D0796479

Edward Batschelet

Introduction to Mathematics for Life Scientists

Third Edition

With 227 Figures

Springer-Verlag Berlin Heidelberg New York 1979

Professor Dr. Edward Batschelet

Mathematisches Institut der Universität Zürich, Switzerland

AMS Subject Classifications (1970)

92-01, 92 A 05, 98 A 35, 98 A 25, 60-01, 93-01, 40-01, 04-01, 15-01, 26-01, 26 A 06, 26 A 09, 26 A 12, 34-01, 35-01

ISBN 3-540-09648-5 3. Auflage Springer-Verlag Berlin Heidelberg New York (broschierte Ausgabe)
ISBN 0-387-09648-5 3rd edition Springer-Verlag New York Heidelberg Berlin (soft cover)

ISBN 3-540-07350-7 2. Auflage Springer-Verlag Berlin Heidelberg New York (broschierte Ausgabe)
ISBN 0-387-07350-7 2nd edition Springer-Verlag New York Heidelberg Berlin (soft cover)

The first edition of this work was originally published by Springer-Verlag Berlin, Heidelberg in 1971 as volume 2 in the series Biomathematics, edited by K. Kricke-berg, R. C. Lewontin, J. Neyman and M. Schreiber

ISBN 3-540-07293-4 2. Auflage Springer-Verlag Berlin · Heidelberg · New York (gebundene Ausgabe)
ISBN 0-387-07293-4 2nd edition Springer-Verlag New York · Heidelberg · Berlin (hard cover)

Library of Congress Cataloging in Publication Data. Batschelet, Edward. Introduction to mathematics for life scientists. "Springer study edition." Bibliography: p. Includes index. 1. Biomathematics. I. Title. QH323.5.B37 1979b./510'.24'574./79-21113

Typesetting and printing: Brühlsche Universitätsdruckerei, Gießen

Preface to the First Edition

A few decades ago mathematics played a modest role in life sciences. Today, however, a great variety of mathematical methods is applied in biology and medicine. Practically every mathematical procedure that is useful in physics, chemistry, engineering, and economics has also found an important application in the life sciences.

The past and present training of life scientists does by no means reflect this development. However, the impact of the fast growing number of applications of mathematical methods makes it indispensable that students in the life sciences are offered a basic training in mathematics, both on the undergraduate and the graduate level. This book is primarily designed as a textbook for an introductory course. Life scientists may also use it as a reference to find mathematical methods suitable to their research problems. Moreover, the book should be appropriate for self-teaching. It will also be a guide for teachers. Numerous references are included to assist the reader in his search for the pertinent literature.

Life scientists are hardly interested in going deeply into mathematics. Therefore, this course differs in many ways from a course offered to mathematicians. Each concept is introduced in an intuitive way. The reader is being kept informed why he is learning a particular method. The relevance of all procedures is proven by examples that have been selected from a wide area of research in the life sciences. It is not intended to distract the student of biology from his main field of activity and to train him as a competent mathematician. The aim is rather to prepare him for an understanding of the basic mathematical operations and to enable him to communicate successfully with a mathematician in case he needs his help.

Many illustrations and some historical notes are inserted to encourage the life scientist who is perhaps somewhat reluctant to be involved with the abstract side of mathematics. Most problems were tested in class. Sections and problems marked with an asterisk are not necessarily more difficult, but may be omitted on first reading.

The book avoids as much as possible the introduction of cookbook mathematics. This requires a somewhat broad presentation. As a consequence no attempt is made to comprise all mathematical methods that are important for life scientists. For instance, computer techniques and statistics are omitted. These two areas can only be presented in special volumes. However, the reader will be prepared for an easier understanding of all topics that could not be covered in this book.

In the beginning I was encouraged to prepare the manuscript by Dr. Sidney R. Galler, Smithsonian Institution. Numerous friends supported the idea and gave me valuable advice and inspiration. I am unable to list all of them. I am very obliged to those biologists who read some chapters and offered valuable criticism and suggestions, especially to Dr. J. P. Hailman, University of Wisconsin, Dr. J. Hegmann and Dr. R. Milkman, both at the University of Iowa, Dr. W. M. Schleidt, University of Maryland.

I gratefully acknowledge the encouragement and considerable support which I received by Dr. Eugene Lukacs, Director of the Statistical Laboratory at Catholic University, Washington, D.C. Some of the more difficult illustrations were made by Mr. C. H. Reinecke with financial support by the Office of Naval Research. I also enjoyed the advice by Dr. V. Ziswiler. The text was carefully typewritten by Mrs. Amelia Miller and Mrs. Phyllis Spathelf for whose patience I wish to express my gratitude. Stylistic, grammatical errors, and other shortcomings were corrected by Dr. Inge F. Christensen and Dr. Maren Brown with great care. I am also indebted to my wife and to Mrs. Eva Minzloff for proofreading and to the staff of the Springer-Verlag for the careful edition.

I would appreciate it if the readers would draw my attention to errors, obscurities and misprints that might still be present in print.

Zurich, October 1971   Edward Batschelet

Preface to the Second Edition

Many users of the first edition complained that the problem section was not large enough. For this reason, numerous problems, both solved and unsolved, were added to the second edition. They are listed at the end of each chapter, but numbered according to sections to facilitate the assignment of problems. At the end of the book the solutions for the odd numbered problems are given.

To make the book self-contained, an appendix with ten numerical tables was added. Chapter 9 was enlarged by a section on methods of integration and Chapter 14 by four sections on determinants and related topics. Many parts of the book were updated and provided with new references. Further, 28 illustrations were added. As a result of the alterations, the size of the book has been enlarged by about a hundred pages.

Many scholars, too many to be listed here, have kindly given me their advice. In addition, I owe particular thanks to Mrs. Alice Peters, Springer-Verlag, for her most valuable recommendations. Dr. Joan Davis edited the text for stylistic errors, and Dr. Armand Wyler carefully checked the problem section. New illustrations were drawn, the manuscript typed, proofs read and reread by Mrs. R. Boller, Mrs. C. Heinzer, Mrs. B. Henop, and by my wife. I had also the unfailing cooperation of the staff of Springer-Verlag. They all deserve my warmest thanks for their help and patience.

Readers are again requested to draw my attention to errors, obscurities and misprints.

Zurich, June 1975   Edward Batschelet

Preface to the Third Edition

Since the second edition appeared in 1975, the units of the "Système international" (SI-units) became legal. To comply with this system, I converted calories into Joules, dynes into Newtons, units of pressure into Pascal, and the unit of viscosity, poise, into Pascalsecond. I also used the opportunity to eliminate some misprints and obscurities. I should like to thank many readers for their help, especially Dr. Tor Gulliksen, University of Oslo, for his substantial contributions.

University of Zürich   Edward Batschelet
July, 1979

Contents

Chapter 4. The Power Function and Related Functions

Chapter 5. Periodic Functions

Chapter 6. Exponential and Logarithmic Functions I

Chapter 7. Graphical Methods

Index of Symbols

Chapter 1

Real Numbers

1.1. Introduction

The purpose of the first chapter is to review some laws and rules of algebra. A selection emphasizing the needs of life scientists will be offered. At the same time we will add important concepts which are usually neglected in textbooks on mathematics.

1.2. Classification and Measurement

Categorizing objects and events is the simplest method of measurement. We assign words, symbols, or numerals to the objects. This is the most primitive kind of a scale. It is called a *nominal scale* (from Latin nomen = name). We also say that we are working at a *nominal level*.

When categorizing biological species, classifying different behavior of animals, or distinguishing among weather conditions, we are working at a nominal level. Likewise an ethologist uses a nominal scale when he categorizes animals as active, inactive, alert, aggressive, submitting. For further examples see Fig. 3.1, 3.3, and 3.4. One may regret the lack of quantitative information, but a clear-cut distinction of objects is already a scientific achievement.

Example 1.2.1. Chemistry is able to distinguish among hundreds of thousands if not millions of different substances. First of all, each substance is either a pure chemical or a mixture of chemicals. Second, among the pure chemicals there is no overlapping. Each one is uniquely defined. Ice, water, and water vapor belong to the same category, whereas a clear distinction is made between different sorts of sugar. This is quite different from everyday language where words such as sugar, salt, wax, alcohol are not well defined.

We obtain a more useful way of scaling if we succeed in *ranking* objects and events. It is easy for us to distinguish two different tones with respect to musical pitch. Not only do we categorize the tones, we are able to rank them as lower or higher. If we are able to rank objects and events we are working with an *ordinal scale* (from Latin ordo = order, rank). We also say that we reach the *ordinal level*.

Example 1.2.2. A medical research worker distinguishes a few categories of *kidney failure* due to excessive consumption of certain chemicals. He may also be able to rank these categories according to severity. It is quite convenient for him to assign the number 0 to "no failure", the number 1 to "weak failure", the number 2 to "modest failure", and the number 3 to "severe failure". He introduces an ordinal scale with *scores* or *rank numbers* 0, 1, 2, 3. Notice that it is not possible to conclude that the increase from weak to modest failure is the same as the increase from modest to severe failure although the numerical difference between the ranks is 1 in either case.

Example 1.2.3. Mineralogists have proposed a *hardness scale* for solid minerals. A specimen S is said to be harder than a specimen T if S scratches T when the two are rubbed together. In this way we can rank the minerals according to hardness. They may be numbered 1, 2, 3, ... with increasing hardness, but there is no way of comparing the intervals between consecutive degrees of hardness.

Example 1.2.4. The intelligence quotient (I.Q.) is an attempt to measure human intellectual performance. The I.Q. is typically a measure at the ordinal level. To what extent ranking of intelligence can be done in a one-dimensional way is a matter of controversy, however.

We reach an even higher level of scaling when intervals become meaningful. Consider the Celsius (°C) temperature scale (formerly called centigrade scale). The zero point 0° C is quite arbitrarily set as the freezing point of water under one atmosphere pressure. But 1° C, 2° C, 3° C, etc. do not simply mean ordered temperatures as with an ordinal scale. An interval of 1° C has a true physical meaning: The mercury of a thermometer rises by the same amount when the temperature increases by 1° C.

Whenever we have a scale with meaningful intervals we call it an *interval scale*. We also say that we are working at an *interval level*.

Example 1.2.5. *Altitude* is typically associated with an interval scale. The reference point, altitude zero, is arbitrary. It could be chosen as floor level, ground level, sea level, etc. But various altitudes have well-defined differences that can be expressed in meters. The intervals do not depend on the choice of the reference level.

Example 1.2.6. *Time t* is measured by an interval scale. There is no natural phenomenon which would indicate a universal reference point

$t = 0$. We always choose a zero point suitable for our particular need. But intervals between time instances have an absolute physical meaning.

Example 1.2.7. *Electric potential* measured in volts can only be defined versus an arbitrary potential zero, but a difference in electric potential, e.g. along a nerve fibre, has a precise physical meaning. Therefore, electric potential is measured by an interval scale.

Example 1.2.8. *Directions* in a plane are measured by angles with an arbitrary zero direction. If north is chosen as the zero direction, and if the angles increase in the clockwise sense, directions are measured by their *azimuth*. The interval between azimuth $0°$ and $90°$ is the same as between $90°$ and $180°$, namely a right angle. Thus directions are determined at an interval level.

There are scales which are even more useful than those at a nominal, ordinal, or interval level. Consider the weight of a body. Here we do not have to set an arbitrary zero point. Weight zero is quite a natural reference point. For this reason, it makes sense to say that one animal weighs twice as much as another one or that the weight increases by two percent. Since the ratio of two weights has a true meaning, we call such a scale a *ratio scale*. We also refer to a *ratio level*.

Notice how foolish it would be to say that a body of $20°$ C is twice as warm as a body of $10°$ C. Similarly, it makes no sense to state that time has advanced 50% or that one direction is ten times as "great" as another direction. Therefore, at an interval level, ratios need not have a proper meaning.

Example 1.2.9. Length, height, thickness, and volume of a body are all measured on ratio scales. However, we have to discriminate between the height of a body and the altitude of a location which is not on a ratio scale (see Example 1.2.5). Height may be interpreted as a difference between two altitudes and, therefore, has a clear reference point, the difference zero. Altitude, on the other hand, has an arbitrary reference point. In the everyday language, altitude is referred usually to sea level, and this level is considered to be absolute. Only under this condition does it make sense to call Mount Everest twice as high as Mount Rainier.

Example 1.2.10. *Temperature* has a lowest point, the temperature of approximately $-273°$ C. It is therefore possible to define a ratio scale for the temperature. Thus by using the intervals of the Celsius scale the Kelvin scale was created. Here $273°$ K means the freezing point and $373°$ the boiling point of water under one atmosphere pressure. With degrees Kelvin, it makes sense to say that $180°$ K is 80% higher than $100°$ K.

Summary. *For adequate use of algebraic rules we have to be aware of different types of measurement. The main properties are listed in the following box:*

level of measurement:	properties:
nominal	no ranking indicated
ordinal	ranking possible, intervals not defined
interval	ranking possible, intervals defined, reference point arbitrary, ratios are meaningless
ratio	ranking possible, intervals defined, reference point is absolute, ratios and percentages are meaningful

This important distinction among several levels of measurement is due to
S. S. Stevens (1951).
For a more detailed treatment, see Hammond and Householder (1963).

1.3. A Problem with Percentages

Does $20\% + 20\% = 44\%$? Certainly not. And yet, numerical work of beginners may result in a paradox like this. Assume that a biologist is studying the growth of a foal. When he starts his investigation the foal weighs 50 kg. The number is simplified for ease of presentation. Within a month the weight increases by 20%, that is, 1/5 of 50 kg, and reaches 60 kg. Assume that in a second month the weight increases again by 20%; then we are inclined to say "the total increase is $20\% + 20\% = 40\%$".

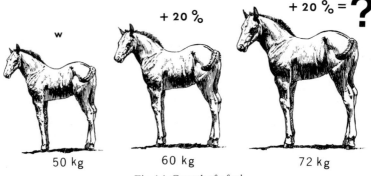

Fig. 1.1. Growth of a foal

However, the calculation leads to a different result: 20% of 60 kg is 12 kg so that the final weight is 72 kg which is 22 kg or 44% more than the original weight of 50 kg (see Fig. 1.1).

Intuitively the two gains have to be added. But an addition that is intuitively clear need not be an addition in the strict sense of algebra. A correct treatment of this example will result in a multiplication rather than an addition. Let w denote the initial weight of the foal. After a month the weight will increase to $w + 0.2\,w$ which is $1.2\,w$. Hence, to obtain the weight after a monthly period we have to multiply by 1.2. The same operation is applied again for the second monthly period resulting in $1.2 \times 1.2\,w = 1.44\,w = w + \dfrac{44}{100}\,w$.

We get a deeper insight in the preceding example if we slightly generalize it. Let w be again the initial quantity[1], e.g. a weight or a length, and p be the percentage of increase. Then the increase is

$$\frac{p}{100} \cdot w .$$

The increased quantity is therefore

$$w + \frac{p}{100} \cdot w = w\left(1 + \frac{p}{100}\right). \tag{1.3.1}$$

To get from the initial quantity w to the increased quantity we have to multiply w by the factor $1 + \dfrac{p}{100}$. For a second increase with the same percentage we have to multiply again by $1 + \dfrac{p}{100}$. Thus we get

$$\left[w\left(1 + \frac{p}{100}\right)\right]\left(1 + \frac{p}{100}\right) \tag{1.3.2}$$

for which we simply write

$$w\left(1 + \frac{p}{100}\right)^2 . \tag{1.3.3}$$

The same operation may be applied over and over again. After the elapse of n time intervals we get for the weight

$$w\left(1 + \frac{p}{100}\right)^n . \tag{1.3.4}$$

This result will be used in Chapter 6 for the introduction to exponential functions.

---

[1] Letters signifying quantities were used for the first time around 1600. This change may appear to us as a trivial improvement. In reality it has caused such a rapid development of algebra that it may be called an explosion of knowledge.

Problem 1.3.1. The White Nile above the dam at Jebel Aulia was infested by a weed known as water hyacinth. In 1958 the plant covered only 12 km^2, but the yearly increase was 50%. a) What area was covered $1, 2, \ldots, n$ years later? b) How long did it take until the full area of backwater measuring 200 km^2 was covered?

Solution: a) Area $A = 12 \times 1.5^n$ km^2 $(n = 1, 2, \ldots)$

             b) $12 \times 1.5^n = 200$ (km^2)

$$n \cdot \log 1.5 = \log(200/12), \quad n = 6.9 \text{ years.}$$

Hence the backwater was fully covered by 1965. For more details see Holm *et al.* (1969).

1.4. Proper and Improper Use of Percentages

As we pointed out in Section 1.2, percentages can only be applied for quantities that are measured on a *ratio scale*.

In the everyday language, as well as in newspapers, magazines, announcements, changes are often described in terms of percentages, although this may not be adequate. Statements such as a 10% increase in the power of cleaning or a 40% drop in the noise level of an engine can hardly be viewed as serious. It is therefore important to carefully study a variety of situations.

Example 1.4.1. It makes little sense to talk about a 40% increase in *productivity* of a laboratory as long as the scale is not specified. Does 40% refer to the number of man-hours, to the amount of laboratory equipment in use, or to the scientific papers published? Assume that the idea is to talk about a 40% improvement of the quality of scientific papers. Then we can hardly accept this idea since quality is best measured on an ordinal scale. Even if 40% refers to the number of publications, we should be suspicious. To be sure, "number of publications" is at the ratio level, but if, in absolute numbers, the increase is from 5 to 7 publications, the ratio 40% indicates an accuracy which is by no means reached.

Example 1.4.2. When a physician observes 6 new cases of tuberculosis in October and 8 new cases in November, then it is arithmetically valid to figure out an increase of $33\frac{1}{3}$%. But this number is misleading. The occurrence of new cases is subject to *random fluctuations*. The ups and downs in a monthly sequence such as 6, 8, 5, 9, 5, 7, etc., do not mean that the risk of infection is changing. Larger deviations would be necessary to signify a true change. The claim that a change has occurred must be supported by a statistical test of the significance of the observed deviations. Even in case of significance one should avoid a percentage, unless the error is small enough. Otherwise an accuracy is suggested that does not exist. If, for instance, the true percentage could range between 10% and 50%, it is ridiculous to report $33\frac{1}{3}$%.

Example 1.4.3. Assume a machine for mass production has an output of 8 defective items among 100 items produced. If the rate can be reduced to 6:100, this may be called a 2% improvement. It is more impressive, however, to describe it as a 25% improvement. All one has to do is to take the 8 defective items as 100%.

Example 1.4.4. *A 50% increase and a subsequent 50% decrease do not cancel each other.* If, for instance, the crop of an apple tree increases from 120 kg to 180 kg within a year, then we call this a 50% increase. A subsequent 50% decrease, however, is usually related to 180 kg as 100%. Thus the crop is reduced to 90 kg instead of the original 120 kg.

Example 1.4.5. A ratio scale may be mathematically and physically meaningful without being biologically adequate. Take for instance our hearing capacity. Children can recognize acoustic waves of frequencies up to 20,000 Hz. The upper limit decreases with age[2]. A forty-year-old person can hardly hear beyond 16,000 Hz. It may strike him that he has lost a range of 4,000 Hz, that is 20% of the original capability. However, this is the wrong way to judge our ear. The interval between two tones has to be measured by the ratio of their frequencies. Hence, the loss is measured by the ratio $20,000 : 16,000 = 5 : 4$. In music this means a small interval, called a *third* (the same interval as from do to mi). Compared with the ten octaves that we can hear, the loss is irrelevant for our life. We obtain a better scale for musical pitch by replacing the frequencies with their logarithms (see Section 6.6).

1.5. Algebraic Laws

To improve our capabilities, but also to prepare ourselves for Boolean algebra in Chapter 2, we have to review the algebraic laws which we all have frequently, but most often unconsciously applied since childhood.

Easiest are the two commutative laws, one for addition, the other one for multiplication:

$$a + b = b + a \quad \text{(commutative law of addition)} \tag{1.5.1}$$

$$ab = ba \quad \text{(commutative law of multiplication)} \tag{1.5.2}$$

---

[2] Hertz (Hz) is the unit of frequency and means cycles per second. Heinrich Rudolf Hertz (1857—1894) was the German physicist who discovered the electromagnetic waves.

In words: *The order in which we add or multiply two numbers may be interchanged.*

Example 1.5.1. If an animal loses weight in a first period by 30% and in a second period by 10%, will we obtain the same result if we interchange the order of reduction? The answer is yes. Indeed, if w denotes the original weight or 100%, the decrease in the first period is expressed by multiplying by the factor $1 - \dfrac{30}{100} = \dfrac{7}{10}$. To figure the second reduction we multiply the result by $1 - \dfrac{10}{100} = \dfrac{9}{10}$. Thus, the final weight is

$$w \cdot \frac{7}{10} \cdot \frac{9}{10},$$

a result which remains unchanged by applying the commutative law of multiplication.

As an introduction to the associative laws consider the addition

$$28 + 13 + 87.$$

One way of carrying out the calculation is to add $28 + 13 = 41$ and in a second step to calculate $41 + 87 = 128$. This procedure may be indicated by parentheses as follows

$$(28 + 13) + 87.$$

There is, however, a smarter way to obtain the result. We first carry out the addition $13 + 87 = 100$ which leads to a round number. The second step, $28 + 100 = 128$, is then particularly simple. This time we followed the rule

$$28 + (13 + 87).$$

Similarly, when multiplying

$$658 \times 2 \times 5,$$

there are two ways without touching the order of the three factors. Either

$$(658 \times 2) \times 5 = 1316 \times 5 = 6580$$

or

$$658 \times (2 \times 5) = 658 \times 10 = 6580.$$

Needless to say that the second way is much faster. In general, the rules are

$$(a+b)+c = a+(b+c) \quad \text{(associative law of addition)} \qquad (1.5.3)$$

$$(ab)\,c = a(bc) \qquad\qquad \text{(associative law of multiplication)} \qquad (1.5.4)$$

In words: *If more than two numbers have to be added or to be multiplied, it does not matter which two of the numbers we add or multiply first.*

Example 1.5.2. We return to formula (1.3.2). There it is quite inconvenient to multiply w and $1 + \dfrac{p}{100}$ first, as indicated by the brackets. It facilitates our task that we are allowed to multiply the second and third factor first which leads to formula (1.3.3).

There is a fifth law which combines addition and multiplication. When carrying out the calculation

$$(4 \times 32) + (4 \times 68)$$

we can do it in a complicated way, that is,

$$4 \times 32 = 128\,, \quad 4 \times 68 = 272\,, \quad 128 + 272 = 400\,,$$

or in an easy way, that is,

$$4 \times (32 + 68) = 4 \times 100 = 400\,.$$

The two versions lead to the same result due to the following law:

$$ab + ac = a(b+c) \quad \text{(distributive law)} \qquad (1.5.5)$$

In words: *Instead of multiplying two numbers by a common factor and adding the products, we may first add the two numbers and then multiply their sum by the factor.*

For an example see also formula (1.3.1).

In this connection we should throw a glance at some convenient ways of writing formulas. In

$$ab + c$$

we may have doubts whether the multiplication or the addition has to be performed first. The results would not be the same! However, there is agreement that *multiplication is first* when parentheses are omitted. Thus $ab + c$ actually means

$$(ab) + c .$$

A similar rule is applied when the horizontal line for fractions is replaced by an oblique stroke[3]. We should learn to clearly distinguish among the four expressions

$$1 + p/100 , \quad (1 + p)/100 , \quad p/100 + 1 , \quad p/(100 + 1)$$

which in turn mean

$$1 + \frac{p}{100} , \quad \frac{1 + p}{100} , \quad \frac{p}{100} + 1 , \quad \frac{p}{100 + 1} .$$

Only the first and the third expressions are of equal value. The parentheses are used to indicate the order in which the operations are performed.

In later chapters we will frequently use the product

$$n! = 1 \cdot 2 \cdot 3 \cdot \cdots \cdot n \tag{1.5.6}$$

for any natural number n (read: n factorial). Thus $1! = 1$, $2! = 2$, $3! = 6$, $4! = 24$, etc.

1.6. Relative Numbers

In comparing numbers with each other we frequently have to observe positive and negative signs. Obviously, a temperature of $-5°$ C is quite different from $+5°$ C. We therefore deal with *relative numbers*. The temperature scale gives us a clue how to arrange the relative numbers along a line, the so-called *real number line* (Fig. 1.2).

Fig. 1.2. The real number line

The numbers are represented by points on the line. As a rule positive numbers are depicted to the right of zero, negative numbers to the left. If we are given two different numbers, one of them is located to the right

---

[3] The oblique stroke /, also called solidus, was proposed in 1845 by the English mathematician Augustus de Morgan (1806—1871). For a typist or a printer it is more economical to set $p/100$ than $\frac{p}{100}$.

of the other number, that is, the real numbers are *ordered*. In the case of temperatures, the number to the right represents the higher temperature. In general, we say that the number to the right is *greater* than the number to the left, and, conversely, the number to the left is *less* than the number to the right. Thus $+2$ is greater than -5, and 0 is greater than -10. For "*b* greater than *a*" we write

$$b > a. \tag{1.6.1}$$

For instance, $2 > (-5)$ and $0 > (-10)$. Symmetrically, for "*a* is less than *b*" we write [4]

$$a < b. \tag{1.6.2}$$

For instance, $0 < 7$ and $(-7) < 0$. Notice that the large end of the signs $>$ and $<$ faces the greater number and the tip the smaller number.

The two temperatures $-5°\,C$ and $+5°\,C$ are equally distant from the point $0°\,C$ on the real number line. To express this fact we say that both temperatures have the same *absolute value*. More precisely, the absolute value of a positive number is the number itself, whereas the absolute value of a negative number is the opposite (positive) number. Thus we write for the absolute value [5]

$$|+5| = +5, \quad |-5| = +5.$$

The number zero is neither positive nor negative. We define $|0| = 0$. Hence, the absolute value of a number is positive except for the number zero. When a number is written without sign, it is positive unless it is zero. Thus 5 and $+5$ are the same number.

Clearly, the greater the distance from 0, the greater is the absolute value. Thus $|-5| > |+2|$, whereas $(-5) < (+2)$. Also $|-8| > |-3|$, whereas $(-8) < (-3)$.

Notice also that for arbitrary numbers x and y

$$|x - y| = |y - x|.$$

When three or more numbers are compared with each other, we may write for instance $a < b < c < d < \cdots$. If a variable x is permitted to take values between 0 and 6, say, we write

$$0 < x < 6$$

---

[4] According to Cajori (1928) the symbols $>$ and $<$ were invented in England in 1631 by Thomas Harriot.

[5] The two vertical bars as symbol for the absolute value were introduced by the German mathematician Karl Weierstrass (1815—1897), according to Cajori (1929).

and call 0 the *lower bound* and 6 the *upper bound* of x. For the same relationship we may also write

$$6 > x > 0.$$

However, it is against the rule to use the two different symbols, $>$ and $<$, in the same relationship. Never write $5 < x > y$. Nor is it permitted to reduce a statement such as "$x < 2$ or $x > 5$, but not between" to $2 > x > 5$.

Occasionally, $x \gtrless y$ is used to indicate that x is either greater than y or less than y, but not equal to y. But for such a statement it is better to write

$$x \neq y$$

with the sign $\neq$ meaning "not equal".

If a variable, say y, is allowed to take the value of its upper bound b or its lower bound a, we may write

$$a \leqq y \leqq b. \tag{1.6.3}$$

We say "y is less than or equal to b" and "y is greater than or equal to a". In print the symbols $\geqq$ and $\leqq$ often appear as $\geq$ and $\leq$.

Note that $a + b > c$ means $(a + b) > c$, but the parentheses are usually omitted.

Every relation using the sign $>$ or $<$ is called an *inequality*. Inequalities occur frequently in classification problems. For instance, a patient is classified as diabetic if the concentration c of glucose in the blood exceeds the value of 1.80 g/l an hour after intake of 50 g of glucose. In symbols

$$c > 1.80 \text{ g/l}.$$

The *number zero* plays an exceptional role in division. Divisions such as $0/5 = 0$ or $0/(-7) = 0$ are not problematic, since the inverse operation $0 \times 5 = 0$, $0 \times (-7) = 0$ leads to a correct result. However, when the denominator is zero, the division cannot be performed. $5/0$ is not a number, because no number x would satisfy the inverse statement $x \cdot 0 = 5$. Nor has $0/0$ any meaning, since the inverse statement $x \cdot 0 = 0$ would be correct for an arbitrary number x. We conclude that the denominator is not allowed to be zero. In later chapters we will frequently have quotients, say p/q, and then add "provided that $q \neq 0$". It is worth keeping the following rule in mind[6]:

> Never divide by zero

---

[6] Readers who are familiar with the symbol ∞ signifying infinity might be inclined to write $3/0 = \infty$. However, this is not correct for two reasons. First, ∞ is not a number so that the sign of equality is not applicable. Second, if $3/0 = \infty$ were true, $3/0 = -\infty$ would also be true. This point will be clarified in Section 8.4.

Notice that the equation $x^2 = px$ cannot be properly solved if we divide each side by x. We would lose the possible solution $x = 0$. Proceeding correctly we collect both terms on the left side, factor out x, and decide that either x or $x - p$ has to be zero. Thus

$$x^2 - px = 0,$$
$$x(x - p) = 0,$$
$$x = 0 \quad \text{or} \quad x = p.$$

1.7. Inequalities

Quite similar to an equation such as

$$2x - 5 = x + 3$$

which we have to solve for x, we also find "inequations" like

$$3x - 5 < x + 3 \qquad (1.7.1)$$

in numerous applications. The word "inequation" is hardly ever used despite its appeal. We accept the usual term *inequality*.

The method of solving an inequality is much the same as for an equation. In our problem (1.7.1) we first add 5 on both sides. This will not change the proposed imbalance of the two sides. Thus we get

$$3x < x + 8.$$

Next we subtract from both sides the same amount x and obtain

$$2x < 8.$$

Again we do not disturb the imbalance when dividing both sides by 2. Hence

$$x < 4$$

which solves our problem (1.7.1).

Great care should be taken when we have to multiply or to divide by a negative number. For instance

$$(-6) > (-10) \qquad (1.7.2)$$

is equivalent to $(+6) < (+10)$. Hence, upon multiplication by -1, the inequality sign changes from $>$ to $<$. The general rule is: *When we multiply or divide an inequality by a negative number we have to reverse the sign of inequality.*

Example 1.7.1. Solve the inequality

$$u(u - 1) > p - 4u + u^2 \qquad (1.7.3)$$

for u. First, we remove the parentheses by applying the distributive law (1.5.5). Second, we subtract u^2 from both sides. Third, we add $4u$ on both sides, and finally, divide by 3. Thus we obtain consecutively

$$u^2 - u > p - 4u + u^2,$$

$$-u > p - 4u,$$

$$3u > p,$$

$$u > p/3.$$

We may reach the same solution upon multiplying $-u > p - 4u$ by (-1), then by subtracting u and adding p on both sides. Thus

$$u < -p + 4u \quad \text{(``less than'' sign!)},$$

$$p < 3u,$$

$$p/3 < u.$$

For biological applications of such inequalities see Section 3.7.

When the unknown quantity is squared as in

$$x^2 < 2, \tag{1.7.4}$$

the solution cannot be found by a simple extraction of the square root. We have to observe the plus and minus signs. Thus (1.7.4) leads to

$$-\sqrt{2} < x < \sqrt{2}$$

which means that x is any number on the real axis between the opposite numbers $-\sqrt{2}$ and $+\sqrt{2}$. We may also write $|x| < \sqrt{2}$.

Notice that

$$x^2 > 9 \tag{1.7.5}$$

implies that

$$x > 3 \quad \text{or} \quad x < -3,$$

a result which could be written in the simple form $|x| > 3$. (Warning: Never write $-3 > x > 3$ since x cannot satisfy both inequalities at the same time).

1.8. Mean Values

When a quantity changes with time or when an experiment is repeated several times, we have to condense the amount of data for practical purposes. One way to do this is to calculate averages or mean values.

In this section we will introduce two types of mean values based on algebraic considerations.

When the quantity under consideration is measured on an interval or ratio scale (see Section 1.2), the most common mean value is the *arithmetic mean.* For two measurements x_1 and x_2 we define

$$\bar{x} = \frac{x_1 + x_2}{2}. \qquad (1.8.1)$$

The bar over the x indicates the arithmetic mean.

There are occasions when the arithmetic mean loses its biological significance. Consider the growth of a foal as described in Section 1.3. At equidistant time intervals the foal attains the weights

$$50 \text{ kg}, \qquad 60 \text{ kg}, \qquad 72 \text{ kg}.$$

The sequence was the result of multiplying 50 kg repeatedly by the constant factor 1.2. We feel that 60 kg is the proper mean between 50 kg and 72 kg. However, the arithmetic mean would be (50 kg + 72 kg)/2 = 61 kg. We may try the *geometric mean* which is defined by the formula

$$x_g = \sqrt{x_1 x_2}. \qquad (1.8.2)$$

We obtain $\sqrt{50 \times 72} = \sqrt{3600} = 60$ which is the desired value.

The geometric mean is only applicable if the quantities involved are measured on a ratio scale.

There are other formulas for mean values possible. The choice of the one to be applied depends on the nature of the problem. This is discussed further in Section 6.5. Sometimes the decision has to be based on a statistical analysis.

Notice that a mean value formula cannot be used at the ordinal level. Three different results may have the scores 3, 7, 20. When the somewhat arbitrary scoring system is changed, the same results may take the scores 2, 15, 18. Obviously, a mean value could be manipulated to fall between the first two or likewise between the last two results.

1.9. Summation

Assume that a measurement is repeated n times where n is one of the numbers 2, 3, 4, ... Let $x_1, x_2, x_3, \ldots, x_n$ denote the n measurements. Then the arithmetic mean is

$$\bar{x} = \frac{x_1 + x_2 + \cdots + x_n}{n}. \qquad (1.9.1)$$

It is convenient to abbreviate a sum of similar terms by using the summation sign Σ (the upper-case Greek Sigma)[7]. Each term of the sum is of the form x_i where the variable subscript i stands for one of the numbers 1, 2, ..., n; $i = 1$ is the lowest and $i = n$ the highest value. Now

---

[7] The summation sign came into common use around 1800.

the sum is written $\sum\limits_{i=1}^{i=n} x_i$ or more frequently $\sum\limits_{i=1}^{n} x_i$, in print also $\Sigma_{i=1}^{n} x_i$.

The sum is read "Summation of x sub i, i ranging (or running) from 1 to n". Formula (1.9.1) may then be rewritten in one of the following ways:

$$\bar{x} = \frac{1}{n} \sum_{i=1}^{n} x_i = \sum_{i=1}^{n} x_i/n . \qquad (1.9.2)$$

Notice that an arbitrary letter can be used for the subscript without changing the meaning. For instance,

$$x_1 + x_2 + \cdots + x_n = \sum_{i=1}^{n} x_i = \sum_{k=1}^{n} x_k = \sum_{v=1}^{n} x_v .$$

Usually a research worker is interested not only in a mean value, but also in the deviations from the mean value. Consider the differences

$$x_1 - \bar{x}, \quad x_2 - \bar{x}, ..., \quad x_n - \bar{x}$$

or briefly $x_i - \bar{x}$ for $i = 1, 2, ..., n$. Since some of the values x_i are greater and some are smaller than $\bar{x}$, the differences $x_i - \bar{x}$ take positive as well as negative values. From (1.9.1) we conclude that the *sum of the differences equals zero*:

$$\sum_{i=1}^{n} (x_i - \bar{x}) = (x_1 - \bar{x}) + (x_2 - \bar{x}) + \cdots + (x_n - \bar{x})$$
$$= x_1 + x_2 + \cdots + x_n - n\bar{x} = 0 . \qquad (1.9.3)$$

This formula is very useful for checking the numerical value of $\bar{x}$.

In order to judge the magnitude of a deviation $x_i - \bar{x}$ it does not matter whether the deviation is positive or negative. We introduce the *absolute value* of $x_i - \bar{x}$, in symbols

$$|x_i - \bar{x}| \qquad (1.9.4)$$

with two vertical bars (Section 1.6). Instead of listing all n deviations $|x_i - \bar{x}|$ it is more convenient to consider the arithmetic mean of the deviations, that is,

$$\frac{1}{n} \sum_{i=1}^{n} |x_i - \bar{x}| .$$

The result gives us some idea of the dispersion of the n measurements. Unfortunately, the expression containing absolute values is very hard to manage algebraically. A mathematician would say that it has "nasty properties". To avoid the difficulties it was an excellent idea of Laplace and Gauss[8] around 1800 to replace the absolute values by the squares of the deviations. Indeed, squaring removes the negative sign. For instance $(-5)^2 = +25$. Therefore, the arithmetic mean of the squares of deviation

$$\frac{1}{n} \sum_{i=1}^{n} (x_i - \bar{x})^2 \tag{1.9.5}$$

is used to judge the amount of dispersion. The expression (1.9.5) is called the *variance* of the n measurements $x_1, x_2, \ldots, x_n$. In statistics the denominator n is frequently replaced by $n-1$ for reasons which cannot be explained in this connection.

It is convenient to perform the calculations in columns as shown by the following example:

x_i	$x_i - \bar{x}$	$(x_i - \bar{x})^2$
2.4	-0.3	0.09
3.1	$+0.4$	0.16
2.8	$+0.1$	0.01
2.9	$+0.2$	0.04
2.2	-0.5	0.25
2.8	$+0.1$	0.01
16.2	0	0.56

$$\bar{x} = 16.2/6 = 2.7 \qquad \Sigma (x_i - \bar{x})^2 = 0.56$$

1.10. Powers

In a variety of biomathematical problems and statistical techniques, powers are involved. For example, in calculating the surface and the volume of a spherical cell we use the formulas

$$S = 4\pi r^2, \qquad V = (4/3)\, \pi r^3 \tag{1.10.1}$$

with the second and third powers of the radius and the number $\pi = 3.14 \ldots$. In Section 1.9 we mentioned the sum of squares of deviations. And in formula (1.3.4) we see that a problem with percentages forced us to consider powers.

In a power such as

$$a^n,$$

---

[8] Pierre Simon Laplace (1749—1827), a French mathematician, and Carl Friedrich Gauss (1777—1855), a German mathematician, both contributed to the development of statistics.

a is called the *base* and n the *exponent*, whereas the term *power* is reserved for the full expression. We read the power "a to the power n", but this should not distract us from calling n an exponent.

$$a^n = \text{power} \quad \begin{array}{l} n = \text{exponent} \\[1ex] a = \text{base} \end{array}$$

Powers are useful to rewrite large and small numbers in a convenient form:

$$100 = 10^2 \,,$$
$$1000 = 10^3 \,,$$
$$10000 = 10^4 \,, \text{ etc.}$$

The base is 10 and the exponents are 2, 3, 4, ... It is practical to extend the list in the opposite way and to generalize the concept of power:

$$10 = 10^1 \,,$$
$$1 = 10^0 \,,$$
$$1/10 \;\; = 0.1 \;\; = 10^{-1} \,,$$
$$1/100 \;\; = 0.01 \;\; = 10^{-2} \,,$$
$$1/1000 = 0.001 = 10^{-3} \,, \text{ etc.}$$

Here we have introduced powers with negative exponents.

The statement $10^0 = 1$ has puzzled many people. The left side seems to indicate "a product consisting of no factor", whereas the right side is one and not zero as might be expected. The following argument may help the reader to understand this detail: Start with $10^3 = 1000$. Take one factor 10 away. This is equivalent to dividing by 10 (not a subtraction!). We get $10^2 = 100$. Continue dividing by 10. We obtain $10^1 = 10$, and in the next step $10^0 = 1$. Thus the number 1 results since "taking a factor 10 away" means a division by 10. If we continue to divide by 10, we will further have $10^{-1} = 1/10$, $10^{-2} = 1/100$, etc. These results may also be obtained by applying the rule $10^n/10^m = 10^{n-m}$ and extending it to the case $m = n$ and then to the case $m > n$. Thus, for instance,

$$10^4/10^2 = 10^2 \,, \quad 10^3/10^2 = 10^1 = 10 \,, \quad 10^3/10^3 = 10^0 = 1 \,,$$
$$10^2/10^3 = 10^{-1} = 1/10 \,.$$

For quantities such as 378000 km or 0.0074 mg we should write

$$3.78 \times 10^5 \text{ km} \quad \text{and} \quad 7.4 \times 10^{-3} \text{ mg} \,.$$

The main advantage is that calculations are easier to perform with decimal powers than with numbers consisting of too many decimals[9].

In this connection we may note some frequently used prefixes for multiples[10]:

Symbol		Factor	
G	giga-	1 000 000 000	$= 10^9$
M	mega-	1 000 000	$= 10^6$
k	kilo-	1 000	$= 10^3$
m	milli-	0.001	$= 10^{-3}$
μ	micro-	0.000001	$= 10^{-6}$
n	nano-	0.000 000 001	$= 10^{-9}$

Examples are μg (microgram), μm (micrometer, formerly denoted by μ only), nm (nanometer), MHz (Megahertz $= 10^6$ cycles per second), mHz (Millihertz $= 10^{-3}$ cycles per second).

The unit ppm indicates how many "parts per million" are involved. Thus 1 mg means 1 ppm with respect to 1 kg. Likewise, 1 cm^3 is 1 ppm with respect to 1 m^3.

Negative powers occur frequently. Let x be any base (except 0). Then we find by subsequent division $x^3/x = x^2$, $x^2/x = x = x^1$, $x/x = 1 = x^0$, $1/x = x^{-1}$, $x^{-1}/x = x^{-2}$, $x^{-2}/x = x^{-3}$, etc.

Hence *reciprocals* such as $1/A$ are also written as A^{-1}. Similarly we find m sec^{-1} instead of m/sec for the unit of velocity.

In connection with powers of ten we mention the phrase "*order of magnitude*". When two quantities are said to be of different orders of magnitude, it could either mean that they are not comparable in size or that one quantity is at least ten times bigger than the other one. "Two orders of magnitude" would then mean a factor of about $10^2 = 100$, etc.

Problem 1.10.1. Lampert *et al.* (1969) measured the dry mass of herpes simplex virus particles (strain 11140) by means of an electron microscope. The core weighs 2×10^{-16} g, the empty naked capsid 5×10^{-16} g, the full naked capsid 7×10^{-16} g, and the enveloped nucleocapsid 13×10^{-16} g. Of what order of magnitude is the weight of a complete herpes simplex virus particle?

Solution: Addition of the above approximate values leads to 15×10^{-16} g. Hence the order of magnitude is 10^{-15} g. This is eight orders of magnitude higher than the weight of a single oxygen molecule $(5 \times 10^{-23}$ g).

---

[9] Another reason for this form is to mark the significant digits (see Section 1.12).

[10] Greek gigas = giant, megas = large, mikros = small, nanos = dwarf.

Problem 1.10.2. The oxygen capacity of mammalian blood is about 200 ml per liter of blood. If fully used, this amount of oxygen will produce an energy of 4.2×10^3 J (Joule). (a) By how many degrees Celsius could the temperature of 1 liter blood be raised by this energy? (b) How many meters can a man of 70 kg weight climb until he uses up the oxygen of his 3.5 liters of blood?

Solution. (a) Blood behaves like water, when it is heated: One liter needs the energy of 4.2×10^3 J for an increase in temperature of $1°$ C. And this is just the energy which oxygen in the blood can supply. Hence, the answer is $1°$ C. (b) The energy required to lift a mass of 1 kg by 1 m is 9.81 Nm (Newtonmeter) $= 9.81$ J. Thus for each meter to climb, a man of 70 kg requires an energy of 70×9.81 J/m. The energy contained in 3.5 liter blood is $3.5 \times 4.2 \times 10^3$ J $= 14.7 \times 10^3$ J. This energy is sufficient for climbing

$$\frac{14.7 \times 10^3 \text{ J}}{70 \times 9.81 \text{ J/m}} = 21 \text{ m}.$$

Thus, when climbing, the body has to renew its oxygen content at least every 21 meters (cf. Schmidt-Nielsen, 1972, p. 73).

Problem 1.10.3. An ultrahigh-frequency electromagnetic field ($v = 2.45$ GHz) kills plants and seeds of many species after a short, but heavy exposure. Calculate the wavelength λ which is in the microwave region of the radio-frequency band. (GHz $=$ Gigahertz $= 10^9$ cycles per second, velocity of waves $c = 3.00 \times 10^5$ km/sec.)

Solution:

$v = 2.45 \times 10^9 \text{ sec}^{-1}$

$c = 3.00 \times 10^5 \text{ km sec}^{-1} = 3.00 \times 10^{10} \text{ cm sec}^{-1}$

$\lambda = c/v = (3.00 \times 10^{10} \text{ cm sec}^{-1})/(2.45 \times 10^9 \text{ sec}^{-1}) = 12.2 \text{ cm}.$

(For experimental results see Davis et al., 1971.)

1.11. Fractional Powers

In the previous sections the exponents were always integers, either positive, negative or zero. Now we introduce exponents that are fractions. Using an intuitive approach we consider an equality such as $2^3 = 8$. Let n be one of the integers 2, 3, 4, ... and raise each side of the equality to the n-th power. This gives $(2^3)^n = 8^n$ or by a well-known rule for the power of a power: $2^{3n} = 8^n$. Conversely we reduce the last equality when we divide each exponent by n. Extending the procedure, let us apply the last step also to exponents that are not multiples of n. Consider again $2^3 = 8 = 8^1$ and divide each exponent by n. We obtain

$$2^{3/n} = 8^{1/n}.$$

In the special case of $n = 3$ we get $2^{3/3} = 8^{1/3}$ or $8^{1/3} = 2$. This is the inverse operation of $2^3 = 8$. It is usually written in quite a different manner, namely $\sqrt[3]{8} = 2$. For the same reason we have $9^{1/2} = \sqrt{9} = 3$, $81^{1/4} = \sqrt[4]{81} = 3$, $6^{2/3} = \sqrt[3]{6^2} = \sqrt[3]{36}$. Thus roots and fractional powers are the same thing[11].

All the rules for powers with integers as exponents can be extended to fractional powers. For instance $a^{2/3} \cdot a^{1/2} = a^{2/3 + 1/2} = a^{4/6 + 3/6} = a^{7/6}$. The same operation with the radical sign, namely $\sqrt[3]{a^2} \cdot \sqrt{a}$, would be rather complicated to perform. Also $x^{-1/2}$ is easier to handle than $1/\sqrt{x}$.

For the typist or the printer it is simpler to write $8^{1/3}$ than $\sqrt[3]{8}$. That is why in modern printing the radical sign is disappearing.

Among the roots *square roots* are especially common in biological and medical research work (see Table B). For the beginner the shift of the decimal point causes some difficulties. We know that $4^{1/2} = \sqrt{4} = 2$, $400^{1/2} = 20$, $40000^{1/2} = 200$, etc. The radicands are 4, 400, 40000, etc., such that the decimal point is shifted by *two places* or by a multiple of two. On the other hand, the square roots are 2, 20, 200, etc., with a shift of the decimal point by only *one place*. Similarly we get $0.04^{1/2} = 0.2$, $0.0004^{1/2} = 0.02$, etc. However, for $40^{1/2}$ we get quite different figures from a table, namely $40^{1/2} = 6.32 \dots$. Applying the rule for the decimal point we obtain for instance $4000^{1/2} = 63.2 \dots$ and $0.4^{1/2} = 0.632 \dots$. Similarly, when a table yields $389^{1/2} = 19.72$, we conclude that $3.89^{1/2} = 1.972$, $0.0389^{1/2} = 0.1972$, and $38900^{1/2} = 197.2$. But we have to look up $38.9^{1/2}$ at a different place in the table, either in a column headed $\sqrt{10n}$ or at the entry 3890. There we find $3890^{1/2} = 62.37$, hence $38.9^{1/2} = 6.237$.

Note that the square root of a negative number such as $(-4)^{1/2}$ is not a real number, since the square of a real number, such as $(+2)^2$ and $(-2)^2$, is always greater than or equal to zero. New numbers have to be invented to solve an equation such as $x^2 = -4$. We will deal with them in Chapter 15.

By definition, the square root of a positive number is always positive. Thus $\sqrt{16} = +4$, but $\sqrt{x^2}$ does not simply equal x. In the case of a negative value x, we have to write

$$\sqrt{x^2} = |x| . \tag{1.11.1}$$

---

[11] The origin of the radical sign is not known. It consists of two parts, $\sqrt{}$ and $\overline{}$, the first part indicating the operation, the second part meaning the same as parentheses today. Thus $\overline{a+b}$ stood for $(a+b)$. According to Cajori (1928) the radical sign was widely accepted at the end of the seventeenth century.

Powers such as a^2, x^3 appeared in print for the first time in 1637 in a work by the French philosopher and mathematician René Descartes (1596—1650), also known under his Latin name Cartesius. In 1655 the English mathematician and theologian John Wallis (1616—1703) introduced powers with negative and fractional exponents such as x^{-n} and $x^{1/n}$. These generalized powers were propagated by the English physicist and mathematician Isaac Newton (1642—1727). Nevertheless it took a long time before they became popular on the European continent.

For $\alpha = 120^0$, $\cos \alpha = -0.5$, but $\sqrt{\cos^2 \alpha} = +0.5$. The solution of $x^2 = 200$ is not completely given by $x = \sqrt{200}$, but by $x = \pm \sqrt{200}$.

1.12. Calculations with Approximate Numbers

Research workers are often concerned with the inaccuracy of figures. One source is errors in counting or uncertainties in reading instruments. Another source is the random fluctuations in sampling. Finally, the inevitable rounding-off errors also cause some difficulties.

In this chapter statistical aspects cannot be discussed. Nor will we try to account for all the rules that are useful in dealing with approximate numbers. For a detailed treatment with instructive biological examples we refer the reader to a monograph by Anderson (1965).

Assume a number such as 14.07 has an error not exceeding 0.005, that is, an error not greater than half of the last digit. Then the last decimal 7 is exact, and we say that 14.07 has four *significant digits* (or figures).

The number of significant digits is quite independent of the decimal point. Clearly 140.7 m and 0.1407 km are the same quantities and consequently have the same degree of accuracy. Zeros that are only required to mark the decimal point should not be counted as significant figures. Thus 0.01407 and 0.001407 contain only four significant figures. In numbers such as 14070 and 140700 it is not clear whether the zeros add to the precision or not. To avoid ambiguity we write these numbers in the form 1.407×10^4 and 1.407×10^5. Notice that 3.8×10^2 has two and 3.80×10^2 has three significant digits. Therefore, the two numbers are quite different in accuracy. The error is at most 0.05×10^2 for the first number and at most 0.005×10^2 for the second number. Without powers both numbers would be 380, thus not indicating the accuracy.

In practical applications we are usually forced to carry *one digit beyond the last significant figure*. For instance, when a measurement or a calculation results in 3.47 mg with an error of at most 0.02 mg, the digit 7 is not significant, but dropping this figure and properly rounding off the quantity to 3.5 mg may cause a loss of relevant information. On the other hand, if the error were 0.1 mg, the digit 7 would be meaningless and should be dropped. Then 3.5 mg would have only one significant figure, namely 3, with an additional digit 5 containing valuable information. In another example, when it is known that the quantity 23.8165 mm is subject to an error of 0.2 mm, it would be ridiculous to carry the last three digits. The result should be rounded off to 23.8 mm. Only the first two digits are significant.

Rounding-off is performed according to the following rules. The digits 1, 2, 3, 4 are *rounded down*, that is the preceding figure is left unchanged. The digits 6, 7, 8, 9 are *rounded up*, that is, the preceding figure is increased by 1. For a 5 some "randomness" should be maintained. It can be enforced by the rule: 5 is rounded down whenever the preceding figure is even, and it is rounded up whenever the preceding figure is odd. Thus 4.65 and 4.75 are rounded off to 4.6 and 4.8, respectively.

Occasionally a rounding rule has to be violated, when a side condition is present. Thus $19.36\% + 34.17\% + 46.47\% = 100\%$ may be rounded off to $19.3\% + 34.2\% + 46.5\% = 100\%$ with a minimal violation for 19.36%. But if 46.47% is a term that is known to be less accurate than the other terms, it is preferable to put $19.4\% + 34.2\% + 46.4\% = 100\%$.

When adding or subtracting approximate numbers, we frequently lose significant figures. For instance, $18.7 + (0.814) = 19.5$ and not 19.514. The last digit of 19.5 is not necessarily significant. Similarly we obtain $(0.493) - (0.4871) = 0.006$ with *at most one* significant digit.

For multiplication and division there exists an easy rule of thumb which we state here without proof:

The result of a multiplication or a division has approximately the same number of significant digits as the term with the fewest significant digits.

Thus in $14.04 \times 2.3/39.7 = 0.813$ the term with the fewest significant digits is 2.3. It has one, at most two significant digits. Hence 0.813 has not more significant digits. The result should be rounded off to 0.81 in case the factor 2.3 has only one significant digit.

Occasionally the number of significant digits is increased. This can happen with the arithmetic mean of ten or more measurements if the dispersion is small enough. For reasons that cannot be explained here, the accuracy of the arithmetic mean is higher than that of the single measurements. For instance, assume ten leaves of a tree are selected at random and their lengths measured in cm are:

5.8   6.1   5.7   5.6   6.2   5.8   5.9   6.2   6.0   5.9 .

Each measurement contains two significant digits. The arithmetic mean calculated by formula (1.9.1) should be written 5.92 and has a higher accuracy than two significant digits.

For lengthy computations it cannot always be recommended to round off intermediary results since this might cause an accumulation of errors. A suitable number of additional digits has to be carried, but the final result should be properly rounded off. A typical example of this sort is given in Problems 1.9.9 and 1.9.10 at the end of this chapter.

*1.13. An Application

To illustrate the use of fractional powers in biology, we consider the shape of quadrupeds (Fig. 1.3). Let l denote the length of a quadruped measured from hip to shoulder and h denote the width which is the average thickness of the body measured in the vertical direction. Then the torso is compared with a uniform bar of length l and width h supported at its ends. Gravity causes sagging of the bar. Physics shows that the ratio

$$l : h^{2/3} \tag{1.13.1}$$

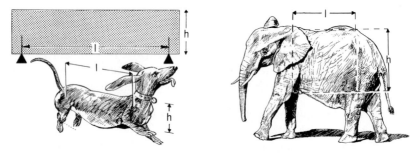

Fig. 1.3. The torso of a quadruped may collapse under gravitation unless its length is limited. The bigger the animal is, the smaller is its length relative to its width

is limited by some value. If the ratio exceeds this value, the bar collapses. For a proof see Rashevsky (1960, Vol. 2, p. 262). Although the torso of an animal is a complicated system consisting of bones, muscles, and tendons and thus quite different from a uniform bar, application of the ratio (1.13.1) indicates roughly the limitation of the length of quadrupeds. Some of the highest values reached by quadrupeds are[12]:

	l	h	$l : h^{2/3}$
Ermine *(Mustela erminea)*	12 cm	4 cm	4.8 : 1
Dachshund	35 cm	12 cm	6.7 : 1
Indian tiger *(Panthera tigris)*	90 cm	45 cm	7.1 : 1
Llama *(Lama glama)*	122 cm	73 cm	7.0 : 1
Indian elephant *(Elephas maximus)*	153 cm	135 cm	5.8 : 1

Most other quadrupeds have a ratio considerably smaller than $7 : 1$. Their torso is less endangered by gravity.

Notice that the ratio depends on the unit of length. If we worked with inches or meters, we would obtain a different ratio.

[12] I am indebted to Mr. H. P. Friedrich for measuring some quadrupeds at the Zoo of Zurich, Switzerland, and to the director of the Zoo, Dr. H. Hediger, for his permission.

As we see from the above table, the shoulder-hip length of a dachshund is almost three times as great as the width. Let us imagine an animal of the same shape but much larger, say $l = 350$ cm and $h = 120$ cm. Then $l : h^{2/3} = 14 : 1$. Such an animal could hardly stand on its feet. The torso would probably touch the ground[13].

The term $h^{2/3}$ can be easily calculated by a pocket calculator. It can also be rewritten by means of a cube root in the following two ways:

$$h^{2/3} = (\sqrt[3]{h})^2 = \sqrt[3]{h^2} .$$

For an elephant with $l = 153$ cm and $h = 135$ cm, we obtain $h^{2/3} = 26.3$. Dropping the units we get $l : h^{2/3} = 153 : 26.3 = 5.8 : 1$ as shown in the above table.

1.14. Survey

The system of real numbers contains the following classes of numbers:

Natural numbers	1, 2, 3, ...
The zero	0
Integers	0, ± 1, ± 2, ± 3, ...
	(the symbol $\pm$ means "plus or minus")
Rational numbers	1/2, 2/3, 8/5, 7/1, $-2/9$, 4.88, -1.5
Irrational numbers[14]	$2^{1/2} = 1.414...$
	$\pi = 3.14159...$
	$\log 2 = 0.30103...$

Notice that natural numbers are a special case of integers and integers a special case of rational numbers.

For real numbers the following laws are valid:

Addition:

commutative law	$a + b = b + a$	
associative law	$a + (b + c) = (a + b) + c$	(1.14.1)

Multiplication:

commutative law	$ab = ba$	
associative law	$a(bc) = (ab) c$	(1.14.2)

---

[13] Advanced knowledge on the stability of animal bodies is offered by Kummer (1959).

[14] A rational number is the result of dividing two integers. The word "rational" is derived from the Latin word "ratio" meaning a fraction. Thus "irrational" is used for a number that cannot be conceived as the fraction of two integers.

Addition and Multiplication:

distributive law $a(b + c) = ab + ac$ (1.14.3)

For the *inverse operations*, subtraction and division, similar laws hold:

$$a - b = (-b) + a$$
$$a - (b - c) = (a - b) - (-c) = a - b + c$$
$$a/b = a \cdot (1/b) = (1/b) \cdot a \quad (b \neq 0) \tag{1.14.4}$$
$$(ab)/(cd) = (a/c) \cdot (b/d) \quad (c \neq 0, d \neq 0)$$
$$(a + b)/c = (a/c) + (b/c) \quad (c \neq 0)$$

Other rules:

$$(a + b)^2 = a^2 + 2ab + b^2 \tag{1.14.5}$$
$$(a - b)^2 = a^2 - 2ab + b^2 \tag{1.14.6}$$
$$(a + b)(a - b) = a^2 - b^2 \tag{1.14.7}$$

For positive or negative integers m and n and for $a \neq 0, b \neq 0$,

$$a^n \cdot a^m = a^{n+m}, \quad a^n/a^m = a^{n-m} \tag{1.14.8}$$
$$(a^n)^m = a^{nm} \tag{1.14.9}$$
$$a^n \cdot b^n = (ab)^n \tag{1.14.10}$$

The formulas (1.14.8) through (1.14.10) are also valid if m and n are fractional numbers and if $a > 0, b > 0$.

Recommended tables for powers, roots, and reciprocals:

Allen (1947), Comrie (1962), Davis and Fisher (1962), Diem *et al.* (1970), Meredith (1967).

Recommended for further reading:

Defares *et al.* (1973), Guelfi (1966), C. A. B. Smith (1966).

Problems for Solution

1.3.2. The earth's surface consists of water (71%) and of land (29%). Two fifths of the land are desert or covered by ice, and one third is pasture, forest or mountainous; the rest is cultivated. What percentage of the total surface of the earth is cultivated?

1.3.3. Each week the world population increases by 1.4 million people. In 1974 the world population was estimated to be 3.8×10^9. Calculate the annual percent increase.

1.3.4. Mexico had a population of 58.4 million people in 1974. Each month the population increased by 175000. Calculate the annual percent increase.

1.3.5. The radium isotope ^{228}Ra loses 9.8% of its intensity of radiation every year. If I_0 denotes the original intensity, what is the intensity after one and two years? Find a formula analogous to (1.3.4) for the intensity I_n after n years.

1.3.6. In a certain tracer method the potassium isotope ^{42}K is used. It loses 5.4% of its intensity every hour. What percentage does it lose within three hours?

1.3.7. Some people are able to taste phenylthiocarbamide as a bitter substance, others find it tasteless. The trait of being a taster or a nontaster is inheritable. In a randomly selected sample the ratio of tasters to nontasters was 1,139 : 461. Calculate the percentages of each group. (Data from Li, 1961, p. 30.)

1.3.8. In a sample from an adult population, geneticists found 219 persons with diabetes mellitus, 380 persons with a mild form of diabetes and 3050 persons with no indication of diabetes. Find the percentages of each group.

1.3.9. 15% of the members of a population were affected by an epidemic disease. 8% of the persons affected died. Calculate the mortality with respect to the entire population.

1.3.10. In a contest 22% of the contestants won an award. Out of the winners 3% got a special distinction. Calculate the percentage of the latter group with respect to the total number of contestants.

1.4.1. Modern medical research tries to develop electrocardiography for automatically diagnosing heart diseases. Assume that it is possible to get a correct diagnosis in 70% of all patients with heart trouble. What meaning may be expressed by a statement such as "by a certain improvement the number of misclassifications is reduced by 20%"? Give two different answers.

1.4.2. Due to the "green revolution" a farmer was able to increase the crop of wheat by 45%. Based on the new figure the next crop was 20% lower. Would the result be the same, if he had first lost 20% and then gained 45%?

1.5.1. Given $a = 4$, $b = 5$, $c = 6$. Calculate $a + b/c$, $(a + b)/c$, $a/b + c$, $a/(b + c)$. Also write the four expressions with the horizontal fraction line.

1.5.2. Calculate $p/q - 3$, $p - q/3$, $p/(q - 3)$, $(p - q)/3$ for $p = 36$, $q = 6$.

1.5.3. Work out the expression $x(x + y) + y(x - y)$ and count how many times the commutative, associative, and distributive laws are applied.

1.5.4. What laws are used by simplifying a) $px \cdot p^2$, b) $m(s + 3) + s(m - 3)$?

1.5.5. Applying the laws of algebra simplify the following calculations:
$$a)\ (17 \times 19) + (13 \times 19),$$
$$b)\ 25 \times 17 \times 4,$$
$$c)\ 33 \times 125 \times 5 \times 8.$$

1.5.6. Calculate a) $(219 \times 67) + (281 \times 67)$, b) $20 \times 3817 \times 5$, c) $4 \times 313 \times 750$ in the most economic way.

1.5.7. Simplify   a) $6!/4!$,   b) $97!/98!$.

1.5.8. Calculate   a) $5!$,   b) $2 + 1/2! + 1/3! + 1/4!$.

1.6.1. Calculate

a) $(-1)^2 + (-2)^2 + (-3)^2$,   b) $(-1)(-2)(-3)$,

c) $(-1)^2 (-2)^2 (-3)^2$,   d) $(+5)(-8)(-200)(+125)$.

1.6.2. Calculate

a) $(-1)^3 + (-2)^3 + (-3)^3$,   b) $(-1)^3(-2)^3(-3)^3$,

c) $1/(-1) + 1/(-2) + 1/(-3)$,   d) $1/(-1)^2 + 1/(-2)^2 + 1/(-3)^2$,

e) $|m - m^2|$ for $m = -5$,

f) $p + |p| + p^2$ for $p = -10$.

1.6.3. Which of the following statements are true and which are false? Correct the false statements.

a) $3/4 > 0.75$,   b) $(-5) < 5$,   c) $(-6) > 5$,

d) $|-5| > 4$,   e) $(-1) < 0 < 1$,   f) $(-5) > (-1) > 0$.

1.6.4. Enumerate the false among the following statements:

a) $(-2) > (-1)$,   b) $|-2| > |-1|$,   c) $|-2| |-1| > 1$,

d) $0.375 < 3/8$,   e) $|-5| < 3$,   f) $5 > |-2| + |-3|$.

1.6.5. For an experiment, mice are seperated into the following weight groups: up to (not including) 20 g, 20 g up to 22 g, 22 g up to 24 g, 24 g up. If w denotes the weight, write the inequalities for w.

1.6.6. A biologist wants to study phytoplankton in relation to salinity of sea water. He decides to subdivide the full range of salinity into three classes, from 3.0% up to and including 3.3%, from 3.3% up to and including 3.5%, and from 3.5% up to 3.8%. Denote the salinity by s and write the three inequalities describing the classes.

1.6.7. Solve the following equations with respect to x:

    a) $8x = x^2$,   b) $px^2 = x$ $(p \neq 0)$,   c) $x = x^3$.

1.6.8. Solve the following equations with respect to q:

    a) $q^2 = 2q$,
    b) $q^3 = 3q^2$,
    c) $a^2 q = q^3$ $(a > 0)$.

1.7.1. Solve the following inequalities:

    a) $x + 4 > 7$,   b) $8 - y > 1$,
    c) $4(u - 5) > 3(5 - u)$,   d) $1 - p^2 \leq (2 - p)(p + 1)$,
    e) $2s^2 < 8$,   f) $3 < t^2$.

1.7.2. a) $z - 4 > 7$,   b) $14 - x < 13$,
    c) $(y + 3)\, 2 < 3(1 - y)$,   d) $x^2 < 16/9$,
    e) $p^2 > 1.44$,   f) $(8 + t)(8 - t) \geq (3 - t)(2 + t)$,
    g) $(u - 3)(2u + 5) < (1 - u)(1 - 2u)$.

1.8.1. The arithmetic mean of 4 and 9 is 6.5. It is greater than the geometric mean which is 6.0. Show that the arithmetic mean of two positive numbers a and b is always greater than or equal to the geometric mean. (Hint: Square both means and reduce the inequality to the obvious statement $(a - b)^2 \geq 0$.)

1.8.2. Show that in the sequence 1, 2, 4, 8, 16, 32, ... each term is the geometric mean of the two neighboring terms.

1.9.1. Write the following sums by means of the summation sign:

    a) $x_1 + x_2 + x_3 + x_4 + x_5$,   b) $z_0 + z_1 + \cdots + z_k$,
    c) $a_3 + a_4 + \cdots + a_n$,   d) $a_1^2 + a_2^2 + a_3^2 + a_4^2$,
    e) $(a_1 + b_1)^2 + (a_2 + b_2)^2 + \cdots + (a_N + b_N)^2$.

1.9.2. Write the following sums without summation sign:

a) $\sum_{i=1}^{5} p_i$, b) $\sum_{k=3}^{5} s_k$, c) $\sum_{j=0}^{n} x_j$,

d) $\sum_{i=1}^{N} x_i^2$, e) $\sum_{k=0}^{4} 2t_k$, f) $\sum_{j=1}^{3} (x_j + y_j)^2$.

1.9.3. Show that $\sum_{i=1}^{n} (x_i + y_i) = \sum_{i=1}^{n} x_i + \sum_{i=1}^{n} y_i$.

1.9.4. Show that $\sum_{k=1}^{m} cu_k = c \sum_{k=1}^{m} u_k$.

1.9.5. Show that $\sum_{j=1}^{n} (x_j + a) = \sum_{j=1}^{n} x_j + na$.

1.9.6. Show that $\sum_{i=1}^{n} (x_i + a)^2 = \sum_{i=1}^{n} x_i^2 + 2a \sum_{i=1}^{n} x_i + na^2$.

1.9.7. Five herring have the following weights x_i in grams:

77.8   75.7   72.3   81.4   76.3.

Find
a) the arithmetic mean $\bar{x}$,
b) the differences $x_i - \bar{x}$,
c) the sum of differences as a check of calculation,
d) the sum of squares of deviations.
(Hint: Use the scheme at the end of Section 1.9.)

1.9.8. Solve the same problem for the following x_i values:

8.3   8.9   9.1   8.5.

1.9.9. By applying formulas (1.14.6) and (1.9.1) show that $\sum_{i=1}^{n} (x_i - \bar{x})^2$
can be rewritten in the following three forms which are often useful in computation:

a) $\Sigma x_i^2 - n\bar{x}^2$, b) $\Sigma x_i^2 - \dfrac{1}{n}(\Sigma x_i)^2$,
c) $\Sigma x_i^2 - \bar{x}(\Sigma x_i)$.

1.9.10. Apply the three formulas of Problem 1.9.9 to the numerical data of problem 1.9.8. What are advantages and disadvantages of the different formulas?

1.10.4. Convert the following powers into fractions:

$$x^{-1} \qquad u^{-2} \qquad (a+b)^{-3} \qquad a^{-2}b^{-3} \qquad (3s-t)^{-1}.$$

1.10.5. Write the following fractions by means of negative powers:

$$1/p \qquad 1/(a+b) \qquad 2/p^5 \qquad 5/(x-z) \qquad (u-v)/(u+v).$$

1.10.6. The 1967–68 eruption of the volcano Halemaumau on Hawaii produced $84 \times 10^6\ m^3$ of lava in 8.2 months. What is the average monthly production? (Data from Swanson, 1972.)

1.10.7. The shorelines of the earth (including lakes and the Arctic and Antarctic regions) total about 440000 km in length (over ten times the earth's perimeter). If everyone of the world population (3.8×10^9 in 1974) decided to possess a portion of this shoreline, how much would he have on the average? (See Inman et al., 1973, p. 26.)

1.10.8. In our civilization each single person needs $60\ m^2$ for housing, $40\ m^2$ for his job, $50\ m^2$ for public buildings and facilities for sports, $90\ m^2$ for traffic, and $4000\ m^2$ for production of his food on the average. Some nations are overpopulated. Take for instance Switzerland with her 6.4 million residents. Arable and habitable land amounts to $11000\ km^2$. For how many people could Switzerland provide adequate space?

1.10.9. Every cm^2 of the earth's surface is loaded by the mass of 1.0 kg of air. The earth's surface amounts to $5.1 \times 10^8\ km^2$.

(a) Calculate the mass of the atmosphere.

(b) What is the mass of O_2? (22% of the total mass is oxygen.)

1.10.10. One km^2 of young forest produces about 2.5×10^5 kg oxygen yearly. What proportion is it of the total mass of atmospheric oxygen above $1\ km^2$ of the earth's surface? (Mass of atmospheric oxygen is from Problem 1.10.9.)

1.10.11. It is estimated that all green plants on the earth (including plankton) produce 0.9×10^{13} kg O_2 yearly. This is the net production which does not include the amount of O_2 consumed by the plants themselves. How many years would it take to build up the oxygen of the atmosphere if no animal life and no fire consumes it? (Use result from Problem 1.10.9.)

1.10.12. Cells in living tissue are remarkably uniform in size. The length of a typical cell is around 3 μm (micrometer). Calculate the volume (in $μm^3$) under the assumption that the shape is spherical.

1.10.13. Wind-generated surface waves dissipate their energy primarily nearshore, mainly in the breaker zone. With an average height of about 1 m, the waves transmit power at a rate of about 10 kW (kilowatt) per meter shoreline.

   (a) How much heat energy will be produced per second (1 kcal/sec = 4.19 kW)?

   (b) If all the heat per meter shoreline went into 100 m^3 of coastal water, how fast would temperature rise? (1 kcal increases temperature of 1 kg water by 1° C. Data from Inman et al., 1973.)

1.10.14. How much do orders of magnitude differ from each other in the following pairs:

   a) Wave length of red light = 0.76 μm,
      Wave length of green light = 0.48 μm (μm = micrometer).

   b) Wave length of blue light = 0.42 μm,
      Length of a tissue cell = 3 μm.

   c) Perimeter of earth = 4.0 × 10^4 km,
      Distance earth-moon = 3.8 × 10^5 km.

   d) Age of earth = 4.5 × 10^9 years,
      Age of hominoids = 10^7 years.

   e) Half-life of ^{14}C = 5.77 × 10^3 years,
      Period of European history = 2500 years.

1.10.15. Each liter of gasoline contains 0.1 to 0.4 g of lead. Assuming that the average consumption of a car ranges from 1200 to 1400 liters per year, how much lead is then dissipated by the 10^6 cars of a large city? Find an upper and a lower bound.

1.10.16. A million liters of fresh water mixed with one liter of mineral oil is unpalatable. What amount of infiltrating mineral oil would suffice to destroy the 1.5 × 10^{10} liters of ground-water which serve as water supply of a town with 100000 people for one year?

1.10.17. Some nuclear power stations use water from a river as coolant. Assume that the station takes 30 m^3 water per second from the river and increases its temperature by 10° C. The total water flow (including the coolant) is 200 m^3/sec. When the heated water mixes with the river, by how many degrees will the temperature of the river rise? (Thermal pollution.)

1.10.18. When cereals are seeded, 5 kg of herbicides are distributed per ha to suppress the growth of weeds. How many metric tons of herbicide would be required to cover an area as large as the state of Iowa (146 × 10^3 km^2). (Note that 1 ha = 10000 m^2.)

1.10.19. Polychlorinated biphenyls (PCB's) are toxic substances and are serious pollutants since they degrade very slowly. They have a wide range of technological applications: heat transfer mechanisms, insulating fluids, cooling systems, large power transformers, automobile tires, brake linings, lubricants, and paints. Some of the PCB's enter the environment. In the North Atlantic the average concentration of PCB's in surface water (0–20 m deep) is 35 ng (nanogram) per liter water. The surface of the North Atlantic measures 5×10^{12} m². Estimate the total amount of PCB's in the surface water. (Data taken from Harvey et al., 1973.)

1.10.20. Measurements throughout 1970 have shown that particles suspended in the air of Chicago have an average mass of 0.45×10^{-6} µg (µg = microgram). The average density of these particles was 86 µg/m³. If all particles were of the same size, how many of them would be in one cubicmeter of air? (Data from Lee, 1972.)

1.10.21. The diameter of an H_2O molecule is approximately 2.5×10^{-10} m. In 1 mole = 18 g of water there are 6.02×10^{23} molecules (Avogadro's number). How long would a "chain" of these molecules be? Compare the result with the distance from earth to sun, which is approximately 1.5×10^8 km.

1.10.22. Spores of ferns float in the atmosphere and are distributed all over the earth by winds. They return to earth only by rain. What mass has a spherical spore of diameter 30 µm, if the density is assumed to be 1.0 g cm⁻³? (Volume of sphere $= 4/3 \cdot \pi r^3$.)

1.10.23. Solve the same problem for a spore with a diameter reduced by 20%.

1.10.24. The mechanical work that is done when one stroke volume of about 100 ml of blood is pumped through the vascular tree against a pressure of about 100 mm Hg is 1.32 J. This energy is transformed to heat by friction. Find the increase in temperature. (Hint: With the energy of 4200 J, one liter of water or blood can be heated by 1° C). This problem is adapted from Randall, 1958, p. 161.

1.10.25. Reduce to lowest terms

a) $x^2 y^2/axy$,

b) $a^3 bc/ab^2 c$,

c) $(a^2 + 2ab)/(2a - a^2)$,

d) $(x^2 - y^2)/(x + y)$.

1.10.26. Reduce to lowest terms

    a) $s^3 t / p s t^4$,
                            b) $A^2 B^2 C^2 / A B^2 C^3$,

    c) $(3pq - p^2)/(6q - 2p)$,
            d) $(m - n)/(m^2 - n^2)$.

1.10.27. Perform the following additions and subtractions:

    a) $(1/x) + (1/y)$,
                     b) $(1/2) - (1/t)$,

    c) $(1/u) - (1/u^2)$,
                 d) $(5/3x^2 y) - (2/xy^2)$.

1.10.28. Perform addition and/or subtraction:

    a) $(1/x^2) - (1/xy)$,
                b) $1/(u - 3) + 1/(u + 3)$,

    c) $(7/a^3 b) + (5/a^2 b^2)$,
         d) $x/(x + 1) - 3x/(x - 2)$.

1.11.1. Write the following roots as fractional powers:

$$\sqrt{7} \quad \sqrt[3]{10} \quad \sqrt{A} \quad \sqrt{a+b} \quad \sqrt[3]{1-x} \quad \frac{1}{\sqrt{3}} \quad \frac{1}{\sqrt[3]{p}}.$$

1.11.2. Write the following powers as roots:

$$8^{1/2} \qquad 2^{3/4} \qquad (p+q)^{1/3} \qquad 5^{-1/2} \qquad x^{-1/4} \qquad (1-z)^{-2/3}$$

1.11.3. Simplify $\sqrt{a} \cdot \sqrt[4]{a} \cdot \sqrt{a^3}$.

1.11.4. Simplify $\sqrt[3]{p^2} \cdot \sqrt{p^3} \cdot \sqrt[5]{p}$.

1.11.5. By what factor does $1773^{1/2}$ differ from $0.1773^{1/2}$?

1.11.6. By means of Table B find the square roots of

        8830,   883,   88.3,   8.83,   0.883,   0.0883.

1.11.7. Let $b > a$. Find $\sqrt{(a-b)^2}$.

1.12.1. Round off the following numbers to one decimal:

        8.06   4.01   18.35   20.85   0.7445   0.1555

1.12.2. Round off the following percentages to integers such that their total remains 100%: 78.3%, 11.4%, 10.3%. Consider 78.3% as less accurate than the other two percentages.

1.12.3. Calculate 25% of 108.52 if a) 25% is a precise percentage; b) 25% is an approximate percentage with an error of not more than 1% (the range is 24% to 26%).

1.12.4. Let 372 have three significant figures. How many significant figures has $372^2 = 138384$? (Hint: Square 371.5 and 372.5 and compare the differences.)

1.12.5. Digital calculators and computers work with a fixed number of digits while the decimal point is floating. By means of the example $a = 2.56$ and $b = 2.54$ and a capacity of only three digits show that the two expressions $a^2 - b^2$ and $(a+b)(a-b)$ do not lead to the same numerical results.

1.12.6. Given three measurements $x_1 = 6.08$, $x_2 = 6.06$, $x_3 = 6.13$ with mean $\bar{x} = 6.09$. Calculations should be performed at each step with three digits only, while the decimal point is floating. Calculate $\Sigma (x_i - \bar{x})^2$ and show that the algebraically equivalent formula $\Sigma x_i^2 - 3\bar{x}^2$ fails to give a useful result.

1.12.7. The difference between a measured value and the exact value of a quantity is called the *absolute error* (For simplicity neglect a possible minus sign). When we divide the absolute error by the exact value, we obtain the *relative error*. It is usually written as a percentage. Find the relative errors in the following examples:

Measured value	Exact value
a) 3.4 µm	3.15 µm
b) 0.88 ml	1.06 ml
c) 80 sec	66.7 sec

1.12.8. Solve the same problem in the examples

Measured value	Exact value
a) 13.5 nm	12.9 nm
b) 125 kHz	108 kHz
c) 0.17 µg	0.21 µg

1.14.1. Work out $(a - b)^2 (a + b)^2$ in the most economic way.

1.14.2. Find the error in the following calculation: Assume that a and b are equal numbers. Then $a = b$, $ab = b^2$, $ab - a^2 = b^2 - a^2$, $a(b - a) = (b + a)(b - a)$, $a = b + a$, $a = a + a$, $a = 2a$, $1 = 2$. (Hint: Read end of Section 1.6.)

Chapter 2

Sets and Symbolic Logic

2.1. "New Mathematics"

What is today called "new mathematics" is a popular expression for a development of mathematics which began early in the last century. For a long time mathematics was considered an area of research dealing with *quantities* such as distances, areas, angles, and weights. Indeed, elementary algebra as well as calculus indicate that countable and measurable items are objects of mathematical operations. However, mathematicians became slowly aware that logical problems, axioms, operations with abstract symbols, and the structure of space are more typical of mathematical thinking. Symbolic logic and set theory were founded before 1850 and abstract algebra before 1900. Today it is a mistake to define mathematics just as the study of quantities.

Applied mathematics was slow in adjusting to the new aspect. A breakthrough came around 1925 when quantum physicists succeeded in applying group theory, an area dealing with abstract elements and an operation on the elements. Since then modern mathematics has rapidly entered other areas of scientific research including the life sciences.

In this chapter we are mainly concerned with sets and operations on sets. At the end of the chapter we will discover a relationship between set theory and logic.

2.2. Sets

The language contains many words to designate a collection of objects. Biologists use categories such as order, family, genus to collect animals and plants that have certain characteristics in common. Economists subdivide the population of a country into different social classes. When statisticians select individuals from a population, they use the word sample. And when they classify measurements, they speak of groups. Psychologists deal with batteries of tests and physicians with syndromes, that is with groups of symptoms. All the words collection, selection, totality, category, class, group, etc., have some common meaning. For a rigorous treatment mathematicians prefer the term *set*.

A set is a collection of any kind of objects – people, animals, plants, phenomena, stimuli, responses, genetic traits, methods, ideas, logical possibilities. *A set is well defined when it is clear whether an object belongs or does not belong to the set.* Ambiguity is not allowed. From now on we will use the term *element* or *member* of a set rather than object. A set may contain a *finite* or an *infinite* number of members. Correspondingly it is called a *finite* or an *infinite* set. A set may contain only a single member. Even the *empty set* or *null set* that contains no member whatsoever turns out to be a useful concept.

Examples

1. It is easy to define the set of patients in a hospital in an unambiguous way, but it is hardly possible to establish the set of patients suffering from rheumatism since the term rheumatism is used in an ambiguous way and since a diagnosis is sometimes doubtful even for a well-defined disease.

2. It is easy to define the set of all plants that produce O_2, since they contain chlorophyll. However, it is difficult if not impossible to define the set of broad-leaved plants. The judgment of "broad-leaved" is subjective and causes ambiguity.

3. The set of all countries located on the Australian continent consists of one single member, namely Australia. The set of all countries located on a polar cap is an empty set.

4. The set of all chemical elements is well defined. It is finite, but not yet completely known. It is also clear what we mean by the set of all chemical compounds although most of them have yet to be discovered. This set is probably infinite.

5. Ecologists are interested in the set of resources and in the set of environmental conditions usually called habitat. Notice that the two sets have common members such as light. This example and subsequent ecological examples are adapted from Patten (1966).

2.3. Notations and Symbols

We denote sets by capital letters such as $A, B, C, \ldots$. The empty set has a standard symbol Ø, a figure zero crossed by a bar. Members of a set are collected by braces. For example

$$S = \{\text{sight, hearing, smell, taste, touch}\}$$

is the set of the traditional five senses.

For the infinite set of *natural* numbers a standard notation is used:

$$\mathbb{N} = \{1, 2, 3, \ldots\} \tag{2.3.1}$$

Similarly the set of all *real* numbers is denoted by $\mathbb{R}$.

To indicate that an object is a member of a set we use the symbol $\in$ which was originally the Greek letter epsilon[1]. Thus

$$a \in T \tag{2.3.2}$$

means that "*a* is a member of *T*" or that "*a* belongs to *T*". The contrary is expressed by $\notin$ meaning "is not a member of" or "does not belong to". For instance

$$5 \in \mathbb{N}, \quad \tfrac{1}{2} \in \mathbb{R}, \quad \tfrac{1}{2} \notin \mathbb{N}.$$

Two sets are said to be *equal*, in symbols

$$A = B, \tag{2.3.3}$$

if they contain exactly the same elements. If x is an element, then $x \in A$ implies $x \in B$ and vice versa. The equality (2.3.3) also means that $x \notin A$ implies $x \notin B$. Since $\{1, 2, 3\}$ and $\{3, 1, 2\}$ have the same members, they are equal sets. Moreover, $\{1, 2, 3\} = \{1, 2, 2, 3\}$ for the same reason.

For a set A containing only elements of a set B, but not necessarily all the members of B we write

$$A \subset B \tag{2.3.4}$$

and say "A is contained in B". A is called a *subset* of B. The symbol $\subset$ reminds us of $<$ for "less than", but has of course a different meaning. The definition of a subset allows us to say that a set is a subset of itself, e.g. B is a subset of B:

$$B \subset B.$$

If we want to say that A is a *proper* subset of B, that is, contains fewer members than B, we may write

$$A \subset B, \quad A \neq B. \tag{2.3.5}$$

The empty set is considered as a subset of every set, that is,

$$\emptyset \subset A. \tag{2.3.6}$$

The same statement as in (2.3.4) may be made by

$$B \supset A. \tag{2.3.7}$$

We read "B contains A". The analogy with $>$ is obvious[2].

---

[1] The symbol $\in$ was proposed by the Italian mathematician Giuseppe Peano (1858—1932).

[2] The symbols $\supset$ and $\subset$ were proposed by the German mathematician Ernst Schröder (1841—1902).

Frequently, sets of any sort of members are represented by *sets of points*. For simplicity of drawing, points in a circle or in a rectangle are used. In this way, the relationship (2.3.4) is graphically represented in Fig. 2.1.

Such a presentation is called a *Venn diagram*[3].

Examples

1. The set of three numbers $\{1, 2, 3\}$ has the following subsets: $\{1, 2, 3\}$ itself and the proper subsets $\{1, 2\}$, $\{1, 3\}$, $\{2, 3\}$, $\{1\}$, $\{2\}$, $\{3\}$, and the empty set $\emptyset$. Notice that $\{3\}$ is not the same as 3. The first symbol

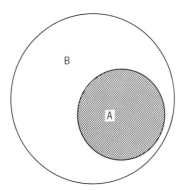

Fig. 2.1. Venn diagram of the relationship "A is contained in B"

designates a set, the second a member. Consequently we write $\{3\} \subset \{1, 2, 3\}$, but $3 \in \{1, 2, 3\}$. Also note $\emptyset \subset \{1, 2, 3\}$, but $0 \notin \{1, 2, 3\}$.

2. Certain minerals constitute a subset of the set of nutrients.

3. For a social worker the blind persons form a subset of all invalids, and the invalids a subset of all handicapped people. This relationship may be symbolically written as $B \subset I \subset H$.

4. There are roughly 5000 species of ants. Each species is a member of one of 170 genera. Each genus is a subset of the family *Formicidae* and this family is a subset of the order *Hymenoptera*.

5. Let A be the set of patients with any type of chest pain and B be the set of patients with Angina pectoris. Then $A \supset B$.

6. Consider quadrilaterals as geometric configurations. Let R be the set of all rectangles (quadrilaterals with four right angles) and P be the set of all parallelograms (quadrilaterals with two pairs of parallel sides). Then $R \subset P$, since each rectangle has two pairs of parallel sides.

---

[3] After the English logician and theologian John Venn (1834—1923). In Venn diagrams, size and shape of the figures are not related to the number of members in each set.

7. The set of natural numbers is a subset of the set of all real numbers. Thus we write

$$\mathbb{N} \subset \mathbb{R}.$$

2.4. Variable Members

For large sets it is inconvenient or impossible to list the members. Instead we try to characterize the members by words or by mathematical statements. For instance, we are unable to enumerate all numbers greater than 5, because this set includes not only integers, but also rational and irrational numbers. Therefore, we introduce a variable member, say x, and characterize the members by stating $x > 5$. The set is then written

$$\{x \,|\, x > 5\}.$$

We read "the set of all numbers x such that x is greater than 5". The vertical bar means "such that" or "given that"[4].

x need not be a number. For instance, let x stand for any chemical compound. Then the set of acids is denoted by

$$\{x \,|\, x = \text{acid}\}.$$

Set theory may be applied to present the solutions of mathematical problems. For instance the equation $x^2 = 4$ has the two solutions $x = 2$ and $x = -2$. These values form the *solution set* of our equation. Hence we may write

$$\{x \,|\, x^2 = 4\} = \{2, -2\}.$$

Similarly we find

$$\{t \,|\, 3t - 4 = 5\} = \{3\}.$$

We add three examples of inequalities and their solutions:

$$\{x \,|\, 2x + 5 < 7\} = \{x \,|\, x < 1\},$$
$$\{y \,|\, y^2 < 4\} = \{y \,|\, -2 < y < 2\} = \{y \,|\, |y| < 2\},$$
$$\{z \,|\, z^{\frac{1}{2}} < 8\} = \{z \,|\, 0 \le z < 64\}.$$

2.5. Complementary Set

When a population is subdivided into two parts, say into sick and healthy individuals, into male and female, or into active and passive we speak of complementary parts or sections. In set theory we would use a slightly different language. Let U be the set whose subdivision is being

---

[4] Some authors prefer a colon to the bar and write for our example $\{x : x > 5\}$.

studied. We call U *the universal set*. Let A be a subset of U, that is $A \subset U$. Then we are interested in those members of U that do not belong to A. They form a new set which is called the *complement of A in U*. We denote this complement by $\bar{A}$ using a horizontal bar[5]. The Venn diagram for $\bar{A}$ is shown in Fig. 2.2.

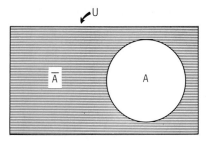

Fig. 2.2. Venn diagram of the complement

Whenever two sets are given, such as U and A, and we form a new set from the given sets, in our case $\bar{A}$, we perform an *operation* on sets. In the following two sections we will define other operations.

In symbolic writing the complement A in U is defined by

$$\bar{A} = \{x \mid x \in U,\ x \notin A\}. \tag{2.5.1}$$

Examples

1. If $U = \{1, 2, 3, 4, 5\}$ and $A = \{2, 5\}$, then $\bar{A} = \{1, 3, 4\}$.

2. Consider all species belonging to the order *Hymenoptera* as the universal set U and the family of *Formicidae* as a subset F of U. Then $\bar{F}$ consists of all species belonging to *Hymenoptera*, but not to *Formicidae*.

2.6. The Union

With two arbitrary sets, say A and B, we can always form a new set, say C, by simply combining the members. We call this new set the *union* and write symbolically

$$C = A \cup B. \tag{2.6.1}$$

We read it "A union B". The symbol $\cup$ reminds us of the letter u in the word "union". C contains exactly those members which are contained in A or in B or in both of them. The same element occurs only once in the union. Fig. 2.3 shows the Venn diagram of the union.

---

[5] There exists unfortunately no standard notation for the complement. Instead of $\bar{A}$ some authors write A', $\tilde{A}$, $\sim A$, A^c or $\complement A$.

The operation $\cup$ resembles somewhat an addition. However, it should be noted that $A \cup A = A$ and, if $B \subset A$, that $B \cup A = A$.

In symbolic writing the union of two sets A and B is defined by

$$A \cup B = \{x \mid x \in A \quad \text{or} \quad x \in B\}. \tag{2.6.2}$$

Notice that the word "or" is used here in the sense "and/or".

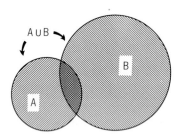

Fig. 2.3. The members of A and B are combined to a new set, called the union

Examples

1. $\{1, 2\} \cup \{1, 5, 6\} = \{1, 2, 5, 6\}$. The member 1 is listed once.

2. Patients suffering from either *Rhinitis atopica* (hay fever, etc.) or from *Urticaria* are considered to be allergic. These patients form two sets. The union consists of all patients having *Rhinitis atopica* or *Urticaria* or both diseases.

3. An ecologist may be interested in the union of the sets "resources" and "habitat" when he studies the totality of influences on a biota.

4. A primitive mammal, the tenrec, makes an ultrasound when the light intensity is changed (stimulus set L) or when a noise is made (stimulus set N). Then $S = L \cup N$ is the set of all known stimuli that produce the ultrasonic response[6].

5. If A is a subset of a universal set U and $\bar{A}$ the complement of A, then the union of A and $\bar{A}$ is the universal set, that is,

$$A \cup \bar{A} = U.$$

2.7. The Intersection

In addition to the operations "take the complement" and "take the union" we will introduce a third operation of great practical value. Consider first two intersecting straight lines. Both lines can be conceived as infinite sets of points. The two sets have one point in common, the point of intersection. More generally, let A and B be any two sets.

---

[6] I am indebted to Dr. Jack P. Hailman, University of Wisconsin, for this example.

We may be interested in the question whether the two sets are over-lapping, that is, whether the two sets have some members in common. We call the set of all common members, whether it is empty or not, the *intersection* of A and B and write

$$D = A \cap B. \tag{2.7.1}$$

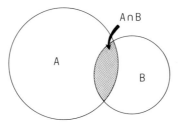

Fig. 2.4. The common members of A and B form a new set, called the intersection

We read "D is equal to A intersection B". If A and B have no common member, then D is an empty set, that is, $A \cap B = \emptyset$. The two sets are then called *disjoint*.

The symbol $\cap$ was not derived from a letter of the alphabet. It was simply chosen in contrast to the symbol $\cup$ for union[7].

The Venn diagram shown in Fig. 2.4 gives an intuitive idea of the new concept.

In symbolic writing the intersection of two sets A and B is defined by

$$A \cap B = \{x \mid x \in A \quad \text{and} \quad x \in B\}. \tag{2.7.2}$$

Notice that the word "and" is used here in the sense "both…and…". The reader should compare (2.7.2) with (2.6.2) and learn that "and" is associated with "intersection" and "or" associated with "union".

Examples

1. $\{3, 4, 5, 6\} \cap \{4, 6, 8\} = \{4, 6\}$, $\{1, 2\} \cap \{3, 4\} = \emptyset$.

2. Two families of the same order are by definition disjoint, that is, they have no species in common. Their intersection is an empty set.

3. Consider lines as infinite sets of points. The intersection of two straight lines in a plane is a single point if the lines have different directions, or the empty set if they are parallel, or a line if the two given lines fall together.

---

[7] The symbols $\cup$ and $\cap$ were proposed by the Italian mathematician Giuseppe Peano in 1888. Occasionally the symbol $\cup$ is called a *cup* and the symbol $\cap$ a *cap*.

4. In an ecosystem let E denote the set of all environmental factors (habitat) and R denote the set of all resources. Then light is a member of both E and R since it acts as stimulus for activity and is also a resource required for the synthesis of chemicals. If l denotes light, we may write $l \in (E \cap R)$ or, as usual, without parentheses $l \in E \cap R$.

5. Consider the MN blood groups. Each individual has the antigen M or the antigen N or both of them. Let U be a population, $\mathcal{M}$ and $\mathcal{N}$ the subsets of individuals having the antigens M and N, respectively[8]. Then $\mathcal{M} \cap \mathcal{N}$ is the subset of all individuals having both antigens.

6. Consider quadrilaterals as geometric configurations. Let R be the set of all rectangles (quadrilaterals with four right angles) and S be the set of all rhombuses (quadrilaterals with four equal sides). Then the squares form the intersection $R \cap S$. Indeed, squares (and only squares) have the property of having four right angles and four equal sides.

7. Among mammalian cells, if I is the set of all insulin-producing cells and A is the set of all antibody-producing cells, then $I \cap A = \emptyset$ because these two specialized molecules are produced by entirely distinct cell types. If, however, we designate by A' the subset of all cells producing antibody of one specificity, and by A'' the subset of all cells producing antibody of some other specificity, then whether $A'' \cap A' = \emptyset$ is a question which has been highly debatable. At present immunologists would probably agree that the intersection is empty for any two distinct specificities.

8. If A is a subset of a universal set U and $\bar{A}$ the complement of A, then

$$A \cap \bar{A} = \emptyset .$$

9. If A is a subset of B, that is $A \subset B$, then $A \cap B = A$.

Still another operation is subtraction of sets. Since there is only a slight analogy to subtraction of numbers, we replace the ordinary minus sign $(-)$ by a slant bar: $\smallsetminus$. We define

$$A \smallsetminus B \qquad\qquad (2.7.3)$$

to be the set of all elements of A which do not belong to B. In symbols:

$$A \smallsetminus B = \{x | x \in A \quad \text{and} \quad x \notin B\} \qquad\qquad (2.7.4)$$

Fig. 2.5 depicts the meaning of $A \smallsetminus B$ and $B \smallsetminus A$ by a Venn diagram.

---

[8] In the same problem different notions should be designated by different symbols. Thus, if M and N signify the antigens, we have to find other symbols for the sets, for instance script letters, $\mathcal{M}$ and $\mathcal{N}$, or letters with asterisks, M^* and N^*, or letters from the Greek alphabet.

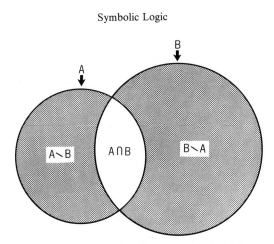

Fig. 2.5. Subtraction of sets. $A \smallsetminus B$ contains all elements of A which do not belong to B

Examples

1. $\{3, 4, 5, 6\} \smallsetminus \{5, 6, 7\} = \{3, 4\}$,
2. The complementary set as defined in Section 2.5 may be written as follows:

$$\bar{A} = U \smallsetminus A .$$

3. Let C be the set of all carnivorous and H be the set of all herbivorous animal species. Then $C \smallsetminus H$ is the set of "purely" carnivores. For instance, lions and tigers belong to $C \smallsetminus H$, bears do not.

*2.8. Symbolic Logic

There are many research workers who believe that logic is part of our common sense and that it needs no further development. In their opinion formal logic is superfluous or, at least, it would not offer additional help in their scientific work. However, these people overlook the ambiguity and the lack of clarity of many statements in science and in daily life. They also forget that scientific thinking becomes more and more complex and calls for a simpler and faster way of presentation.

A story may illustrate the ambiguity involved in communication. A research group working in human genetics was concerned with the relationship between allergic diseases. The head of the group asked his assistant to collect from the hospital's catalog all cases with asthma bronchiale and with atopic rhinitis (hay fever, etc.). On a slip of paper he wrote symbolically $A + \mathrm{Rh}$. The assistant as well as the head was convinced that the task was quite clear. The assistant inspected thousands of cards and made notes about all patients suffering from asthma bronchiale as well as atopic rhinitis, that is from *both* diseases. Unconsciously he interpreted the symbol + in the sense of "as well as". The head of the group, however, meant "and/or" with his symbol +. He wanted all patients listed who suffered from at least one of the diseases. Thus the assistant had to do the time-consuming work over again.

With this section on symbolic logic we pursue two purposes, first to provide an unambiguous language and second to find a universal shorthand, thus avoiding lengthy and wordy sentences. Progress in science was accelerated when relationships such as the equation $x^2 - 2x + 5 = 0$ or the chemical reaction $CO_2 + H_2O \rightarrow H_2CO_3$ could be written without words.

We begin with the term "proposition". There exists no explicit definition. But we call any statement *a proposition* if the statement is either *true or false*. Further distinctions such as "nearly true" or "sometimes false" are excluded. For instance, "the earth has the general shape of a ball" or "the earth is a flat disk" are two propositions. The first of them is true, the second is false. The proposition "127 is a prime number" cannot be quickly verified, but there is no doubt that it is either true or false.

We denote propositions by capital letters such as $P, Q, \ldots$. The symbol $\vee$ is used for the word "or". Then if P and Q are propositions, we generate a new proposition writing

$$P \vee Q \quad \text{(read "}P\text{ or }Q\text{")} . \tag{2.8.1}$$

The new proposition is true if P is true, or if Q is true, or if both P and Q are true. Otherwise $P \vee Q$ is false[9].

For instance, the proposition "Adrenaline is an enzyme or a hormone" is true, since adrenaline is a hormone. But the proposition "Adrenaline is a vitamin or an enzyme" is false, because adrenaline is neither a vitamin nor an enzyme.

To find a bridge from logic to set theory we use an illustration. Let x be a variable meaning any chemical compound and let

$$E: \text{set of all enzymes} .$$

Then we consider the proposition

$$P: \ x \text{ is an enzyme} .$$

---

[9] The symbol $\vee$ was originally the letter v in the Latin word "vel" meaning "or", or more precisely "and/or". The meaning should not be confused with "either – or" which is expressed in Latin by "aut – aut". The symbol $\vee$ came into common use in 1910 and has been since then a standard symbol.

This proposition is true for every $x \in E$, but false for every $x \notin E$. Thus we established the following correspondence:

Sets	Proposition
$\in$	true
$\notin$	false

Now define the sets

$$E: \text{ set of all enzymes},$$
$$H: \text{ set of all hormones},$$

and the corresponding propositions

$$P: x \text{ is an enzyme},$$
$$Q: x \text{ is a hormone}.$$

From (2.6.2) it follows that

$$E \cup H = \{x \,|\, P \vee Q\},$$

or in words: The union of the sets of all enzymes and hormones consists of all compounds x for which the proposition P or Q is true. This shows how "union" and "or" are related with each other. Notice the similar shape of the symbols $\cup$ and $\vee$.

The symbol $\wedge$ is used to designate the logical operation "and" in the sense of "as well as" or "both ... and ...". $\wedge$ is not derived from a letter of the alphabet but was chosen in contrast to the sign $\vee$ for "or"[10]. If P and Q denote propositions,

$$P \wedge Q \quad \text{(read "}P\text{ and }Q\text{")} \tag{2.8.2}$$

is a new proposition which is true if P is true and Q is true. Otherwise, the proposition $P \wedge Q$ is false.

For instance, the proposition "the brown bear is carnivorous and herbivorous" is true, because this bear eats animal food as well as plant food. But the analogous proposition about a lion instead of a bear would be false.

To interconnect again logic with set theory we use an illustration. Let x be a variable meaning any mammal, and let C and H denote the sets of carnivores and of herbivores, respectively. Then we consider the

---

[10] The symbol $\wedge$ for "and" is not yet standard. Some mathematicians prefer a squarelike dot $\bullet$ or the familiar symbol $\&$.

two propositions

$$P:\ x \text{ is carnivorous},$$
$$Q:\ x \text{ is herbivorous}.$$

Then from (2.7.2) it follows that

$$C \cap H = \{x \mid P \wedge Q\},$$

or in words: The intersection of the sets of carnivorous and herbivorous mammals consists of all mammals for which the propositions P and Q are both true. This shows how "intersection" and "and" are related with each other. Notice the similar shape of the symbols $\cap$ and $\wedge$.

*2.9. Negation and Implication

The logical negation is another operation demonstrating the analogy between set theory and symbolic logic. Let P be any proposition. Then

$$\neg P \quad \text{(read "not } P\text{")}$$

signifies a proposition which is false if P is true, and true if P is false[11]. To show how the negation is formed we use an example: The negation of "Birds have feathers" is "It is not true that birds have feathers". The proposition "The ewe is a male sheep" is false, since an ewe is a female sheep. The negation is "The ewe is not a male sheep" which is true. Notice that the negation does not claim that the ewe is a female sheep.

Consider now the universal set U of all vertebrates and let x be a variable meaning any kind of vertebrate, that is, $x \in U$. The class of birds forms a subset, say B, that is $B \subset U$. Let P be the proposition "x has feathers". Then $\neg P$ means "x has no feathers". This proposition corresponds to the complement $\bar{B}$ in U (see Section 2.5). Hence we get the relationship

$$B = \{x \mid P\}, \qquad \bar{B} = \{x \mid \neg P\} \tag{2.9.1}$$

indicating the correspondence between "complement of a set" and "negation of a proposition"[12].

---

[11] The symbol $\sim$ for the logical negation has been used since 1910. Now the symbol $\neg$ is in common use.

[12] Sometimes the language of mathematics uses the negation to avoid listing two or more properties. Thus "a non-negative number" is a convenient phrase to replace "a number which is positive or zero".

The meaning of the logical symbols $\lor$, $\land$, $\neg$ may be summarized by a so-called *truth table*:

Given		It follows		
P	Q	$P \lor Q$	$P \land Q$	$\neg P$
true	true	true	true	false
true	false	true	false	false
false	true	true	false	true
false	false	false	false	true

There is a further analogy between symbolic logic and set theory. The statement "set K is contained in set M", in symbolic writing $K \subset M$, corresponds to "proposition P *implies* proposition Q", in symbolic writing

$$P \Rightarrow Q. \qquad (2.9.2)$$

We illustrate this connection by means of animal species. Define the sets

$$K: \text{set of kangaroos},$$
$$M: \text{set of mammals},$$

and the propositions

$$P: x \text{ is a kangaroo},$$
$$Q: x \text{ is a mammal}.$$

Then $K \subset M$ and by the same token $P \Rightarrow Q$.

The precise meaning of $P \Rightarrow Q$ is: If P is true, then Q is also true. However, if P is false, then Q may be either true or false. This is quite analogous to: If $x \in K$, then $x \in M$. However, if $x \notin K$, then either $x \in M$ or $x \notin M$.

The following example may lead us a step further. Define the propositions

$$P: \text{the diagonals of a quadrilateral bisect each other},$$
$$Q: \text{opposite sides of a quadrilateral are parallel}.$$

From geometry we know that P and Q are both true for a parallelogram. Here P implies Q, but also Q implies P. Thus

$$P \Rightarrow Q \quad \text{and} \quad Q \Rightarrow P$$

for which we simply write

$$P \Leftrightarrow Q. \qquad (2.9.3)$$

The precise meaning of (2.9.3) is: Either P and Q are both true, or P and Q are both false. In other words: P and Q have the same "truth value". The corresponding relationship for sets is the equality. Indeed, $A = B$ means that $x \in A$ implies $x \in B$, and that $x \notin A$ implies $x \notin B$.

The following list summarizes the analogies between sets and propositions:

Sets	Propositions
$x \in A$	P is true
$x \notin A$	P is false
$A \subset B$	$P \Rightarrow Q$
$A = B$	$P \Leftrightarrow Q$
$\bar{A}$	$\neg P$
$A \cup B$	$P \vee Q$
$A \cap B$	$P \wedge Q$

The truth table may serve as an aid in manipulating these relationships. As an example, assume the following propositions:

$$P = \text{mammals feed their young by milk,}$$

$$Q = \text{kangaroos feed their young by milk,}$$

$$R = \text{birds feed their young by milk.}$$

We know that P and Q are true but R is false. What is the truth value of

$$P \vee (Q \wedge R)?$$

Let $Q \wedge R = S$. From the truth table it follows that S is false (that is, that $Q \wedge R$ is false) because R is false. However, $P \vee S$ is true since P is true. Therefore, $P \vee (Q \wedge R)$ is true.

*2.10. Boolean Algebra

In applying set theoretical and logical operations to two or more sets or propositions a peculiar algebra emerges which is called *Boolean algebra*[13]. It is easy to verify the *commutative law* for the operations "union", "intersection", "or", "and", that is

$$A \cup B = B \cup A, \qquad A \cap B = B \cap A,$$
$$P \vee Q \Leftrightarrow Q \vee P, \qquad P \wedge Q \Leftrightarrow Q \wedge P. \tag{2.10.1}$$

---

[13] The inventor is the logician and mathematician George Boole (1815—1864). He was born in England, but lived in Ireland in his later years. Today Boolean algebra is applied to sets, to propositions, and to electrical networks. Attempts are also made to understand complicated neural nets by means of Boolean algebra.

Now we examine the case of three sets. We adapt an example presented in Thrall, Mortimer, Rebman, and Baum (1967, Example PE4)
dealing with blood types. The individuals of a certain population are
tested for the presence of one or more of the antigens A, B, and Rh.
A and B are antigens in the AB0-blood group and Rh is the antigen in
the Rhesus-blood groups. We designate the population as the universal
set U and denote the subsets of individuals carrying the antigens A, B,
Rh by $\mathscr{A}, \mathscr{B}, \mathscr{R}$, respectively. Let u be a variable denoting any individual
of the population, that is $u \in U$. Then in symbolic writing we get

$$\mathscr{A} = \{u \mid u \text{ has antigen A}\},$$
$$\mathscr{B} = \{u \mid u \text{ has antigen B}\},$$
$$\mathscr{R} = \{u \mid u \text{ has antigen Rh}\}.$$

We may ask for those subsets of individuals that carry at least two of the
antigens. These subsets are intersections, namely

$$\mathscr{A} \cap \mathscr{B} = \{u \mid u \text{ has antigens A and B}\},$$
$$\mathscr{A} \cap \mathscr{R} = \{u \mid u \text{ has antigens A and Rh}\},$$
$$\mathscr{B} \cap \mathscr{R} = \{u \mid u \text{ has antigens B and Rh}\}.$$

To obtain the subset of individuals carrying *all three antigens* we
have several possibilities:

$$(\mathscr{A} \cap \mathscr{B}) \cap \mathscr{R} \quad \text{or} \quad (\mathscr{A} \cap \mathscr{R}) \cap \mathscr{B} \quad \text{or} \quad (\mathscr{B} \cap \mathscr{R}) \cap \mathscr{A}.$$

These sets are identical. The result is depicted in the Venn diagram of
Fig. 2.6. We have, therefore, proved the *associative law* for the

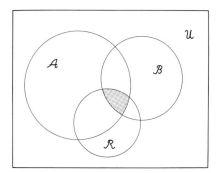

Fig. 2.6. The shaded area represents the intersection of three sets

operation "intersection". In general, if E, F, G denote any three sets, the equality

$$(E \cap F) \cap G = E \cap (F \cap G) \tag{2.10.2}$$

is satisfied. Interpreting the law we may say that the parentheses are unnecessary and can be omitted. Thus for the intersection of three sets we simply write $E \cap F \cap G$.

A corresponding result holds for the union of three sets. First, in combining two sets we get

$$\mathscr{A} \cup \mathscr{B} = \{u \mid u \text{ has antigen A or B}\},$$

$$\mathscr{A} \cup \mathscr{R} = \{u \mid u \text{ has antigen A or Rh}\},$$

$$\mathscr{B} \cup \mathscr{R} = \{u \mid u \text{ has antigen B or Rh}\}.$$

Second, to obtain the subset of individuals carrying at least one of the antigens, that is, A or B or Rh, we have several possibilities:

$$(\mathscr{A} \cup \mathscr{B}) \cup \mathscr{R} \quad \text{or} \quad (\mathscr{A} \cup \mathscr{R}) \cup \mathscr{B} \quad \text{or} \quad \mathscr{A} \cup (\mathscr{B} \cup \mathscr{R}).$$

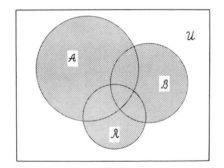

Fig. 2.7. The shaded area represents the union of three sets

These sets are identical as shown in Fig. 2.7. In general, any three sets E, F, G satisfy the *associative law* for the operation "union":

$$(E \cup F) \cup G = E \cup (F \cup G). \tag{2.10.3}$$

As a consequence, the parentheses are superfluous and can be omitted. We simply write $E \cup F \cup G$.

In Section 1.4 we stated the commutative and associative laws for addition and multiplication of numbers. So far the algebra for the operations "intersection" and "union" is quite analogous. Now we try the *distributive law*. For numbers, the law states $a(b+c)=ab+ac$. Notice that addition and multiplication play a different role. By interchanging the two operations we would obtain $a+(bc)=(a+b)(a+c)$ which is wrong, since $a+bc \neq a^2+ac+ab+bc$. For sets we do not know if intersection and union will correspond to multiplication and addition. Hence, we have to try both alternatives. Consider first

$$E\cup(F\cap G)=(E\cup F)\cap(E\cup G). \qquad (2.10.4)$$

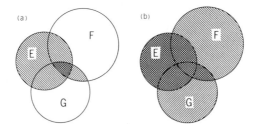

Fig. 2.8. Proof of the distributive law (2.10.4) by means of a Venn diagram

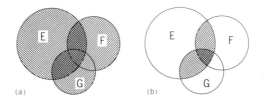

Fig. 2.9. Proof of the distributive law (2.10.5) by means of a Venn diagram

For a proof we look at Fig. 2.8. In (a) the union of E and $(F\cap G)$ is shown. This corresponds to the left side of Eq. (2.10.4). In (b) the intersection of $(E\cup F)$ and $(E\cup G)$ is depicted by the doubly hatched area. This corresponds to the right side of the equation. The two results are equal which completes the proof.

Second, we consider the other possibility:

$$E\cap(F\cup G)=(E\cap F)\cup(E\cap G). \qquad (2.10.5)$$

The proof is shown in Fig. 2.9. In (a) the doubly hatched area is the intersection of E and $(F\cup G)$ which is the left side of the equation. In (b) the union of $(E\cap F)$ and $(E\cap G)$ is depicted by the hatched area which corresponds to the right side of the equation. Both results are equal so that (2.10.5) is proved.

The fact that *both distributive laws are valid* is a remarkable property of Boolean algebra. Together with the properties $A \cup A = A$ and $A \cap A = A$ this shows that Boolean algebra is quite different from the ordinary algebra of numbers.

For the corresponding laws for propositions we refer the reader to problem 2.10.5 at the end of this chapter.

In Fig. 2.6 we see that the population of U is subdivided into eight disjoint sets. They correspond to eight mutually exclusive classes of blood types. All of them can be expressed with the symbols for intersection and complement. In biomedical writing, presence of A and B is symbolized by AB, absence of A and B as 0, presence of Rh by Rh^+, and absence of Rh by Rh^-. With these notations we obtain the following classes of blood types:

$$\mathscr{A} \cap \mathscr{B} \cap \mathscr{R} = \{u \mid \text{type (AB, Rh}^+)\} ,$$
$$\mathscr{A} \cap \mathscr{B} \cap \overline{\mathscr{R}} = \{u \mid \text{type (AB, Rh}^-)\} ,$$
$$\mathscr{A} \cap \overline{\mathscr{B}} \cap \mathscr{R} = \{u \mid \text{type (A, Rh}^+)\} ,$$
$$\mathscr{A} \cap \overline{\mathscr{B}} \cap \overline{\mathscr{R}} = \{u \mid \text{type (A, Rh}^-)\} ,$$
$$\overline{\mathscr{A}} \cap \mathscr{B} \cap \mathscr{R} = \{u \mid \text{type (B, Rh}^+)\} ,$$
$$\overline{\mathscr{A}} \cap \mathscr{B} \cap \overline{\mathscr{R}} = \{u \mid \text{type (B, Rh}^-)\} ,$$
$$\overline{\mathscr{A}} \cap \overline{\mathscr{B}} \cap \mathscr{R} = \{u \mid \text{type (0, Rh}^+)\} ,$$
$$\overline{\mathscr{A}} \cap \overline{\mathscr{B}} \cap \overline{\mathscr{R}} = \{u \mid \text{type (0, Rh}^-)\} .$$

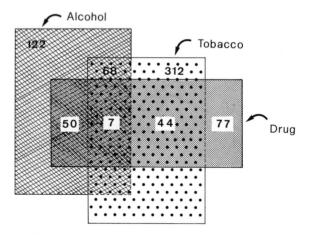

Fig. 2.10. Venn diagram representing the frequencies of alcohol, tobacco, and drug addiction

These equalities may be proved by means of the Venn diagram. A similar classification in ecology is presented in Patten (1966, p. 594).

The graphical presentation of complex data may differ from one author to another. We will conclude this section with a behavioral study: 680 addicts were asked whether they are dependent on alcohol, tobacco, a drug, or some combination. 122 of them were alcohol addicts, 312 tobacco addicts, and 77 drug addicts exclusively. 68 were simultaneously addicts of alcohol and tobacco, 50 of alcohol and drugs, and 44 of tobacco and drugs. Only 7 participated in all three sorts of addiction. The result is shown in Fig. 2.10.

Recommended for further reading: Arbib (1964), Feinstein (1967), George (1961), Hays (1963), Kerlinger (1964), Lefort (1967), Nahikian (1964), Rashevsky (1960, Vol. 2), Stibitz (1966), Tarski (1965).

Problems for Solution

2.3.1. Consider quadrilaterals as geometric configurations. Let U be the set of all quadrilaterals, P be the subset of parallelograms, R the subset of rhombuses, T the subset of rectangles, and S the subset of squares. What are the relationships among these sets?

2.3.2. Consider the set of all possible scales on a straight line (Section 1.2). Denote the set of all nominal scales by N, the set of all ordinal scales by O, the set of all interval scales by I, and the set of all ratio scales by R. Which of these sets are subsets of others?

2.4.1. What is the relationship between the two sets of numbers, $C = \{x \mid x > 5\}$ and $D = \{x \mid x \geq 5\}$?

2.4.2. What is the relationship between the sets
$$A = \{t \mid 0 < t < 3\} \quad \text{and} \quad B = \{\tfrac{1}{2}, \tfrac{3}{2}, \tfrac{5}{2}\}?$$

2.4.3. Find the solution sets of

a) $\{x \mid x + 7 < 12\}$, b) $\{y \mid 3y = 15 - y\}$,
c) $\{x \mid x = 3(x + 3) - x\}$, d) $\{u \mid u^2 = 10\}$,
e) $\{x \mid 3x - 5 \geq x + 7\}$, f) $\{t \mid t^2 > 0\}$.

2.4.4. Find the solution set of

   a) $\{x|(x-3)(x+2)=x^2\}$ b) $\{t|2t+3=2(t-1)\}$
   c) $\{x|x^2<16\}$ d) $\{y|y^2>9\}$.
   e) $\{x|x^2-4=(x+2)(x-2)\}$.

Notice that a solution set can be an empty set or consist of an infinite number of values.

2.5.1. A litter of five mice may contain $0, 1, 2, \ldots$, or 5 males, the rest females. We denote the six possibilities by $a_0, a_1, \ldots, a_5$ and collect them to a set U. Write down the two sets

A: at least four mice are male ,

B: at most three mice are female ,

and find $\bar{A}$ and $\bar{B}$.

2.5.2. By proper definition men may be assigned to one of the following categories:

a: tall and slim,

b: tall and fat,

c: short and slim,

d: short and fat.

These four categories constitute a set $U=\{a,b,c,d\}$. Form the subsets tall, short, slim, fat and find the corresponding complementary sets.

2.7.1. In a plane we draw a circle and a straight line. Interpret both lines as point sets and list three possibilities for the intersection of the two lines.

2.7.2. Interpret a straight line and the surface of a sphere as two point sets. Enumerate all possibilities for the intersection of the two sets.

2.7.3. Interpret planes in the three-dimensional space as point sets. Distinguish three cases for the intersection of two planes. What cases are possible for three planes?

2.7.4. We consider the points of a plane as the universal set U. In this plane we draw a circle. Let K be the set of points on the circle, I be the point set inside, and A be the point set outside. Find all possible relationships among these sets.

2.7.5. In a study of AB0 blood-groups, 6000 Chinese were tested. 2527 had the antigen A, 2234 the antigen B, and 1846 no antigen. How many individuals had both antigens? (Adapted from Li, 1958, p. 48).

2.7.6. Let A and B be subsets of the universal set U and assume that $A \cap B \neq \emptyset$. By using a Venn diagram show that

$$(A \cap \bar{B}) \cup (B \cap \bar{A})$$

is the set of all members that belong either to A or to B, but not to both sets A and B.

2.7.7. Show that the same set as in Problem 2.7.6. can be written as

$$(A \smallsetminus B) \cup (B \smallsetminus A).$$

2.7.8. Consider MN blood groups. An individual has either an M antigen or an N antigen or both of them. Let U be a given population and $\mathcal{M}$ and $\mathcal{N}$ be the subsets of all individuals with antigen M and N, respectively. Using a Venn diagram describe the subsets $\bar{\mathcal{M}}, \bar{\mathcal{N}}, \mathcal{M} \cap \mathcal{N}, \mathcal{M} \cup \mathcal{N}, \bar{\mathcal{M}} \cap \mathcal{N}, \bar{\mathcal{M}} \cup \mathcal{N}$.

2.7.9. In a plane we consider two points P_1 and P_2. We form two sets: Set A_1 contains all straight lines through P_1, and A_2 all straight lines through P_2. Find $A_1 \cap A_2$. (Hint: The elements of A_1 and A_2 are straight lines, not points. Distinguish between the cases $P_1 = P_2$ and $P_1 \neq P_2$.)

2.7.10. In the three-dimensional space we consider two straight lines g_1 and g_2. They may be equal, parallel, intersecting, or skew. We form two sets: Set A_1 contains all planes through g_1 and set A_2 all planes through g_2. Find $A_1 \cap A_2$.

*2.10.1. Electrical switching circuits are used in the study of neurons in the brain. In Fig. 2.11(a) and (b), two basic circuits are depicted. In Fig. 2.11(a) the switches are said to be in series. An electric current can only flow if switch a as well as b is closed. Let A be the proposition "a is closed" and B the proposition "b is closed". Then the condition for flow is $A \wedge B$. In Fig. 2.11(b) the switches are said to be in parallel. The current can flow if switch a or switch b or both switches are closed. Here the condition for flow is $A \vee B$. Find the corresponding statements in Fig. 2.11(c) and (d).

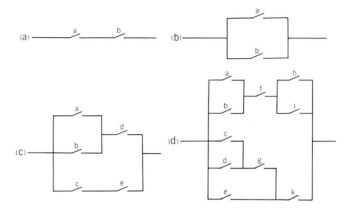

Fig. 2.11. Boolean algebra is applied to electrical switching circuits

*2.10.2. Find electrical networks which represent the propositions

a) $A \wedge (B \vee C)$,   b) $(A \vee B) \wedge (C \vee D)$,

c) $(A \wedge B) \vee (C \wedge D)$,   d) $(A \wedge B) \vee (C \vee D)$.

*2.10.3. Transform

a) $(x > 0) \wedge (y > 0) \Rightarrow x + y > 0$,

b) $(x < 0) \wedge (y < 0) \Rightarrow xy > 0$,

c) $x^2 + y^2 > 0 \qquad \Rightarrow (x \neq 0) \vee (y \neq 0)$

into verbal statements.

*2.10.4. Let a, b, c, d be the sides of a quadrilateral and $\alpha, \beta, \gamma, \delta$ its angles, both in a clockwise or counter-clockwise order. Define the propositions

$$P_1: a \parallel c, \qquad P_2: b \parallel d,$$
$$Q_1: \alpha = \gamma, \qquad Q_2: \beta = \delta,$$
$$R_1: a = c, \qquad R_2: b = d.$$

Show that

a) $P_1 \wedge P_2 \Leftrightarrow R_1 \wedge R_2$,

b) $Q_1 \wedge Q_2 \Leftrightarrow P_1 \wedge P_2$.

*2.10.5. Using a truth table (Section 2.9) prove that for arbitrary propositions P, Q, R the following rules are valid:

a) $\qquad P \vee P \Leftrightarrow P, \qquad P \wedge P \Leftrightarrow P$,

b) $(P \wedge Q) \wedge R \Leftrightarrow P \wedge (Q \wedge R)$,

$(P \vee Q) \vee R \Leftrightarrow P \vee (Q \vee R)$,   (associative laws)

c) $P \vee (Q \wedge R) \Leftrightarrow (P \vee Q) \wedge (P \vee R)$,

$P \wedge (Q \vee R) \Leftrightarrow (P \wedge Q) \vee (P \wedge R)$ (distributive laws).

Chapter 3

Relations and Functions

3.1. Introduction

In the life sciences not all relationships are of a quantitative nature. Although future scientists may be able to understand cells, viruses, genes, antibodies, etc. in terms of molecules, their structure is so complex that a description will be more or less qualitative. The study of interconnections between cells, either by chemical or by electrical exchange, calls for mathematical tools that are not simply formulas. How an organ operates, its response to a stimulus, or how an individual behaves can hardly be expressed by numbers alone. Therefore, in this chapter we give definitions that are wide enough to comprise qualitative as well as quantitative properties.

Using set theory we will give a mathematical definition of what is generally called a *relation*. The notion of a function will appear as a special case. The chapter will end with a review of the linear function.

3.2. Product Sets

We consider first an introductory example. The AB0 blood groups are theoretically explained by three genes or alleles at the same gene locus. We denote the alleles by a, b, o. The alleles a and b are dominant over o. Individuals carrying the gene combination oo have neither antigen A nor B in their blood. With the combination aa or ao the blood contains the antigen A, with the combinations bb or bo it contains the antigen B, and with the combination ab the blood contains both antigens. To get all possibilities for genetical recombination we introduce the set of alleles

$$S = \{a, b, o\} .$$

Each gamete (egg or sperm cell) carries one of the three alleles. We represent the possible alleles of the sperm cell by points on a horizontal line and the possible alleles of the egg cell by points on a vertical line (see Fig. 3.1). Then we draw parallel lines through each series of points and obtain nine points of intersection, the so-called *lattice points*. Each lattice point represents a possible recombination of alleles in a zygote (fertilized cell). There are nine different types of recombinations, namely the *ordered pairs*

$$
\begin{array}{ccc}
(a, a) & (a, b) & (a, o) \\
(b, a) & (b, b) & (b, o) \\
(o, a) & (o, b) & (o, o).
\end{array}
\qquad (3.2.1)
$$

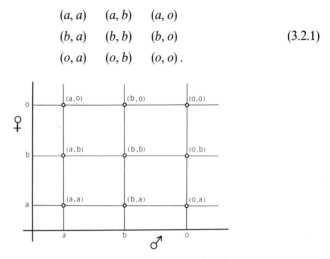

Fig. 3.1. Representation of all possible gene combinations in a zygote

One of the genetic laws says that the order of the two alleles is irrelevant. For instance, individuals with (a, b) and (b, a) cannot be distinguished from each other. However, when a similar representation is applied to other objects of investigation, the order may well be of importance.

The pairs in (3.2.1) form a new set which was generated by an *operation* on the original set $S = \{a, b, o\}$. The new set is called a *product set* or a *Cartesian product*[1] and denoted by $S \times S$. We read "S cross S", and by definition we may write

$$
\begin{aligned}
S \times S = \{&(a, a),\ (a, b),\ (a, o),\ (b, a),\ (b, b),\\
&(b, o),\ (o, a),\ (o, b),\ (o, o)\}.
\end{aligned}
\qquad (3.2.2)
$$

The new operation on sets should not be confused with a multiplication of numbers.

---

[1] Cartesius is the Latin name for the French philosopher and mathematician René Descartes (1596—1650). He invented the coordinate system with two axes. Today the coordinate system is viewed as a special application of the concept of Cartesian product.

In the preceding example we studied the product set of *finite* sets. With continuously varying quantities, however, a set may be infinite. Being concerned with fever we are interested in the relationship between body temperature and pulse rate. Both quantities can be measured with high accuracy. Let t be the temperature in degrees Celsius ranging from 35° in healthy to 41° in ill persons, and let f be the pulse frequency ranging from 50 to 150 cycles per minute. Both quantities vary continuously. Now we form the two sets

$$A = \{t \mid 35° \le t \le 41°\},$$
$$B = \{f \mid 50 \le f \le 150\}.$$

Combining temperature and pulse frequency we obtain the ordered pairs (t, f) which constitute the product set $A \times B$. This set is infinite.

The best known application of product sets is the Cartesian plane. Let $\mathbb{R}$ denote the set of real numbers. Then by definition the Cartesian plane is the set of all ordered pairs (x, y) with real numbers x and y, namely the product set

$$\mathbb{R} \times \mathbb{R} = \{(x, y) \mid x \in \mathbb{R} \text{ and } y \in \mathbb{R}\}. \tag{3.2.3}$$

The graphical presentation consists of two perpendicular axes intersecting each other at the point with coordinates $x = 0$, $y = 0$. But neither the right angle nor the particular point of intersection are true requirements. In most applications the units on the two axes are different. We also call the Cartesian plane a *two-dimensional space* and denote it by

$$\mathbb{R}^2 = \mathbb{R} \times \mathbb{R}, \tag{3.2.4}$$

whereas $\mathbb{R}$ may be interpreted as a one-dimensional space.

Notice that a pair (x, y) is ordered if we distinguish (x, y) from (y, x). Two ordered pairs (x_1, y_1) and (x_2, y_2) are equal if $x_1 = x_2$ and $y_1 = y_2$; otherwise they are unequal.

The first number is called *abscissa*, the second number *ordinate*, and both are called *coordinates*. The x and the y axes are graphs of the set $\mathbb{R}$ of real numbers. The two sets are sometimes denoted by X and Y. Both axes together form a *coordinate system*. Each ordered pair is represented by a point in the Cartesian plane. Two different pairs correspond to two different points. The converse is also true: Two different points are associated with two different ordered pairs of numbers. Such a relationship is called a *one-to-one correspondence* and is symbolically written $1 - 1$. For brevity (x, y) is frequently called a point (x, y), although a pair of numbers and a point are by no means identical. The point $(0, 0)$ is called the *origin* of the coordinate system. For some peculiarities of presentation see Fig. 3.2.

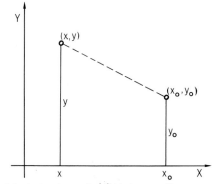

Fig. 3.2. A point (x, y) is depicted together with its coordinates x and y. The letter x is placed at the point $(x, 0)$ just below the point (x, y) and the letter y is usually placed at the center of a vertical line connecting the points (x, y) and $(x, 0)$

Summary. *A product set $A \times B$ consists of all ordered pairs (a, b) such that a is a member of the set A and b a member of the set B. The Cartesian plane consists of all points (x, y), that is, of all members of the product set $\mathbb{R} \times \mathbb{R}$ where $\mathbb{R}$ is the set of real members.*

3.3. Relations

An example may lead us to the definition of the term "relation". We discuss the interaction between neurons (nerve cells in the brain). In Fig. 3.3a we consider the set

$$S = \{a, b, c, d, e, f\}$$

of six neurons labeled in arbitrary order (cf. Thrall, Mortimer, Rebman, and Baum, 1967, Model CE 1). Either a neuron is able or is not able to send an impulse directly to another neuron. Biologically an impulse is transmitted by a rather complicated mechanism consisting of dendrites, axon and synapse, a mechanism which is schematically depicted by ———C——— in Fig. 3.3a. The impulse goes only in one direction: Neuron a can send to neurons b and d, but cannot receive impulses from b and d, etc. The relationship "a can send an impulse to b" is mathematically represented by an ordered pair (a, b). The product set $S \times S$ with its 36 members contains all imaginable connections of the six neurons. The existing connections form a subset of $S \times S$ and are shown by heavy dots in Fig. 3.3b. We denote the subset by R. It contains only the following ten members:

$$R = \{(a, b), \ (a, d), \ (c, a), \ (b, d), \ (b, c),$$
$$(c, d), \ (d, e), \ (c, f), \ (d, f), \ (f, e)\} \, .$$

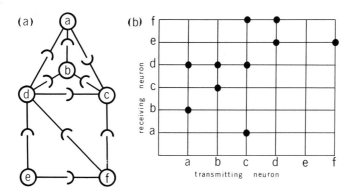

Fig. 3.3. The interconnection of neurons presented as a relation within a product set

According to Section 2.3 we write

$$R \subset S \times S .$$ (3.3.1)

The subset R reveals the relationship between the neurons. In mathematics R is simply called a *relation* in the product set $S \times S$. To state that an ordered pair (a, b) actually occurs in the relation R we also write

$$a\,R\,b$$

and read it "a is R-related to b". In our example $a\,R\,d$, $c\,R\,d$, $b\,R\,d$, etc. But $b\,R\,a$, $c\,R\,e$, $f\,R\,f$ would be wrong.

Another example of a relation is taken from physics. Let

$$A = \{H, He, Li, Be, B, C, N, O, ...\}$$

be the set of chemical elements ·ordered according to their atomic number and let

$$\mathbb{N} = \{1, 2, 3, 4, ...\}$$

be the set of all imaginable atomic weights (rounded-off to integers). Then an ordered pair such as $(He, 4)$ means a helium isotope of weight 4, the same as ^4He in standard physical notation. All isotopes existing in nature or artificially produced in the laboratory form a subset of the product set $A \times \mathbb{N}$. Again, this subset is called a relation in the product set $A \times \mathbb{N}$. Part of this relation is depicted in Fig. 3.4.

A relation can also be determined by a mathematical formula. For instance,

$$x^2 + 3y^2 = 21$$ (3.3.2)

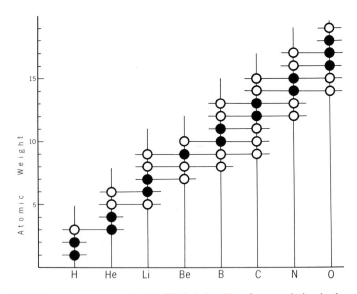

Fig. 3.4. Stable isotopes are marked by filled circles. They form a relation in the product set $A \times \mathbb{N}$. Unfilled circles mark unstable isotopes. They form another relation. Both types of circles together form a third relation

has an infinite number of ordered pairs (x, y) as solutions. Among them are $(3, -2), (3, 2), (-3, 2), (-3, -2), (0, \sqrt{7}), (0, -\sqrt{7}), (\sqrt{21}, 0), (1, \sqrt{20/3})$. The set

$$R = \{(x, y) | x \in \mathbb{R}, y \in \mathbb{R}, x^2 + 3y^2 = 21\} \qquad (3.3.3)$$

is a subset of $\mathbb{R} \times \mathbb{R}$ and hence a relation in the Cartesian plane. A graph of R is shown in Fig. 3.5.

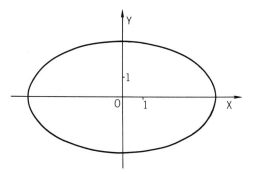

Fig. 3.5. Graph of the relation (3.3.3). The points constitute an ellipse

Similarly, an inequality such as

$$y > x$$

defines a subset of $\mathbb{R} \times \mathbb{R}$ which is again called a relation. In symbolic writing the relation is

$$R = \{(x, y) | x \in \mathbb{R}, y \in \mathbb{R}, y > x\},$$

but in "every day language" of mathematics we simply say "the relation $y > x$". A graph of this relation is shown in Fig. 3.6.

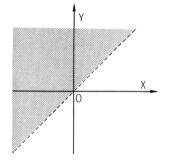

Fig. 3.6. Graph of the relation $y > x$. The shaded area is the subset of points with y greater than x

Summary. *Let $A \times B$ be a product set consisting of all ordered pairs (a, b) where a is a member of A and b is a member of B. Then any subset of $A \times B$ is called a relation.*

3.4. Functions

Consider a human population, say a set U of persons, and the set of fingerprints of these persons, say F. Each person has ten fingerprints. Since the relationship of fingerprints and persons is of practical interest, we introduce ordered pairs

(fingerprint, person).

These pairs are members of the product set $F \times U$. Certain fingerprints belong to certain persons so that such pairs form a subset of $F \times U$. We call this subset a relation as in the preceding section. In our example the relation has a remarkable property: *Each fingerprint x is associated* *with exactly one person y.* Given a fingerprint we can uniquely identify the person. Such a relation is called a *mapping* or a *function*[2].

Quite similarly, when we observe a group of people in sunlight, each person has exactly one shadow. Thus, we may state that the relationship is a function. One variable is the person x, the other variable the shadow y. Both variables are of the nominal type (see Chapter 1).

With each biological species there is associated a typical number of chromosomes (pathological cases disregarded). Therefore, this association is called a function. Here, the species is a variable of the nominal type but the number of chromosomes a quantitative variable.

Notice in all of these cases that the uniqueness of association is in one direction only. Each person y has more than one fingerprint x; a given shadow y may belong to more than one person x (if, for example, they are standing close together); a chromosome number y may be characteristic of a large number of species x. To speak of a function, therefore, we require uniqueness of association in only one direction, as expressed by the ordered pairs (fingerprint, person); (person, shadow); (species, chromosome number).

When both variables x and y are quantitative, either at the interval or the ratio level, we obtain more familiar examples of functions. For example, the pressure of a gas is a function of its temperature provided that the volume is kept constant.

A very special kind of function is given by a formula, for instance by $y = x^2$. With each value of a numerical variable x there is associated only one number y. This association is therefore a function. It is customary to say that y is a quadratic function of x. Notice, however, that the term "function" means a relationship, not the special variable y.

We introduced a function as a special sort of relation. But not all relations are functions. For instance, in Fig. 3.4 the same chemical element can have different atomic weights. Therefore, atomic weight is not a function of chemical element. Similarly, we recognize in Figs. 3.5 and 3.6 that more than one y value is associated with a single x value.

In neither case do we have a function.

In general, let x and y denote any two variables of a nominal, ordinal, interval, or ratio level. The variable x may stand for the members of a set A and y for the members of a set B, symbolically $x \in A$,

---

[2] Both words need some explanation. *Mapping* was originally used for depicting a landscape. To each object of a landscape there corresponds a *unique* mark on a map. In modern mathematics the word "mapping" is used for any kind of unique association.

The word *"function"* has quite a different history. In the 17th century mathematicians conceived the idea of *variable quantities.* A dependent variable was called a function of an independent variable. It was given by a formula such as $y = x + 3$ or $y = a\sqrt{x}$, etc. During the last hundred years the notion of a function was more and more generalized. Today x and y may denote not only numbers but members of any kind of sets.

$y \in B$. Assume that to each x there is associated a unique y, then we call this relation a function. We write

$$f : x \mapsto y \tag{3.4.1a}$$

or

$$x \mapsto f(x) \tag{3.4.1b}$$

or

$$y = f(x). \tag{3.4.1c}$$

The first two notations are modern, the third is traditional[3].

Notice that $x \mapsto f(x)$ and $t \mapsto f(t)$ mean the same mapping and, hence, the same function. The choice of letter for the independent variable is irrelevant.

The concept of mapping is not restricted to a Cartesian coordinate system. An abstract idea of mapping is shown in Fig. 3.7.

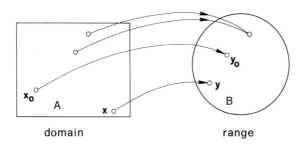

Fig. 3.7. Mapping of elements x from domain A into elements y of range B

x is called the *independent*, y the *dependent variable*. The set A of the x variable is called the *domain* of the function. If the set B is defined in such a way that all members of B are associated with members of A, then B is called the *range* of the function.

Example 3.4.1. Fig. 3.8 depicts a piece of a curve. The ordinate y of a point of this curve is uniquely associated with an abscissa x. Hence, the curve defines a function. We may denote the association by $y = f(x)$. The curve is called the *graph* of the function. The domain A and the range B are also shown in the figure.

---

[3] The symbol $\mapsto$ for mapping has a meaning which is quite different from logical implication ($\Rightarrow$, Section 2.9) and from convergence ($\rightarrow$, Section 8.1). In the older literature the same symbol $\rightarrow$ is used for all three operations.

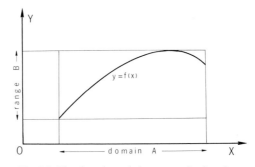

Fig. 3.8. The domain and the range of a function

Example 3.4.2. With each electromagnetic wave of wave length λ (Greek lambda) taken from the approximate domain

$$A = \{\lambda \,|\, 3.8 \times 10^{-7}\,\text{m} < \lambda < 7.8 \times 10^{-7}\,\text{m}\}$$

the human eye perceives a specific spectral color which may or may not have a name. This association is a function. The dependent variable, the color, is a quality of the ordinal level. The range is the infinite set of all spectral colors.

Example 3.4.3. Some functions may be defined by a formula such as $y = ax + b$, $y = 1/x$, $y = x^{1/2}$. Each of these formulas associates a unique y value with a given x value. In each case we should specify a domain A and a range B.

For $y = ax + b$, the variable x can take on any real value. Thus, we may identify the domain A with the set $\mathbb{R}$ of all real numbers. In applications to the natural sciences, however, the domain A is usually a finite interval since x cannot be arbitrarily small or large.

In the case of $y = 1/x$ any real value qualifies for x except $x = 0$ since we cannot divide by zero. Thus, the domain A can be any set on the x axis which does not contain the point $x = 0$.

For the formula $y = x^{\frac{1}{2}}$ we have to exclude all negative numbers x if we are to avoid imaginary numbers. Hence, the domain A is the set $\{x \,|\, x \geq 0\}$ or any subset of it.

The graphs of these three functions are lines which we will study in future sections.

Notice that not every formula constitutes a function. As we already know, formula (3.3.2) defines a relation, but y is not uniquely associated with x.

Example 3.4.4. The graph of a function of numerical variables is not always a curve. The domain as well as the range may consist of isolated points. This is for instance true for an empirical function with a finite number of measurements $y_1, y_2, \ldots, y_n$ which are associated with given values $x_1, x_2, \ldots, x_n$, respectively. The domain as well as the range consist of a finite number of members as long as we do not interpolate.

Example 3.4.5. Let the domain be $A = \mathbb{N} = \{1, 2, 3, \ldots\}$. With each $x \in \mathbb{N}$ we associate the largest prime factor of x and denote it by y. This is a mapping. We may write

$$f : x \mapsto y$$

or $y = f(x)$. No formula exists for this function. The range is $B = \mathbb{N}$. The mapping is depicted in Fig. 3.9.

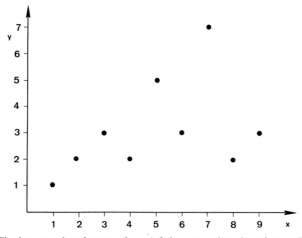

Fig. 3.9. The largest prime factors of $x = 1, 2, 3, \ldots$ are plotted against x. The domain as well as the range is $\mathbb{N}$.

Summary. *Three things are required to establish a function, namely a set A, a set B, and a rule whereby each member of A is associated with exactly one member of B. We say that each member of the domain A is mapped into a member of the range B.*

Natural scientists tend to believe that some *causality* must be involved in the term "function". For instance, the temperature of a solution may vary as a function of energy release. However, we should be cautious by introducing a cause into the concept of function. Sometimes the cause relationship is not well established. Even if the cause is known, the result need not lead to a unique association of y with x. Indeed, a repetition of an experiment could lead to different y values for the same x. The main reason, however, why causality fits poorly into the concept of function is the following: For many functions there exists a so-called inverse function, but causality is one-sided and cannot be reversed. If, for instance, $y = x^2$, then $x = y^{\frac{1}{2}}$. Thus, "y is a function of x" as well as "x is a function of y", but if y is "caused" by the amount of x, the converse cannot be true.

Finally, even if the modern definition of a function is accepted, the word "function" is often used in an *ambiguous* way. Strictly speaking, phrases such as "y is a function of x" and "the function varies between 0 and 1" are wrong. The function is a relationship and cannot have numerical values. Moreover, as an established relationship a function cannot vary. Nevertheless, such phrases are widely used in the everyday scientific language and can hardly be eradicated. When a misunderstanding is unlikely, we will sometimes accept a more liberal use of the word function.

3.5. A Special Linear Function

One of the easiest functions is given by the formula

$$y = ax \tag{3.5.1}$$

where a is a constant number. In the Cartesian plane this function is represented by a straight line through the origin. The fact that the graph is a straight line accounts for the term *linear function*.

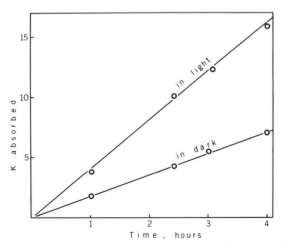

Fig. 3.10. Potassium absorption by leaf tissue of *Zea mays* in light and in dark. Adapted from Rains (1967)

As an illustrative example we shall discuss the absorption of potassium (K) by leaf tissue of *Zea mays* (corn) as a function of time. We follow a report by Rains (1967). The independent variable is the time t, measured in hours. The dependent variable y is the amount of absorbed potassium, measured in μMoles per unit weight of leaf tissue (which is not specified here). The function $y = at$ fits the data very well for a domain $\{t \mid 0 \leq t \leq 4\}$. When the experiment is performed in darkness, the constant takes on the value $a = 1.8$ μMoles per unit weight per hour. If, however, the tissue is illuminated (by light intensity of roughly 2×10^4 lumen/m^2), then the constant turns out to be $a = 4.0$ μMoles per unit weight per hour. The constant a is called the *rate of absorption*. Hence the result of the experiment may be summarized as follows: The rate of absorption in the light is about twice the rate in the dark. The two straight lines are shown in Fig. 3.10.

Formula (3.5.1) means that y is proportional to x. For this relationship the symbol $\propto$ is sometimes used. Thus if the type of function is more important than the specific value of a, we write

$$y \propto x \tag{3.5.2}$$

and read "y is proportional to x"[4].

The constant proportion $y/x = a$ is called a *growth rate*, especially if the independent variable is time. Thus we speak of a rate of reaction, a rate of absorption, a rate of mutation, etc. In the Cartesian plane the constant a plays the role of a *slope*. If $a = 0$, the straight line coincides with the x axis. For $a > 0$ the straight line ascends from left to right and for $a < 0$ it descends. Some caution is required: A large value of a does not necessarily mean that the straight line is steep. All depends on the units we choose on the x and the y axes. The same linear relationship can be depicted in different ways. The two diagrams in Fig. 3.11 represent exactly the same functions as in Fig. 3.10, but in 3.11a the reader is under the impression that the two absorption rates are small, whereas Fig. 3.11b evokes the opposite impression. *The impression of steepness is highly subjective.* It is a well-known trick for advertisement or for political purposes to manipulate the steepness. A humorous as well as instructive example is presented in Huff (1954) and reproduced in Fig. 3.12.

In this connection we conclude that the *angle of inclination* of the straight line has no meaning in general. There are a few exceptions. In *calibration*, for instance, x means a variable measure read on the scale of an instrument, whereas y is the corresponding exact measure given

---

[4] The proportionality sign $\propto$ is very practical, but not widely used. Mathematicians apply it rarely. However, the symbol is quite popular in biophysics.

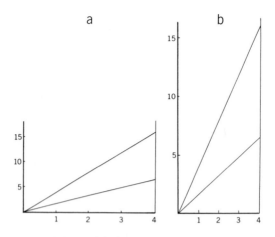

Fig. 3.11. The steepness of the straight lines depends on the choice of the units on the
x and the y axes

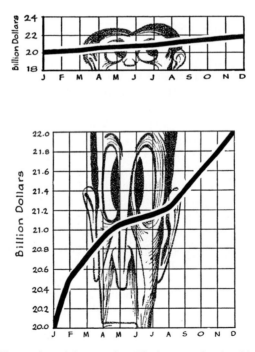

Fig. 3.12. Two different plots of the same fact: The increase of national income in a year.
From Huff (1954, p. 62)

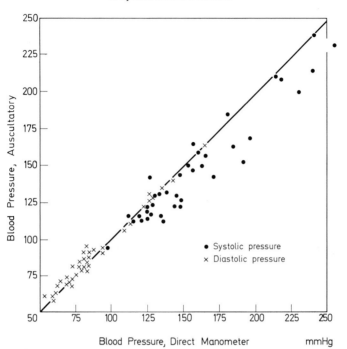

Fig. 3.13. The blood pressure measured in two different ways. Let x and y denote the measurements at the same individual. Without errors of measurements we would have $y = x$. The graph is a straight line with $\alpha = 45°$ as angle of inclination (from Sunderman and Boerner, 1949, after a paper by Steele)

by a suitable norm. Here, x and y are written in the same units and, by convention, also plotted on the same scale. Slightly more general is the case where two experimental procedures are compared with each other and corresponding results x and y are plotted against each other. In such cases the angle of inclination may be quite relevant. An example is shown in Fig. 3.13.

Provided that x and y are plotted in the same units, the angle α of inclination can be determined from the equation

$$\tan \alpha = a \qquad (3.5.3)$$

by means of a table. For a positive slope, that is $a > 0$, α ranges from $0°$ to $90°$. For a negative slope, that is, $a < 0$, α ranges from $90°$ to $180°$ (the lower and upper bounds are excluded). For the definition of tan see formula (5.5.6).

3.6. The General Linear Function

We consider now the equation

$$y = ax + b.\qquad(3.6.1)$$

The right side consists of two terms, the linear part ax discussed in the previous section and the constant b. Correspondingly, the graph of the function can be obtained from the straight line $y = ax$ passing through the origin by adding the (positive or negative) constant b to each ordinate (see Fig. 3.14). This procedure leads to a *parallel straight line*

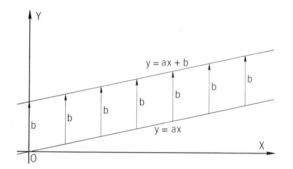

Fig. 3.14. Construction of the straight line of the equation $y = ax + b$ using the graph of $y = ax$

which we also get when we shift the straight line $y = ax$ in the y direction by the constant b. The point of intersection with the y axis is $(0, b)$, and b is therefore called the *y-intercept*. The y-intercept is seen in a graph only if the x axis intersects the y axis at the point $(0, 0)$.

The constant b does not contribute to the slope of the straight line. The slope of (3.6.1) is the same as that of $y = ax$, namely a. We get a deeper insight in the meaning of slope when we write the equation of the straight line in a different and often more practical way. Let (x_0, y_0) be an arbitrary, but fixed point of the straight line, and let (x, y) be a variable point of the same line (see Fig. 3.15). The differences

$$\varDelta x = x - x_0, \qquad \varDelta y = y - y_0\qquad(3.6.2)$$

are called the *increments*[5] of x and y. They can be positive, negative, or zero. Assume now that $\varDelta x \neq 0$. Then from Fig. 3.15 we see that the

---

[5] $\varDelta$ is the upper case Greek letter delta. This symbol is used in mathematics to indicate a difference.Cf. the use of $\varDelta$ in Section 4.4. In formulas, $\varDelta x$ should not be confused with a multiplication of two numbers $\varDelta \cdot x$.

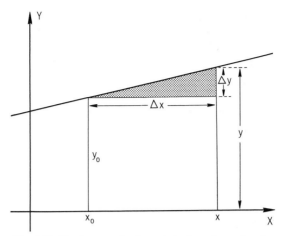

Fig. 3.15. The increments Δx and Δy for a linear function

slope a of the straight line is

$$a = \Delta y/\Delta x \,, \qquad (3.6.3)$$

no matter how large or how small Δx is chosen. In other words: While the variable point (x, y) is moving along the straight line, we get always the same ratio $\Delta y/\Delta x$ unless the variable point (x, y) coincides with the fixed point (x_0, y_0). For $\Delta x = 0$ formula (3.6.3) breaks down, since we cannot divide by zero.

We may also say that Δy is *proportional to* Δx and write

$$\Delta y \propto \Delta x \,.$$

Formula (3.6.3) will play a major role in Chapter 9 on calculus.

By means of (3.6.2), formula (3.6.3) can be rewritten in the form

$$\frac{y - y_0}{x - x_0} = a \qquad (x \neq x_0) \,. \qquad (3.6.4)$$

This is an equation which restricts the variable point (x, y) to lie on the straight line through the point (x_0, y_0) and with slope a. Hence, (3.6.4) is another form of the Eq. (3.6.1). On rearranging the terms in (3.6.4) and comparing with (3.6.1) we obtain for the constant term

$$b = y_0 - ax_0 \,.$$

Example 3.6.1. Given the linear function $y = (3.2)x + 30.5$. Plot the graph of the function.

Solution: The graph is a straight line with slope $a = 3.2$ and y-intercept $b = 30.5$. Plotting a graph causes some difficulties since a and b are both large values. It is reasonable to choose different scales on the two axes, e.g. setting the x-unit five times as large as the y-unit. Second, we may cut the y axis to lower the straight line (Fig. 3.16). Let B be the point of intersection with the y axis. From there we draw a segment of length $\Delta x = 1$ (in the x-unit) to the right. Let the point we have reached be denoted by C. From Eq. (3.6.3) we get $\Delta y = a\Delta x = 3.2$. Then from C upward we plot a segment of length $\Delta y = 3.2$ (in the y-unit). With the two segments Δx and Δy the graph is determined.

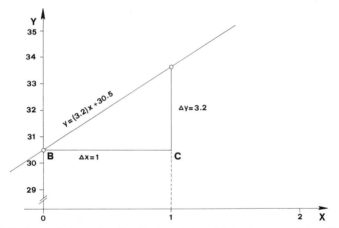

Fig. 3.16. Graph of a linear function. By $\Delta x = 1$, $\Delta y = 3.2$ the slope $a = 3.2$ is easily plotted

Example 3.6.2. In a study of the reduction of 2,6-dichlorophenol-indophenol by light, McNaughton (1967) reports that the photochemical reaction in *Typha latifolia* (broad-leaved cattail) is the more efficient, the higher the altitude is at which plants of this species grow. More precisely, McNaughton found that the so-called Hill activity is an almost linear function of the frost-free period at the place where the plants live. Fig. 3.17 is a diagram of the straight line fitted to the data[6]. Let us first determine the equation. We denote the frost-free period by x (in days) and the Hill activity by y (in Hill units $=$ μMoles of 2,6-dichlorophenol-indophenol per mg of chlorophyll per minute). On the straight line we choose two points that are not close together, say

$$x_0 = 100 \text{ days} \qquad y_0 = 42 \text{ Hill units}$$
$$x_1 = 300 \text{ days} \qquad y_1 = 21 \text{ Hill units}$$

---

[6] For an analytic method of fitting a straight line to a scatter diagram, see Section 12.3.

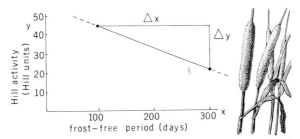

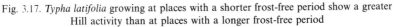

Fig. 3.17. *Typha latifolia* growing at places with a shorter frost-free period show a greater Hill activity than at places with a longer frost-free period

(read with the highest accuracy that the graph allows). Using (3.6.2) and (3.6.3), where we replace (x, y) by (x_1, y_1), we obtain

$$\Delta x = x_1 - x_0 = 200 \text{ days (positive)}$$

$$\Delta y = y_1 - y_0 = -21 \text{ Hill units (negative)}$$

$$a = -21/200 \text{ Hill units/day} = -0.105 \text{ Hill units/day}.$$

Now, formula (3.6.4) furnishes the equation of the straight line (the units are omitted for simplicity)

$$\frac{y - 42}{x - 100} = -0.105$$

or, on rearranging the equation,

$$y = (-0.105)\, x + 52.5. \tag{3.6.5}$$

We may check the result by comparing the constant term (that is 52.5 Hill units) with the y-intercept in Fig. 3.17.

The converse problem, that is to plot the diagram, is solved in the reverse order. Given is the Eq. (3.6.5). Then we determine the co-ordinates of an arbitrary point of the straight line, for simplicity the point $(0, b)$, in our example $(0, 52.5)$. We plot this point and from there we proceed in the direction given by the slope a. Knowing that $\Delta y = a \Delta x$ we choose an arbitrary step Δx to the right and from there another step up or down with the corresponding value of Δy. In our example we may choose $\Delta x = 300$ days (for good accuracy Δx should be at least one half of the width of the diagram). The corresponding value of Δy is

$$\Delta y = a\, \Delta x = (-0.105) \times 300 = -31.5 \text{ (Hill units)}.$$

Hence, from the point $(0, 52.5)$ we proceed $\Delta x = 300$ steps to the right and $\Delta y = -31.5$ steps in the perpendicular direction (down). This leads us to a second point of the straight line.

A slightly different way to plot the straight line is based on calculating the coordinates of *two* points. In our example we may choose $x_0 = 100$ days and $x_1 = 300$ days. From (3.6.5) we get $y_0 = 42$ and $y_1 = 21$ (Hill units). With the two points (x_0, y_0) and (x_1, y_1) the straight line is determined.

As in Section 3.5 we have to warn the reader that the slope a has in general nothing to do with the angle of inclination. Only in the special case where both x and y are measured and plotted in the same units to the same scale, the angle α of inclination may be meaningful. It can be determined from (3.5.3).

For completeness, the domain of a function ought to be specified. In our example, the relationship between Hill activity and frost-free period is only valid for $\{x \mid 50 < x \leq 365\}$. For $x < 50$ the relationship is not supported by the experiment.

*3.7. Linear Relations

In applications of mathematics linear functions often occur *implicitly*. An example shall serve as illustration. It is known that 100 g of dried soybeans contain 35 g of protein and 100 g of dried lentils contain 26 g of protein. Men of average size living in a moderate climate need 70 g of protein in their daily food [7]. Assume a man wants to provide for these 70 g of protein by eating soybeans and/or lentils. Let x be the amount of soybeans and y be the amount of lentils daily (x and y measured in units of 100 g). What is the relationship between x and y?

The protein taken with soybeans is $35x$ and with lentils $26y$ a day (both measured in g). The total daily amount of protein is 70 g. Hence we obtain the equation

$$35x + 26y = 70. \qquad (3.7.1)$$

By rearranging the terms we can express y as a function of x:

$$y = -\frac{35}{26}x + \frac{70}{26} \quad \text{or} \quad y = (-1.35)x + 2.69. \qquad (3.7.2)$$

We call an equation such as (3.7.1) an *implicit function* and an equation such as (3.7.2) an *explicit function*.

A listing of all possible pairs (x, y) that satisfy (3.7.1) or equivalently (3.7.2) is impossible, since there is an infinite number of them even if we observe the obvious side conditions $x \geq 0$ and $y \geq 0$. But a diagram offers a good survey of all possibilities. The function (3.7.2) is represented by a straight line with y-intercept 2.69 and slope -1.35 (see

---

[7] The figures were taken from Diem *et al.* (1970).

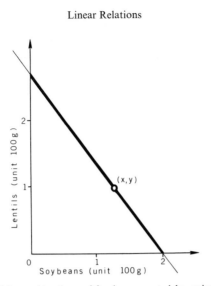

Fig. 3.18. Possible combinations of food represented by points of a segment

Fig. 3.18). The segment satisfying the side conditions $x \geq 0$, $y \geq 0$ is shown by a heavy line. All suitable pairs (x, y) are located on this segment.

For preparing the diagram it is not necessary to express y explicitly as a function of x. We could immediately use Eq. (3.7.1). For $x = 0$ we get $y = 70/26 = 2.69$, and for $y = 0$ we get $x = 70/35 = 2.00$. The latter amount is the x-intercept.

Generalizing this example we consider the *linear equation*

$$Ax + By + C = 0 \qquad (3.7.3)$$

which is equivalent to the explicit formula

$$y = -\frac{A}{B}x - \frac{C}{B} \quad \text{or} \quad y = ax + b \qquad (3.7.4)$$

provided that $B \neq 0$. If $B = 0$, y cannot be a function of x.

Let us now slightly generalize the previous example. Assume that a man wants to eat soybeans and lentils to get *at least* 70 g of protein a day. Then we have to replace the sign of equality in (3.7.1) by the sign for "greater than or equal to". Thus we obtain the *inequality* (also called *inequation*)

$$35x + 26y \geq 70 . \qquad (3.7.5)$$

Inequalities of this type usually occur in problems of "mixing" two kinds of "opportunities", such as mixing nutrition, supplies, plankton, bacteria cultures, etc.

The question is the same as before: What are the possible pairs (x, y) that satisfy the inequality (3.7.5)? To find a practical way we first solve the inequality (3.7.5) with respect to y. From both sides we subtract $35x$ and then divide by 26 in the same way as we would proceed for an equation:

$$35x + 26y \geq 70$$
$$-35x \qquad\quad = -35x$$

$$26y \geq -35x + 70$$
$$\frac{26y}{26} \geq \frac{-35x + 70}{26} .$$

Thus we obtain the explicit inequality

$$y \geq (-1.35)\, x + 2.69 . \tag{3.7.6}$$

The inequality (3.7.6) differs from the Eq. (3.7.2) only by the sign $\geq$. Whereas the Eq. (3.7.2) associates a unique value of y with a specific value of x, this is no longer true for the inequality (3.7.6). Therefore, y is not a function of x, but according to Section 3.3 the inequality may still be called a relation. More specifically we call (3.7.6) a *linear relation*.

From what we just said it is easy to plot the relation given by (3.7.6). In Fig. 3.18 we add to the points on the straight line all those points that have a higher value of y. Thus we get the set of all points above the straight line. Such a set is called a *half-plane*. For our specific problem of nutrition we have also to observe the inequalities $x \geq 0$ and $y \geq 0$. The admissible points lie in the shaded area of Fig. 3.19.

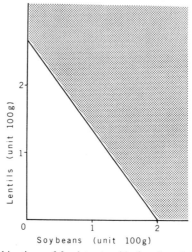

Fig. 3.19. Possible combinations of food to provide for the minimum required amount of protein

Biological problems of this type are sometimes more complex. Instead of only one inequality, more of them have to be satisfied. In addition some optimal property may be required. Problems of this type are today treated under the heading *linear programming*. We conclude this section with a typical example. In a laboratory there are two bacteria counters available. Counter C_1 can be operated by a graduate student who earns $\$2.00$ per hour. On the average he is able to count six samples an hour. Counter C_2 is faster, but also more sophisticated. Only a well trained person earning $\$5.00$ an hour can operate it. With the same precision as C_1, counter C_2 allows ten countings an hour. Given are 1000 samples to be counted within a time period not exceeding 80 hours. How long should each of the two counters be used in order to perform the task at a minimum cost?

Let x denote the number of hours counter C_1 is operated, and y the corresponding number for C_2. Then we get the following table:

Counter	Samples counted per hour	Wages per hour	Number of hours in operation
C_1	6	$\$2.00$	x
C_2	10	$\$5.00$	y

Since the task should be performed within 80 hours, we obtain the inequalities

$$0 \leq x \leq 80,$$
$$0 \leq y \leq 80. \tag{3.7.7}$$

C_1 counts $6x$ and C_2 counts $10y$ samples. Altogether they count

$$6x + 10y = 1000 \tag{3.7.8}$$

samples. The cost for operating C_1 is $2x$, and for C_2 the cost is $5y$ (both amounts in dollars). Hence the total cost is

$$2x + 5y, \tag{3.7.9}$$

and this amount should be minimized.

We solve the problems by means of a diagram. In Fig. 3.20 the shaded square contains the points (x, y) that satisfy the inequalities (3.7.7). In addition, the linear Eq. (3.7.8) has to be fulfilled which means that the points (x, y) lie on the corresponding straight line. To satisfy all conditions the admissible points have to be located on the *intersection* of the square (shaded) and the straight line. This intersection is

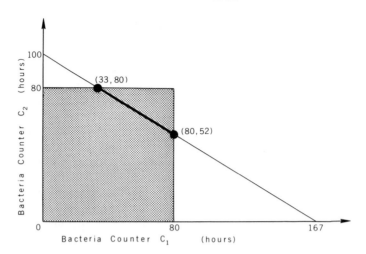

Fig. 3.20. Construction of the optimal point in the use of bacteria counters

plotted as a heavy segment. We calculate the coordinates of the end-points. For the upper endpoint we know that $y = 80$, and from (3.7.8) we obtain $x = 33$ (rounded off to the nearest integer). For the lower endpoint we find $x = 80$ and $y = 52$. Now we calculate the cost for the operation of the counters along the segment. From (3.7.8) we get $y = (-0.6)x + 100$. With this value of y the total cost (3.7.9) becomes

$$2x + 5y = 2x + 5((-0.6)x + 100) = 500 - x .$$

Hence, the cost is a linear function of x. When we move from the upper to the lower endpoint, that is when x increases from 33 to 80, the total cost decreases from 467 to 420 dollars. Thus the final result is: The expense is minimized if counter C_1 is used 80 hours and counter C_2 is used 52 hours.

Recommended for Further Reading: Defares et al. (1973), Grossman and Turner (1974), Guelfi (1966), Hays (1963), Kerlinger (1964), Lefort (1967), Nahikian (1964), C.A.B.Smith (1966), Stibitz (1966).

Problems for Solution

3.2.1. Determine the product set of $A = \{0, 1\}$ and $B = \{0, 2, 4\}$ and plot a graph. Show that $A \times B \neq B \times A$ which means that the commutative law is not valid for the cross product.

3.2.2. Let $S = \{a, A\}$, $T = \{b, B\}$, $U = \{c, C\}$. Define ordered triples by the cross product $S \times T \times U$.

3.3.1. Let $A = \{0, 1, 2, 3, 4\}$. The inequality $x + y \geq 3$ defines a relation in the product set $A \times A$. Find a graphical representation of this relation. How many of the 25 pairs (x, y) satisfy the inequality?

3.3.2. In the pedigree shown in Fig. 3.21 the squares are symbols for males and the circles symbols for females. The individuals are denoted by $a, b, c, ..., n$. Establish the following relations using graph paper:

a) x is brother or sister of y,

b) x is a descendant of y,

c) x is an ancestor of y.

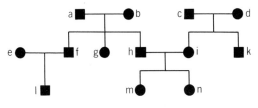

Fig. 3.21.

3.3.3. In a marine ecosystem phytoplankton is either eaten up by zooplankton, or by omnivores, or it decays. Zooplankton is eaten up by omnivores, or by carnivores, or it decays. Finally, omnivores and carnivores eat themselves up or they decay. With this situation in mind define a relation in the product set $S \times S$ where $S = \{$phytoplankton, zooplankton, omnivores, carnivores, decay$\}$ and find a suitable graph (cf. Patten, 1966, p. 595).

3.3.4. The Pacific gooseneck barnacle (*Pollicipes polymerus*), a crustacean attached to rocks, has a primitive circulatory system. Hemolymph is pumped through the body by an unidentified muscle. The contraction frequency is strongly dependent on body temperature. If body temperature (°C) is plotted on a horizontal line and the contraction frequency (beats per minute) on a vertical axis, a relation is given by all points in a triangle whose vertices are:

°C	Beats per minute
7	1
23	15
25	50

Plot the relation and decide whether the following points belong to the relation: $(15°, 9 \text{ b.p.m.})$, $(15°, 25 \text{ b.p.m.})$, $(24°, 20 \text{ b.p.m.})$. Data are taken from Fyhn *et al.* (1973).

3.3.5. Five persons checked their physical strength on an ascending mountain trail. They measured the time which they needed to climb from a certain altitude to a higher level:

Person No.	Classification	Altitude		Time needed
		Lower level	Higher level	
1	very strong	800 m	2200 m	2 h 42 min
2	not trained	1200 m	2100 m	3 h 6 min
3	very strong	1200 m	2900 m	3 h 30 min
4	fairly strong	1500 m	2350 m	2 h 12 min
5	not trained	950 m	2250 m	4 h 6 min

Calculate the "power of climbing" in m h^{-1} (meter vertical distance per hour) for each person and establish a relation by plotting the "power of climbing" versus classification of strength.

3.4.1. Twenty experimental mice numbered $1, 2, ..., 20$ are tested whether they react to a certain dose of strychnine. We associate the number one with a mouse if it reacts positively, otherwise the number zero. This association is a function. Why? Determine the domain and the range of this function.

3.4.2. Twelve persons numbered $1, 2, ..., 12$ are classified right-handed, left-handed, or "both-handed". This association establishes a function. What are domain and range of this function?

3.4.3. In which of the following cases is it correct to say that x-members are mapped into y-members or, equivalently, that the association of y with x is a function?
   a) $x = $ pulse rate, $y = $ body temperature of a particular patient, x and y being measured several times,
   b) $x = $ triangle, $y = $ area of the triangle,
   c) $x = $ frequency of electromagnetic wave, $y = $ spectral color,
   d) $x = $ section of a road, $y = $ average speed of an automobile on a particular trip,
   e) $x = $ speed of a particular car, $y = $ shortest stopping distance, x and y being measured under different road conditions.

3.4.4. Let $x \in \mathbb{N} = \{1, 2, 3, ...\}$ and y be the remainder 0, 1, 2 or 3 which results after dividing x by 4. Why is the association of y with x a function? Plot a diagram of this function and specify its domain and range.

3.5.1. According to Timoféeff-Ressovsky and Zimmer (1947, p. 36), the number of sex related mutations in *Drosophila melanogaster* increases almost linearly with the dose of X-rays provided that the dose does not exceed 6 kR (kilo-Roentgen). Let x denote the dose measured in kR and y be the mutation rate (percentage). For dose 0 no mutation is observed. With a dose of 3 kR the mutation rate is 8.4%. Plot a diagram and establish the equation for y and x. What is the domain and the range of the function? Is the angle of inclination meaningful?

3.5.2. The pressure of water is proportional to its depth. Let d denote the depth (meters) and p the pressure (atmospheres). In seawater the following measurement was made: $d = 98.0$ m, $p = 10.21$ atm. Express p in terms of d. If an error in p of less than 1 % is negligible, the domain of the function is approximately $\{d \mid 0 \le d \le 1000\}$. What is the corresponding range?

3.5.3. A leaping animal, such as a cat, a porpoise, or a flea, falls in such a way that the vertical speed of its center of gravity increases by 9.81 m/sec every second. What is the equation of the vertical speed, if the time zero is chosen at the instant when the vertical speed is zero (at the point of culmination)? What are the vertical speeds for $t = 0.1$ sec, 0.2 sec, etc.? Plot a diagram. Interpret the values of the function for negative values of t.

3.5.4. The rate of oxygen consumption of the lung is measured in ml min^{-1}. The larger this rate is, the more extended the alveolar surface area, A, must be. Since oxygen consumption is a diffusion process, the consumption rate, Q, is expected to be proportional to the alveolar surface area. Find the arithmetic mean of Q/A from the following data (adapted from Tenney and Remmers, 1963):

	$A(\mathrm{m}^2)$	$Q(\mathrm{ml\ min}^{-1})$
Rat	0.4	2.0
Rabbit	6.5	18
Man	62	210
Porpoise	110	1000
Cow	300	1400
Whale	1700	12000

3.5.5. The earth is an approximate sphere of circumference 40000 km. Assume that a wire is spanned around the equator of such a sphere. Now we enlarge the required length of 40000 km by 10 m and span the wire again so that a gap of constant width is left between earth and wire. Will a mouse be able to slip between wire and earth?

3.6.3. Plot the function $y = 8.8x - 20$ using a unit on the y axis that is only one tenth of the unit on the x axis.

3.6.4. Plot the function $y = 0.35x + 25$ with equal units on the x and y axis, but with an "interrupted" y axis in order to save space.

3.6.5. Plot the straight lines given by the equation $y - 3 = a(x - 2)$ for $a = 0$, 0.6, 1.3, 2.8, -0.9, -1.6. (Hint: Compare with Eq. (3.6.4). To plot the slope choose an arbitrary Δx, for instance $\Delta x = 10$, determine Δy, and proceed Δx steps to the right and Δy steps up or down.)

3.6.6. Without any calculation, plot the graphs of $y - 5 = a(x - 2)$ for

$$\text{a) } a = 2.5, \qquad \text{b) } a = 3.5, \qquad \text{c) } a = -3.5.$$

Use different scales for the x and y axes, such as setting the x-unit twice as large as the y-unit. (Hint: Use Eq. (3.6.4). To plot the slope, let $\Delta x = 1$.)

3.6.7. A linear function $Q = Q(t)$ assumes the value $Q_1 = 88.3$ mg at time instant $t_1 = 14$ sec and the value $Q_2 = 89.6$ mg at $t_2 = 39$ sec. Find the rate of growth and establish the linear function.

3.6.8. As a rule biological experiments are time consuming. The cost consists of two parts:

1. a fixed amount for purchase of instruments, equipment, animals or plants,

2. a variable amount increasing daily by a salary rate for the experimenter and/or his assistant, and the cost rate for food, supplies and power. Let f be the fixed amount, r the daily rate of increase, t the time in days, and F the total cost. Find the equation expressing F in terms of t. Check that $\Delta F/\Delta t$ is constant.

3.6.9. If a helical spring is under the influence of a force, its length is a linear function of the force unless the force exceeds a certain bound (Hooke's law). Let F be the force (measured in Newtons) and l be the length of the spring (cm). l_0 denotes the initial length when no force acts. Let $a = \Delta l/\Delta F$ be the rate of increase. Express l in terms of F.

3.6.10. The carbon dioxide concentration in the weather-active range of 9 to 12 km height was 313 ppm in 1960 and 321 ppm in 1970. There was a monotone increase. Using linear extrapolation estimate the CO_2 concentration for the years 1980, 1990, and 2000 (ppm = parts per million).

3.6.11. Are the following three points located on the same straight line?

$$x_0 = 1.5 \qquad x_1 = 4.5 \qquad x_2 = 12.0$$
$$y_0 = -2.0 \qquad y_1 = 2.5 \qquad y_2 = 13.75$$

(Hint: Calculate $\Delta y/\Delta x$ for different pairs of points.)

3.6.12. Let

$$x_0 = 2, \qquad x_1 = 6, \qquad x_2 = 9,$$
$$y_0 = 7.5, \qquad y_1 = 5.5, \qquad y_2 = w.$$

Using $\Delta y/\Delta x$ find w such that the three points (x_0, y_0), (x_1, y_1), (x_2, y_2) are on the same straight line.

3.6.13. Simpson, Roe, and Lewontin (1960, p. 218) report that in females of the snake *Lampropeltis polyzona* the total length y is a linear function of the tail length x with great accuracy. The domain is the interval from 30 mm to 200 mm, and the range is the interval from 200 mm to 1400 mm. The following two points are given:

$$x_0 = 60 \text{ mm} \qquad x_1 = 140 \text{ mm}$$
$$y_0 = 455 \text{ mm} \qquad y_1 = 1050 \text{ mm} .$$

Find the equation of y as a function of x and plot a diagram with suitable units for x and y. Is the angle of inclination meaningful?

3.6.14. A copper rod, part of an instrument, is exposed to different temperatures. Its length l is almost a linear function of the temperature t provided that $t < 150°$ (Celsius). Find the equation for l using the following measurements:

$$t_1 = 15°, \ l_1 = 76.45 \text{ cm} \quad \text{and} \quad t_2 = 100°, \ l_2 = 76.56 \text{ cm}.$$

3.6.15. The temperature on the Celsius scale, denoted by x, and the same temperature on the Fahrenheit scale, denoted by y, are connected by the linear relation $5y - 9x = 160$. Express y as a function of x and plot the function. Prepare a conversion table for $x = 36.0°, 36.1°, 36.2°, \ldots, 37.0°$.

3.6.16. In the lungs, the air reaches body temperature. The exhaled air is below body temperature since it is cooled at the walls of the nose. Measurements were made on cactus wrens (small desert birds). For the ambient temperature the domain was $\{T_A | 12° < T_A < 30°\}$. The exhaled air temperature T_E depends linearly on the ambient temperature T_A:

$$T_E = 8.51 + 0.756\, T_A \qquad \text{(empirical function)}.$$

Plot a graph of this function and determine the range. (Data from Schmidt-Nielsen, 1972, p. 9.)

3.6.17. A major pollutant produced by burning fossil fuels is sulfur dioxide (SO_2). An investigation in Oslo, Norway, has shown that the number N of deaths per week is a linear function of the mean concentration C of SO_2 measured in $\mu g/m^3$. The empirical function is $N = 94 + (0.031)\, C$. The domain is $\{C | 50 < C < 700\}$. a) Plot a graph of this function. b) Calculate the range. (Data from Wilson, 1972.)

3.6.18. By burning fossil fuels man releases 200 million metric tons of poisonous carbon monoxide (CO) into the atmosphere each year. Nevertheless the concentration of CO remains between 0.04 and 0.90 ppm (parts per million) in the ambient air. The main reason is that microorganisms in the soil absorb CO rapidly and convert it to carbon dioxide and/or methane. In an experiment with 10 liters of air and some potting soil, 1443 μg of CO were reduced to 47 μg within 3 hrs. The decrease was linear. a) Plot the result in a rectangular coordinate system. b) Find the equation for the decreasing mass of CO. (Data from Inman, R. E., et al., 1971.)

3.6.19. The ability of a person to discriminate spectral colors is measured by establishing at each frequency v_0 the closest frequency that can be distinguished, say v_1. The absolute value of the difference, that is, $\delta = |v_1 - v_0|$ may be plotted as ordinate with the reference frequency v_0 as abscissa. δ is minimal at frequencies near the boundaries between two hues (such as blue, green, red) and maximal at frequencies near the center of hues. Do δ and v_0 constitute a relation? a function? a linear function?[8]

*3.7.1. Solve the following inequalities with respect to y and plot the relations in the Cartesian plane:

$$\text{a)}\ x + y < 5 \qquad \text{b)}\ x - y < 5$$
$$\text{c)}\ x + 2y > 8 \qquad \text{d)}\ 3x - 3y < 10.$$

---

[8] Suggested by Dr. Jack P. Hailman, University of Wisconsin. v is the Greek letter nu.

*3.7.2. Given the two inequalities

$$A: \ x > y \qquad B: \ 3x < y.$$

Find graphically all pairs (x, y) for which

          a) A as well as B,

          b) A and/or B,

          c) either A or B

are satisfied.

*3.7.3. Find graphically the intersection and the union of the sets

$$\{(x, y)|x + y - 5 > 0\} \quad \text{and} \quad \{(x, y)|x - 2y + 2 > 0\}.$$

*3.7.4. Assume an adult needs at least 300 g of carbohydrates in his daily food. What possibilities does he have if he wants to fulfill this condition by eating a combined food consisting of potatoes and soybeans? 100 g of raw potatoes contain 19 g and 100 g of dried soybeans 35 g of carbohydrates (data from Diem et al., 1970). Show the result in a diagram.

*3.7.5. Find the solution of the problem at the end of Section 3.7 assuming that the wages are $ 2.00 and $ 3.00 for the operators of bacteria counters C_1 and C_2, respectively.

*3.7.6. An area of at most 10 ha should be planted with wheat and potatoes. To avoid monoculture it is required that at most 70% of the area actually cultivated be reserved for wheat or for potatoes. Let x be the number of hectares planted with wheat and y the corresponding number for potatoes. Find graphically all pairs (x, y) that fulfill these conditions.

*3.7.7. In a developing country an area of at most 10 km^2 ($= 1000$ ha) is being planted with potatoes and corn. To fight against insects and against plant diseases it is required that not more than 70% of the area actually used be planted with potatoes or corn. The crop of potatoes is estimated to be 8.3 t (metric tons) per ha and of corn 7.7 t per ha. Each kilogram of raw potatoes contains 20 g of protein whereas the corresponding amount for fresh corn is 32 g (data from Diem et al., 1970). It is required that the total amount of protein in the crop is not less than 200 t. If x denotes the number of km^2 planted with potatoes and y the corresponding number for corn, what are the pairs (x, y) that satisfy all conditions? Solve the problem graphically.

Chapter 4

The Power Function and Related Functions

4.1. Definitions

In Section 3.5 we studied the special linear function $y = ax$, that is, y is proportional to x. The variable x may be written as the first power x^1, and then the linear function appears to be a particular case of

$$y = ax^n. \tag{4.1.1}$$

where a and n are constant numbers. A function of this type is called a *power function*. (4.1.1) says that y is proportional to x^n, a property which we may write as

$$y \propto x^n \tag{4.1.2}$$

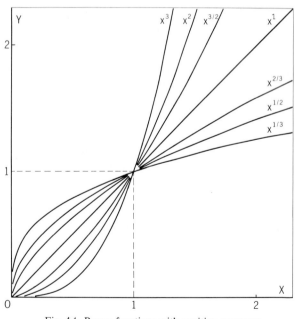

Fig. 4.1. Power functions with positive exponent

using a symbol introduced in formula (3.5.2). The behavior of a power function is mainly determined by the exponent n. Fig. 4.1 gives a survey of some power functions with positive exponents n. For simplicity we choose $a = 1$ and restrict the domain to $x \geq 0$. The greater n is, the faster y increases.

In (4.1.1) the constants a and n characterize a power function, that is, specific values of a and n belong to one and only one power function. In general, constants with such a property are called *parameters* of a function. We also say that (4.1.1) is a *two-parametric* function. Another example of a *two-parametric* function is $y = ax + b$ with the slope a and the y-intercept b as parameters.

In this chapter we will deal not only with power functions, but also with polynomials which are sums of power functions.

4.2. Examples of Power Functions

As a first application we study a jumping or leaping animal such as a flea, a cat, or a porpoise. Galilei's law[1] states that the center of gravity is subject to the same acceleration for all bodies provided that the air resistance can be neglected. The movement of the center of gravity is described by this law no matter what action the animal performs during the jump. Consider the highest level reached by the center of gravity. Let $t = 0$ be the time instant when the highest level is reached and let s denote the vertical distance of the center of gravity measured from its highest level (see Fig. 4.2). Then

$$s = \frac{g}{2} t^2 , \tag{4.2.1}$$

where g is the acceleration of the center of gravity. The vertical distance is proportional to the square of the time. If t is measured in sec and s in cm, then $g = 981$ cm/sec^2 for the surface of the earth.

From (4.2.1) we derive the table:

t (sec)		0	0.1	0.2	0.3	0.4 ...
s (cm)	$\dfrac{981}{2}$ times	0	0.01	0.04	0.09	0.16 ...

While the time increases in constant steps of 0.1 sec, the vertical distance increases as the squares $1^2 = 1$, $2^2 = 4$, $3^2 = 9$, $4^2 = 16$, etc., that is, with acceleration. Formula (4.2.1) defines a power function. Since the exponent

---
[1] Galileo Galilei, Italian physicist and astronomer (1564–1642).

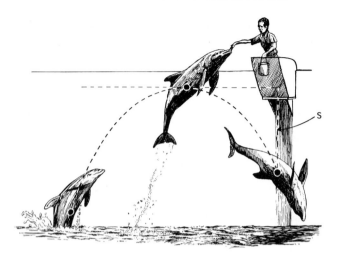

Fig. 4.2. A leaping porpoise. The center of gravity describes a parabola no matter how the porpoise acts during the motion

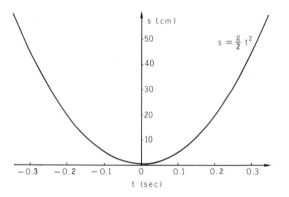

Fig. 4.3. The graph of the vertical distance as a function of time is a parabola

is $n = 2$, the function is also called a function of the *second degree* or a *quadratic function*. We may include negative values of t, that is, consider the time before the animal reaches culmination. Then s is still the vertical distance from the level of culmination. The graph of the function is a *quadratic parabola* (see Fig. 4.3).

Biophysics furnishes important applications of power functions. Consider a unicellular being of spherical shape. Let r be the radius of the sphere. Then the surface[2] S and the volume V of the cell are given by

$$S = 4\pi r^2, \tag{4.2.2}$$

$$V = \frac{4}{3}\pi r^3 \tag{4.2.3}$$

with $\pi = 3.14\ldots$. The surface S is a quadratic function of r. If the radius grows by a factor 2, 3, etc., the surface increases by a factor 4, 9, etc. (see Fig. 4.4). The volume V is a cubic function of r. Thus if the radius

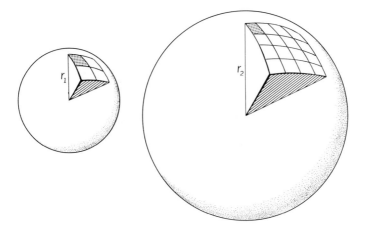

Fig. 4.4. Two spheres of different radius. If $r_2 = 2r_1$, the surface is four times as large

is doubled, the volume increases by a factor 8. When a living cell grows, its volume increases faster than its surface. Conversely, as long as the cell is small in size, the surface is relatively large. Assume that the various physical and chemical properties of the cell components remain unaltered while the cell grows. Then the flow of chemicals, such as O_2 and CO_2, through the surface increases at a slower pace than the metabolic capacity of the cell, which is proportional to the volume. The same is true for the light energy received through the surface. A proper balance between flow through the surface and metabolism inside is only possible for a cell that is not too large and not too small. On one hand, if the radius is less than a lower critical value, say $r < r_1$, the metabolism presumably collapses since the volume is not sufficient relative to the

---

[2] Unfortunately the word surface is used ambiguously. It could mean the set of all points that form the face of a solid, a meaning that is expressed by the notion *boundary* in modern mathematics. A second meaning of the word surface is the measure of the boundary, that is, a number.

efficiency of the surface. On the other hand, for a radius greater than an upper critical value, say $r > r_2$, the chemical exchange and the energy flow through the surface may break down since the surface is too small relative to the capacity of metabolism. Only for $r_1 < r < r_2$ is the cell competitive and able to survive in the struggle for existence. Future theoretical research may furnish numerical values for r_1 and r_2 and thus suggest why unicellular beings are of the size observed in nature.

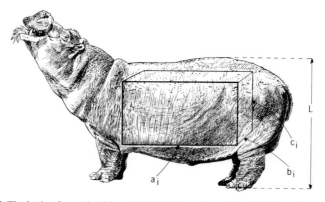

Fig. 4.5. The body of an animal is subdivided into rectangular solids in order to show that the volume is proportional to the third power of a linear dimension

For plants and animals with a shape that is more complicated than a sphere, there is still an easy geometric relationship between surface and volume of the body. For better understanding we introduce the term of a *linear dimension*. The total length of a body, its height, or its width are linear dimensions. Also any distance between well-defined points of a body such as the shoulder-hip distance may serve as a linear dimension. In the special case of a spherically shaped cell we used the radius r as a linear dimension. Now we consider bodies of different size, but exactly the same shape. Such bodies are *similar* in the geometric sense. Let l be a suitably chosen linear dimension. Then we will express the volume as a function of l. For this purpose we subdivide the body into a large number of cubes of different sizes (or of rectangular solids as in Fig. 4.5). Let V be the total volume of these cubes. It is close to the volume of the body. We can perform the subdivision in such a refined way that the error of approximation is arbitrarily small.

We number the cubes by $i = 1, 2, 3, \ldots$ and denote the side of the ith cube by s_i. Then

$$V = \sum_i s_i^3 . \tag{4.2.4}$$

Now we increase all linear dimensions of the body by the same factor, say by $c > 1$. If we denote the new sizes by primes, we obtain

$$l' = cl, \quad s_i' = cs_i \quad (i = 1, 2, 3, \ldots),$$

$$V' = \Sigma(s_i')^3 = \Sigma(cs_i)^3 = \Sigma c^3 s_i^3 = c^3 \Sigma s_i^3. \tag{4.2.5}$$

It follows that

$$V' = c^3 V, \tag{4.2.6}$$

that is, the *volume has to be multiplied by the third power of c*.

We may reformulate the result by considering l, s_i and V as *variable quantities*. For each value of a linear dimension l, the sides s_i and the volume V are uniquely determined. They may be considered as functions of l. Since s_i grows proportional to l we may write

$$s_i \propto l \quad (i = 1, 2, 3, \ldots).$$

But, as formula (4.2.5) indicates, the volume grows as s_i^3. Therefore, V is proportional to s_i^3 and in turn to l^3. Thus we may write

$$\boxed{V \propto l^3}. \tag{4.2.7}$$

In words: *The volume of a body of any shape is proportional to the cube of any of its linear dimensions.*

Similarly, we may subdivide the *surface* of a body into a large number of squares of different sizes. If S denotes the total surface, we obtain by the same method as before the result

$$\boxed{S \propto l^2}. \tag{4.2.8}$$

In words: *The surface of a body of any shape is proportional to the square of any of its linear dimensions.*

Special applications of formulas (4.2.7) and (4.2.8) are found in the formulas (4.2.2) and (4.2.3) for the sphere. Here, the radius r is used as linear dimension.

Multicellular plants and animals of a size considerably larger than a single cell cannot exist unless they have a sufficiently large surface. This follows from our discussion of a spherical cell. Gills, lungs, intestines, kidneys for animals, roots and leaves for plants are devices for an increased surface in order to stimulate diffusion of molecules or absorption of energy. All these devices are appropriate as long as the body size remains within certain limits.

As an illustration take a mouse. A giant mouse, geometrically similar in shape to an ordinary mouse but with a linear dimension ten times greater, could hardly survive. Its volume has grown $10^3 = 1000$ times, whereas the surface (including the inner surface of lungs, intestines, etc.) has increased by the factor $10^2 = 100$ only. Thus, gas exchange, food resorption, and the renal functions would not suffice. Conversely, a dwarf mouse, geometrically similar in shape to an ordinary mouse but with a linear dimension one tenth as great, is also handicapped. The surface is $1/100$ that of the ordinary mouse, whereas the volume has dropped to $1/1000$. Metabolism, also decreased to $1/1000$, would be unable to maintain the necessary functions, especially since a mouse is a warm-blooded animal and has to compensate for heat loss which is proportional to the surface.

It is an experimental fact that muscle force is approximately proportional to the number of muscle fibers and, therefore, proportional to l^2 where l is a linear dimension of the muscle. The energy produced by a muscle is "force × segment" and thus proportional to $l^2 \times l = l^3$. Consider now a flea *(Ctenocephalis canis)*. A leaping flea can reach a height nearly 200 times its own height. The energy required to do this is approximated by the formula "weight × height of leap". Imagine a giant flea, geometrically similar in shape but 10 times larger in its linear dimension. The energy required to jump 200 times as high as its own height would be $10^3 \times 10 = 10^4$ times as much as for the ordinary flea. This, however, cannot be done by the giant flea since its muscle strength has increased by a factor 10^3 only. Thus, it could reach a height of only 20 times its own height. By the same token, a "supergiant" flea, 100 times larger in its linear dimension than an ordinary flea, could leap to a height that is only twice its own height despite a tremendous leaping muscle.

Slightly generalizing this example we see that animals in running away from predators or in searching for food consume *energies that are approximately proportional to the fourth power of their linear dimensions*, whereas the energy supply is proportional to the third power only. We conclude that an animal species is able to compete with other species, only if the body size is subject to certain restrictions. For more examples of this type see Schmidt-Nielsen (1972), Slijper (1967), J. M. Smith (1968), and Thompson (1917, 1961).

In general, if the linear dimensions of an organism are changed, the physical and chemical properties will change. They can be kept constant only over a small range of increase or decrease. Jonathan Swift's six-inch tall Lilliputians are most entertaining. In *Gulliver's Travels* they are able to perform like humans. However, they remain fiction. Biology cannot work that way.

So far we have considered power functions with exponents $n = 2$, 3, or 4. There are also biological applications of power functions with *fractional exponents*. However, they will be treated either in connection with a double-logarithmic coordinate system (Section 7.3) or in connection with allometric growth (Section 11.6).

Problem 4.2.1. Assume that a crystal grows in such a way that all linear dimensions increase by 18%. What are the percent increases of the crystal's surface S and volume V?

Solution: The factor c in Eq. (4.2.5) is

$$c = 1 + 18/100 = 1.18 .$$

Hence,

$$S' = c^2 S , \qquad c^2 = 1.39 ,$$

increase of surface $= 39\%$,

$$V' = c^3 V , \qquad c^3 = 1.64 ,$$

increase of volume $= 64\%$.

4.3. Polynomials

Sometimes the quadratic function $y = ax^2$ appears in a slightly generalized form. Linear terms, say $bx + c$, are added. Thus we get the *general quadratic function*

$$y = ax^2 + bx + c \quad (a \neq 0) \tag{4.3.1}$$

with three parameters a, b, c. We also call the right-hand expression a *polynomial in x of the second degree*.

Polynomials of higher degree are occasionally used in biology. The expression

$$a_0 + a_1 x + a_2 x^2 + a_3 x^3 + \cdots + a_n x^n \tag{4.3.2}$$

is called a polynomial in x of the *n-th* degree with parameters a_i $(i = 0, 1, 2, ..., n)$ if $a_n \neq 0$ and if all exponents of x are positive integers. We will confine ourselves to second degree polynomials.

We know that a graph of the special quadratic function $y = ax^2$ is a quadratic *parabola* with vertex at the origin. It is a pleasant property of the general quadratic function that its graph is also a quadratic parabola. Only now, the vertex need not be at the origin. We show this fact with a numerical example (Fig. 4.6). Consider

$$y = \frac{1}{2} x^2 . \tag{4.3.3}$$

We shift each point (x, y) of the parabola three units to the right and two units down. A point (x, y) moves then into the new point $(x + 3, y - 2)$. We denote its new coordinates by x' and y' (x prime, y prime). Thus we get

$$x' = x + 3, \qquad y' = y - 2. \tag{4.3.4}$$

In this manner we want to shift all points of the parabola. The vertex is shifted to the point $(3, -2)$. To obtain the equation of the shifted parabola we solve the Eq. (4.3.4) with respect to x and y:

$$x = x' - 3, \qquad y = y' + 2. \tag{4.3.5}$$

Then, in (4.3.3) we replace x and y by the expressions (4.3.5). Thus we get

$$y' + 2 = \frac{1}{2}(x' - 3)^2$$

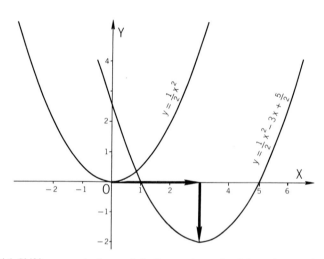

Fig. 4.6. Shifting a quadratic parabola three units to the right and two units down

or

$$y' = \frac{1}{2}x'^2 - 3x' + \frac{5}{2}.$$

Finally, we drop the primes for simplicity. Then the quadratic function becomes

$$y = \frac{1}{2}x^2 - 3x + \frac{5}{2} \tag{4.3.6}$$

which is of the form (4.3.1). Notice that the quadratic term has not changed, but that a linear expression $-3x + 5/2$ has been added.

4.4. Differences

For tabulating and checking the values of a polynomial, we recommend the use of differences. We explain the method using the special polynomial (4.3.6). For $x = 0, 1, 2, 3, 4$ it is easy to calculate y. We get

x	y
0	2.5
1	0
2	-1.5
3	-2
4	-1.5

These values may be compared with the corresponding coordinates in Fig. 4.6. Now we form the differences of consecutive y-values. If the y-values are denoted by $y_0, y_1, y_2, \ldots$ and the differences by

$$\Delta y_0, \Delta y_1, \Delta y_2, \ldots,$$

we obtain [3]

$$\Delta y_0 = y_1 - y_0 = 0 - 2.5 = -2.5$$
$$\Delta y_1 = y_2 - y_1 = (-1.5) - 0 = -1.5$$
$$\Delta y_2 = y_3 - y_2 = (-2) - (-1.5) = -0.5$$
$$\Delta y_3 = y_4 - y_3 = (-1.5) - (-2) = +0.5.$$

We see that Δy increases from -2.5 to 0.5. To get further insight, we form the "differences of the differences". They are denoted by $\Delta^2 y$ and are called second differences [4]. In our example we get

$$\Delta^2 y_0 = \Delta y_1 - \Delta y_0 = (-1.5) - (-2.5) = +1$$
$$\Delta^2 y_1 = \Delta y_2 - \Delta y_1 = (-0.5) - (-1.5) = +1$$
$$\Delta^2 y_2 = \Delta y_3 - \Delta y_2 = (+0.5) - (-0.5) = +1.$$

---

[3] The upper case Greek letter delta is generally used to symbolize a difference. The symbol Δ stands in front of the variable from which a difference has to be formed. It indicates an operation. For this reason, mathematicians call Δ a *difference operator*.

[4] In a second difference, the operator Δ is applied twice. We may write $\Delta(\Delta y)$, but it is convenient to abbreviate this expression by $\Delta^2 y$, much in the same way as we abbreviate a multiplication $a \cdot a = a^2$. The term $\Delta^2 y$ is read "delta two y". Correspondingly, differences of higher order are $\Delta(\Delta^2 y) = \Delta^3 y$, $\Delta(\Delta^3 y) = \Delta^4 y$, etc.

Hence, the second differences remain constant. The question arises whether this occurred by chance, or whether there exists a rule for quadratic functions. We will show in what follows that the second alternative is true. Let

$$y = ax^2 + bx + c \qquad (4.4.1)$$

be a quadratic function. We choose any fixed value of x and denote the corresponding value of y by y_0. Then we increase x by 1. Thus we obtain the new value

$$y_1 = a(x+1)^2 + b(x+1) + c$$

$$= ax^2 + 2ax + a + bx + b + c.$$

The difference is

$$\Delta y_0 = y_1 - y_0 = 2ax + a + b. \qquad (4.4.2)$$

If we consider x as a variable and write Δy instead of Δy_0, we see that Δy is a *linear* function of x. For this new function we again form a difference:

$$\Delta y_0 = 2ax + a + b,$$

$$\Delta y_1 = 2a(x+1) + a + b = 2ax + 3a + b,$$

$$\Delta^2 y_0 = \Delta y_1 - \Delta y_0 = 2a.$$

This second difference is independent of x and, therefore, a constant. For all values of x we get

$$\Delta^2 y = 2a. \qquad (4.4.3)$$

Summarizing we have the following rule:

For a quadratic polynomial, the first differences form a linear function, the second differences remain constant. In our numerical example, we have $a = \frac{1}{2}$. Hence, $\Delta^2 y = 2a = 1$ in agreement with the values of $\Delta^2 y_0$, $\Delta^2 y_1$, $\Delta^2 y_2$ obtained above.

The property of constant second differences may be used not only for checking numerical tables of quadratic functions but also for extending the tables. The procedure is shown in the following table:

x	y	Δy	$\Delta^2 y$
0	+ 2.5		
		−2.5	
1	0		+1.0
		−1.5	
2	− 1.5		+1.0
		−0.5	
3	− 2		+1.0
		+0.5	
4	− 1.5		+1.0
		+1.5	
5	0		+1.0
		+2.5	
6	+ 2.5		+1.0
		+3.5	
7	+ 6.0		+1.0
		+4.5	
8	+10.5		...
		...	
...	...		

We first extend the column headed by $\Delta^2 y$. Then we add $\Delta^2 y$ to the last value of Δy, in our example, $(+0.5) + 1 = +1.5$. Then we add the result to the last value of y, in the example, $(-1.5) + 1.5 = 0$. Thus we obtain a new value of y (in the example $y = 0$ for $x = 5$). The procedure may be repeated as many times as desired. It is much faster than recalculating the polynomial over and over again. Notice that no squaring and no multiplication are required. If more digits are involved, an adding machine can perform all calculations.

4.5. An Application

We consider the *flow of blood in a blood vessel*. A piece of an artery or of a vein may be conceived as a cylindric tube of constant width. Assume that the cross section is a circle of radius R. Like any liquid the blood has an inner friction which is called viscosity and denoted

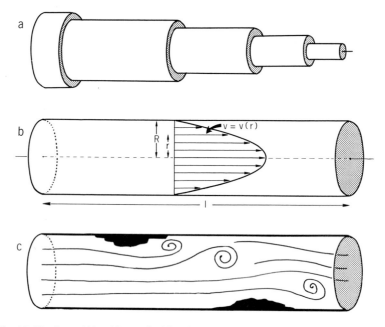

Fig. 4.7. The flow of blood in a cylindric tube. (a) and (b) laminar flow. (c) turbulent flow

by η (Greek letter eta). The viscosity is measured in Pa s (Pascal-second), one of the SI-units[5] which equals $kg\, s^{-1}\, m^{-1}$. There is also friction at the walls of the tube. Immediately at the wall the velocity of the blood is zero. The speed is highest along the center axis of the tube. If the velocity does not exceed a certain critical value, the flow is *laminar*, that is, all particles of the liquid move parallel to the tube and the velocity increases regularly from zero at the wall toward the center (see Fig. 4.7a, b). We may think of an infinite number of cylindrical laminae (Latin word for layers) which move like the tubes of a telescope. The velocity increases toward the center. If, however, the speed exceeds the critical value, for instance in a vessel that is partially occluded, the flow is *turbulent* and sound can be heard (Fig. 4.7c).

Now we assume laminar flow. Let r be the distance of any point of the liquid from the axis. Then the velocity v is a function of r. We may write $v = v(r)$. The domain of the function is the interval $0 \leqq r \leqq R$. The function was experimentally discovered by J. L. Poiseuille. Later it could

---

[5] Up to 1975, the unity of viscosity was "poise", after Jean Louis Poiseuille (1799–1869) who was a French physiologist and physician. He discovered the law of laminar flow. The new unit for pressure is Pa (Pascal) = $kg\, s^{-2}\, m^{-1}$, after Blaise Pascal (1623–1662), a French mathematician, physicist, and philosopher.

be derived in a theoretical way. The velocity $v(\text{m s}^{-1})$ is

$$v = \frac{p}{4\eta l}(R^2 - r^2) \tag{4.5.1}$$

where l denotes the length of the tube (meters) and p the pressure difference (measured in Pascal). The radii R and r are measured in meters. Clearly, $v = 0$ for $r = R$. For $r = 0$ the velocity reaches its maximum. Thus the range of the function is $0 \leq v \leq pR^2/4\eta l$. See Beier (1962, p. 337–363), McDonald (1960), Randall (1962, p. 210 to 223), Ruch and Patton (1965), Thrall, Mortimer, Rebman, and Baum (1967, Example OA 4).

To get more insight into Poiseuille's law we study a numerical example which is chosen to be as realistic as possible. Take arterial blood with its high concentration of O_2 bound to hemoglobin. For human blood its viscosity is somewhat lower than that of venous blood, on the average $\eta = 0.0027$ Pa s (from Diem *et al.*, 1970, p. 558). The blood is assumed to flow through an arteriole (wide arterial capillary) of length $l = 0.02$ m and radius $R = 8 \times 10^{-5}$ m. At one end the pressure is assumed to be higher than at the other end such that the difference is $p = 400$ Pa (≈ 3 mm mercury). Then (4.5.1) becomes

$$v = \frac{400}{4 \times 0.0027 \times 0.02}(64 \times 10^{-10} - r^2)$$

or

$$v = 1.185 \times 10^{-2} - (1.85 \times 10^6)r^2 \tag{4.5.2}$$

measured in m s^{-1}.

According to the definition, v is a quadratic function of r. Hence the graph is a section of a quadratic parabola (see Fig. 4.8 and 4.7b). The peak velocity is 0.0185 m s^{-1}.

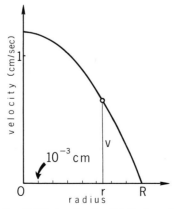

Fig. 4.8. Graph of the function (4.5.2) showing the distribution of velocities in laminar flow

4.6. Quadratic Equations

At the end of this chapter we deal with a question which frequently arises in connection with quadratic functions: For what values of x is $y = ax^2 + bx + c = 0$? Or, where does the quadratic parabola intersect the x axis? To answer this question we have to solve the *quadratic equation*

$$ax^2 + bx + c = 0 \qquad (4.6.1)$$

with $a \neq 0$. Eq. (4.6.1) is also called the *standard form* of a quadratic equation since all terms are on the left side in an expanded form.

It is not worthwhile to study the way the solution is found. Rather it is economic to learn the solution by heart[6]:

$$x = \frac{-b \pm (b^2 - 4ac)^{1/2}}{2a}. \qquad (4.6.2)$$

The expression the square-root of which must be taken is called the *discriminant* of the quadratic equation. We denote it by D:

$$D = b^2 - 4ac. \qquad (4.6.3)$$

If $D < 0$, the square root of D is not a real number so that no real solutions exist. The *condition for real solutions* is, therefore,

$$D \geqq 0. \qquad (4.6.4)$$

For $D > 0$ there exist two different solutions x_1 and x_2. But if $D = 0$, the two solutions fall together in one single solution $x_1 = x_2$. The two values x_1 and x_2 are also called *zeros* of the polynomial $ax^2 + bx + c$ or *roots* of the quadratic Eq. (4.6.1).

Often the roots are required to factor a quadratic expression. In the equation

$$(x - x_1)(x - x_2) = 0 \qquad (4.6.5)$$

$x = x_1$ is a root since the first factor vanishes and $x = x_2$ another root since the second factor vanishes. Working out the product we obtain

$$x^2 - (x_1 + x_2)x + x_1 x_2 = 0. \qquad (4.6.6)$$

Comparing the result with Eq. (4.6.1) we conclude that

$$x_1 + x_2 = -b/a, \qquad x_1 x_2 = c/a. \qquad (4.6.7)$$

---

[6] For readers who are curious to know the proof, the main steps are indicated here: The quadratic Eq. (4.6.1) is rewritten in the form $\left(x + \dfrac{b}{2a}\right)^2 = \dfrac{b^2 - 4ac}{4a^2}$. Then square roots are taken on both sides of this equation.

Thus, in the standard form of a quadratic equation, the *sum of the roots* is $-b/a$ and the *product of the roots* is c/a.

Problem 4.6.1. Let r be a given number. Find a quadratic equation whose roots are

$$x_1 = 2r, \qquad x_2 = r - 1.$$

Solution: Applying Eq. (4.6.5) we get

$$(x - 2r)(x - (r - 1)) = 0,$$
$$x^2 - (r - 1)x - 2rx + 2r(r - 1) = 0,$$
$$x^2 - (3r - 1)x + 2r(r - 1) = 0.$$

This is the required equation, written in the standard form.

Problem 4.6.2. Let k be a given number. Discuss the character of the roots of the quadratic equation

$$3y^2 - ky + 3 = 0.$$

Solution: According to Eq. (4.6.3) the discriminant is

$$D = k^2 - 36.$$

The roots are real numbers if $D \geq 0$, that is, $|k| \geq 6$. In the cases $k = 6$ or -6, the two roots are identical.

Problem 4.6.3. Kramer (1964, p. 517) studied a genetical problem of recombination of chromosomes. Let p be the recombination fraction with the limitation $0 \leq p \leq \frac{1}{2}$, and let $1 - p$ be the nonrecombination fraction. In that study the behavior of the product $p(1 - p)$ had to be considered.

Solution: Let

$$y = p(1 - p). \qquad (4.6.8)$$

This is a quadratic function of p with domain $0 \leq p \leq \frac{1}{2}$. It follows that $y \geq 0$. For a given value of y the question arises whether there exists a corresponding value of p and, if so, how to find p. Equation (4.6.8) leads to a quadratic equation. By rearranging the terms we find the standard form

$$p^2 - p + y = 0. \qquad (4.6.9)$$

Comparing with (4.6.1) we identify x by p, a by 1, b by -1, and c by y. From (4.6.2) we obtain

$$p = \frac{1 \pm (1 - 4y)^{1/2}}{2}.$$

A real solution exists only if $1-4y \geq 0$ which implies $y \leq \frac{1}{4}$. Since p is limited by $0 \leq p \leq \frac{1}{2}$, one of the two solutions (the one with $+$) is not admissible unless $y = \frac{1}{4}$ in which case the two roots are identical. Hence the final solution of (4.6.9) is

$$p = \frac{1-(1-4y)^{1/2}}{2}. \tag{4.6.10}$$

A wealth of quadratic equations applied to genetic problems can be found in Jacquard (1974).

Recommended for further reading: Defares et al. (1973), Guelfi (1966), Lefort (1967), C. A. B. Smith (1966).

Problems for Solution

4.1.1. For the power function $y = ax^2$, choose the domain consisting of all real numbers. Plot the function for the parameter values $a = 1, 2, 0.5, 0.1, -1, -2, -3$. Discuss the range of the function.

4.1.2. Plot the power function $y = ax^3$ for the domain $\{x | -2 \leq x \leq 2\}$ and the parameter values $a = 1, 0.2, -1$. Discuss the range of the function.

4.1.3. Plot the set $\{(x, y) | y > x^2$ and $y < 4 - x\}$.

4.1.4. Solve the simultaneous inequalities $P^2 < Q + 1$ and $P < Q$ graphically. (Hint: Let $P = x$ and $Q = y$.)

4.2.2. Assume that all linear dimensions of an animal increase by 12%. Then the animal will have the same shape. How do the surface, the volume, and the weight (under the assumption of a constant specific gravity) increase? Give the percentage of increase.

4.2.3. Converse problem: All linear dimensions (height, length, diameter, etc.) have grown at a uniform rate such that the volume increases by 60%. By what percentage does the surface increase?

4.2.4. Consider a spherical cell of volume V and surface S. Express V as a function of S. What type of function is it? How does doubling S influence V? (Hint: In formulas (4.2.2) and (4.2.3) eliminate the radius r).

4.3.1. Shift the graph of $y = 6x - x^2$ one unit to the left and two units upward. Perform the transformation also algebraically.

4.3.2. Shift the graph of $y = 2/(x+2)$ five units to the right and one unit upward. Perform the transformation also algebraically.

4.4.1. For $x = -1, 0, 1$ the quantity $y = 8 - 3x + x^2$ takes on the values 12, 8, 6, respectively. Using first and second differences only, find the functional values for $x = 2, 3, \ldots, 10$.

4.4.2. Given the function $y = 3 + x - \frac{1}{4}x^2$. Calculate y for $x = -2, -1, 0, +1, \ldots, +7$ by using first and second differences. Plot a diagram.

4.4.3. Show that for the cubic function $f(x) = x^3$ the third differences are constant. Begin with numerical values for $x = -1, 0, +1, +2, \ldots, +6$. Then generalize showing that the first difference $\Delta f(x) = f(x+1) - f(x)$ is a quadratic function of x, the second difference $\Delta^2 f(x)$ is a linear function of x, and that the third difference $\Delta^3 f(x)$ is a constant.

4.5.1. For the function given by formula (4.5.2) calculate the velocity v of blood for $r = 0\,\text{cm}, 10^{-3}\,\text{cm}, 2 \times 10^{-3}\,\text{cm}, 3 \times 10^{-3}, \ldots, 8 \times 10^{-3}\,\text{cm}$ and compare the result with Fig. 4.8. Check the values of v by means of first and second differences.

4.6.4. Convert the following quadratic equations in x into the standard form:

a) $(3-x)(4+x) = 6x^2$,   b) $2x + 6/x = 5$,

c) $(p-x)x = 2px - q$.

4.6.5. Solve the "incomplete" quadratic equations in x:

a) $x^2 - 2px = 0$,   b) $qx^2 = 3x$,   c) $rx^2 + (r+3)x = 0$.

(Hint: factor x.)

4.6.6. For $a > 0$ solve the following equations with respect to z:

a) $az^2 = z$,   b) $az = 1/z$,

c) $az + z^2 = 0$,   d) $a/z - 2z = 0$.

4.6.7. Solve $y^4 - y^2 = 12$ (Hint: Substitute $y^2 = x$.)

4.6.8. Wright (1964, p. 27) in an analysis of Mendelian heredity was led to the equation $4x^2 - 2x - 1 = 0$. Find the roots.

4.6.9. Solve $u - 2\sqrt{u} = 3$. (Hint: Substitute $\sqrt{u} = x$.)

4.6.10. Fisher (1965, p. 131), discussing the breeding of animals with a long pregnancy and only one offspring at birth, considers the equation $8\lambda^2 - 8\lambda + 1 = 0$ where λ is the Greek letter lambda. Solve this equation.

4.6.11. Find the standard form of a quadratic equation with roots $x_1 = 7, x_2 = -3s$.

4.6.12. Find the standard form of a quadratic equation with roots $x_1 = t^2, x_2 = -t/2$.

4.6.13. The sum of two unknown numbers is 6, their product 4. Determine the numbers to an accuracy of two decimals.

4.6.14. The sum and the product of two unknown numbers are both equal to k where $k \geq 4$ is a given quantity. Find the two numbers.

4.6.15. Discuss the character of the roots in the quadratic equation
$$9x^2 - (m - 3) x + 1 = 0 .$$

4.6.16. For what values of k are the roots of
$$(3x + k)^2 = 4(k + x)$$
real?

4.6.17. Li (1958, p. 216) in an analysis of decreasing heterozygosis studies the quadratic equation in λ:
$$2N\lambda^2 - 2(N - 1) \lambda - 1 = 0 .$$

Show that the equation has two different real roots for all values of the constant N except for $N = 0$.

4.6.18. For what values of p are the roots of
$$(p - 2) \mu^2 = 2(p - 1) \mu + 2$$
real? Discuss the case $p = 2$ separately.

4.7.1. For several kilometers a highway has only one lane with no opportunity of passing. The "bottleneck" is crowded all day. If cars drive at 100 km/h, they have to observe a safe distance of $s = 70$ m from car to car. Police reduces the speed to 50 km/h with safe distance of $s = 25$ m. If everybody keeps the safe distance, which of the two speeds allows the greater number of cars per hour? (Distances between cars are measured from midpoint to midpoint.)

4.7.2. On a lake of area A (square meters) motor boats are permitted to cruise at a speed v (m/sec). Assume that the lake is crowded and that each boat requires a strip of length $s_1 = t_1 v$ and width $s_2 = t_2 v$ where t_1 and t_2 are given time length necessary for safe maneuvering. What is the maximum admissible number of boats? (Numerical example: $t_1 = 10$ sec, $t_2 = 5$ sec, $A = 1$ km^2.)

Chapter 5

Periodic Functions

5.1. Introduction and Definition

This chapter deals with some mathematical tools that are required for the study of *biological rhythms*. The best known rhythms are seasonal variations, menstruation, daily cycles, breathing and heart beat. It is typical of rhythms that the same or nearly the same pattern is repeated from cycle to cycle. Phenomena of this type are also called *periodic*.

Fig. 5.1 depicts an electrocardiogram with several cycles. The curve may be interpreted as the graph of a function $y = f(t)$ with the time t as independent variable and the voltage y plotted perpendicularly versus the time axis. Whereas a real electrocardiogram is not exactly periodic, Fig. 5.1 shows a somewhat idealized behavior: In consecutive time intervals of equal length the same curve repeats itself. The constant interval is called the *period*[1]. We denote it by l. Each point on the curve represents a particular *phase* of the rhythm. Points that differ by one, two, or more periods are said to be in equal phase.

A less intuitive, but mathematically precise definition of a periodic function is as follows: Let x be any value for which the function $y = f(x)$ is determined, that is, let x belong to the domain of the function. Let l denote a constant positive number. Assume that $x + l$, $x + 2l$, $x + 3l$, ...

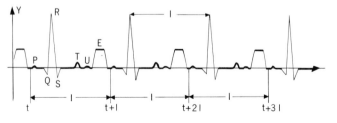

Fig. 5.1. The scheme of an electrocardiogram

---

[1] The word "period" has many different meanings. In mathematics it is exclusively used in the sense of an interval required to complete a cycle.

also belong to the domain. The values of y at these points of the x axis are given by $f(x)$, $f(x+l)$, $f(x+2l)$, etc. Then the function $y = f(x)$ is called periodic with period l if

$$f(x) = f(x+l) = f(x+2l) = \cdots \tag{5.1.1}$$

is valid for all possible values of x.

Some of the simpler periodic functions can be generated by a rotating wheel. Fig. 5.2 shows some *cycloids* that are curves generated by a wheel rolling along a line. The trigonometric functions such as the sine and the

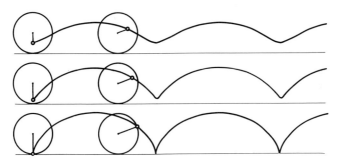

Fig. 5.2. Cycloids as illustration for periodic curves. They are traced by a point on a wheel rolling along a straight line

cosine may also be generated by a rotating wheel. Before we can define them we have to introduce angles and polar coordinates in the coming two sections. The treatment of polar coordinates will lead us to another sort of biological application, namely to the study of *animal orientation*. Periodicity in *leaf-arrangement* will be mentioned in Section 8.5. For an application to *population dynamics* see Fig. 11.10 and 11.12.

5.2. Angles

An angle is a measure of the amount of rotation. For a precise definition we consider a rectangular coordinate system with x and y axes and an origin O. We introduce a half-line h and assume that it originally coincides with the positive x axis (see Fig. 5.3). Now we rotate the half-line h in the counter-clockwise direction keeping the point O fixed. When h coincides for the first time with the positive y axis, then h is said to form a *right angle* with the positive x axis[2]. On subdividing this rotation into

---

[2] The word "angle" has a double meaning. If we speak of a right angle, we mean the configuration. In saying "an angle is 90°" we mean the measure of the angle.

ninety equal steps we get a traditional unit for measuring angles, the *degree*. An angle α between $0°$ and $90°$, that is $0° < \alpha < 90°$, is called *acute*.

Now we continue the rotation of the half-line h. When h coincides for the first time with the negative x axis, then h is said to form a *straight angle* with the positive x axis ($\alpha = 180°$). Angles in size between the right and the straight angle are called *obtuse*. They satisfy the inequality $90° < \alpha < 180°$. There is no upper bound in the amount of rotation. An angle of $360°$ is the measure for a full rotation. We may easily go beyond that and consider angles such as $450°$, $720°$, etc.

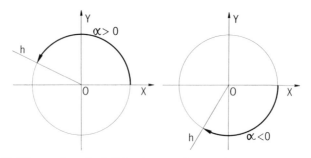

Fig. 5.3. The rotation of a half-line h generating positive and negative angles

For many applications it is essential to distinguish between the two directions of rotation. On rotating the half-line h from its original position on the positive x axis in the clockwise direction we generate *negative angles*. When, for instance, h coincides for the first time with the negative y axis, it forms the angle $\alpha = -90°$ with the positive x axis. $-360°$ is the measure for a full rotation in the clockwise direction.

The subdivision of a right angle into 90 degrees is a rather arbitrary procedure and often inconvenient for the study of trigonometric functions. Let us, therefore, introduce another unit for the angle, the *radian*, which is more natural. For this purpose we have to assume that on the x and the y axes the *same unit of length* is chosen (Fig. 5.4). We draw a circle around the origin whose radius is of unit length. This particular circle is called the *unit circle*. We again rotate the half-line h from the positive x axis in the counter-clockwise direction. Let A and B be the points of intersection of h with the unit circle before and after the rotation. Then we measure the angle by the *arc* from A to B. Since the circumference of a circle of radius r is $2\pi r$, and since the unit circle has radius $r = 1$, the full rotation is measured by the number $2\pi = 6.28318\ldots$. We say that this angle is 2π radians. By comparison with degrees we may write

$$360° = 2\pi \text{ radians}. \tag{5.2.1}$$

In mathematics the unit radian is often dropped and the angle treated as a pure number. Hence it is also correct to write $360° = 2\pi$. The straight angle is measured by $2\pi/2 = \pi = 3.14...$ and the right angle by $2\pi/4 = \pi/2 = 1.57....$ From (5.2.1) it follows that

$$1° = 2\pi/360 = 0.01745 \ldots \text{ (radians)} \tag{5.2.2}$$

and

$$1 \text{ radian} = 360°/2\pi = 57.295° \ldots . \tag{5.2.3}$$

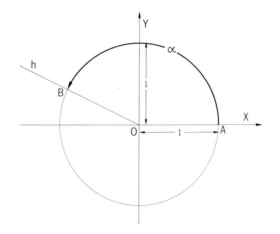

Fig. 5.4. An angle may be measured by the arc on the unit circle

The arc of this angle is equal to the radius. Conversion of degrees into radians and conversely is usually done by tables (see recommendations at the end of Chapter 5).

It may surprise the reader that the result in (5.2.3) is not given in degrees, minutes, and seconds. One minute is 1/60 degree and one second is 1/60 minute. It is rather tedious to work with minutes and seconds. They are also impractical for computers. Therefore, they are more and more being abandoned in favor of decimal fractions (cf. Stibitz, 1966, p. 181). It is much easier to perform calculations with an angle of 7.315°, say, than with the old-fashioned $7°18'54''$. Newer trigonometric tables are made up in the decimal system (see recommendations at the end of Chapter 5).

For clockwise directed angles, the minus sign should be observed. Thus, for instance, $-90° = -\pi/2$ radians. Any real number can be interpreted as an angle measured in radians.

5.3. Polar Coordinates

K. von Frisch discovered the language of honey bees. When a bee has found a plentiful source of food, she flies back to the hive, makes known the smell, and dances on the vertical surface of a honeycomb. If the food is more than a hundred meters away, the bee runs straight ahead for a short distance, returns in a semicircle to the starting point, again runs through the straight stretch, describes a semicircle in the opposite direction, and so on in regular alternation. On the straight part of the run, the bee is vigorously wagging her abdomen. A path vertically

Fig. 5.5. The tail-wagging dance. Four followers are receiving the message (from von Frisch, 1967, p. 57). By the dance the polar coordinates of the food source are transmitted

upwards means that the food is in the direction of the sun. If it is 30° to the right of the vertical, it signifies that the source is 30° to the right of the sun, etc. (Fig. 5.5). The distance is mainly signaled by the number of runs per time unit. For instance, for a distance of 1000 m the explorer bee performs about 18 runs per minute (for more details see von Frisch, 1967).

From this brief description of the language we learn that bees do not use rectangular x- and y-coordinates. Instead they work with an angle and a distance or, in other words, with *polar coordinates*. In animal behavior, particularly in the study of orientation of birds and of fish, polar coordinates play a major role (for the mathematics used in animal orientation see Batschelet (1965)).

Fig. 5.6 illustrates the relationship between polar and rectangular coordinates. We assume again that x and y are measured in the same unit of length. Let P be an arbitrary point in the plane, but different from the origin O. Then the position of P is uniquely determined by its distance r from O and by an angle α between the positive x axis and the line OP. The numbers r and α are called the *polar coordinates* of P. r is the *polar distance* and α the *polar angle*. The origin O plays a somewhat exceptional role: Its position is already determined by the statement $r = 0$, and the polar angle is not defined.

The positive x axis characterized by $\alpha = 0$ is called the *polar axis*.

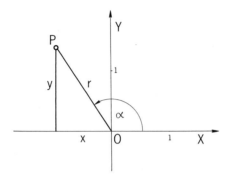

Fig. 5.6. Rectangular and polar coordinates of a point P

For polar coordinates there is usually no need to use angles greater than $360°$ or 2π radians. Instead of $\alpha = 390°$, say, we could as well operate with $\alpha = 30°$. Therefore, some limitation is practical, for instance

$$0° \leqq \alpha < 360° \quad \text{or} \quad 0 \leqq \alpha < 2\pi \quad \text{(radians)} . \qquad (5.3.1)$$

In some applications another way of limitation is preferred: By using negative angles we may assume that

$$- 180° \leqq \alpha < 180° \quad \text{or} \quad - \pi \leqq \alpha < \pi \quad \text{(radians)} . \qquad (5.3.2)$$

Here the absolute value of α cannot exceed $180°$ or π radians. With either of the two limitations (5.3.1) or (5.3.2), a point P (that does not coincide with O) determines its polar coordinates uniquely.

The conversion of polar coordinates into rectangular coordinates and conversely requires trigonometric functions and will be treated in Section 5.5.

5.4. Sine and Cosine

Let α be an angle, positive or negative, not limited in size. Let P be the point with polar coordinates $r = 1$ and α (see Fig. 5.7). As in Section 5.2, the point P lies on the unit circle. Let x and y be the rectangular coordinates of P. They are both uniquely determined by α. Then cosine and sine of α are defined to be the functions

$$x = \cos\alpha,$$
$$y = \sin\alpha.$$

(5.4.1)

With each value of α, the cosine associates a unique value x and the sine a unique value y. The domain of these two functions is the set $\mathbb{R}$ of all real numbers: $\alpha \in \mathbb{R}$. The values of x and y, however, reach a maximum, namely $+1$, and a minimum, namely -1. For instance,

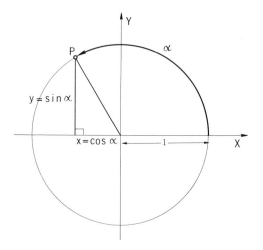

Fig. 5.7. The definition of sine and cosine for arbitrary values of α

$\cos 0° = +1$, $\cos 180° = -1$, $\sin 90° = +1$, and $\sin(-90°) = -1$. Hence the range of the two functions is the interval from -1 to $+1$:

$$-1 \leq \cos\alpha \leq +1,$$
$$-1 \leq \sin\alpha \leq +1.$$

(5.4.2)

If we add the angle of a full rotation, $360°$ or 2π radians, to any angle α, then the point P with polar angle α keeps its original position. Therefore, x and y remain unchanged, and from (5.4.1) it follows that

$$\cos(\alpha + 360°) = \cos\alpha,$$
$$\sin(\alpha + 360°) = \sin\alpha.$$

(5.4.3)

Hence, according to Eq. (5.1.1), the functions $\cos\alpha$ and $\sin\alpha$ are *periodic* with *period* $360°$ or 2π radians. A graph of the two functions is shown in Fig. 5.8.

It may help the intuition when we compare the unit circle with a rotating wheel. The wheel is assumed to have a fixed center and to rotate with constant speed. Let P be a point on the circumference of the wheel. The projection of P on the x axis is by definition a point with $x = \cos\alpha$ where α is the polar angle of P. While P is rotating with constant speed, the projection is oscillating between the points with $x = +1$ and

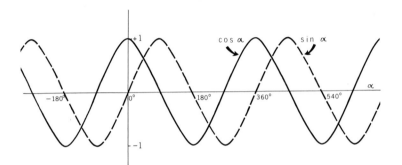

Fig. 5.8. Graphs of the functions $\cos\alpha$ and $\sin\alpha$

$x = -1$. Similarly, the projection of P on the y axis is a point with ordinate $y = \sin\alpha$, and this point oscillates within the interval from $y = +1$ to $y = -1$ (see Fig. 5.9).

Most tables of trigonometric functions list the values of cosine and sine only for the interval from $0°$ to $90°$ (see Table C of the Appendix). All other values have to be derived. This is done by means of the following relations:

$$\cos(-\alpha) = \cos(360° - \alpha) = +\cos\alpha ,$$
$$\sin(-\alpha) = \sin(360° - \alpha) = -\sin\alpha , \tag{5.4.4}$$

$$\cos(180° - \alpha) = -\cos\alpha , \qquad \sin(180° - \alpha) = +\sin\alpha ,$$
$$\cos(180° + \alpha) = -\cos\alpha , \qquad \sin(180° + \alpha) = -\sin\alpha . \tag{5.4.5}$$

All these formulas follow directly from the definitions of cosine and sine. Fig. 5.10 may help the reader find the details.

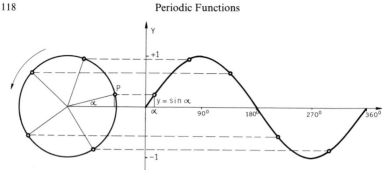

Fig. 5.9. When P rotates with constant speed, its projection on the y axis oscillates as the
values of $y = \sin \alpha$

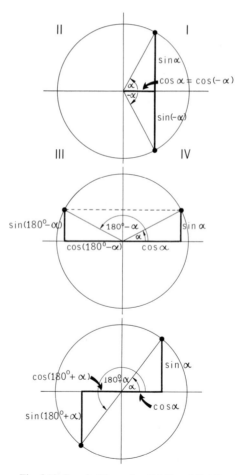

Fig. 5.10. Proof of formulas (5.4.4) and (5.4.5)

If we subdivide the plane into the four quadrants I, II, III, IV in the counter-clockwise direction, the trigonometric functions have the following signs:

Table 5.1

Quadrant	Interval	cosine	sine
I	$0° < \alpha < 90°$	+	+
II	$90° < \alpha < 180°$	−	+
III	$180° < \alpha < 270°$	−	−
IV	$270° < \alpha < 360°$	+	−

For functional values of $\sin \alpha$, $\cos \alpha$, $\tan \alpha$ see Table C at the end of the book.

5.5. Conversion of Polar Coordinates

Now we are prepared to obtain the conversion of polar coordinates into rectangular coordinates. In Fig. 5.11, a point P is plotted in the plane with polar coordinates r, α and rectangular coordinates x, y. It is assumed that P does not coincide with the origin O. Let P' be the point where the line OP intersects the unit circle, and let x', y' be the rectangular coordinates of P'. Since P' has polar coordinates $1, \alpha$, it follows from definition (5.4.1) of cosine and sine that

$$x' = \cos \alpha , \qquad y' = \sin \alpha . \tag{5.5.1}$$

Moreover, the two right triangles with legs x, y and x', y', respectively, are similar. Hence

$$x : r = x' : 1 , \qquad y : r = y' : 1 . \tag{5.5.2}$$

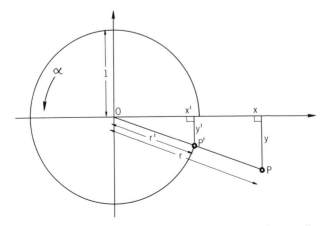

Fig. 5.11. The conversion of polar coordinates into rectangular coordinates

In these proportions x, y and x', y' can take negative values, but x has always the same sign as x' and y the same sign as y'. From (5.5.2) we conclude that

$$x = rx', \qquad y = ry'$$

and hence from (5.5.1) it follows that

$$x = r\cos\alpha, \qquad y = r\sin\alpha. \tag{5.5.3}$$

This is the solution of our conversion problem.

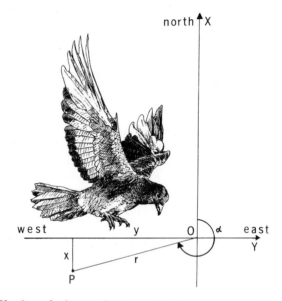

Fig. 5.12. Homing of pigeons. Polar coordinates are converted into rectangular coordinates

Example 5.5.1. In an experiment on orientation and navigation, some pigeons were released 72 km away from their loft. If we consider the loft as the center of a polar coordinate system, the point of release had an azimuth of 241° (azimuth = angle measured clockwise from the north direction to the point of release). How many kilometers is the point of release southward and how many westward of the loft?

To solve this problem we first adjust a rectangular coordinate system to the geographical map, assuming for simplicity that the surface of the earth is flat (Fig. 5.12). The origin O is at the loft. The positive x axis points northward and the positive y axis eastward. Let P be the point of release. Then $r = 72$ km and $\alpha = 241°$ are the polar coordinates of P.

The problem is now equivalent to finding the rectangular coordinates x, y of P. From formulas (5.5.3) we obtain

$$x = 72 \cos 241°, \qquad y = 72 \sin 241° \quad (\text{km}).$$

Now $241° = 180° + 61°$. Using (5.4.5) we get $\cos(180° + 61°) = -\cos 61°$ and $\sin(180° + 61°) = -\sin 61°$. From a table we find $\cos 61° = 0.485$ and $\sin 61° = 0.875$. Hence

$$x = 72(-0.485) = -34.9, \qquad y = 72(-0.875) = -63.0 \quad (\text{km}).$$

The point of release is therefore 34.9 km southward and 63.0 km westward of the loft.

The inverse problem consists of finding the polar coordinates when the rectangular coordinates are given. We solve this problem in three steps.

First, we apply the Pythagorean theorem to the triangle with legs x, y in Fig. 5.12:

$$r^2 = x^2 + y^2. \tag{5.5.4}$$

By square rooting we calculate the polar distance r. Second, from (5.5.3) we obtain

$$\cos \alpha = x/r, \qquad \sin \alpha = y/r. \tag{5.5.5}$$

Third, from a table of trigonometric functions we find the polar angle α. The proper quadrant can be determined by Table 5.1.

The origin O plays an exceptional role. From $x = 0$ and $y = 0$ it follows that $r = 0$, and hence (5.5.5) cannot be applied. Thus a polar angle is not defined.

Example 5.5.2. We consider again an experiment on orientation and navigation in pigeons. Assume that the point of release is 23.5 km southward and 65.7 km eastward from the loft. We adjust the same coordinate system as in Fig. 5.12. Thus we get $x = -23.5$ and $y = +65.7$ (km). What are the polar coordinates?

Formula (5.5.4) yields

$$r^2 = (-23.5)^2 + (+65.7)^2 = 4869, \qquad r = 69.8 \quad (\text{km}).$$

From (5.5.5) we obtain

$$\cos \alpha = (-23.5)/69.8 = -0.337, \qquad \sin \alpha = (+65.7)/69.8 = +0.941.$$

Table 5.1 indicates that $90° < \alpha < 180°$. For the acute angle $180° - \alpha = \alpha'$, say, we get $\cos \alpha' = 0.337$ and $\sin \alpha' = 0.941$. In a table of trigonometric functions we find $\alpha' = 70.2°$. Hence $\alpha = 180° - 70.2° = 109.8°$. The polar distance r and the polar angle α (azimuth) have thus been calculated.

Some writers recommend a slightly different procedure. They use the function $\tan \alpha$ which is defined in terms of cosine and sine by

$$\tan \alpha = \frac{\sin \alpha}{\cos \alpha} \qquad (5.5.6)$$

provided that $\cos \alpha \neq 0$. The two formulas (5.5.5) are replaced by the single formula

$$\tan \alpha = y/x \qquad (x \neq 0). \qquad (5.5.7)$$

There are advantages in using (5.5.7) to determine α in that only a single trigonometric function occurs and r need not be known. However, there is also a serious disadvantage. Before we can detect it, we have to study a property of the function $\tan \alpha$.

From (5.5.6) and (5.4.5) it follows that

$$\tan(180° - \alpha) = \frac{\sin(180° - \alpha)}{\cos(180° - \alpha)} = \frac{\sin \alpha}{-\cos \alpha} = -\tan \alpha, \qquad (5.5.8)$$

$$\tan(180° + \alpha) = \frac{\sin(180° + \alpha)}{\cos(180° + \alpha)} = \frac{-\sin \alpha}{-\cos \alpha} = \tan \alpha. \qquad (5.5.9)$$

The latter formula proves that the function $\tan \alpha$ is *periodic* with period $\alpha = 180°$ or π radians. Thus the function $\tan \alpha$ repeats its pattern already after half the period as compared with $\sin \alpha$ and $\cos \alpha$. This property can also be seen from a graph of $\tan \alpha$ as shown in Fig. 5.13.

Now we are prepared to discuss the disadvantage of using (5.5.7). This equation does not determine α uniquely. If α is a solution of (5.5.7),

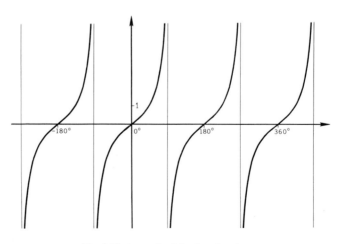

Fig. 5.13. A graph of the function $\tan \alpha$

$\alpha + 180°$ is also a solution because of (5.5.9). For example, the equation $\tan\alpha = 0.2679$ has two solutions, $\alpha = 15°$ and $\alpha = 195°$. The ambiguity can only be removed by observing the signs of x and y or by drawing a graph.

Nevertheless, formula (5.5.7) provides a good check of calculation. In Example 5.5.1 we obtain for instance

$$\tan\alpha = \tan 241° = \tan 61° = y/x = (-63.0)/(-34.9) = 1.805\,.$$

On the other hand, we get from a table $\tan 61° = 1.804$ which corresponds well to our calculated value.

Similarly, in Example 5.5.2 we obtain

$$\tan\alpha = \tan 109.8° = -\tan 70.2° = y/x = (+65.7)/(-23.5) = -2.80\,.$$

On the other hand, we get from a table $\tan 70.2° = 2.79$. The difference can be explained by rounding-off errors.

5.6. Right Triangles

We denote the two acute angles by α and β, the leg adjacent to α by x, and the leg adjacent to β by y, and the hypotenuse by r. As Fig. 5.14 shows, this is a special case of what we previously treated. Formula (5.5.5) leads to

$$\cos\alpha = x/r = \frac{\text{leg adjacent to } \alpha}{\text{hypotenuse}} = \sin\beta\,,$$

$$\sin\alpha = y/r = \frac{\text{leg opposite } \alpha}{\text{hypotenuse}} = \cos\beta\,. \tag{5.6.1}$$

From (5.5.7) we obtain

$$\tan\alpha = y/x = \frac{\text{leg opposite } \alpha}{\text{leg adjacent to } \alpha}\,. \tag{5.6.2}$$

Also sometimes used is

$$\cot\alpha = 1/\tan\alpha. \tag{5.6.3}$$

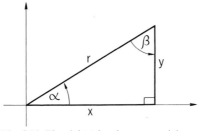

Fig. 5.14. The right triangle as a special case

Example 5.6.1. Let I be the intensity of sunlight measured in units of energy per cm^2 per sec. If a leaf forms an angle $\alpha < 90°$ with the incident rays, the intensity is reduced. By what factor?

The question may be answered by inspecting Fig. 5.15. Let A_1 (cm^2) be the area of cross section of those rays that hit the leaves. Further, let A_2 be the area of the leaf. The incident light energy per second is IA_1. When distributed evenly over the leaf, the energy per sec per cm^2, I', is then

$$I' = IA_1/A_2 .$$

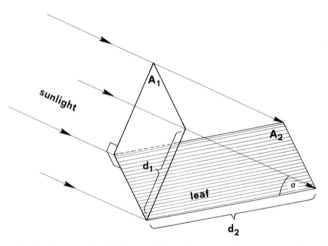

Fig. 5.15. A leaf of area A_2 is hit by sunlight of cross section A_1. The light intensity is reduced by the factor A_1/A_2

Hence, A_1/A_2 is the factor of reduction. It may be expressed by the two segments d_1 and d_2 shown in Fig. 5.15 and by α:

$$A_1/A_2 = d_1/d_2 = \sin\alpha .$$

Example 5.6.2. Some muscles do not simply consist of parallel fibres. Rather, the fibres are short and inclined to a tendon in their center. Such muscles are called *pinnate* (meaning featherlike). Cross sections of a pinnate muscle are shown in Fig. 5.16 (adapted from Alexander, 1968, p. 15).

We may ask by how much the tendon is moved when the length of each muscle fibre, λ, is reduced to λ'. To answer the question, let y denote the distance between the bones to which the muscle is attached. Furthermore, let α and α' be the angle between fibres and tendon in the

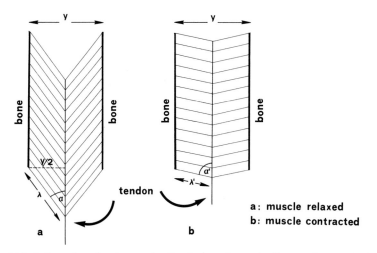

Fig. 5.16. Action of a pinnate muscle. Shortening of muscle fibers and movement of tendon are in different directions. Magnitudes are connected by trigonometric functions

relaxed and contracted stage, respectively. Then

$$\sin\alpha = \frac{y/2}{\lambda}, \qquad \sin\alpha' = \frac{y/2}{\lambda'}.$$

The distance, d, the tendon moves is

$$d = \lambda\cos\alpha - \lambda'\cos\alpha'.$$

Eliminating λ and λ' we obtain

$$d = y/2\left(\frac{\cos\alpha}{\sin\alpha} - \frac{\cos\alpha'}{\sin\alpha'}\right) = y/2(\cot\alpha - \cot\alpha').$$

The distance d is smaller than for a muscle with parallel fibres, but a pinnate muscle is considerably stronger.

Example 5.6.3. When we observe a *cylindrical tissue structure* under the microscope, we most likely see an *oblique cut*. Fig. 5.17a depicts a side view of a fiber whose cross section is a circle with diameter d. Fig. 5.17b shows an oblique cut as viewed under the microscope. It is easy to measure the diameter d. In addition, we can also measure two distances a and b which are explained by Fig. 5.17b. We are interested in the thickness t of the layer under the microscope. The meaning of t becomes clear from Fig. 5.17c which shows a side view of the layer.

To calculate the thickness t in terms of the observed quantities a, b, and d, we introduce two right triangles in Fig. 5.17c. In triangle ACE,

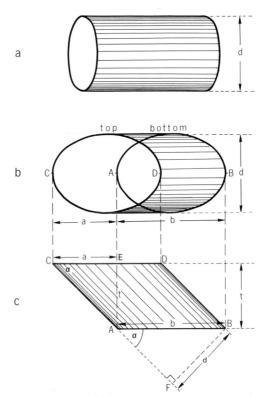

Fig. 5.17. An oblique cut of a cylindrical tissue structure as observed under the microscope. The thickness t of the layer in relation to the diameter d of the cylinder

the right angle is at E, and the legs are a and t. If α denotes the acute angle at C, we obtain from formula (5.6.2)

$$t = a \tan \alpha. \tag{5.6.4}$$

Here we have to eliminate the unknown angle α. For this purpose we consider the second triangle ABF. The diameter d of the cylinder forms one leg joining B and F, and b is the hypotenuse. By the Pythagorean theorem we get for the other leg

$$\overline{AF} = \sqrt{b^2 - d^2}. \tag{5.6.5}$$

Since AB is assumed to be parallel to CD, the acute angle at A is the same as the angle α at C. It follows from the right triangle ABF that

$$\tan \alpha = \frac{d}{\sqrt{b^2 - d^2}}. \tag{5.6.6}$$

Finally, by combining formulas (5.6.4) and (5.6.6) we obtain the desired formula

$$t = \frac{ad}{\sqrt{b^2 - d^2}}. \tag{5.6.7}$$

Example 5.6.4. Fig. 5.18 shows a human femur[3]. The axis of the femoral shaft is GE. The knee-joint determines the condylar axis GH. The two axes span the frontal plane which is vertical when the person is standing upright. The axis EB of the femoral neck does not fall into the frontal plane. Instead, together with the line GE it spans another plane, the antetorsion plane. The new plane forms an acute angle α with the frontal plane. It is called the antetorsion angle. This angle α and the obtuse angle β between the lines GE and EB play an important role in orthopedics. Various afflictions such as a hip-joint dislocation may be caused by improper angles.

In a patient the angles α and β cannot be measured directly. X-ray pictures do not show the true angles since they are projections of the

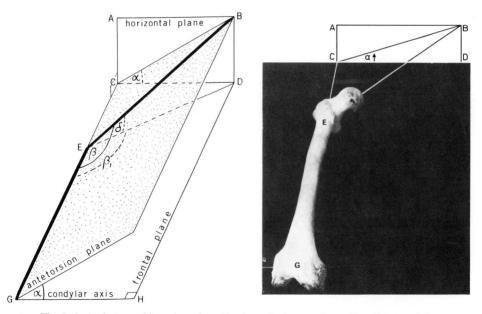

Fig. 5.18. A photographic and a schematic view of a human femur. For diseases of the hip-joint, improper size of the angles α and β may be responsible. The photograph is reproduced from Morscher (1967). The drawing was made with minor changes after the same paper

---

[3] I am indebted to Dr. E. Morscher, University of Basel, Switzerland for his kind permission to reproduce his figures.

femur. For instance, when the patient is lying with horizontal condylar axis, the X-ray being vertical, the axis EB of the femoral neck is projected into a line ED in the frontal plane. The angle β_1 between ED and GE differs slightly from β. A correction is required. In order to obtain exact formulas we introduce the plane $ABCD$ which is horizontal when the person is standing upright. From the right triangles CBE and CDE we get

$$\tan(180^\circ - \beta) = BC/CE, \quad \tan(180^\circ - \beta_1) = CD/CE.$$

From this we deduce by means of formula (5.5.8)

$$\tan \beta_1/\tan \beta = \frac{-CD/CE}{-BC/CE} = \frac{CD}{BC}.$$

On the other hand, from the right triangle CDB we get $\cos \alpha = CD/BC$. Combining the results we obtain the exact relationship between β_1 and β:

$$\tan \beta_1/\tan \beta = \cos \alpha. \tag{5.6.8}$$

Sometimes the angle δ between the axis EB of the femoral neck and the frontal plane is considered. δ should not be confused with α. For the exact relationship between the two angles see Problem 5.6.4 at the end of this chapter.

Example 5.6.5. Another biological application refers to the control of equilibrium in the guppy *(Lebistes reticulatus)*. It is well known that the guppy uses not only gravitation but also the incident light to adjust its upright position in water. Fig. 5.19 depicts the behavior when light enters vertically or from the side. We imagine that the guppy is under the influence of two forces, one force F, vertical, caused by gravitation, and the other force L, parallel to the incident rays, caused by light

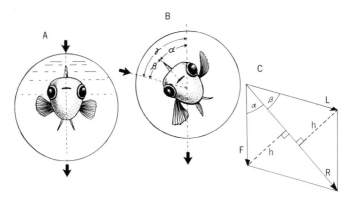

Fig. 5.19. Control of equilibrium in the guppy by gravitational force and by light. The figure is adapted from Lang (1967)

sensation. These two forces generate a resultant force R which determines the upright position of the guppy.

Let α and β be the angles between F and R and between L and R, respectively. Then $\gamma = \alpha + \beta$ is the angle between F and L. The problem consists in finding a relation between α and β in terms of F and L. In Fig. 5.19c we may express the height h in two ways:

$$h = F \sin \alpha \quad \text{and} \quad h = L \sin \beta . \qquad (5.6.9)$$

Eliminating h we find

$$\frac{\sin \alpha}{\sin \beta} = \frac{L}{F} . \qquad (5.6.10)$$

This is the desired relationship. It may be used for calculating the ratio L/F when α and β have been measured[4].

5.7. Trigonometric Relations

Among the hundreds of identities for trigonometric functions there are a few that are essential for a basic background in mathematics. Formulas that occur frequently should be learned by heart.

Application of the Pythagorean theorem to Fig. 5.7 yields

$$(\cos \alpha)^2 + (\sin \alpha)^2 = 1 \qquad (5.7.1)$$

for any angle α. Thus $\cos \alpha$ and $\sin \alpha$ are numerically interrelated. It is customary to write $\cos^2 \alpha$ and $\sin^2 \alpha$ instead of $(\cos \alpha)^2$ and $(\sin \alpha)^2$, respectively.

The following two formulas will be proven in Section 15.4 by means of complex numbers:

$$\begin{aligned} \cos(\alpha + \beta) &= \cos \alpha \cos \beta - \sin \alpha \sin \beta , \\ \sin(\alpha + \beta) &= \sin \alpha \cos \beta + \cos \alpha \sin \beta . \end{aligned} \qquad (5.7.2)$$

These formulas show how the cosine and the sine of the sum of two angles can be resolved. Sometimes two angles have to be subtracted instead of added. All we have to do in formulas (5.7.2) is replace β with $-\beta$. Then from (5.4.4) we get

$$\begin{aligned} \cos(\alpha - \beta) &= \cos \alpha \cos \beta + \sin \alpha \sin \beta , \\ \sin(\alpha - \beta) &= \sin \alpha \cos \beta - \cos \alpha \sin \beta . \end{aligned} \qquad (5.7.3)$$

---

[4] The relationship (5.6.10) may also be derived by applying the law of sine. This law is skipped here.

For $\alpha = \beta$ formulas (5.7.2) become

$$\cos 2\alpha = \cos^2 \alpha - \sin^2 \alpha ,$$
$$\sin 2\alpha = 2 \sin \alpha \cos \alpha .$$

(5.7.4)

For later purposes we also need the formulas

$$\sin \varphi + \sin \psi = 2 \sin \frac{\varphi + \psi}{2} \cos \frac{\varphi - \psi}{2} ,$$

(5.7.5)

$$\sin \varphi - \sin \psi = 2 \cos \frac{\varphi + \psi}{2} \sin \frac{\varphi - \psi}{2} ,$$

(5.7.6)

$$\cos \varphi + \cos \psi = 2 \cos \frac{\varphi + \psi}{2} \cos \frac{\varphi - \psi}{2} ,$$

(5.7.7)

$$\cos \varphi - \cos \psi = - 2 \sin \frac{\varphi + \psi}{2} \sin \frac{\varphi - \psi}{2} ,$$

(5.7.8)

written with the Greek letters φ (phi) and ψ (psi). These formulas are derived from (5.7.2) and (5.7.3) by adding or subtracting the equations and by replacing $\alpha + \beta$ with φ and $\alpha - \beta$ with ψ.

*5.8. Polar Graphs

In plotting a periodic function $y = f(x)$ it is somewhat redundant to show the same cycle over and over again. A more natural way of presenting periodic functions is by a *polar graph* or a *polar diagram*. For simplicity we choose a new unit for x such that the period l becomes $360°$ or 2π radians corresponding to a full rotation. In this way we have introduced a new variable which may be interpreted as an angle; let us denote it by α. The introduction of a new variable is called a *transformation*. The relationship between α and x is given by one of the following so-called *transformation equations*:

$$\alpha = \frac{360°}{l} x \quad \text{or} \quad \alpha = \frac{2\pi}{l} x$$

(5.8.1)

depending on whether degrees or radians are preferred. As x increases from 0 to l, α increases from $0°$ to $360°$ or from 0 to 2π radians, respectively. The inverse transformation is given by

$$x = \frac{l}{360°} \alpha \quad \text{or} \quad x = \frac{l}{2\pi} \alpha ,$$

(5.8.2)

respectively.

In the original function $y = f(x)$, we first replace x by the expression (5.8.2). Second, we interpret y as the polar distance r. Transforming x and replacing y with r finally leads to a new equation which we write $r = g(\alpha)$. When α increases from $0°$ to $360°$ (or from 0 to 2π radians), a closed curve appears. With α taking on higher and higher values, the curve repeats itself.

We call α the *phase angle* since its value determines the actual phase within a cycle.

Since it is assumed that r is nonnegative, that is, positive or zero, polar graphs are suitable only for those periodic functions for which y avoids negative values.

Example 5.8.1. Consider the periodic function

$$y = a(1 + \cos x) \quad (a > 0). \tag{5.8.3}$$

As plotted along a horizontal x axis, the graph is a sine wave with period 2π. The values of y oscillate between 0 and $2a$. To avoid redundancy in plotting, we switch to polar coordinates. We interpret x as a polar angle α and y as a polar distance r. Thus we rewrite formula (5.8.3) in the form

$$r = a(1 + \cos \alpha) \quad (a > 0). \tag{5.8.4}$$

For simplicity we choose a as our unit of length. Then for equidistant values of α, say for $\alpha = 0°, 15°, 30°, 45°, \ldots, 360°$, we determine $1 + \cos \alpha$ by means of a table. Finally, we plot r versus α in a polar coordinate system. The result is shown in Fig. 5.20 a. The curve is called a cardioid.

Fig. 5.20 exhibits six examples of periodic functions together with their polar graphs. Absolute values were taken in cases (d), (e), and (f) to avoid negative values. The coefficient a is an arbitrary positive constant.

The curve (a) is named *cardioid* because of its alleged resemblance to a heart. Curve (d) looks somewhat similar to curve (a), but is not similar in the strict geometric sense. Curve (c) is an *ellipse* for all values of the constant e from the interval $-1 < e < +1$. The origin O of the polar coordinate system coincides with one of the two focuses.

All six functions have period $360°$ or 2π radians, but for curves (e) and (f) there exist also smaller periods. They are $120°$ for curve (e) and $72°$ for curve (f).

The curves of Fig. 5.20 are pleasant because of their symmetries. Each one is axially symmetric. The curves (a) through (d) have a horizontal axis of symmetry. The curve (e) has three and the curve (f) even five axes of symmetry. For more polar diagrams of periodic functions see Thompson (1961, p. 282–284, or an earlier edition). Statistical applications of polar diagrams are offered in Batschelet (1965).

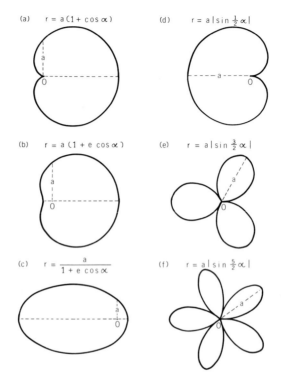

(a) $r = a(1 + \cos \alpha)$

(b) $r = a(1 + e \cos \alpha)$

(c) $r = \dfrac{a}{1 + e \cos \alpha}$

(d) $r = a\left|\sin \tfrac{1}{2}\alpha\right|$

(e) $r = a\left|\sin \tfrac{3}{2}\alpha\right|$

(f) $r = a\left|\sin \tfrac{5}{2}\alpha\right|$

Fig. 5.20. Some periodic functions with their polar diagrams. In (b) and (c), the constant e is $e = 0.8$

Nonperiodic functions can also be plotted in a polar diagram, but the curves are then not closed. If the value of y is steadily increasing, its polar diagram is called a *spiral*. Some spirals will be treated in Section 6.7.

*5.9. Trigonometric Polynomials

Data concerning biological rhythms can often be approximated by sine curves. Fig. 5.22 is an example. In order to fit the data, the function $y = \cos \alpha$ has to be adjusted in four steps[5].

First, the independent variable is the time t with known or unknown period l. The phase angle α has to be transformed into the time t. This is done by formula (5.8.1) wherein we replace x by t:

$$\alpha = \frac{360°}{l} t \quad \text{or} \quad \alpha = \frac{2\pi}{l} t .\qquad (5.9.1)$$

---

[5] Whether $\cos \alpha$ or $\sin \alpha$ is applied, is mathematically equivalent. Both functions are graphically represented by sine curves as shown in Fig. 5.8.

The coefficient is $360°/l$ or $2\pi/l$ depending on whether we measure α in degrees or in radians. In the following we simply write ω for this coefficient (ω = Greek letter omega):

$$\omega = 360°/l \quad \text{or} \quad \omega = 2\pi/l. \tag{5.9.2}$$

ω may be considered as a scale factor which introduces a new unit on the horizontal axis. Sometimes ω is called *angular frequency*, because ω indicates how often the period l is contained in a full rotation of the angle α.

The transformation (5.9.1) is now written $\alpha = \omega t$. It takes the function $y = \cos \alpha$ into the form

$$y = \cos \omega t. \tag{5.9.3}$$

A graph of this function is shown in Fig. 5.21a.

Second, suppose our curve is such that there is no peak at $t = 0$. So assume that a peak occurs at time t_0 where $0 < t_0 < l$. This calls for shifting the curve in the direction of the t axis by the amount t_0. It simply means that we add to the old variable t the constant t_0 so that the new variable, say t', is

$$t' = t + t_0. \tag{5.9.4}$$

The inverse transformation is $t = t' - t_0$. Thus (5.9.3) turns into $y = \cos \omega(t' - t_0)$. For simplicity we drop the prime and write

$$y = \cos \omega(t - t_0). \tag{5.9.5}$$

Fig. 5.21b depicts the new function. The constant t_0 indicates the position of the first peak to the right of the origin. For this property, t_0 is called *acrophase*[6].

Third, we have to adjust the function for a different *amplitude*. By amplitude we mean the greatest deviation of an oscillation from its center. The graphs of (5.9.3) and (5.9.5) have amplitude 1. If, instead, the amplitude is a known or unknown constant $c \neq 1$, we have to multiply the cosine by c. The positive constant c may be interpreted as a scale factor for the y axis. (5.9.5) turns into

$$y = c \cos \omega(t - t_0). \tag{5.9.6}$$

For a graph see Fig. 5.21c.

Fourth, y could oscillate around a value c_0 which may be different from 0. This calls for shifting the curve in the direction of the y axis by the amount c_0. It simply means that we have to add c_0 to all values of y.

---

[6] The term "acrophase" was coined by Dr. F. Halberg, University of Minnesota, Minneapolis.

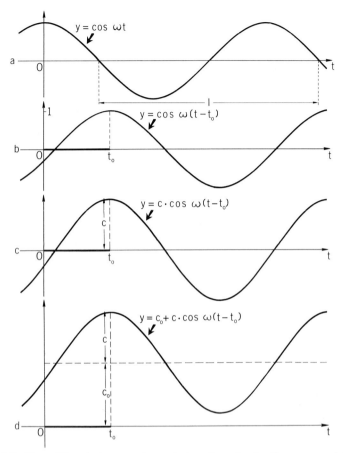

Fig. 5.21. Four different steps are shown to transform the function $y = \cos\alpha$ into the function (5.9.7)

Thus we finally obtain

$$y = c_0 + c \cos\omega(t - t_0) . \qquad (5.9.7)$$

A graph is shown in Fig. 5.21d. c_0 is called *mean level* and sometimes *mesor*.

It might be wise to make a brief check: For $t = t_0$ we get $y = c_0 + c \cos 0° = c_0 + c$ which is the highest value that y can take. This is in agreement with our assumption that t_0 is the phase of a peak. For $t = t_0 + l$ we obtain $\omega(t - t_0) = \omega l = 360°$ and $y = c_0 + c \cos 360° = c_0 + c$. Hence, after the elapse of one period, y takes again its maximum value.

The function (5.9.7) contains the four parameters c_0, c, ω, t_0. When data are given which are subject to random fluctuations, a statistical

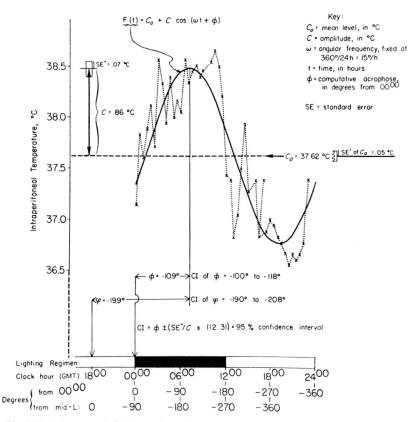

Fig. 5.22. A sine curve is fitted to telemetric measurements of intraperitoneal temperatures of an adult female rat. The rat is subject to a special lighting regimen which influences the acrophase. Data were obtained in preparation for a space shot (Reprinted from Halberg, 1969)

procedure is required to estimate the parameters. It would go far beyond the scope of this presentation to deal with estimation procedures. We refer the reader to Bliss and Blevins (1959), Halberg, Engeli, Hamburger, and Hillman (1965), Halberg, Reinberg (1967), and Halberg, Tong, Johnson (1967), Pavlidis (1973). The result of such an estimation procedure is shown in Fig. 5.22.

In certain cases ordinary statistical procedures fail to be applicable. Then a rather advanced method, the so-called *time series analysis*, has to be used. For treatments that do not require high level mathematics we refer the reader to Stumpff (1937), Blume (1965), Kendall and Stuart (1966). For a bibliography on the vast literature on biological rhythms see Sollberger (1965).

The function (5.9.7) is sometimes written in a different form. From (5.7.3) it follows that

$$\cos\omega(t - t_0) = \cos(\omega t - \omega t_0) = \cos\omega t \cos\omega t_0 + \sin\omega t \sin\omega t_0 .$$

The factors $\cos\omega t_0$ and $\sin\omega t_0$ are constants. Hence (5.9.7) becomes

$$y = c_0 + a \cos\omega t + b \sin\omega t \qquad (5.9.8)$$

with $a = c \cos\omega t_0$ and $b = c \sin\omega t_0$. In this form, the amplitude c and the acrophase t_0 do not appear explicitly. Instead new parameters a and b have been introduced.

Occasionally two or more sine curves are *superposed*. We begin with a special case which is important for the interference of light. We assume that two sine curves have the same amplitude and the same angular frequency, but that they may differ in acrophase. The equations are

$$y = c \cos\omega(t - t_0) \quad \text{and} \quad y = c \cos\omega(t - t_1) .$$

Superposition means that we add the two ordinates. Writing again y for the new ordinate we obtain

$$y = c \cos\omega(t - t_0) + c \cos\omega(t - t_1) .$$

Applying formula (5.7.7) we get

$$y = 2c \cos\omega \frac{t_1 - t_0}{2} \cos\omega\left(t - \frac{t_0 + t_1}{2}\right).$$

Abbreviating the constant $2c \cos\omega(t_1 - t_0)/2$ by C and the term $(t_0 + t_1)/2$ by $\bar{t}$ we obtain

$$y = C \cos\omega(t - \bar{t}) . \qquad (5.9.9)$$

Hence the graph is again a sine curve. Its amplitude is C and its acrophase $\bar{t}$. For $t_0 = t_1$ we get $C = 2c$, which means doubling of the amplitude. On the other hand, for a phase difference of half the period, that is, for $t_1 - t_0 = l/2$, we obtain

$$\omega \frac{t_1 - t_0}{2} = \frac{360°}{l} \frac{l}{4} = 90°$$

and, therefore, $C = 2c \cos 90° = 0$. Superposition of two such functions leads to the trivial function $y = 0$. The two cases are depicted in Fig. 5.23.

If the process of superposing sine waves is interpreted as interference of light waves, we get maximum amplification for $t_0 = t_1$ and extinction for $t_1 - t_0 = l/2$. The beautiful colors of a peacock are generated by interference of daylight and not by pigments in the feathers. Indeed, photographs of a feather made by an electron microscope reveal a fine

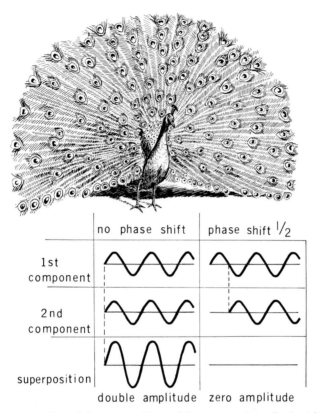

Fig. 5.23. Superposition of sine curves with equal frequency and amplitude. A biological application is found in the colors of a peacock. They are generated by interference of light and not by pigments

regular structure which causes amplification or extinction of light waves depending on their frequencies.

Superposition of sine waves with different frequencies, amplitudes, and phases may lead to periodic curves that do not necessarily resemble a sine curve. Thus we may add terms of the form

$$c_i \cos \omega_i (t - t_i) = a_i \cos \omega_i t + b_i \sin \omega_i t \quad (i = 1, 2, 3, \ldots).$$

In order to get periodicity we have to choose the angular frequencies ω_i as multiples of a certain ω, that is, ω, 2ω, 3ω, 4ω, etc. This assumption corresponds to periods l, $l/2$, $l/3$, $l/4$, etc. Thus we are led to a function

$$y = a_0 + a_1 \cos \omega t + a_2 \cos 2\omega t + \cdots + a_n \cos n\omega t$$
$$+ b_1 \sin \omega t + b_2 \sin 2\omega t + \cdots + b_n \sin n\omega t$$

$$(5.9.10)$$

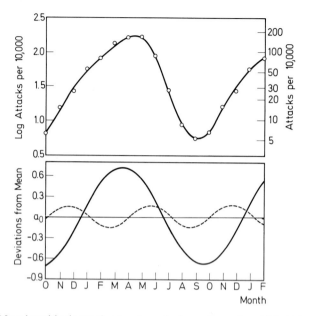

Fig. 5.24. Mean logarithmic attack rates of measles for each month and fitted trigonometric polynomial (upper diagram). Separate plot of the components (lower diagram). Reprinted from Bliss and Blevins (1959)

with $2n + 1$ terms and period l. For any natural number n such a function is called a *trigonometric polynomial*. Fitting periodic data by a trigonometric polynomial is called *harmonic analysis* or *Fourier analysis*[7]. The application of trigonometric polynomials in the life sciences is described in Bliss (1970), Bliss and Blevins (1959), and Waterman (1963). An application from medicine is shown in Fig. 5.24.

Another application related to vibrations of coupled systems is presented in Fig. 5.25.

We conclude the chapter by mentioning a famous mathematical result: *Any continuous or reasonably discontinuous periodic function can be represented by a trigonometric polynomial* (5.9.10) *with any degree of accuracy*. To obtain a high accuracy, the number of terms has to be correspondingly large. When n tends to infinity, the trigonometric polynomial turns into a *Fourier series*.

---

[7] Jean Baptiste Joseph Fourier (1768–1830), French physicist and mathematician.

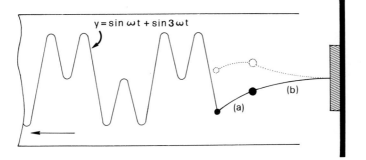

Fig. 5.25. Two vibrating systems, (a) and (b), are coupled. It is assumed that the frequency of system (a) is three times as high as the frequency of system (b). The oscillations are recorded on a moving tape

Recommended tables:

For trigonometric functions:
   a) from 0° to 90° in steps of 0.001°: Salzer and Levine (1962);
   b) from 0° to 90° in steps of 0.1°: Selby (1968), Meredith (1967);
   c) from 0° to 360° in steps of 1° or
      from 0 h to 24 h in steps of 4 min: Batschelet (1965).

For conversion of degrees into radians: Allen (1947), Meredith (1967).

Recommended for further reading: Defares *et al.* (1973), Guelfi (1966), Lefort (1967), C. A. B. Smith (1966), Sollberger (1965).

Problems for Solution

5.1.1. Let x be any natural number $1, 2, 3, \dots$ and y be the remainder after dividing x by 5. Thus, if $x = 19$, then $y = 4$. Show that y is a periodic function with period $l = 5$. Is $l' = 10$ also a period?

5.1.2. Let x be multiples of 3, that is, 3, 6, 9, 12, $\dots$ and y be the remainder after dividing x by 5. Is y a periodic function of x? If yes, what is the period? Plot a graph of the function.

5.2.1. Express the angles 30°, 45°, 60°, 120°, 135°, 270°, 450° in radians (Example: $90° = \pi/2$ radians).

5.2.2. Let φ take on the values 0.71, 1.65, and 2.90 (radians). Convert these values into degrees ($\varphi =$ Greek phi).

5.4.1. Verify graphically that
   a) $\sin 156° = \sin 24°$,   b) $\sin 240° = -\sin 120° = -\sin 60°$,
   c) $\cos(-70°) = \cos 70°$,   d) $\cos 105° = -\cos 75°$.

5.4.2. By means of a table find numerical values for
   a) $\sin(-37.5°)$,   b) $\cos 110°$,
   c) $\tan 128°$,   d) $\cos(-68.1°)$,
   e) $\sin 395°$,   f) $\tan(-65.6°)$.

5.5.1. A dancing explorer bee signals a polar angle of 75° counter-clockwise versus the sun's azimuth and a distance of 550 m. Convert this information into rectangular coordinates. (Read Section 5.3.)

5.5.2. An explorer bee discovers a source of honey at noon. This source is located 850 m east and 1200 m south from the hive. What polar coordinates will the bee signal?

5.5.3. In a behavioral experiment three turtles were released at a point O and later observed resting. To measure their position, a polar axis was chosen with origin O and oriented toward east. Polar angles were measured counter-clockwise. The polar coordinates of the three turtles were:

$$r_1 = 27.5 \text{ m} \qquad\qquad \alpha_1 = 73°$$
$$r_2 = 18.7 \text{ m} \qquad\qquad \alpha_2 = 165°$$
$$r_3 = 31.3 \text{ m} \qquad\qquad \alpha_3 = 106° .$$

Convert these polar coordinates into rectangular coordinates assuming that the origin is at O, that the x axis points eastward and the y axis northward.

5.5.4. Assume the same situation as in the previous problem. This time the rectangular coordinates are given:

$$x_1 = 20.5 \text{ m} \qquad x_2 = -2.8 \text{ m} \qquad x_3 = 15.3 \text{ m}$$
$$y_1 = 3.8 \text{ m} \qquad y_2 = 15.6 \text{ m} \qquad y_3 = -8.5 \text{ m} .$$

Find the corresponding polar coordinates.

5.5.5. Determine the following set of angles:
   a) $\{\alpha | \sin \alpha = 0.500, 0 < \alpha < 360°\}$,
   b) $\{\beta | \sin \beta \geq 0.500, \; 0 < \beta < 360°\}$.

5.5.6. Determine the following set of angles:
   a) $\{\varphi | \cos \varphi > 0.914, \; -90° < \varphi < 90°\}$,
   b) $\{\theta | 0 < \tan \theta < 1.60, \; 0 < \theta < 360°\}$.

5.5.7. The slope of a straight line in a rectangular coordinate system is measured by $a = \Delta y/\Delta x$ (cf. Section 3.6). Assume that Δx and Δy are measured and plotted in the same unit of length. It is then meaningful to ask for the angle of inclination. This is the angle α by which we have to rotate the x axis in the counter-clockwise direction until it coincides (for the first time) with the given straight line. The angle of inclination is limited by $0° \leq \alpha < 180°$. We are given

    a) $\Delta x =$ 15 cm, $\Delta y =$ 7 cm
    b) $\Delta x =$ 8 cm, $\Delta y = -6$ cm
    c) $\Delta x = -5$ cm, $\Delta y =$ 3 cm
    d) $\Delta x = -5$ cm, $\Delta y = -9$ cm .

Find the angle of inclination graphically and then numerically using formula (3.5.3).

5.6.1. A vertical rod of length 2 m casts a shadow on a horizontal plane. The rays of sunlight have an inclination $\theta = 67°$ to the horizontal plane. a) How long is the shadow? b) How long would the shadow be, if the rod were horizontal and the plane vertical facing the sun?

5.6.2. Angles of inclination, α, for mountain trails range from $0°$ to $15°$ if without steps, and from $15°$ to $45°$ if with steps. What are the boundaries of the slope, that is, of $\tan \alpha$? When the average angle of inclination is $\alpha = 12°$, how many kilometers of horizontal distance has a mountaineer to hike in order to climb 1200 m vertical distance?

5.6.3. During wet periods, a thin layer of water is present on fallen leaves and other litter on the soil surface. On this film bacteria, protozoans, fungi, and spores are floated onto higher levels (Bandoni and Koske, 1974). If the angle of inclination, α, is $10°$, $30°$, or $50°$ and the distance moved upward, d, is 5 cm, calculate the difference, h, in level reached by the microorganisms.

5.6.4. In a study of the human femur the two angles α and δ are introduced as shown in Fig. 5.18. Prove that $\tan \delta/\tan \alpha = \cos(\beta_1 - 90°)$.

5.7.1. In a table we find $\sin 37.4° = 0.6074$ and $\cos 37.4° = 0.7944$. With these values check the equality $\sin^2 \alpha + \cos^2 \alpha = 1$.

5.7.2. Using formulas (5.7.1) and (5.7.4) show that
$$\cos 2\alpha = 2 \cos^2 \alpha - 1 = 1 - 2 \sin^2 \alpha .$$

*5.8.1. Find the equation $r = f(\alpha)$ in polar coordinates for the line $y = 2$.

*5.8.2. Plot the polar graph of
   a) $r = |\sin \alpha|$ b) $r = |\sin 2\alpha|$.

*5.9.1. Plot $y = 1 + 2 \cos 2\pi t$ in rectangular coordinates.

*5.9.2. A certain biorhythm can be approximately described by the formula

$$y = 2.5 + 1.5 \cos \frac{360°}{24\,\text{h}} (t - 5\,\text{h}) \qquad (\text{h} = \text{hours}) .$$

Plot the function for $t = 0, 1, 2, \ldots, 24\,\text{h}$. What is the period, the amplitude, and the phase of maximum y value (acrophase)?

*5.9.3. Plot $y = \cos x + \cos 2x$.

*5.9.4. Plot $y = \sin x + \frac{1}{2} \sin 2x$.

Chapter 6

Exponential and Logarithmic Functions I

6.1. Sequences

We began the first chapter considering the growth of a foal. It was assumed that the weight increases at a rate of 20% during consecutive time intervals of equal length. Let w be the initial weight and p be the rate of growth. Then the weights at the end of 0, 1, 2, ... time intervals are

$$w, \ w\left(1+\frac{p}{100}\right), \ w\left(1+\frac{p}{100}\right)^2, \ w\left(1+\frac{p}{100}\right)^3, \dots .$$

With the abbreviating notation

$$q = 1 + \frac{p}{100} \tag{6.1.1}$$

the weights are

$$w, \ wq, \ wq^2, \ wq^3, \ wq^4, \dots . \tag{6.1.2}$$

An ordered arrangement of values such as (6.1.2) is an example of a *sequence*. To get a definition of a sequence we begin with the set of natural numbers

$$1, 2, 3, \dots$$

and choose this set as the domain of a function $y = f(x)$. Then the y values, arranged in the same order as the natural numbers, are

$$f(1), \ f(2), \ f(3), \dots .$$

This arrangement is called a *sequence*.

Sometimes the domain is $\{0, 1, 2, \dots\}$, that is, it also contains 0, or it is finite, say $\{5, 6, 7, 8, 9\}$, or we prefer the negative integers, say $\{-1, -2, \dots\}$.

Therefore, a sequence may be conceived as a special type of function: The domain (values of x) consists of consecutive integers and the range (values of $y = f(x)$) of correspondingly ordered values. Our Example (6.1.2) could be written as a function:

$$y = wq^x, \quad x \in \{0, 1, 2, \dots\} . \tag{6.1.3}$$

We take another example: Under ideal conditions a cell may subdivide into two cells in a certain time interval. The new cells subdivide again after the elapse of the same time interval, etc. We assume that the time is measured in units that coincide with the time interval needed for a cell division. Let t denote the time and N the number of cells. Then we obtain the sequence 1, 2, 4, 8, 16, ... or

$$N = 2^t, \quad t \in \{0, 1, 2, ...\} . \tag{6.1.4}$$

Still another sequence arises when a radioactive substance decays. The carbon isotope ^{14}C has a half-life of 5760 years. This means that the number N of ^{14}C atoms in the substance is reduced to $N/2$ after the elapse of 5760 years. Thus if t denotes the time measured in units of 5760 years and N_0 the number of ^{14}C atoms in the substance at time $t = 0$, we get the sequence

$$N = N_0 \cdot 2^{-t}, \quad t \in \{0, 1, 2, ...\} . \tag{6.1.5}$$

We notice that in all three examples the independent variable is an *exponent* of a power. Whenever this occurs, the sequence is called geometric. The general form of a geometric sequence is $a, aq, aq^2, aq^3, ...$ or

$$y = aq^x, \quad x \in \{0, 1, 2, ...\} . \tag{6.1.6}$$

Two consecutive terms of a geometric sequence have a constant ratio, in our notation q. Let aq^n and aq^{n+1} be two consecutive terms. Then we get indeed $aq^{n+1}/aq^n = q$. The number q is called the *common ratio*.

Geometric sequences are strongly related to the *geometric mean* which we introduced in formula (1.8.2). We consider three consecutive terms of (6.1.6), say

$$aq^{n-1}, \quad aq^n, \quad aq^{n+1} .$$

Then we take the geometric mean of the first and the third term and obtain

$$\sqrt{aq^{n-1} \cdot aq^{n+1}} = \sqrt{a^2 q^{2n}} = aq^n ,$$

that is, we get the second term. In other words: *Every term of a geometric sequence is the geometric mean of the preceding and the following term.*

Geometric sequences may be compared with *arithmetic sequences* which are of the form

$$b, \quad a+b, \quad 2a+b, \quad 3a+b, \quad 4a+b, ...$$

or

$$y = ax + b, \quad x \in \{0, 1, 2, ...\} . \tag{6.1.7}$$

This may be interpreted as a *linear function* with a domain consisting of consecutive integers. The arithmetic sequence has properties similar to those of the geometric sequence:

Two consecutive terms of an arithmetic sequence differ by a constant value, in our notation by the quantity a. This number is called the *common difference*.

We consider three consecutive terms of an arithmetic sequence, say

$$a(n-1)+b, \quad an+b, \quad a(n+1)+b.$$

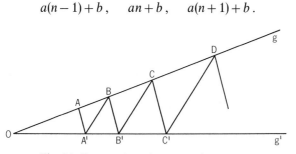

Fig. 6.1. Construction of a geometric sequence

The second term, $an+b$, is the arithmetic mean of the other two terms. In other words: *Every term of an arithmetic sequence is the arithmetic mean of the preceding and the following term.*

For a better intuitive understanding we compare geometric with arithmetic sequences by graphical presentations. We begin with Fig. 6.1. An angle with vertex O and sides g and g' is intersected by two sets of parallel lines, $AA' \parallel BB' \parallel CC'$ etc. and $A'B \parallel B'C \parallel C'D$ etc. Since the triangles OAA' and OBB' are similar, we get for the lengths of the corresponding sides

$$OA/OB = OA'/OB'.$$

In the same way we get from the similar triangles OBA' and OCB'

$$OA'/OB' = OB/OC.$$

Combining the two results, we obtain

$$OA/OB = OB/OC$$

so that OA, OB, OC form three consecutive terms of a geometric sequence. The argument can be easily extended to more than three terms. The segments OA, OB, OC, ... of Fig. 6.1 are interpreted in Fig. 6.2a as ordinates $y_1, y_2, y_3, ...$ on a y axis.

On the other hand, if we replace the sides g and g' of an angle by two parallel lines and apply then the same procedure, we obtain an arithmetic sequence $y_1, y_2, \ldots$ as shown in Fig. 6.2 b. This presentation of geometric and arithmetic sequences is adapted from Gebelein and Heite (1951, p. 32).

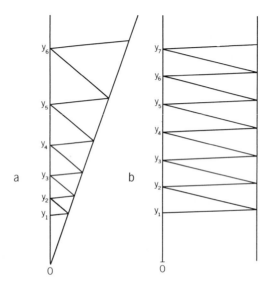

Fig. 6.2. Geometric and arithmetic sequences constructed by two sets of parallel lines

6.2. The Exponential Function

An animal does not grow in steps. It grows continuously. Therefore, we may ask whether the sequence (6.1.3) has any meaning if x does not only take on integers but also fractional or even real numbers. Mathematically speaking, we try to replace the domain $\{0, 1, 2, \ldots\}$ by the set $\mathbb{R}$ of all real numbers. To begin with, we extend the sequence $2, 2^2, 2^3, 2^4, \ldots$ in the opposite direction: $2^0 = 1$, $2^{-1} = \frac{1}{2}$, $2^{-2} = 1/4, \ldots$. Thus negative integers as exponents do not cause any difficulty. Next we consider fractional exponents, for instance $2^{0.5} = 2^{\frac{1}{2}}$. With the radical sign this is $\sqrt{2} = 1.414 \ldots$. Similarly, $2^{0.25} = 2^{1/4} = \sqrt[4]{2} = 1.189 \ldots, 2^{0.75} = 2^{3/4} = \sqrt[4]{8} = 1.681 \ldots, 2^{1.25} = 2^{5/4} = \sqrt[4]{32} = 2.378 \ldots$, etc. The values that we have found so far are plotted in Fig. 6.3. With tables of square and cube roots we could quickly find many more values of the function $y = 2^x$. As long as x is a rational number, the function $y = 2^x$ is well defined. When x is an irrational number such as $x = \sqrt{2}$ or $x = \pi$, it is not so easy to define 2^x (for a definition see Section 10.5). At this stage, however, the reader may take it for granted that a definition is possible and that the function $y = 2^x$ is represented by the smooth line drawn in Fig. 6.3 for all real values of x, that is, for $x \in \mathbb{R}$.

The function $y = 2^x$ and, more generally, the function

$$y = aq^x, \quad q > 0, \quad x \in \mathbb{R} \tag{6.2.1}$$

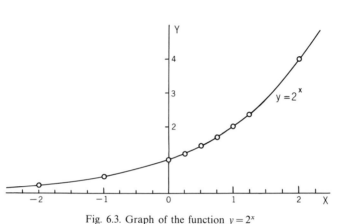

Fig. 6.3. Graph of the function $y = 2^x$

with two parameters a and q is called an *exponential function*. The assumption $q > 0$ is required since values such as 0^{-1} and $(-1)^{\frac{1}{2}}$ are either not defined or are not real numbers. The name "exponential function" is derived from the property that the independent variable is an exponent. The reader is warned not to confuse an exponential function with a power function $y = ax^n$. In a power function the exponent is a constant and the independent variable is the *base* of a power. The reader is invited to compare the graphs of some power functions in Fig. 4.1 with the graph of a typical exponential function in Fig. 6.3.

As we have seen, a geometric sequence is only a special case of an exponential function in that the domain is restricted to consecutive integers. It would be desirable to abandon the historic word "geometric" in this connection and to speak of an *exponential sequence*. By the same token, an arithmetic sequence should rather be called a *linear sequence* (cf. formula (6.1.7)).

In Chapter 10, we will give a definition of an exponential function which avoids the difficulty of x being an irrational number. There, by means of the calculus, we will define the natural logarithm and then find a way to the exponential function e^x with base $e = 2.718 \dots$.

6.3. Inverse Functions

We introduced the exponential function $y = aq^x$ in connection with growth, and the special function $N = 2^t$ in connection with doubling the number of cells. The following question arises frequently: We are given the value of y or N, that is, the value of the dependent variable. At what time of growth or cell division does the function reach the given value? Thus we are asked to calculate the value of the independent variable x or t. In trying to answer this question it would be in vain to

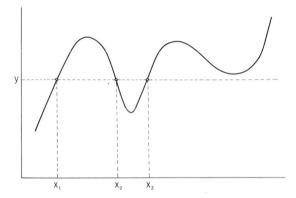

Fig. 6.4. Graph of a function which is not monotone. Therefore, the inverse function does not exist. To a given y there correspond several x values

apply the operations of addition, subtraction, multiplication, division, or root extraction. A new operation is required, the use of logarithms. Before we enter this area, we introduce the notion of an *inverse function*.

For a linear function $y = ax + b$ it is easy to find x in terms of y. We get $x = (y - b)/a$ provided that $a \neq 0$. Here y appears as the independent and x as the dependent variable. For each value of y, the quantity x is uniquely determined. Hence, x is a function of y, called the *inverse function*.

In the case of a power function $y = ax^n$, we can also solve the equation with respect to x. We obtain $x = (y/a)^{1/n}$ provided that $y/a > 0$. However, there may be ambiguity. Consider, for example, the quadratic function $y = x^2$. For a given positive value of y, we find two different solutions, $x = \sqrt{y}$ and $x = -\sqrt{y}$. Due to this ambiguity, x cannot be called a function of y unless we accept some restriction. Fig. 6.4 depicts the graph of a function whereby the problem of finding x, given y, leads to even greater ambiguity.

To avoid the difficulty, we have to exclude functions whose values go up and down. We will restrict ourselves to functions that are either increasing or decreasing but not both. A glance at Fig. 6.3 shows that $y = 2^x$ increases for all values of x. We say that this function is *monotone increasing* since $2^{x_2} > 2^{x_1}$ whenever $x_2 > x_1$.

In general, let the domain of a function $y = f(x)$ be an *interval* on the x axis and let x_1 and x_2 be two values in this interval. *If the property*

$$f(x_2) > f(x_1) \tag{6.3.1}$$

holds whenever $x_2 > x_1$, the function is called monotone increasing.

Similarly, *if the property*

$$f(x_2) < f(x_1) \tag{6.3.2}$$

holds whenever $x_2 > x_1$, the function is called monotone decreasing[1].

A function is monotone if it is either monotone increasing or monotone decreasing.

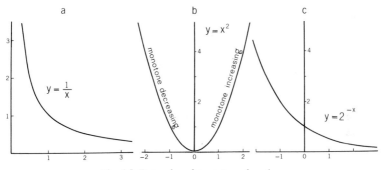

Fig. 6.5. Examples of monotone functions

Example 6.3.1. The linear function $y = ax + b$ is monotone increasing for all values of x (that is, $x \in \mathbb{R}$) if $a > 0$, and monotone decreasing if $a < 0$. In the first case the slope is positive, in the second case negative.

The power function $y = 1/x = x^{-1}$ is monotone decreasing for $x > 0$ and $x < 0$. In Fig. 6.5a the function is depicted for the domain $x > 0$.

---

[1] Sometimes it is important to distinguish between *weakly* and *strictly monotone*. A function satisfying $f(x_2) \geq f(x_1)$ or $f(x_2) \leq f(x_1)$ whenever $x_2 > x_1$, is called monotone in the weak sense. Such a function can remain constant in some intervals. In the above text and later we use the word monotone in the strict sense.

The word "monotone" is used in mathematics as adjective and as adverb. The words "monotonic" and "monotonically" would sound more familiar.

The power function $y = x^2$ is monotone increasing for the domain $x \geq 0$. But for $x \leq 0$, $y = x^2$ is monotone decreasing. With no restriction for the domain, $y = x^2$ is not monotone at all (see Fig. 6.5 b).

The exponential function $y = 2^{-x}$ is monotone decreasing for all values of x (that is, for $x \in \mathbb{R}$). A graph is shown in Fig. 6.5 c.

Example 6.3.2. The trigonometric function $y = \sin \alpha$ is monotone increasing, monotone decreasing or not monotone, depending on the choice of its domain. For instance, if $-90° < \alpha < 90°$, the sine is monotone increasing. For $90° < \alpha < 270°$ it is monotone decreasing. But for $0° < \alpha < 180°$ the sine is not monotone (cf. Fig. 5.8).

Example 6.3.3. As a rule, with increasing sea level, the temperature of ambient air gradually decreases. However, the drop need not be monotone. In winter time, for instance, we frequently observe *temperature inversion*: In mountains above the fog, the sunshine may heat the air well above the freezing point, while the temperature beneath remains below $0°$ C (Fig. 6.6).

Now we are prepared to understand an important result: *To each monotone function $y = f(x)$ there exists an inverse function.* The proof is illustrated in Fig. 6.7 a—c. In Fig. 6.7 a the function is monotone increasing in a certain interval of the x axis. It is assumed that the function does not make any jump. Under this condition the range of the function is an interval on the y axis. If y is a value from the range, the line through the point $(0, y)$ parallel to the x axis hits the curve at only one point. Thus the x value corresponding to y is uniquely determined. The same is true

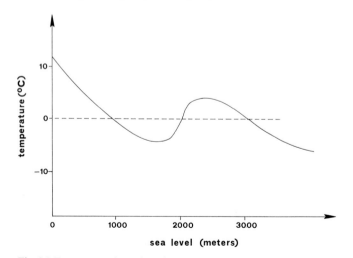

Fig. 6.6. Temperature inversion above fog. The function is not monotone

for a monotone decreasing function without jump as shown in Fig. 6.7b. Finally in Fig. 6.7c it is assumed that the function has a jump at $x = x_0$ (discontinuity). Whereas the domain is an interval of the x axis, the range consists of two nonadjacent intervals. But, again, if y belongs to the range, there exists only one value of x that is associated with y. Thus, in all these cases, x is uniquely determined by y. The inverse function exists indeed.

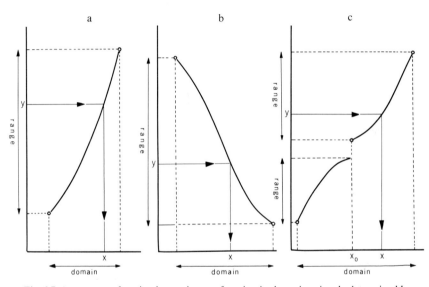

Fig. 6.7. A monotone function has an inverse function in that x is uniquely determined by y

Let $y = f(x)$ be a monotone function. Then we may denote the inverse function by $x = g(y)$. Whereas $y = f(x)$ maps the domain into the range, $x = g(y)$ maps the range into the domain. Often the inverse function of $y = f(x)$ is designated by $x = f^{-1}(y)$. The superscript -1 should not be confused with an exponent, even though this symbol was chosen because of some analogy with a reciprocal.

Sometimes the inverse function is studied quite independently of the original function. Then it is convenient to change notation and to use x again for the independent and y for the dependent variable. Thus, instead of $x = g(y)$ we would write $y = g(x)$. As a consequence, the x and y axes, as well as range and domain, have to be interchanged. This procedure removes the graph of the original function and puts it in a new position. In the special case where x and y are measured in the same unit of length, the displacement may be interpreted as a reflection about the line $y = x$ (see Fig. 6.8).

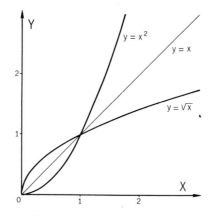

Fig. 6.8. The function $y = x^2$ for $x \geq 0$ and its inverse function $y = \sqrt{x}$. On both axes the same unit of length is chosen. Then the two curves are reflections of each other about the line $y = x$

Example 6.3.4. $y = x^2$ with domain $x \geq 0$ and range $y \geq 0$. The inverse function is $y = \sqrt{x}$ with domain $x \geq 0$ and range $y \geq 0$ (see Fig. 6.8).

$y = 1/x$ with domain $x > 0$ and range $y > 0$. The inverse function is identical with the original function.

Example 6.3.5. $y = \sin \alpha$ with domain $-90° \leq \alpha \leq +90°$ and range $-1 \leq y \leq +1$. The inverse function is called the *arcsine function* and we write $\alpha = \arcsin x$ or $\alpha = \sin^{-1} x$ with domain $-1 \leq x \leq +1$ and range $-90° \leq \alpha \leq +90°$. In $\sin^{-1} x$ the superscript is not an exponent, and $\sin^{-1} x$ should not be confused with $1/\sin x$.

Example 6.3.6. $y = \tan \alpha$ with domain $-90° \leq \alpha \leq +90°$. As shown in Fig. 5.13 $\tan \alpha$ is monotone increasing. The inverse function is $\alpha = \arctan x$. This function is defined for all $x \in \mathbb{R}$, but α is restricted to the interval $-90° \leq \alpha \leq +90°$.

Notice that the inverse function of a monotone increasing function (without jump) is also increasing. Monotone decreasing functions have an analogous property.

6.4. The Logarithmic Functions

The function $y = 2^x$ with domain $\mathbb{R}$ (set of all real numbers) and range $y > 0$ is monotone increasing (cf. Fig. 6.3). The same is true for every exponential function $y = aq^x$ with $a > 0$ and $q > 1$. Therefore, the inverse function exists. It is called a *logarithmic function*. In the particular

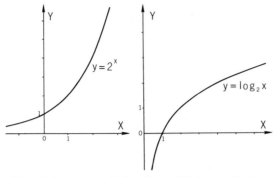

Fig. 6.9. An exponential function and its inverse function

case of $y = 2^x$ we write for the inverse function $x = \log_2 y$ or, on interchanging x and y,

$$y = \log_2 x. \qquad (6.4.1)$$

The original as well as the inverse function are plotted in Fig. 6.9. The domain of the logarithmic function (6.4.1) is the interval $x > 0$, that is, we can take logarithms only of positive numbers. The range is $\mathbb{R}$, that is, a logarithm can take on every real number. The inverse statement of $2^3 = 8$ is $\log_2 8 = 3$. Similarly, $\log_2(1/16) = -4$ stems from $2^{-4} = 1/16$.

In the following we will concentrate on the *common logarithm*[2]. Its base is 10. Instead of $\log_{10}$ we simply write log. We list a few statements together with the inverse statements (the decimal fractions are rounded off):

$$
\begin{array}{llll}
10^3 & = 1000, & \log 1000 & = 3 \\
10^{2.30103} & = 200, & \log\ 200 & = 2.30103 \\
10^2 & = 100, & \log\ 100 & = 2 \\
10^{1.30103} & = 20, & \log\ 20 & = 1.30103 \\
10^1 & = 10, & \log\ 10 & = 1 \\
10^{0.30103} & = 2, & \log\ 2 & = 0.30103 \\
10^0 & = 1, & \log\ 1 & = 0 \\
10^{-0.69897} & = 0.2, & \log\ 0.2 & = -0.69897 = 0.30103 - 1 \\
10^{-1} & = 0.1, & \log\ 0.1 & = -1.
\end{array}
$$

To each positive number there corresponds a unique logarithm. The original positive number is called the *antilogarithm*. Thus the antilogarithm of 0.30103 is 2. It is also true that to each logarithm there

---

[2] The natural logarithm with base $e = 2.71828 \ldots$ will be introduced in Chapter 10.

corresponds a unique antilogarithm. Hence, there is *a one-to-one mapping between logarithms and antilogarithms.*

We will derive four *basic rules* for logarithms:

$$\log ab = \log a + \log b \qquad (a>0,\ b>0) \qquad (6.4.2)$$

$$\log 1/a = -\log a \qquad\qquad (a>0) \qquad\qquad (6.4.3)$$

$$\log a/b = \log a - \log b \qquad (a>0,\ b>0) \qquad (6.4.4)$$

$$\log a^n = n\log a \qquad\qquad (a>0,\ n\in\mathbb{R}). \qquad (6.4.5)$$

The proofs are based on some well-known rules for powers. Let

$$a = 10^u, \quad \text{that is,} \quad u = \log a,$$

$$b = 10^v, \quad \text{that is,} \quad v = \log b.$$

Then

$$ab = 10^u \cdot 10^v = 10^{u+v}$$

which implies $\log ab = u + v = \log a + \log b$ as stated in (6.4.2). Moreover,

$$1/a = 1/10^u = 10^{-u}$$

which implies $\log 1/a = -u = -\log a$ as stated in (6.4.3). From (6.4.2) and (6.4.3) it follows that

$$\log a/b = \log a(1/b) = \log a + \log 1/b = \log a - \log b$$

which proves (6.4.4). Finally

$$a^n = (10^u)^n = 10^{nu}$$

which implies $\log a^n = nu = n\log a$. This proves (6.4.5).

Notice that (6.4.5) includes fractional and negative powers since n can be any real number. Thus, for instance,

$$\log \sqrt[3]{a} = \log a^{1/3} = (1/3)\log a.$$

Formula (6.4.2) reduces the multiplication of two numbers a and b to a simple addition. The idea is applied in the *slide rule*, a handy and inexpensive instrument. Notice, on the other hand, that the logarithm of a sum, say $\log(a+b)$, cannot be expressed in terms of $\log a$ and $\log b$.

The numerical value of a common logarithm consists of two parts called *characteristic* and *mantissa*. In a logarithmic table we find the mantissa which is the fractional part of a logarithm. For instance, in a four digit table the mantissa of $\log 73$ is 8633. We know that $\log 73$ falls between $\log 10 = 1$ and $\log 100 = 2$. Hence the integral part or the characteristic is 1 and $\log 73 = 1.8633$. Changing the decimal point only influences the characteristic, not the mantissa. Indeed, $\log 730 = \log 10 \times 73 = \log 10 + \log 73 = 1 + 1.8633 = 2.8633$. Similarly, $\log 7300 = 3.8633$, $\log 73000 = 4.8633$, etc. In the opposite direction we get $\log 7.3 = \log 73/10 = \log 73 - \log 10 = 1.8633 - 1 = 0.8633$. Similarly, $\log 0.73 = 0.8633 - 1$, $\log 0.073 = 0.8633 - 2$, $\log 0.0073 = 0.8633 - 3$, etc.

Four-place values of common logarithms are given in Table D of the Appendix.

Logarithms convert a *geometric mean*, say $(AB)^{1/2}$, into an *arithmetic mean* of $\log A$ and $\log B$:

$$\log(AB)^{1/2} = \tfrac{1}{2}(\log A + \log B).$$

Similarly, taking logarithms of all terms of

$$a, aq, aq^2, aq^3, \dots$$

and letting $\log a = c$, $\log q = d$, we get

$$c, c+d, c+2d, c+3d, \dots,$$

that is, a *geometric or exponential sequence is converted to an arithmetic or linear sequence by applying logarithms*.

Finally, we determine the inverse function of

$$y = aq^x (a > 0, q > 0, q \neq 1). \tag{6.4.6}$$

Taking logarithms on both sides we obtain

$$\log y = \log a + x \cdot \log q$$

and from this

$$x = (\log y - \log a)/\log q. \tag{6.4.7}$$

6.5. Applications

Problem 6.5.1. The quantity of *timber* in a young forest grows almost exponentially. We may assume that the yearly rate is 3.5%. What increase is expected within ten years?

Answer: To answer this question we calculate first the factor $q = 1 + p/100$ with $p = 3.5\%$ (see formula (6.1.1)). We obtain $q = 1.035$. As explained in Section 1.3, the growth in ten years is determined by the factor q^{10}. To get this value numerically we apply logarithms:

$$\log q^{10} = 10 \log q = 10 \log 1.035 = 10 (0.01494) = 0.1494.$$

The antilogarithm is 1.41. Hence the quantity of timber has increased in ten years by 41%.

Problem 6.5.2. We assume again that the growth rate of timber is $p = 3.5\%$. How many years will it take until the quantity of timber has doubled?

Answer: Let n be the unknown number of years and $q = 1 + p/100$. Then we obtain the equation

$$q^n = 2. \tag{6.5.1}$$

Such an equation is called an *exponential equation*, since the exponent is the unknown quantity.

$$\log q^n = n \log q = \log 2, \quad n = \log 2 / \log q.$$

With $q = 1.035$ we obtain $n = 0.30103/0.01494 = 20.1$. Hence it takes a little over 20 years until the quantity of timber has doubled.

Problem 6.5.3. Calloway (1965) suggested that the *ages of experimental animals* should be known as exactly as possible, because biological forms change rapidly with age, especially in the early periods of life. So far as feasible, experiments ought to be done on groups of animals of different ages.

A geometric or exponential sequence of ages seems to be most appropriate. Find such a sequence, if the youngest age is $A = 3$ weeks and if the constant factor is chosen as $q = 1.5$.

Solution: The sequence is

$$A, Aq, Aq^2, Aq^3, \dots \tag{6.5.2}$$

or

$$3.0 \quad 4.5 \quad 6.8 \quad 10.1 \quad 15.2 \quad 22.8 \dots \text{ weeks}.$$

It is quite different from an arithmetic or linear sequence which, beginning with the same terms 3 and 4.5 weeks, would be

$$3 \quad 4.5 \quad 6.0 \quad 7.5 \quad 9.0 \quad 10.5 \dots \text{ weeks}.$$

Problem 6.5.4. When the ages of experimental animals are chosen such that they form a geometric or exponential sequence, the following problem may arise: Given the number of age groups, the lowest and the highest age, what are the terms of the sequence?

Solution: Let A be the age of the youngest group, B the age of the oldest group, and n be the number of groups. With the unknown factor q the sequence is

$$A, \quad Aq, \quad Aq^2, \dots, Aq^{n-1}. \tag{6.5.3}$$

The last exponent is $n - 1$, because the first term may be written Aq^0 and there are n terms altogether. The last term was assumed to be B.

Hence we get the *exponential equation* (cf. Adler, 1966)

$$A q^{n-1} = B .$$ (6.5.4)

The equation is solved by logarithms. With formulas (6.4.2) and (6.4.5) the Eq. (6.5.4) becomes

$$\log B = \log A q^{n-1} = \log A + (n-1) \log q .$$

Hence

$$\log q = \frac{\log B - \log A}{n-1} .$$ (6.5.5)

A numerical example may illustrate the procedure. Given $A = 4$ weeks, $B = 10$ weeks, and $n = 5$. Then

$$\log q = (\log 10 - \log 4)/4 = (1 - 0.60206)/4 = 0.09948 .$$

The antilogarithm is $q = 1.257$. To obtain the sequence (6.5.3) numerically it is easier to work with $\log q$. The logarithms taken of the terms in (6.5.3) form the linear sequence

$$\log A , \quad \log A + \log q , \quad \log A + 2 \log q , \dots , \log A + (n-1) \log q .$$ (6.5.6)

The numerical values are

$$0.60206 \quad 0.70154 \quad 0.80102 \quad 0.90050 \quad 0.99998 .$$

The last term differs slightly from the exact value $\log B = 1$. The difference can be explained by rounding-off errors. The antilogarithms are

$$4 \quad 5.03 \quad 6.32 \quad 7.95 \quad 10 \text{ weeks} .$$

This is the desired sequence of ages.

*6.6. Scaling

One of the important biological applications of exponential functions and of logarithms goes back to 1846 when E. H. Weber[3] studied the *response* of humans to *physical stimuli*.

We assume that a person holds a weight of 20 g (grams) in his hand and that he is tested for the ability to distinguish between this weight and a slightly higher weight. Experiments show that a person is not able to discriminate between 20.5 g and 20 g, but that he finds 21 g to be heavier than 20 g most of the time. The required increase of stimulus is 1 g. With an initial weight of 40 g, the result is quite different. A person cannot reliably discriminate between 41 g and 40 g. The increase should

---

[3] Ernest Heinrich Weber (1795–1878), German anatomist and physiologist.

be 2 g instead of 1 g. Similarly, experience shows that 63 g can be discriminated from 60 g, 84 g from 80 g, and 105 g from 100 g, but that the interval cannot be reduced. From these figures it follows that discrimination is possible if the magnitude of stimulation is increased by one twentieth or 5% of the original value.

Analogous results were found for sound and light reception, as well as for smell and taste. In general, let s be the magnitude of a measurable stimulus and Δs the increase just required for discrimination. Then the ratio

$$r = \frac{\Delta s}{s} \tag{6.6.1}$$

is *constant*, that is, it does not depend on s. In other words: *Noticeable differences in sensation occur when the increase of stimulus is a constant percentage of the stimulus itself.* This is *Weber's law*. The following list of approximate ratios $\Delta s/s$ may illustrate the sensitivity of human senses:

Visual brightness   1 : 50 ($s =$ light intensity)

Tone   1 : 10 ($s =$ sound intensity)

Smell for rubber   1 : 8   ($s =$ number of molecules)

Taste for saline solution   1 : 4   ($s =$ concentration of solution).

It should be emphasized, however, that Weber's law is at best a good approximation to reality. The law fails to be valid when the magnitude s of stimulation is either too small or too large. As is the case for most natural laws, Weber's law is only valid within a certain domain of s. For a more advanced discussion see Luce and Galanter (1963a, p. 193 to 206).

Strongly related to the problem of discrimination is the problem of *scaling*. As a rule, sensations are not measurable. Nevertheless, it is of great practical value to scale responses to stimuli. In 1860 Fechner[4] proposed a method of scaling which is based on Weber's law. Let $r = \Delta s/s$ be the constant ratio of Weber's law and let s_0 be a fixed value of s. Then we calculate the nearest noticeably higher stimulus:

$$s_1 = s_0 + \Delta s_0 = s_0 + \frac{\Delta s_0}{s_0} s_0 = s_0 + r s_0 = s_0(1 + r).$$

Abbreviating the factor $1 + r$ by q we obtain

$$s_1 = s_0 q.$$

---

[4] Gustav Theodor Fechner (1801–1887), German physicist and philosopher.

In our example of weight lifting we had $s_0 = 20$ g, $r = 1/20$. Hence $q = 1.05$, $s_1 = 21$ g.

The next higher distinguishable stimulus is $s_1 q = s_0 q \cdot q = s_0 q^2$, etc. Hence, *noticeably different stimuli follow each other in a geometric or exponential sequence*:

$$s_0, \quad s_0 q, \quad s_0 q^2, \quad s_0 q^3, \dots . \tag{6.6.2}$$

The general term is

$$s_n = s_0 q^n \quad (n = 0, 1, 2, \dots). \tag{6.6.3}$$

On the other hand, it is quite clear that the corresponding sensation does not follow a geometric sequence. We feel that *sensation proceeds in equal steps* and that it should be represented by an *arithmetic or linear sequence*. Thus we get the following table:

Stimulus	Level of sensation
s_0	0
$s_1 = s_0 q$	1
$s_2 = s_0 q^2$	2
$s_3 = s_0 q^3$	3
....	....
$s_n = s_0 q^n$	n

We may consider n as a function of s_n, actually as the *inverse function* of (6.6.3):

$$n = (\log s_n - \log s_0)/\log q . \tag{6.6.4}$$

To simplify we write s instead of s_n, denote $1/\log q$ by A and $-\log s_0/\log q$ by B. Hence, a suitable measure for the level of sensation is

$$n = A \log s + B . \tag{6.6.5}$$

By means of (6.6.5) we may judge the level of sensation not only by an ordinal scale, but even by an *interval scale* (see Section 1.2). Indeed, differences in the level of sensation are not arbitrary. They can be meaningfully expressed in terms of measurable stimuli.

For scaling sensation we are not restricted to Formula (6.6.5). We may take any multiple of n and add an arbitrary constant. The new quantity would still be a linear function of $\log s$. Hence, in general, if M denotes a suitable quantity for scaling sensation, we get

$$M = a \log s + b . \tag{6.6.6}$$

The constants a and b can be chosen freely, subject to the restriction $a \neq 0$. This formula is known as the *psychophysical law of Weber-Fechner*. It is not a law in the usual sense inasmuch as it fails to relate two or more measurable quantities to each other. Formula (6.6.6) should rather be interpreted as a suitable *definition* of M. *In the absence of an objective measure of sensation, Fechner's formula is a useful way of scaling sensation*[5].

Example 6.6.1. We consider the subjective impression of *loudness*. Experienced sensation is not proportional to the physical intensity of sound. Let I denote this intensity (Watt/m^2). Then, according to (6.6.6), loudness L is scaled by

$$L = a \log I + b.$$

The constants a and b can be chosen arbitrarily. It is customary, however, to fix a and b in the following way: At a frequency of 1000 Hz (Hz = Hertz = cycles per second) the threshold of audibility or lowest intensity that can be heard is nearly $I_0 = 10^{-12}$ Watt/m^2. Then L is made equal to

$$L = 10(\log I - \log I_0) = 10 \log(I/I_0). \tag{6.6.7}$$

The unit of L is called decibel[6] and abbreviated by dB. For $I = I_0$, L is zero decibel.

For a tone of any frequency other than 1000 Hz, formula (6.6.7) and the unit dB cannot be used for the human ear, since the ear is not equally sensitive to tones of different frequencies. For any tone deviating from 1000 Hz, for a mixture of tones, whether harmonious or noisy, a subjective manner of scaling is required. "Normal" observers match the loudness of a given sound with a tone of 1000 Hz frequency. When this tone is scaled L decibels, it is said that the given sound has L *phons*. For instance, ordinary conversation has 60 phons. This means that an equally loud tone of 1000 Hz (judged subjectively) has 60 dB (measured objectively). For comparison we mention that a whisper has roughly 10 phons, a quiet automobile 40 phons, a loud orchestra 80 phons, and thunder 120 phons. For more details on scaling loudness see Stuhlman (1943, p. 287 ff) and Randall (1958, p. 36 and 182–190).

---

[5] Based on a large number of psychophysical experiments S. S. Stevens and his coworkers claim that power functions are more suitable for a psychophysical law than Fechner's logarithmic formula. For details see e.g. S. S. Stevens (1970). A thorough discussion of the problem of scaling can be found in Luce (1959, Chap. 2) or in Luce and Galanter (1963 b).

[6] 1 decibel = 1/10 Bel in honor of Alexander Graham Bell (1847–1922), born in Scotland; American scientist who is best known for his invention of the telephone.

Example 6.6.2. Consider three tones which are equally spaced on the frequency scale, for instance the three tones with frequencies

$$300\,\text{Hz}, \quad 600\,\text{Hz}, \quad 900\,\text{Hz}.$$

People who hear these tones agree unanimously that the interval between the second and the third tone is considerably smaller than between the first and the second tone. Therefore, our sensation for *musical pitch* is not proportional to frequency. To get two consecutive intervals which are perceived as equal in size we have to adopt a geometric sequence. For instance, the three tones with frequencies

$$300\,\text{Hz}, \quad 600\,\text{Hz}, \quad 1200\,\text{Hz}$$

proceed in intervals of octaves. A proper scale for musical pitch is based on the *logarithm of frequency*, quite in agreement with the law by Weber-Fechner.

Example 6.6.3. This example is known to astronomers. The experienced *brightness* of a star is by no means proportional to the light energy received by the eye. Again we have a linear relationship between brightness and the logarithm of light intensity. A standard formula is

$$m = c - 2.5 \log I \tag{6.6.8}$$

where I is the light intensity and c a constant determined by the unit in which I is measured. m is called the *apparent magnitude* of a star. The brightest star, Sirius, has magnitude -1.6, Vega has magnitude 0.1, and Betelgeuse 0.9. The clumsiness of a negative magnitude might have been avoided if the magnitude zero had been placed better.

Example 6.6.4. Perhaps the application of the Weber-Fechner law that is especially important for life scientists is the *dose-response relationship in biological assay*. When a certain dose of a chemical (drug, poison, vitamin, hormone, etc.) is administered to an animal, the response or reaction, whether it is qualitative or quantitative, cannot be linearly related to the dose. If, for instance, a dose of 10 mg is increased by 5 mg, it is likely that the response will vary. If, however, a dose of 100 mg is increased by 5 mg, the response will hardly change. In both cases the increase is 5 mg, but in the first case we have 50%, in the second case only 5% of the original dose. It is this *rate of increase* which is relevant. Therefore, the law of Weber-Fechner is applicable. Although the response is usually of a qualitative nature and cannot be measured, it is assumed that *the response is linearly dependent on the logarithm of the dose*.

As a consequence, when animals are tested for their response to doses of different levels, *the doses should form a geometric or exponential sequence*. Let d_0 be the lowest dose and let q be a convenient factor greater than one. Then the chemical is administered in doses

$$d_0, \quad d_0 q, \quad d_0 q^2, \quad d_0 q^3, \dots. \tag{6.6.9}$$

A sequence such as 10 mg, 20 mg, 30 mg, 40 mg, etc. is not suitable for doses. Beginning with the same terms it should be 10 mg, 20 mg, 40 mg, 80 mg, etc. whereby $d_0 = 10$ mg and $q = 2$.

Example 6.6.5. We conclude this section with an application of logarithms to chemistry. The hydronium concentration, denoted by $[H_3O^+]$ is an important factor in living tissue as well as in the soil where plants grow. A high concentration ranging from about 10^{-2} to almost 10^{-7} mol/l is present in *acid* solutions. For distilled water $[H_3O^+] = 10^{-7}$ mol/l approximately, a concentration which is called *neutral*. Lower concentration down to about 10^{-12} mol/l indicate an *alkaline* solution or a *base*. Because of this wide range, a logarithmic scale for judging the hydronium concentration appears to be more practical. We may use the exponents of the above powers and get rid of the minus sign. Thus the quantity

$$pH = -\log[H_3O^+] \tag{6.6.10}$$

is introduced. For a neutral solution $pH = 7$, for an acid solution $pH < 7$ and for an alkaline solution $pH > 7$. For the human brain it was found that the cerebrospinal fluid has $[H_3O^+] = 4.8 \times 10^{-8}$ mol/l on the average. Hence $pH = -\log(4.8 \times 10^{-8}) = -(\log 4.8 + \log 10^{-8}) = -(0.68 - 8) = 7.32$ (this value is from Diem *et al.*, 1970, p. 635).

*6.7. Spirals

Molluscan shells, ammonites, the arrangement of seeds in sunflowers and of scales in pine cones all indicate that spirals occur frequently in nature. We restrict ourselves to those spirals that are plane curves and are generated as follows: We assume that $y = f(x)$ is a monotone increasing function (see Section 6.3). Instead of a graph in a rectangular coordinate system we choose a *polar graph*. For this purpose we interpret x as the polar angle and y as the polar distance and put $x = \alpha$ and $y = r$ (see Section 5.8). Then the graph of $r = f(\alpha)$ is called a *spiral*.

Perhaps the simplest example of a spiral is given by a *linear function*

$$r = a\alpha + b \quad (a > 0, \quad b \geq 0) \tag{6.7.1}$$

with domain $\alpha \geq 0$ and range $r \geq b$. As α grows, r increases linearly. For each full angle completed by α, the polar distance r has increased by $2\pi a$ provided that α is measured in radians. Hence successive windings or whorls of the spiral have a constant width from each other. Such a curve

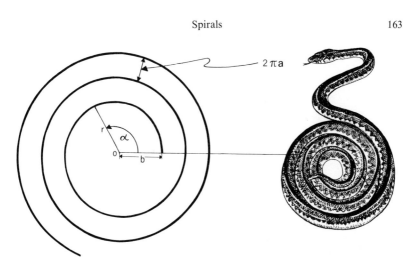

Fig.6.10. A spiral of Archimedes. The polar distance r grows linearly as a function of the polar angle α

is called a *spiral of Archimedes*[7] (see Fig. 6.10). An intuitive idea gives a coiled rope or a coiled snake. We may also think of a revolving gramophone disk on which the needle approaches the center with constant speed.

Another famous spiral is the *logarithmic spiral*. It is the polar graph of an *exponential function*

$$r = aq^\alpha \quad (a > 0, \quad q > 1) \tag{6.7.2}$$

with domain $\alpha \in \mathbb{R}$ and range $r > 0$. With increasing α, the polar distance grows faster than for a spiral of Archimedes (see Fig. 6.11). The name "logarithmic spiral" stems from the fact that $\log r$ is a linear function of α. Indeed, from (6.7.2) it follows that $\log r = \log a + \alpha \log q$.

From a rectangular graph we know that $y = aq^x$ with $q > 1$ tends asymptotically to zero as x takes on smaller and smaller negative values (cf. Fig. 6.3). It follows that the logarithmic spiral winds infinitely many times around the origin O without ever reaching O when α decreases over negative values.

Another property of the logarithmic spiral can be derived from a triangle $OA_1 A_2$ depicted in Fig. 6.11. Let α_1 and α_2 be two polar angles with a fixed difference $\Delta\alpha = \alpha_2 - \alpha_1$, and $r_1 = OA_1$, $r_2 = OA_2$ the corresponding polar distances. Then we conclude from Eq. (6.7.2) that

$$r_2/r_1 = q^{\alpha_2}/q^{\alpha_1} = q^{\alpha_2 - \alpha_1} = q^{\Delta\alpha}. \tag{6.7.3}$$

---

[7] Archimedes of Syracuse (Sicily), mathematician and physicist, died 212 B. C.

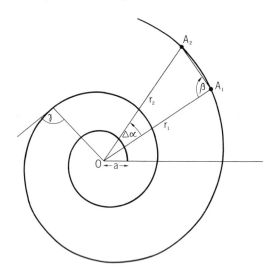

Fig. 6.11. A logarithmic spiral. The polar distance r is an exponential function of the polar angle α

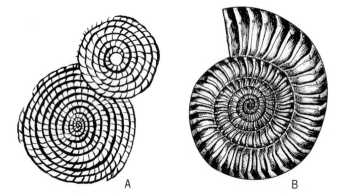

Fig. 6.12. Nummulites (a) and ammonites (b) are examples of two different kinds of spirals (nummulites redrawn from Schaub, 1966)

The ratio r_2/r_1 is determined by the choice of $\Delta\alpha$. Hence, no matter how big α_1 and α_2 are, the triangle $OA_1 A_2$ has always the same shape and, therefore, the angle $\beta = \sphericalangle OA_1 A_2$ is constant. With this property in mind, many points of a logarithmic spiral can be quickly constructed (see problems 6.7.1 and 6.7.2).

When $\Delta\alpha$ tends to zero, the line joining A_1 and A_2 tends to the tangent of the spiral and β tends to a limiting value which is denoted by γ in Fig. 6.11. Hence *the logarithmic spiral cuts the radii at a constant angle.*

As there are infinitely many monotone functions, there are also infinitely many spirals of different shapes. For curve fitting, however, the spiral of Archimedes and the logarithmic spiral are usually sufficient. In Fig. 6.12 two fossils are shown. Nummulites (A) resemble the spiral of Archimedes and ammonites (B) the logarithmic spiral.

For more details or other biological applications of spirals see Bourret, Lincoln, and Carpenter (1969), Dormer (1972), Fraenkel and Gunn (1961, Fig. 48), Kleerekoper (1972), Steinhaus (1960, p. 139–145), Thompson (1917, p. 493 ff., or 1961, p. 172 ff.), Winfree (1972).

Recommended tables:

For common logarithms: four digits, Diem (1970), five digits, Meredith (1967), six digits, Allen (1947).

For common antilogarithms: four digits, Diem (1970), ten digits, National Bureau of Standards (1953).

Recommended for further reading:

Lefort (1967), C. A. B. Smith (1966).

Problems for Solution

6.1.1. Among the following sequences identify those that are arithmetic and/or geometric:

a) 7, 11, 15, 19, ...	f) 0.3, 0.03, 0.003. ...
b) 6.5, 5, 3.5, 2, ...	g) 18, 20, 21, 23, 25, 26, ...
c) 2, 6, 18, 54, 162, ...	h) 1, 4, 9, 16, 25, ...
d) 4, 4, 4, 4, ...	i) 1/2, 1/3, 1/4, 1/5, ...
e) $(-1/2)$, $(-3/2)$, $(-5/2)$, ...	j) 4, 2, 1, 0.5, 0.25, ...

6.1.2. Add three more terms to the following geometric sequences:

a) 2, 10, 50,	d) 1, $(-1/2)$, 1/4,
b) 1, 1/3, 1/9,	e) (-6), 2, $(-2/3)$,
c) 220, 22, 2.2,	f) 75, 15, 3.

6.1.3. A culture medium is infected with N_0 bacteria. The bacteria cells divide every two hours. How many bacteria will be in the medium 24 h later? At what time had the number of bacteria reached 25% of the previous total?

6.1.4. Assume that each female rabbit of a colony gives birth to three female rabbits. How many female rabbits of the tenth generation will be descendants of a single rabbit of the first generation?

6.1.5. In a tracer method the potassium isotope ^{42}K is used for labeling. The half-life of ^{42}K is 12.5 h. If N_0 is the original number of atoms, how many are expected to remain after the elapse of two days and two hours? How many hours will it take until only $(1/1024) N_0$ atoms remain?

6.1.6. A female moth *(Tinea pellionella)* lays nearly 150 eggs. In one year there may live up to five generations. Each larva eats about 20 mg of wool. Assume that 2/3 of the eggs die and that 50% of the remaining moths are females. Estimate the amount of wool that may be destroyed by the descendants of one female within a year. (The first female belongs to the first generation.)

6.2.1. Plot the graph of the exponential function $y = a \cdot 2^x$ for the parameter values $a = 2, 0.5, -0.5$.

6.2.2. Plot the exponential function $y = q^x$ for the parameter values $q = 2, 1, 0.5$.

6.2.3. Find the increment $\Delta y = f(x+1) - f(x)$ for the exponential function $y = f(x) = aq^x$ and show that this increment is also an exponential function of x.

6.2.4. Modify the previous problem by defining $\Delta y = f(x+h) - f(x)$ with a given number h.

6.3.1. Draw graphs of the following functions and decide which of them are monotone functions and which are not:
a) $y = 3 - x$ with domain $x \in \mathbb{R}$,
b) $y = 4 - x^2$ with domain $-2 \leq x \leq +2$,
c) $y = \frac{1}{4} x^2 + x + 1$ with domain $x > -2$,
d) $y = \cos x$ with domain $0 \leq x \leq \pi$,
e) $y = \cos x$ with domain $-\pi/2 \leq x \leq \pi/2$.

6.3.2. Show that for the following functions inverse functions exist. Find the explicit expressions for these inverse functions.
a) $y = -2x + 3$ with domain $x \in \mathbb{R}$,
b) $y = x^2 + 2$ with domain $x \geq 0$,
c) $y = x^2 + 2$ with domain $x \leq 0$,
d) $y = 1/x^2$ with domain $x > 0$,
e) $y = 1 + (1/x)$ with domain $x > 0$.

6.4.1. Find the inverse functions of the following exponential functions:

a) $y = 2^x$,   c) $r = \frac{1}{2} \cdot 5^t$,

b) $y = a \cdot 10^x$ $(a > 0)$,   d) $Q = 2w^x$ $(w > 0)$.

What are the widest possible domains and ranges for these functions and their inverses?

6.4.2. Solve the same problem for

a) $w = (0.5)^s$,   c) $y = aq^{3x}$ $(a > 0, \; q > 0)$,

b) $M = a(0.1)^n$ $(a > 0)$,   d) $z = r^{2s-3}$ $(r > 0)$.

6.5.5. The world's population in 1970 is estimated to be 3.7×10^9 persons. The yearly growth rate is approximately 2%. Under the assumption that the current growth rate remains constant, how large would the world's population be in the years 1980, 1990, and 2000?

6.5.6. The population of India amounted to 550 million in 1972. The annual growth rate is 2.4%. Under the assumption that the growth rate remains constant, find a) a formula to represent the population size, b) the year when India's population size will reach 860 million, that is, the presumed population size of China in 1972.

6.5.7. Nevada has the fastest growing population of all states of the USA. The population increased from 291000 in 1960 to 480000 in 1970. Assuming exponential growth, what is a) the annual percent increase, b) the doubling time?

6.5.8. In Western Europe 25% of the required energy was supplied by derivatives of mineral oil in 1950. This proportion increased to 70% on 1970. Assuming exponential growth, what is the factor q of annual increase?

6.5.9. The outcome of a certain experiment with mice is expected to be age dependent. A first group of mice is three weeks old, a second group five weeks old. What are the ages of two more groups, if a geometric sequence of ages is required?

6.5.10. Same situation as in the previous problem. This time only three groups of experimental mice are planned. The youngest group is three weeks old, the oldest 10 weeks old. What age is desirable for the middle group?

6.5.11. To test the content of vitamin A in carrots, pieces of this vegetable are fed to vitamin A deficient rats. The dose levels are arranged in a geometric sequence. If 20 g and 50 g are the first two doses of the sequence, how does the sequence continue?

6.5.12. Same situation as in the previous problem. Assume that six different dose levels are planned. The lowest dose is 20 g, the highest 200 g. Find the sequence of doses.

6.5.13. On his 13th birthday a child was 112 cm tall, on his 14th birthday 121 cm tall. Assume a geometric monthly growth rate. What is this rate?

*6.6.1. Two tones of 1000 Hz frequency are assumed to have loudness 30 dB and 45 dB. What follows for the ratio of their physical intensities I_1 and I_2?

*6.6.2. By the chromatic scale the interval of an octave is subdivided into 12 equal intervals. This means that 13 tones follow each other such that their frequencies form a geometric sequence and that the last frequency is twice as high as the first frequency. What is the ratio of frequencies for two consecutive tones?

*6.6.3. Find pH if   a) $[H_3O^+] = 3.7 \times 10^{-5}$,
        b) $[H_3O^+] = 8.1 \times 10^{-8}$,   c) $[H_3O^+] = 0.27 \times 10^{-7}$ mol/l.

*6.6.4. For human blood the pH falls between 7.37 and 7.44. Find the corresponding bounds for $[H_3O^+]$.

*6.6.5. The intensity of sound decreases inversely to the square of the distance from the source ($I = c/r^2$). How does loudness, as defined by Eq. (6.6.7), depend on distance?

*6.6.6. Luminance L is a measure for the brightness of visible light. It is expressed in units "candela per square meter" ($= \text{cd m}^{-2}$). Luminance in daily life ranges mainly from 1 to 100 cd m^{-2}. To test Weber's law (Section 6.6), ΔL was measured versus L:

L (cd m^{-2})	ΔL (cd m^{-2})
0.616	0.013
1.23	0.023
2.47	0.042
6.16	0.097
12.3	0.21
24.7	0.39
61.6	1.13
123	2.5

Calculate $r = \Delta L/L$ for each measurement and determine the arithmetic mean $\bar{r}$ and the sum of squares of deviations $\Sigma (r - \bar{r})^2$.

*6.7.1. Let O be the origin of a polar coordinate system. From the polar axis (denoted by h_0) we rotate a half-line in steps of a fixed acute angle α. Thus we obtain the half-lines h_1, h_2, h_3, ... On h_0 we choose a point A_0 (different from O) and put $OA_0 = a$. Then at A_0 we draw a line perpendicular to h_0 and denote its intersection with h_1 by A_1. From A_1 we proceed with a line perpendicular to h_1 and denote its intersection with h_2 by A_2, etc. Thus we get the broken line $A_0 A_1 A_2 A_3 \ldots$ (Fig. 6.13). Find OA_1, OA_2, etc. in terms of a and α and prove that the points $A_0, A_1, A_2, \ldots$ lie on a logarithmic spiral.

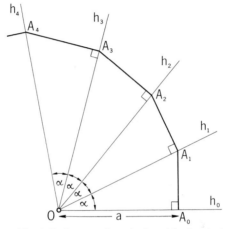

Fig. 6.13. Construction of a logarithmic spiral

*6.7.2. Generalize the previous problem, replacing the right angles with an angle $\beta \neq 90°$. Show that OA_{n+1}/OA_n is constant for $n = 0, 1, 2, \ldots$ and that, therefore, $A_0, A_1, A_2, \ldots$ lie on a logarithmic spiral.

*6.7.3. A moth attracted by a light flies toward the light keeping a constant angle of $120°$ to the light rays. Thus the insect spirals inward instead of maintaining a straight course (menotaxis). Sketch such a spiral. What sort of spiral is it?

Chapter 7

Graphical Methods

7.1. Nonlinear Scales

Our eye is very quick in recognizing pattern and shape. Information contained in mathematical relationships can often be presented in graphical form for better understanding.

Unless high accuracy is required, graphical representation, such as diagrams, histograms and nomograms may replace lengthy tables of functions. Sometimes even tedious numerical calculations can be avoided by employing graphical tools.

We are already familiar with rectangular and polar graphs. They are advantageous in that they clearly show intervals where a function is increasing or decreasing, where maxima and minima are located, etc. But there are also disadvantages. Sometimes appropriate space for the range of a function is lacking. Or the relationship of more than two variables should be presented in the plane. In such cases *functional* or *nonlinear scales* help considerably.

To introduce the idea we consider a monotone function, for example the exponential function $y = 10^x$ with a graph plotted in Fig. 7.1. Now we want to economize the space required for representation: Starting with a particular value of the y axis we move parallel to the x axis. Then from the intersection with the graph we move parallel to the y axis until we reach the x axis. There we mark the particular y value just opposite the corresponding x value. Repeating this procedure we get a *double scale*, the y scale adjacent to the x scale. We may compare the result with a "river" and its two "banks". On one bank we have the x, on the other bank the y scale. The x scale is *linear*, that is, equally spaced[1]. The y scale, however, is in general *nonlinear*. In our example the y scale is called a *logarithmic scale* since $x = \log y$.

A double scale fills only a one-dimensional space. Yet it contains the same information as a graph plotted in a two-dimensional space. Therefore, a double scale is space saving and economic.

---

[1] A rigorous definition of a linear scale is based on the following condition: The distance between successive points of the scale is equal for equal increments of the variable.

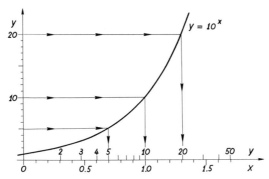

Fig. 7.1. Transition from a graph of a function to a nonlinear scale. Construction of the logarithmic scale

For preparing a double scale it is essential to assume that there is a one-to-one correspondence between the x and the y values. This property is guaranteed by assuming that $y = f(x)$ is monotone (cf. Section 6.3).

A double scale can be more quickly constructed when the function and its inverse are tabulated. In our example of $y = 10^x$, the inverse function is the common logarithm $x = \log y$. A brief table follows:

y	$x = \log y$	y	$x = \log y$	y	$x = \log y$
1	0	6	0.778	20	1.301
2	0.301	7	0.845	30	1.477
3	0.477	8	0.903	40	1.602
4	0.602	9	0.954	50	1.699
5	0.699	10	1.000	etc.	

On one side of a line we plot a linear scale for x, on the other side we insert the y values at their proper positions (see Fig. 7.2). The double scale is the same as in Fig. 7.1.

Logarithmic scales are very practical for representing quantities with a large range such as paleontological ages, micro- and macroscopic dimensions, or frequencies of electromagnetic waves. A highly efficient

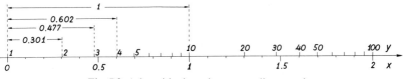

Fig. 7.2. A logarithmic scale versus a linear scale

aid for multiplication and division based on the logarithmic scale is the *slide rule*.

To construct a double scale for $y = f(x)$, the following steps are required:

1. If necessary, reduce the domain of $y = f(x)$ so that the mapping is *monotone*.

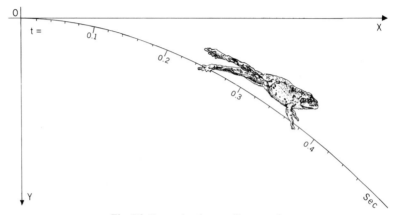

Fig. 7.3. Example of a curvilinear scale

2. Decide whether x or y should be plotted on a *linear scale*.

3. If x is chosen to be plotted on a linear scale, tabulate y in *equal steps* and find the corresponding x-values. Otherwise interchange x and y.

4. On one "bank" of a straight line draw the linear x-scale. On the other "bank" plot the corresponding y-values (or x and y interchanged).

Step 3 is essential. If the wrong variable is tabulated in equal steps, the method fails.

Linear and nonlinear scales are sometimes placed on curves, especially in nomograms. An example may serve to explain how *curvilinear scales* are constructed. Let x be a quantity that increases linearly as a function of time, and y be a simultaneous quadratic function of time. For a leaping animal, x may be interpreted as the horizontal, y as the vertical component of the trajectory (Fig. 7.3). We assume that

$$x = at, \quad y = ct^2 \tag{7.1.1}$$

with certain constants a and c. Then we choose a sequence of t values, say $t = 0.1, 0.2, 0.3, \ldots$ sec, and calculate the corresponding values of x and y. In a rectangular coordinate system we obtain a sequence of points labeled with the t values (Fig. 7.3). The points lie on a smooth curve and form a curvilinear scale.

Should the t values be of no interest, then t could be eliminated from (7.1.1). We get $t = x/a$ and $y = c(x/a)^2 = (c/a^2)\,x^2$. Hence, y is a quadratic function of x and the curve a quadratic parabola. The original equations (7.1.1) are called the *parametric equations of the curve* and the variable time t the *parameter*. In this connection the word "parameter" signifies an independent variable which determines two or more other variables[2].

7.2. Semilogarithmic Plot

When the x axis or the y axis, but not both axes, bears a logarithmic scale, the coordinate system is called *semilogarithmic*.

Example 7.2.1. We consider a dose-response relationship reported by Copp, Cockcroft, and Kueh (1967). The objective was to prove that the hormone calcitonin is produced in the ultimobranchial glands (and not in the thyroid glands) in *Squalus suckleyi* (small shark, dogfish) and in *Gallus domestica* (chicken). Extracts from the ultimobranchial glands were administered to rats in several dose levels. The response was measured in terms of an area between the plasma calcium curves of treated and untreated rats. Fig. 7.4 exhibits the result. *The dose levels*

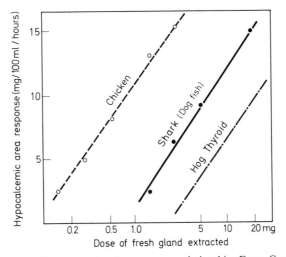

Fig. 7.4. Semilogarithmic plot of a dose-response relationship. From Copp, Cockcroft, and Kueh (1967)

---

[2] In mathematics the word "parameter" is used in different ways. For another meaning of this word see Section 4.1.

are plotted on a logarithmic scale because the logarithm of a dose rather than the dose itself is biologically relevant (cf. Example 6.6.4). The response can be approximated by a linear function of the logarithmic dose.

Example 7.2.2. Now we illustrate the use of a logarithmic scale in a case where one variable covers a *wide range*. A linear presentation would be virtually impossible. The example is taken from Strehler (1963,

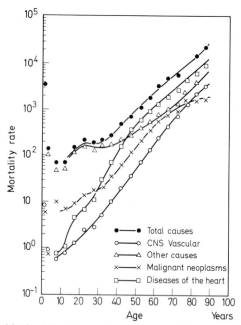

Fig. 7.5. Semilogarithmic plot of the age dependent mortality rate for 100,000 persons in the United States around 1960. From Strehler (1963, p. 114)

p. 114). The mortality rate measured in incidence per 100,000 of population varies between about 10^0 and 10^4 as a function of the age (Fig. 7.5).

Occasionally the functional values range from 0 to a high value. If the value 0 is actually taken, an ordinary logarithmic scale could not be applied since $\log 0$ does not exist[3]. In this case we recommend adding the number 1 to all values of y and then taking the logarithm, that is, plotting the quantity

$$Y = \log(y + 1).\tag{7.2.1}$$

_____

[3] As y tends to zero over positive numbers, $x = \log y$ tends to $-\infty$ as can be seen from Fig. 7.1.

This transformation was proposed by Bartlett (1947). An example is offered in problem 7.2.7 at the end of this chapter.

There is still another reason for using a semilogarithmic coordinate system. When an exponential function

$$y = aq^x \quad (a > 0, \quad q > 0) \tag{7.2.2}$$

with domain $x \in \mathbb{R}$ and range $y > 0$ has to be plotted, it is advantageous to apply the *logarithmic transformation*. First we obtain from (7.2.2)

$$\log y = \log a + x \log q. \tag{7.2.3}$$

Second, we introduce *new variables and constants*. It is customary to denote them with capital letters:

$$Y = \log y, \quad A = \log q, \quad B = \log a. \tag{7.2.4}$$

Then (7.2.3) turns into the *linear function*

$$Y = Ax + B. \tag{7.2.5}$$

Of course, it is much simpler to draw a graph of (7.2.5) than a graph of the original function (7.2.2). Moreover, curve fitting to empirical data and the discussion of deviations are greatly facilitated.

Example 7.2.3. To illustrate such a transformation we consider a method of determining the volume of blood plasma in a human or an animal body. A known quantity of thiosulfate is injected into the blood stream. If the thiosulfate mixed homogeneously with the blood plasma without any loss to other parts of the body, it would be easy to calculate the blood volume from the concentration of thiosulfate in the plasma. Instead, the substance is continuously excreted by the kidney, whereas mixing of thiosulfate with plasma is hardly completed before the elapse of 10 minutes. To cope with this problem several measurements of thiosulfate concentration are made, and a graphical method is applied to extrapolate backward. Following Randall (1958 or 1962, p. 67) we consider an experiment: Into an animal, 0.5 g (grams) of thiosulfate was injected. Ten minutes after the injection and at successive 10-minute intervals the following plasma concentrations of thiosulfate were obtained:

$$44 \quad 38 \quad 33 \quad 28 \quad 25 \text{ mg/100 ml}.$$

As usual the excretion by the kidney leads to an exponential decrease of concentration. Therefore, we plot the data in a semilogarithmic coordinate system (Fig. 7.6). The dots lie on a straight line with considerable accuracy in agreement with formula (7.2.5). We are interested in the concentration as it was before the kidney began to excrete thiosulfate.

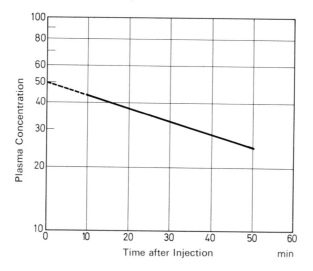

Fig. 7.6. Semilogarithmic plot of plasma concentration after injection of 0.5 g of thiosulfate. Adapted from Randall (1958, p. 68)

Hence, we extrapolate the data back to obtain plasma concentration at
the time of injection. Thus we get 50 mg/100 ml. Let V be the total
plasma volume of the animal. Then

$$0.5 \text{ g} : V = 50 \text{ mg} : 100 \text{ ml}$$

and

$$V = \frac{0.5 \text{ g} \times 100 \text{ ml}}{50 \text{ mg}} = 1000 \text{ ml} = 1.0 \text{ l}.$$

7.3. Double-Logarithmic Plot

When both axes of a coordinate system bear logarithmic scales, it is
called *double-logarithmic*. The term log-log plot is also used. Let

$$X = \log x, \quad Y = \log y. \tag{7.3.1}$$

Then Fig. 7.7 illustrates three ways of presentation.

In Fig. 7.7a both axes bear adjacent scales. The X and Y scales, both
linear, are first drawn. Then the logarithmic scales for x and y are added
as explained in Section 7.1. The values of x and y as well as their logarithms X and Y are shown. Because of the many details this presentation
is scarcely used in print.

Fig. 7.7b exhibits only the logarithmic scales. When the graduation
is emphasized by lines drawn parallel to the axes, we speak of double-

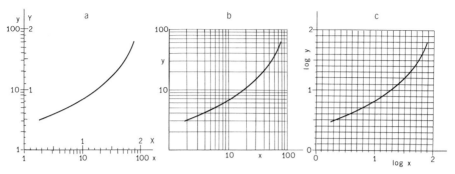

Fig. 7.7. Three presentations of a double-logarithmic coordinate system

logarithmic paper. Dropping the linear scales is rather popular and frequently seen in print, but it is not always practical for research work. Neither plotting of measurements nor reading of coordinates is easy.

Finally, Fig. 7.7c is less spectacular but perhaps more economic than the other methods. Ordinary graph paper suffices. By a table of logarithms, values for $\log x$ and $\log y$ are determined and then plotted on linear scales.

Double-logarithmic plots are employed for a variety of reasons: wide range of x and y, dose-response relationships, etc. A major application is the graphical representation of a *power function*:

$$y = ax^n \quad (a > 0, \quad n \in \mathbb{R}) \tag{7.3.2}$$

with domain $x > 0$ and range $y > 0$. We apply the *logarithmic transformation*. First, (7.3.2) yields

$$\log y = \log a + n \log x.$$

Second, new variables and constants are introduced by

$$Y = \log y, \quad X = \log x, \quad B = \log a. \tag{7.3.3}$$

Thus (7.3.1) is transformed into

$$Y = nX + B, \tag{7.3.4}$$

that is, into a linear relation. Hence, in a double-logarithmic plot the exponent n of a power function appears as the slope of a straight line. Examples with various values of n, including fractions and negative numbers, are offered in Problem 7.3.4 at the end of this chapter.

Example 7.3.1. As a first biological application we study the relationship between the length and weight of *Heterodon nasicus* (western hognose

snake). Assume that all snakes of this species, whether young or old, have
the same shape and the same specific gravity. Then, following Section 4.2,
their weight W should be proportional to the cube of their length L,
that is,

$$W \propto L^3 \quad \text{or} \quad W = aL^3 \tag{7.3.5}$$

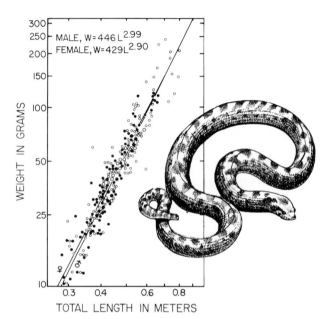

Fig. 7.8. Regression lines of weight versus length fitted to a sample of 158 male and 167
female western hognose snakes (*Heterodon nasicus*) from Harvey County, Kansas. Males
are represented by filled and females by open circles. Small circles represent a single record
and larger symbols represent from two to four records. Both the abscissa and ordinate
are in logarithmic scale. From D. R. Platt (1969)

with a certain constant a. Applying logarithms we obtain

$$\log W = \log a + 3 \log L . \tag{7.3.6}$$

This is a linear relationship between $\log L$ and $\log W$. Hence, a double-
logarithmic plot is a straight line with slope 3. Actual measurements
show that the dots representing individual snakes deviate slightly from a
straight line (see Fig. 7.8)[4].

---

[4] D. R. Platt (1969) applied a statistical method to fit a straight line to his data. The
result is an exponent 2.99 for males and 2.90 for females. Both exponents are very close
to the theoretically predicted value 3.

Example 7.3.2. We plot the mean daily heat production of warm blooded animals against the average body weight (Fig. 7.9). If logarithmic scales for both quantities are used, the dots are very close to a straight line. This indicates that the relationship can be well approximated by a power function. Other instructive examples are presented in Jerison (1970) and in Schmidt-Nielsen (1972). For the use of double-logarithmic plots in allometric growth see Section 11.6.

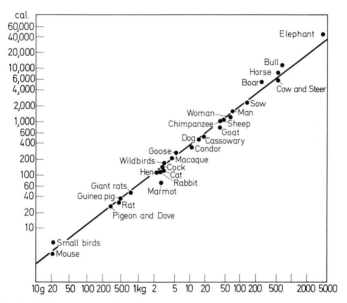

Fig. 7.9. The daily heat production of warm blooded animals plotted against the body weight on logarithmic scales (after Benedict, 1938, from Davson, 1964, p. 221)

Summarizing the main results of Sections 7.2 and 7.3 we formulate a rule which is worthwhile to be learned by heart:

For an *exponential function* we apply *semilogarithmic*, but for a *power function* a *double-logarithmic* plot. In each case the graph of the function is a *straight line*.

7.4. Triangular Charts

We begin this section with a geometric proposition: *Let P be a point in an equilateral triangle. Then the sum of the distances of P from the three sides is equal to the height of the triangle.* In other words: The sum of the distances remains constant when we move the point P from any position to any other position inside the triangle.

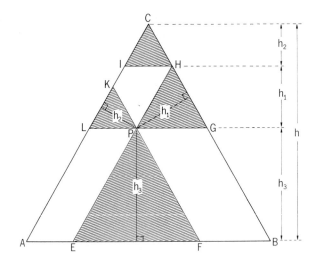

Fig. 7.10. The sum of three distances h_1, h_2, h_3 of a point P from the sides of an equilateral triangle is equal to h, the height of the triangle

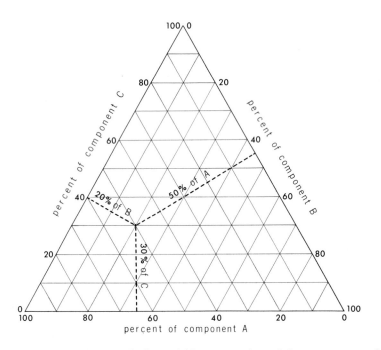

Fig. 7.11. Trilinear chart for plotting variable concentrations of three components of a substance

For a proof we look at Fig. 7.10. Let A, B, C be the vertices of the triangle such that $AB = BC = CA$. We denote the perpendicular distances of P from the sides by h_1, h_2, h_3 and the height of the triangle ABC by h. Through P we draw three lines: $LG \parallel AB$, $FK \parallel BC$, and $HE \parallel CA$. Also we draw $HI \parallel AB$. These lines together with the sides of the triangle form four equilateral triangles PEF, PGH, PKL and CIH, since all angles are equal to 60°. Now, h_1 is height of triangle PGH, h_2 height of triangle PKL, and h_3 height of triangle PEF. Since $HE \parallel CA$, triangle CIH has also height h_2. Arranging the three heights such that they are all perpendicular to AB we get immediately

$$h_1 + h_2 + h_3 = h, \quad \text{Q.E.D.} \tag{7.4.1}$$

Therefore, whenever three variables, say x, y, z, add up to a constant value, a *trilinear chart* is very convenient. The constant is plotted as height of an equilateral triangle. The sides are interpreted as *axes* and the distances x, y, z from the sides as *triangular coordinates*.

Fig. 7.11 shows how such a chart is constructed. The three variables may be the concentration of three components A, B, C of a substance. They add up to 100%.

Example 7.4.1. A biological application is shown in Fig. 7.12. In the tentacles of *Hydra attenuata Pall.* (a fresh-water polyp) there are three kinds of nematocysts (thread cells): the relatively large stenoteles, the

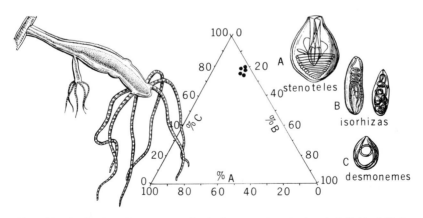

Fig. 7.12. Distribution of nematocysts in the five tentacles of a female individual *Hydra attenuata Pall.*

isorhizas, and the desmonemes, both somewhat smaller. All three kinds were counted in the five tentacles of a female individual[5]. The results are:

Tentacle No.	Stenoteles freq.	%	Isorhizas freq.	%	Desmonemes freq.	%
1	124	5.6	334	15.0	1768	79.4
2	124	7.2	298	17.4	1297	75.4
3	123	6.7	302	16.5	1411	76.8
4	114	5.8	310	15.6	1556	78.6
5	148	8.4	319	18.1	1293	73.5

As we have three components, it is quite convenient to plot the percentages in a trilinear chart (Fig. 7.12)[6].

Example 7.4.2. Interesting applications of trilinear charts are also known in genetics. Let A and a be two genes at the same locus. Individuals are then either homozygotes AA, aa, or heterozygotes Aa. Let p_{AA}, p_{aa}, p_{Aa} be the percentages of the three genotypes in a population. Then, by definition

$$p_{AA} + p_{aa} + p_{Aa} = 100\% . \tag{7.4.2}$$

Each population is represented by a dot in a trilinear chart whose height is 100% and whose coordinates are p_{AA}, p_{aa}, p_{Aa}. Under the assumption

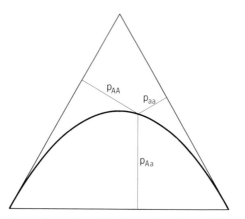

Fig. 7.13. The percentages of genotypes AA, Aa, aa plotted on a trilinear chart. Panmictic populations are represented by points falling on a parabola

---

[5] I am indebted for the data to Mr. Adrian Zumstein, University of Zurich, Switzerland.
[6] The drawing of a hydra is made after Engelhardt (1962) and the drawing of tentacles after Brohmer (1964).

of panmixia (random mating) the Hardy-Weinberg law states that

$$p_{Aa}^2 = 4p_{AA}p_{aa} . \tag{7.4.3}$$

It can be shown that populations satisfying (7.4.3) are represented by points located on a parabola (Fig. 7.13). For details and for advanced use of trilinear charts see Cannings and Edwards (1968), de Finetti (1926), Jacquard (1974), Li (1958), Levene, Pavlovsky, and Dobzhansky (1954), Schaffer and Mettler (1970), Turner (1970), Wright (1969, p. 39, 133, 139).

For other applications of trilinear charts in the life sciences see Goldin *et al.* (1968), King *et al.* (1972), Ostrander *et al.* (1969), and Taylor *et al.* (1971).

*7.5. Nomography

There are techniques for *representing a functional relationship among three variables in a plane*, thus avoiding the three-dimensional space. For example, let w be the geometric mean of two quantities u and v, that is,

$$w = \sqrt{uv} \quad (u > 0, \quad v > 0). \tag{7.5.1}$$

Squaring this equation and taking logarithms we obtain

$$2 \log w = \log u + \log v . \tag{7.5.2}$$

With new variables

$$U = \log u, \quad V = \log v, \quad W = \log w, \tag{7.5.3}$$

the equation becomes

$$2W = U + V. \tag{7.5.4}$$

W is the arithmetic mean of U and V.

We may employ a rectangular coordinate system with a U and a V axis. Then for each fixed value of W, (7.5.4) is the equation of a straight line. For different values of W we get different straight lines. They are mutually parallel (Fig. 7.14a). When we are given two values U and V, we find the point (U, V) falling on a particular straight line characterized by the corresponding value of W.

Using (7.5.3) we may replace the linear scales with logarithmic scales, thus obtaining a graphical representation of the geometric mean (Fig. 7.14b).

Charts of the type shown in Fig. 7.14 are called *Cartesian charts* or *concurrency nomograms*. The lines $W = $ const may also be interpreted as contour lines on a map. Therefore, concurrency nomograms are sometimes called *topographic charts*.

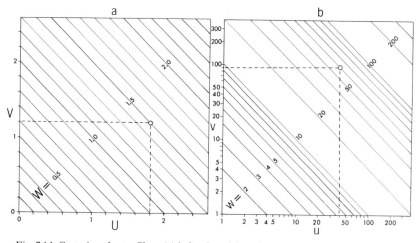

Fig. 7.14. Cartesian charts. Chart (a) is for the arithmetic mean $W = (U + V)/2$. Numerical example: $U = 1.8$, $V = 1.2$, $W = 1.5$. Chart (b) is for the geometric mean $w = \sqrt{uv}$. Numerical example: $u = 40$, $v = 90$, $w = 60$

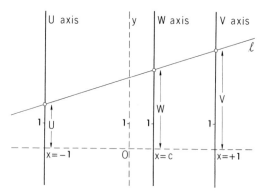

Fig. 7.15. Construction of an alignment nomogram with three parallel lines

Under favorable conditions concurrency nomograms or topographic charts can be replaced with so-called *alignment nomograms*. They are usually more convenient. In the simplest form such a nomogram consists of three parallel straight lines which bear scales for three variables u, v, w.

One way of constructing alignment nomograms is based on the use of an auxiliary x, y coordinate system. The dashed lines in Fig. 7.15 are the x and y axes. The units on the two axes need not be the same. Through the points $(-1, 0)$ and $(+1, 0)$ we draw lines parallel to the y axis. They serve as linear U and V axes. We also draw a third axis parallel to the

y axis through a point $(c, 0)$ where c is an arbitrary but fixed number $(c \neq 1, c \neq -1)$. We call it the W axis. All three axes are chosen such that they are linear with points $U = 0$, $V = 0$, $W = 0$ on the x axis and that the unit is the same as on the y axis.

Now, let l be a straight line that intersects the U, V, W axes. The points of intersection are

$$(-1, U) \quad \text{on the } U \text{ axis},$$

$$(+1, V) \quad \text{on the } V \text{ axis},$$

$$(c, W) \quad \text{on the } W \text{ axis}.$$

The slope a of l can be calculated in different ways by the formula $a = \Delta y / \Delta x$. For the points of intersection of l with the V and W axes we get $a = (V - W)/(1 - c)$, and for the points of intersection with the U and V axes $a = (V - U)/2$. Equating the results we obtain

$$\frac{V - W}{1 - c} = \frac{V - U}{2} \tag{7.5.5}$$

or, after cross multiplication and rearrangement of terms,

$$U(1 - c) + V(1 + c) = 2 W. \tag{7.5.6}$$

The equation establishes a linear relationship among the variables U, V, and W. Every linear relationship can be adapted to formula (7.5.6). This permits construction of an alignment nomogram.

As an example, we return to formula (7.5.4) which defines W as the arithmetic mean of U and V. In order to adjust this formula to (7.5.6) we simply let $c = 0$. Hence, the W axis coincides with the y axis. The nomogram is shown in Fig. 7.16a. For simplicity the x and y axes are not shown.

By replacing the linear scales with logarithmic scales following (7.5.3), we obtain an alignment nomogram for the geometric mean (Fig. 7.16b).

One of the three axes may be *curvilinear*. In order to construct such a nomogram we slightly generalize the previous concept. In Fig. 7.17 we assume that the curve bearing the values of w is represented in parametric form

$$x = f(w)$$
$$y = g(w) \tag{7.5.7}$$

with parameter w as explained in Section 7.1. In the same way as we obtained (7.5.5) and (7.5.6), we now get

$$\frac{V - y}{1 - x} = \frac{V - U}{2} \tag{7.5.8}$$

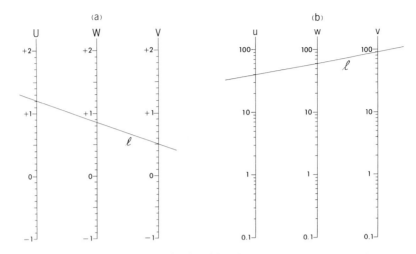

Fig. 7.16. Alignment nomograms (a) for the arithmetic mean $W = (U + V)/2$ and (b) for the geometric mean $w = \sqrt{uv}$. When two values u and v are given, the corresponding points on the u and v axes are connected by a straight line l. At the intersection of l with the w axis we find the geometric mean $w = \sqrt{uv}$. Numerical example: $u = 40$, $v = 90$, $w = 60$

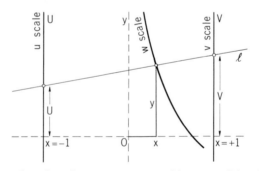

Fig. 7.17. Construction of an alignment nomogram with two parallel and one curvilinear axes. U and V are scaled in the same manner as y, whereas u and v are represented, in general, by non-linear scales

and

$$U(1 - x) + V(1 + x) = 2y. \qquad (7.5.9)$$

Replacing x and y by the functions (7.5.7) and U and V by certain monotone functions

$$U = h(u), \qquad V = k(v), \qquad (7.5.10)$$

we derive from (7.5.9) the formula

$$h(u)\,(1 - f(w)) + k(v)\,(1 + f(w)) = 2g(w). \qquad (7.5.11)$$

Hence, *if we can write a functional relationship among three variables* *u, v, w in the form* (7.5.11), *there exists an alignment nomogram of the type shown in Fig. 7.17.*

Example 7.5.1. Consider the *quadratic equation* in w

$$w^2 + uw + v = 0 . \tag{7.5.12}$$

When the coefficients u and v are given, we want to determine graphically the positive roots of this equation. We may rewrite (7.5.12) in the form

$$uwF + vF = - w^2 F \tag{7.5.13}$$

with an *arbitrary factor F*. In order to adapt (7.5.13) to (7.5.11) we let

$$h(u) = u , \quad k(v) = v ,$$

$$1 - f(w) = wF , \quad 1 + f(w) = F , \quad 2g(w) = - w^2 F .$$

Eliminating F we get

$$f(w) = \frac{1 - w}{1 + w} , \quad g(w) = - \frac{w^2}{1 + w} . \tag{7.5.14}$$

These functions of w specify the parametric Eq. (7.5.7). For various positive values of w we calculate x and y. Plotting the points (x, y) and

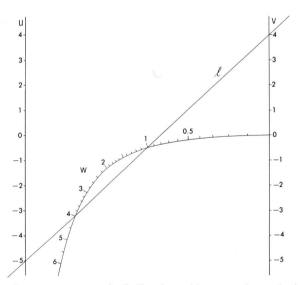

Fig. 7.18. An alignment nomogram for finding the positive roots of a quadratic equation $w^2 + uw + v = 0$. Numerical example: $w^2 - 5w + 4 = 0$. The line connecting the points $u = - 5$, $v = 4$ intersects the w scale at $w_1 = 1$ and $w_2 = 4$. Thus the equation has two positive roots

labeling these points with the corresponding values of w we obtain the curvilinear scale depicted in Fig. 7.18. For simplicity, the auxiliary x and y axes are not shown.

For more advanced alignment nomograms, especially in the area of life sciences, see Batschelet *et al.* (1952), Boag (1949), Comroe (1965), Consolazio *et al.* (1963), Diem *et al.* (1970, p. 528–569), Hamilton (1947), Hennig (1967), Levens (1959, p. 252, 254, 260), Peters (1969, p. 36), Reinke *et al.* (1969), Ricci (1967, p. 260), Schuler *et al.* (1969), Severinghaus *et al.* (1956), Singh *et al.* (1971), C.A.B.Smith (1966, p. 170), Sunderman *et al.* (1949, p. 96, 97), Swerdloff *et al.* (1967), Thews (1971), Waterman *et al.* (1965, p. 10–12).

*7.6. Pictorial Views

In the previous section we have shown how functional relations among three variables can be represented in a plane. This procedure of reducing three to two dimensions is primarily chosen for economic reasons. However, there are other considerations, such as clarity and quick intuitive understanding, which favor *perspective or pictorial views* of surfaces and curves. In addition, three-dimensional objects as studied in morphology also call for a representation that is as close as possible to our everyday experience. A special case of perspective views occurs

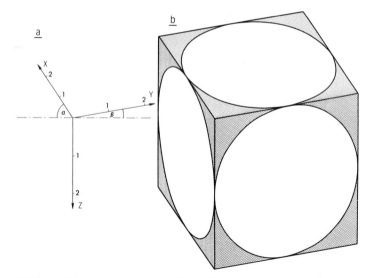

Fig. 7.19. Parallel-perspective presentation. Three mutually perpendicular axes are depicted by three lines in a plane. Lines parallel in space remain parallel in the drawing. The unit of length may be shortened in some direction. Circles are depicted by ellipses

when a landscape is depicted. Then the drawing is usually called a *block diagram*.

There are a variety of ways to obtain good and precise pictorial views. We begin with the *parallel-perspective presentation*. We introduce three mutually perpendicular axes and denote them by x, y and z. One of them, say the z axis, is depicted by a vertical line. For representing the x and y axes we choose lines that form acute angles α and β with a horizontal line (Fig. 7.19 a). The unit of length may be plotted in its natural size or may be suitably shortened. In Fig. 7.19 a we have assumed no shortening on the z and the y axes but some reduction on the x axis. The ratio of scales is $\frac{2}{3} : 1 : 1$. Lines that are parallel in space *remain parallel* in the

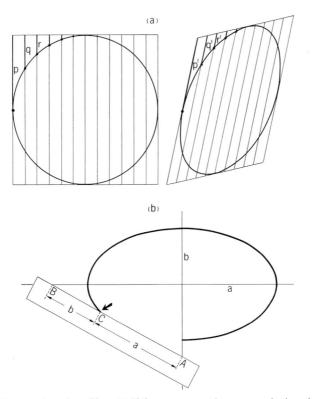

Fig. 7.20. Construction of an ellipse. (a) If the axes are not known, we obtain points of the ellipse by reducing segments $p, q, r, \ldots$ located on lines parallel to the sides of the original square. The new segments $p', q', r', \ldots$ are then proportional to $p, q, r, \ldots$. (b) If the axes are known, we prepare a strip of paper and plot on a straight edge two adjacent segments a and b in the size of the half-axes. Then the strip is moved in such a way that one endpoint, say A, remains on a vertical axis and the other endpoint, say B, remains on a horizontal axis. Then the point C which separates a and b moves along the ellipse

parallel-perspective illustration. With these rules in mind a cube is depicted in Fig. 7.19 b. Each point in space can be exactly placed in the drawing as long as we know its rectangular coordinates x, y, z. We have simply to proceed from an origin by x steps in the x direction, by y steps in the y direction, and by z steps in the z direction, always observing the proper unit of length.

Circles are represented by *ellipses*. Fig. 7.19 b shows the pictorial view of three circles that are located on the surface of a cube. There should be no concern about constructing an ellipse. Fig. 7.20 explains the practical procedure.

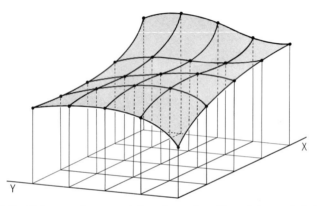

Fig. 7.21. Parallel-perspective presentation of a surface. The points are plotted vertically above a grid pattern. The points are connected by smooth curves

For plotting a surface a *grid pattern* is quite useful. In the x, y plane we draw equidistant lines parallel to the axes, that is, we prepare a pictorial view of ordinary graph paper (Fig. 7.21). Then from each mesh point we plot the height of the surface. Finally we connect the points thus obtained by smooth curves in the direction of the x and/or the y axes. This procedure gives a precise and often a good intuitive view of a curved surface.

For simplicity, two special cases of parallel-perspective presentation are often chosen:

a) The *isometric drawing*. Here, $\alpha = \beta = 30°$. No shortening of the unit of length is required (ratio of scales $1:1:1$). An example is shown in Fig. 7.22.

b) *The oblique view*. Here α is zero and β is chosen between 30° and 45°. The units along the x and z axes are not shortened, whereas the unit on the y axis is reduced. The ratio of scales is $1:c:1$ with c between 0.5

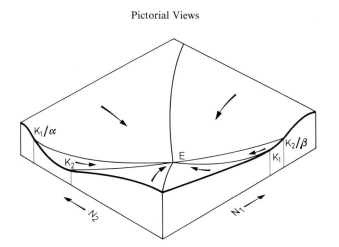

Fig. 7.22. Isometric drawing of the interaction between two competing species. N_1 and N_2 are the number of individuals of species No. 1 and 2. K_1, K_2 denote saturation values and α, β competition coefficients. The model was proposed by L. B. Slobodkin. The figure is taken from Slobodkin (1961, p. 66)

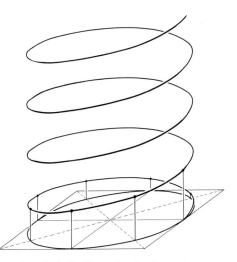

Fig. 7.23. Oblique view of a helix. The ratio of scales on the x, y, z axes is $1 : \frac{2}{3} : 1$

and 0.8. Fig. 7.23 shows an oblique view of a *helix*. The projection of this three-dimensional curve by rays parallel to the z axis into the x, y plane is a circle with radius r. The parametric equations of this circle are

$$x = r \cos \varphi$$

$$y = r \sin \varphi$$

$$(7.6.1)$$

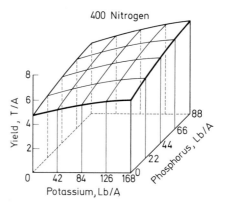

Fig. 7.24. Oblique view of a response surface which was calculated from data by statistical methods. Three fertilizers were applied: a constant amount of nitrogen (400 pounds per acre) and variable amounts of potassium and phosphorus. The yield is measured in tons per acre. From Platt and Griffiths (1964, p. 49) after Welch, Adams, and Carmon (1963)

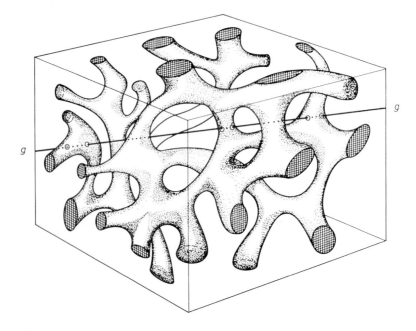

Fig. 7.25. Two-point perspective view of a complicated surface. The size of this surface may be estimated by the number of intersections with a straight line g. The figure is taken from Fischmeister (1967, p. 224)

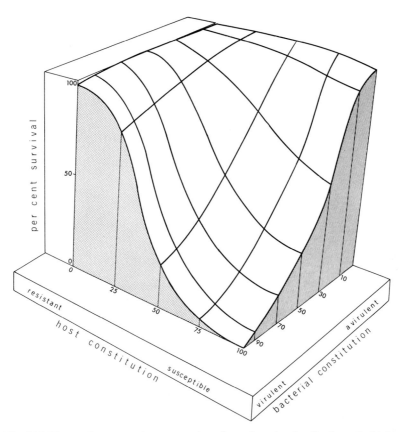

Fig. 7.26. Three-point perspective presentation of a surface showing the theoretical inter-
relation of genotypes of host and pathogen in controlling the severity of a disease.
Redrawn from Gowen (1952, p. 287)

with variable parameter φ (for the notion of parametric equations see
Section 7.1). The angle φ may take on any real value and may be measured either in degrees or in radians. As φ increases, the coordinate z
increases proportionally. Hence, another parametric equation is

$$z = k\varphi \tag{7.6.2}$$

with a certain constant $k > 0$. When φ increases by $360°$ or 2π radians,
the helix climbs by $k \cdot 360°$ or $k \cdot 2\pi$, respectively. (7.6.1) together with
(7.6.2) form the parametric equations of the helix. For an isometric
drawing of a helix see Fig. 8.9.

Another example of an oblique view is shown in Fig. 7.24.

The parallel-perspective presentation is not always satisfactory. As seen with our eye, parallel lines converge into one point (cf. the two rails of a straight railroad track). Hence, particularly if lengthy segments on parallel lines have to be depicted, a so-called *angular-perspective presentation* is preferable. We begin with a special case: The x and z axes as well as lines parallel to these axes are drawn as in an oblique view, but the lines parallel to the y axis converge into one point (which is not

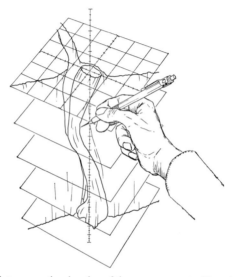

Fig. 7.27. Three-point perspective drawing of tissue components. The relation of successive sections is shown. For accurate drawing, a grid pattern is required in each section. The figure is from Mitchell and Thaemert (1965, p. 1480)

necessarily part of the illustration). Such a presentation is called *one-point perspective*. An example is offered in Fig. 9.27.

When convergence is chosen in two directions, the presentation is called *two-point* perspective. Fig. 7.25 is an application.

An ordinary camera usually generates a *three-point perspective* view of lines that are parallel to three different directions. It may be recommended for a more sophisticated way of presentation. Fig. 7.26 and 7.27 are examples.

In an angular-perspective view circles are still represented by ellipses so that the construction shown in Fig. 7.20 is applicable. However, shortening can no longer be performed at a constant ratio. In general, angular-perspective drawings are more time consuming than parallel-

perspective drawings. A labor saving method is the use of a printed
perspective chart that can be purchased. Fig. 7.28 depicts the principle.

Computers with attached plotters are able to draw the more complicated pictorial views. They are even able to rotate a three-dimensional
pattern and to replot it in various perspective views (see Rosenfeld, 1969).
Programming, however, is time consuming and the process expensive.
If hand drawing is feasible, it should be used since it is fast and inexpensive.

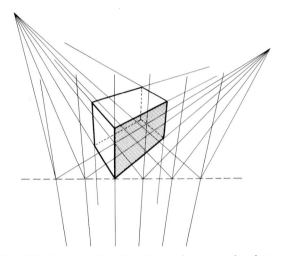

Fig. 7.28. Converging lines in a three-point perspective chart

For more pictorial views of biological objects and of relations see
Chiarappa *et al.* (1972), Dingle (1972), Elias (1971), Elias and Hennig
(1967), Patrick *et al.* (1973), Weibel and Elias (1967). For block diagrams
see Jenks and Brown (1966).

Recommended for further reading. D. S. Davis (1962), Karsten (1925),
Levens (1959, 1962 and 1965), Schmid (1954), C. A. B. Smith (1966),
Worthing and Geffner (1959).

Problems for Solution

7.1.1. Plot a double scale (or adjacent scales) for the function $y = x^{1/2}$
with scale values $x = 0$, 0.5, 1.0, ..., 4.0 and $y = 0$, 0.2, 0.4, ..., 2.0.
Plot x on a linear scale.

7.1.2. Same problem, but y should be plotted on a linear scale. Which of the two double scales seems to be more practical?

7.1.3. Plot a double scale (or adjacent scales) for $y = \sin\alpha$ with scale values $\alpha = 0°, 10°, ..., 90°$ and $y = 0, 0.1, 0.2, ..., 1.0$ whereby the y scale is linear.

7.1.4. Solve the same problem for $y = \tan\alpha$, but restrict the angle to the interval from $0°$ to $45°$.

7.1.5. Draw a curvilinear scale by means of the parametric equations $x = 2\cos\alpha$, $y = \sin\alpha$ for the scale values $\alpha = 0°, 15°, 30°, ..., 360°$.

7.1.6. Given the parametric equations $x = (\theta/90°)\cos\theta$, $y = (\theta/90°)\sin\theta$. Draw a curvilinear scale for the values $\theta = 0°, 15°, 30°, ..., 360°$.

7.2.1. The lead content of human blood, C_B measured in μg/100 ml, increases with the mean lead content of ambient air, C_L measured in μg/m³, according to formula

$$C_B = 60 \log C_L - 20$$

for the domain $\{C_L | 5 < C_L < 100\}$. Plot this function by means of a straight line.

7.2.2. Part of a sample of Mycoplasmatales virus, known as MVL 1, can be deactivated by ultraviolet irradiation (wave length 257 nanometers). The percentage p of virus not yet deactivated follows the empirical law

$$\log p = 2 - (8.50 \times 10^{-4})\, d$$

where d is the dose of ultraviolet light measured in 10^{-7} J mm^{-2}. Draw the graph by means of a semilogarithmic coordinate system. The domain is $\{d | 0 < d < 1600\}$. (Adapted from Liss et al., 1971.)

7.2.3. To detect LSD (d-lysergic acid diethylamide) in human urine a method was developed by which traces of LSD inhibit a certain chemical binding. Between the amount of LSD in pg (pico-gram $= 10^{-12}$ g) and the percent inhibition the following empirical relation was established:

LSD (pg)	Inhibition (%)
20	14
30	29
40	39
70	65
100	80

Plot the percent inhibition versus the logarithm of LSD-quantity and fit a straight line by eye (semilogarithmic plot). (Adapted from Taunton-Rigby *et al.*, 1973.)

7.2.4. The biogenic amine "serotonin" is associated with emotional stability in man. To measure traces of serotonin a method was established which is based on the inhibition of a certain chemical process. The following data is available:

Serotonin (ng)	Inhibition (%)
1.2	19
3.6	36
12	60
33	84

$(ng = Nanogram = 10^{-9} g)$

Prepare a semilogarithmic plot with log (serotonin) on the horizontal axis. Fit a straight line by eye and estimate the serotonin quantity that results in 50% inhibition. (Adapted from Peskar *et al.*, 1973.)

7.2.5. Prepare a semilogarithmic plot of the following exponential functions:

a) $y = 2^x$,　　　　b) $y = 5 \cdot 2^x$,　　　c) $y = (1.25)^x$,

d) $y = (0.45)(1.25)^x$,　　e) $y = (\frac{1}{2})^x$,　　f) $y = (0.33)^x$.

7.2.6. Prepare a semilogarithmic plot for the exponential growth function

$$w = w(n) = a \left(1 + \frac{p}{100}\right)^n$$

which we considered in Section 1.3. As domain for the independent variable n choose the interval from 0 to 50. For simplicity, let $a = 1$, and treat the cases $p = 1\%$, 2%, 3%.

7.2.7. We are given the following empirical function:

$x =$	1.5	2.5	3.5	4.5	5.5	6.5	7.5 (mg)
$y =$	0	0.47	2.8	27.4	83	18.3	12.4 (min) .

Because of the large range of y, a semilogarithmic plot is required. Use the modified transformation $Y = \log(y + 1)$, since y can take on zero.

7.2.8. In world running records, let x be the distance in meters and t be the record time in seconds. From the World Almanac the following table can be extracted:

x	t
90	7.8
200	20.0
400	44.8
800	105.1
1 600	231.1
3 200	499.8
4 800	772.4
9 600	1607.0
16 000	2832.8
24 000	4368.2

Let $v = x/t$ be the average speed in m/sec. Plot $1/v$ versus $\log x$ in a rectangular coordinate system and fit the best straight line by inspection. From the diagram derive an empirical law for v as a function of x. (Adapted from Good, 1971.)

7.2.9. The enzymatic activity of catalase is lost during exposure to sunlight in the presence of oxygen. Let y be the concentration of catalase (in µg/10 ml) as a function of the time t (in minutes). Then we have approximately

$$y = aq^t \quad (0 \leq t \leq 80)$$

with certain parameters a and q. Estimate a and q graphically using the following data (from Mitchell and Anderson, 1965):

t (minutes)	0	10	30	50	60	70	80
y (µg/10 ml)	121	74	30	12	6.7	3.7	2.0

(Hint: Use a semilogarithmic plot).

7.3.1. To investigate mutation caused by low doses of neutron radiation, stamen hair of a herb from genus *Tradescantia* was used. Normally the flowers are blue. Pink cells indicate the presence of mutant chromosomes. The absorbed dose d is measured in Joule/kg. Let m be the average mutation frequency per stamen hair. Then $\log m$ is a linear function of $\log d$ over the domain $\{d | 10^{-4} < d < 10^{-1}\}$.

Find the formula for $\log m$ when the following data are given:

d (J/kg)	m
0.001	0.0030
0.05	0.16

(Adapted from Sparrow *et al.*, 1972.)

7.3.2. For several species of animals the energy cost of running was measured. Let E be the energy cost to transport 1 g of body weight over 1 km distance (in $J\,g^{-1}\,km^{-1}$) and let M be the body mass. Some empirical data are:

Animal	M (g)	$E(J\,g^{-1}\,km^{-1})$
White mouse	21	54
Ground squirrel	236	15
White rat	384	18
Dog (small)	2.6×10^3	7.1
Dog (big)	1.8×10^4	3.9
Sheep	3.9×10^4	2.4
Horse	5.8×10^5	0.63

Plot $\log E$ versus $\log M$ and fit a straight line by inspection. Estimate the slope of this line. (Adapted from Schmidt-Nielsen, 1972a.)

7.3.3. The larvae of the gypsy moth (*Porthetria dispar* L.) are serious defoliators of forest and orchard trees. For mating, males are lured by females using a sex attractant pheromone. This pheromone is cis-7,8-epoxy-2-methyloctadecane and may be used in an attempt to control the population. Males were caught in certain traps. The following relationship was found in a field experiment:

x = amount pheromone μg	N = number of males trapped
0.1	3
1	7
10	11
100	20

$\log N$ increases almost linearly with $\log x$. Check this property by a double-logarithmic plot. (Adapted from Beroza *et al.*, 1972.)

7.3.4. For $x > 0$ represent the following power functions by straight
lines using a double-logarithmic plot:

a) $y = x^2$ b) $y = \frac{1}{2} x$ c) $y = 4/x$

d) $y = x^{-2}$ e) $y = 6.5 \, x^2$ f) $y = 2 \, x^{\frac{1}{2}}$

g) $y = x^{3/2}$ h) $y = x^{-\frac{1}{4}}$.

7.3.5. We are given three empirical functions:

a) x	y	b) x	y	c) x	y
1.5	1.778	1.5	5.217	1.5	10.87
2.5	2.611	2.5	7.328	2.5	5.047
4.0	4.642	4.0	10.03	4.0	2.495
5.0	6.813	5.0	11.64	5.0	1.782
6.5	12.11	6.5	13.87	6.5	1.201

Plot all three functions first in a semilogarithmic coordinate system and second in a double-logarithmic coordinate system. By
inspection decide which of the three functions is best fitted by
a power function or by an exponential function.

7.4.1. The plasma of human blood contains on the average 92% water,
6% proteins, and 2% electrolyte (the percentages refer to the
weight). Represent this data by a dot in a trilinear chart. How
does the dot change its position when the first component (water)
is successively reduced to 70% and to 40%?

7.4.2. To classify deep-sea fossils of Foraminifera (protozoans with
shells), not only their morphology but also the content of amino
acids in the shells may be used. The following amino acids were
considered: Alanine (Ala), Aspartic acid (Asp), Glutamic acid
(Glu), Glycine (Gly), Proline (Pro), Serine (Ser), Threonine (Thr),
Valine (Val). Three factors turned out to be efficient for dis-
criminating among species:

Factor 1: percentage of Ala, Pro, Val,
Factor 2: percentage of Asp, Thr,
Factor 3: percentage of Glu, Gly, Ser.

Some results are listed as follows:

Species	Factor			Total
	1	2	3	
Globoquadrina dutertrei (recent)	27%	35%	38%	100%
Globoquadrina altispira (extinct)	39%	32%	29%	100%
Globoquadrina dehiscens (extinct)	41%	27%	32%	100%

Plot the three species as dots in a triangular chart. (Adapted
from King *et al.*, 1972.)

7.4.3. Let A and a be two different alleles which may or may not be present at a certain locus. Let p and q be the relative frequencies of A and a, resp., in a population ($0 \le p \le 1$, $0 \le q \le 1$, $p + q = 1$). Under the assumption of panmixia (random mating), the three genotypes AA, Aa, aa occur with relative frequencies p^2, $2pq$, q^2, resp., according to the Hardy-Weinberg law. Plot these three frequencies in a trilinear chart for the following numerical values:

a) $p = 0.2$, $q = 0.8$,   b) $p = 0.4$, $q = 0.6$,
c) $p = 0.6$, $q = 0.4$,   d) $p = 0.8$, $q = 0.2$,
e) $p = 1$, $q = 0$.

7.4.4. Find an alternative proof for formula (7.4.1). Instead of drawing parallels to the three sides, draw the lines AP, BP, CP. Express the areas of triangles BPC, APC, APB in terms of h_1, h_2, h_3 and sum up these areas.

*7.5.1. Draw a concurrency nomogram for the volume of a right circular cylinder, $V = \pi r^2 h$ with $\pi = 3.14$. Use logarithms to get straight lines for constant values of V.

*7.5.2. Find an alignment nomogram with three parallel scales
a) for the weight of a cube, $W = \varrho a^3$ where ϱ denotes the specific gravity,
b) for the volume of a right circular cylinder, $V = \pi r^2 h$ with $\pi = 3.14$.
(Hint: In both cases take logarithms and adapt the result to formula (7.5.6)).

*7.5.3. Find an alignment nomogram with three parallel scales for the exponential growth
$$y = (1 + p/100)^x = q^x.$$
The exponent x takes values from 1 to 50 and the base q takes values from 1.01 to 1.10 which correspond to the interval of p from 1% to 10%. This nomogram is very useful in studies of population growth. (Hint: Take logarithms twice. Let $U = \log x - 1.5$, $V = \log(\log q) + 1.5$, and $W = \frac{1}{2} \log(\log y)$. Apply formula (7.5.6)).

*7.6.1. A rectangular solid has length 8 cm, width 5 cm and height 3 cm. Prepare a) an isometric drawing, b) an oblique view, c) a one-point perspective view, d) a three-point perspective view.

*7.6.2. Find an isometric drawing of a helix.

*7.6.3. Prepare a block diagram of the exponential function $z = q^x$ with $q = 1, 2, 3, 4$ and x ranging between 0 and 2. The z axis should not be shortened, while shortening the q and x axes is optional.

Chapter 8

Limits

8.1. Limits of Sequences

The purpose of this chapter is to make differential and integral calculus understandable. With the study of limits we will obtain a most powerful tool for defining such concepts as (instantaneous) growth rate, rate of decay, reaction rate, diffusion rate and their counterparts, total amount of growth, of decay, etc.

Let n be any natural number, that is, let $n \in \{1, 2, 3, \ldots\}$. We consider any function of n and denote such a function either by $f(n)$ or by a_n. The set of functional values $f(1), f(2), f(3), \ldots$ or $a_1, a_2, a_3, \ldots$ arranged in their natural order is called a *sequence* (cf. Section 6.1). Some examples may illustrate the notion:

$$a_n = 3n \qquad \text{generates the sequence} \quad 3, 6, 9, 12, 15, \ldots \qquad (8.1.1)$$

$$a_n = \frac{3}{n} \qquad \text{generates the sequence} \quad 3, \frac{3}{2}, 1, \frac{3}{4}, \frac{3}{5}, \frac{3}{6}, \frac{3}{7}, \ldots \qquad (8.1.2)$$

$$a_n = \frac{n-1}{n+1} \quad \text{generates the sequence} \quad 0, \frac{1}{3}, \frac{2}{4}, \frac{3}{5}, \frac{4}{6}, \frac{5}{7}, \ldots \qquad (8.1.3)$$

The usual notation for such a sequence is $\{a_n\}$.

In the theory of limits we are interested in the behavior of a sequence when n takes larger and larger values or, in other words, when n tends to infinity. With increasing n the sequence (8.1.1) also tends to infinity. On the other hand, the sequences (8.1.2) and (8.1.3) behave quite differently. In both cases we have

$$0 \leq a_n \leq 3 .$$

We say that the sequences remain *bounded*. A *lower bound* for a_n is 0 and an *upper bound* is 3.

In the history of mathematics the term "*infinite*" was obscure for a long period. On one hand, there were people claiming that infinity could be treated as a number and, consequently, could be involved in algebraic operations. On the other hand, working with infinity has led to several contradictions. Bolzano[1] was able to rationalize the previously obscure notion of infinity. His concept is based on easy relations. Nevertheless, it takes some time to become familiar with his definition:

A sequence $\{a_n\}$ is said to tend or to diverge to infinity, if for any given $C > 0$ it is possible to find a natural number n_C such that

$$a_n > C \tag{8.1.4}$$

holds whenever $n > n_C$. In symbols[2]

$$a_n \to \infty . \tag{8.1.5}$$

In modern mathematics, ∞ is not a number, and no algebraic operation is defined for this symbol. In connection with sequences, "infinite" simply means that a_n, from a certain n on, exceeds any bound C, no matter how big C is chosen.

The reader unfamiliar with this way of thinking may study the following example: Let the term be $a_n = n/100$. We choose $C = 10^5$. Then $a_n > C$ for every $n > 10^7$. Hence, for the particular value of C there exists $n_C = 10^7$ such that $a_n > C$ whenever $n > n_C$. Had we chosen $C = 10^{10}$ or any higher value, we could find another n_C, namely $n_C = 10^{12}$ or correspondingly higher. Therefore, the sequence $\{a_n\}$ tends to infinity.

Similarly, if a sequence such as $-1, -4, -9, -16, \ldots$ with the term $a_n = -n^2$ becomes smaller than an arbitrarily low bound $-C$, we say that $\{a_n\}$ tends to *minus infinity* and write

$$a_n \to -\infty . \tag{8.1.6}$$

Now we return to sequence (8.1.2). As $n \to \infty$, we see that $a_n = 3/n$ becomes smaller and smaller without ever reaching the value zero. Nevertheless, $3/n$ approaches zero so that we are inclined to say that $3/n$ tends to zero. The precise mathematical meaning of this statement may be formulated after Bolzano:

A sequence $\{a_n\}$ is said to tend or to converge to zero, if for any given $\varepsilon > 0$ it is possible to find a natural number n_ε such that

$$|a_n| < \varepsilon \tag{8.1.7}$$

holds whenever $n > n_\varepsilon$. In symbols[3]

$$a_n \to 0 \quad \text{or} \quad \lim_{n \to \infty} a_n = 0 . \tag{8.1.8}$$

---

[1] Bernhard Bolzano (1781–1848). Bohemian theologian, logician, and mathematician.

[2] The symbol ∞ for infinity was proposed by the English mathematician and theologian John Wallis (1616–1703), according to Cajori (1929, p. 44).

[3] The Greek letter ε (epsilon) has become a standard notation for an arbitrarily small positive number. The symbol $\lim_{n \to \infty}$ is read "limit as n tends to infinity". For the printer it is easier to write $\lim_{n \to \infty}$.

Example 8.1.1. Let $a_n = 3/n$. If we choose $\varepsilon = 10^{-5}$, then, for every $n > 3 \cdot 10^5$, we get $a_n < \varepsilon$. Hence, for the particular choice of ε we have $n_\varepsilon = 3 \cdot 10^5$. Had we chosen $\varepsilon = 10^{-10}$ or smaller, we would get another n_ε, namely $n_\varepsilon = 3 \cdot 10^{10}$ or correspondingly higher. Therefore, $\{a_n\}$ tends to zero. The behavior of this sequence is depicted in Fig. 8.1.

Fig. 8.1. The term $a_n = 3/n$ tends to zero. The points representing a_n accumulate in the "neighborhood" of the point zero without ever reaching this point

Example 8.1.2. The two vertical bars in Eq. (8.1.7) signify the absolute value of a_n (see Section 1.6). This operation is required for a sequence which approaches zero from the negative side or even from both sides. Let $a_n = (-1)^n (1/n)$, that is,

$$a_1 = -1, \; a_2 = +\tfrac{1}{2}, \; a_3 = -\tfrac{1}{3}, \; a_4 = +\tfrac{1}{4}, \dots.$$

Then a_n approaches the limiting value 0 from both sides. For every $\varepsilon > 0$, there exists a certain n_ε so that

$$-\varepsilon < a_n < \varepsilon$$

for all $n > n_\varepsilon$. With the presentation in Fig. 8.2 we may say that a_n falls into an ε-strip whenever $n > n_\varepsilon$.

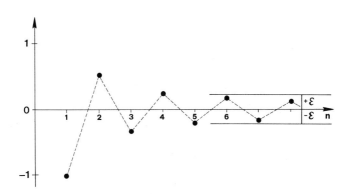

Fig. 8.2. Graph of a sequence a_n approaching a limit. From a certain n on, the dots remain in the ε-strip

Example 8.1.3. Consider the sequence

$$\frac{1}{3}, \frac{1}{4}, \frac{1}{5}, \frac{1}{3}, \frac{1}{6}, \frac{1}{7}, \frac{1}{3}, \frac{1}{8}, \frac{1}{9}, \frac{1}{3}, \frac{1}{10}, \ldots$$

Most terms tend to zero, but every third term is $\frac{1}{3}$. If we choose $\varepsilon = \frac{1}{4}$ there will be terms not satisfying Eq. (8.1.7) no matter how large n is. Therefore, the sequence does not converge to zero. As we will see, it has no limit whatsoever.

Example 8.1.4. Let $a_n = (n-1)/(n+1)$. When n becomes large, say $n = 1000$, we have $a_n = 999/1001$. We guess that $\{a_n\}$ tends to 1, that is

$$\frac{n-1}{n+1} \to 1 \quad \text{or} \quad \lim_{n \to \infty} \frac{n-1}{n+1} = 1.$$

In order to verify the statement we introduce a modified sequence $\{b_n\}$ where

$$b_n = a_n - 1 = \frac{n-1}{n+1} - 1 = -\frac{2}{n+1}.$$

As n tends to infinity, $\{b_n\}$ tends to zero, since $|b_n| < \varepsilon$ for any arbitrarily small positive number ε whenever n is large enough. It follows that a_n deviates from 1 by an arbitrarily small number for large values of n.

In general, the case of a finite limit can be reduced to the special case where the limit is zero. We say that *a sequence $\{a_n\}$ tends to a finite limit A if the modified sequence $\{b_n\}$ with term*

$$b_n = a_n - A \tag{8.1.9}$$

converges to zero.

Two convergent sequences may fulfill a peculiar relationship. For example, let $a_n = 2 - \dfrac{5}{n}$ and $c_n = 2 - \dfrac{1}{n}$. Here

$$a_n < c_n$$

for every natural number n. But

$$\lim_{n \to \infty} a_n = \lim_{n \to \infty} c_n = 2.$$

Hence, two sequences can converge to the same limit, although the members of one sequence are smaller than the corresponding members of the other sequence. Therefore, from $a_n < c_n$ we cannot conclude that $\lim_{n \to \infty} a_n < \lim_{n \to \infty} c_n$. The proper conclusion is

$$\lim_{n \to \infty} a_n \leq \lim_{n \to \infty} c_n \tag{8.1.10}$$

which allows for the possibility of an equal sign.

In the rest of this section we will learn how to determine limits of some sequences. Clearly, sequences such as

$$\{a_n = n^2\}, \{a_n = n^3\}, \ldots$$

tend to infinity. Conversely,

$$\left\{a_n = \frac{1}{n^2}\right\}, \left\{a_n = \frac{1}{n^3}\right\}, \ldots$$

tend to zero. When we are given a more complicated sequence $\{a_n\}$, for example with

$$a_n = \frac{100 - 5n + 3n^2}{8 + 10n + 2n^2},$$

we may find a limit by reducing the problem to simpler terms. For this purpose we divide the numerator and the denominator by the highest occuring power, in our example by n^2. We obtain

$$a_n = \frac{\dfrac{100}{n^2} - \dfrac{5}{n} + 3}{\dfrac{8}{n^2} + \dfrac{10}{n} + 2}.$$

Here, $100/n^2$, $-5/n$, $8/n^2$, and $10/n$ tend to zero. Hence, the numerator tends to 3, the denominator to 2, and the whole fraction to 3/2. Therefore,

$$\lim_{n \to \infty} a_n = 3/2.$$

Similarly we get

$$\lim_{n \to \infty} \frac{n+1}{n-2} = \lim_{n \to \infty} \frac{1 + 1/n}{1 - 2/n} = 1,$$

$$\lim_{n \to \infty} \frac{n^3 - 1000}{5n^3 + 20n^2} = \lim_{n \to \infty} \frac{1 - 1000/n^3}{5 + 20/n} = 1/5,$$

$$\lim_{n \to \infty} \frac{10 + n}{2 + n^2} = \lim_{n \to \infty} \frac{10/n^2 + 1/n}{2/n^2 + 1} = 0.$$

A sequence such as

$$1, \frac{1}{1/2}, \frac{1}{1/3}, \frac{1}{1/4}, \frac{1}{1/5}, \ldots$$

creates a particular problem. The denominator tends to zero. One is tempted to think that the sequence tends to 1/0. However, 1/0 has no mathematical meaning as we pointed out at the end of Section 1.6. We get a correct result if we rewrite the general term $a_n = \dfrac{1}{1/n}$ in the simple

form $a_n = n$. Clearly, $\{a_n\}$ diverges to infinity. Other examples of divergence are

$$\frac{1}{(1/n)-(1/n)^2} = \frac{n}{1-(1/n)} \to \infty,$$

$$\frac{2/n^2+1}{10/n^2+1/n} = \frac{2/n+n}{10/n+1} \to \infty,$$

$$\frac{n^2-100}{n+100} = \frac{n-100/n}{1+100/n} \to \infty.$$

In Section 6.1 we introduced the geometric or exponential sequence with the term

$$a_n = aq^n. \tag{8.1.11}$$

We may now study its limiting behavior. This depends only on the factor q^n. Several cases have to be distinguished. For $q=1$ we get $q^n=1$ for all values of n and hence, $q^n \to 1$. For $q>1$ the sequence q^n is monotone increasing. Because of $\log q > 0$ we see that $\log q^n = n \cdot \log q$ tends to infinity. Therefore, $\{q^n\}$ tends also to infinity. For $0 < q < 1$, however, the sequence $\{q^n\}$ is monotone decreasing. The reciprocal $1/q = q_1$, say, is greater than one. Thus $1/q^n = q_1^n$ tends to infinity. Therefore, $\{q^n\}$ tends to zero.

For negative values of the base q the sign of the sequence $\{q^n\}$ alternates. For $-1 < q < 0$ the sequence tends to zero from the positive as well as from the negative side. For $q \leq -1$, however, the sequence does not tend to a single finite or infinite limit. We simply say that $\{q^n\}$ *diverges*.

The results may be summarized as follows: As n tends to infinity

$$\begin{array}{lll} q^n \to \infty & \text{for} & q>1, \\ q^n \to 1 & \text{for} & q=1, \\ q^n \to 0 & \text{for} & |q|<1, \\ q^n \ \text{diverges for} & q \leq -1. \end{array} \tag{8.1.12}$$

Example 8.1.5. A biological application may illustrate the case $q = -\frac{1}{2}$. In man, and in fruitflies, there are two kinds of *sex chromosomes*, the X and the Y chromosome. An XX individual is female and an XY individual is male. Let A and a be two alleles at a certain locus of the X chromosome for which there is no corresponding locus of the Y

chromosome. Therefore, males can only be of genotype $A \cdot$ or $a \cdot$, while females may be one of three genotypes $AA, Aa,$ and aa. The theory shows that under the assumption of panmixia (random mating) the frequencies of A and a oscillate from generation to generation. The oscillation is damped so that the frequencies quickly converge to certain limits (see Li, 1958, p. 59–68). For the ease of presentation we skip the proof and confine ourselves to a numerical example. Let q_1 be the initial frequency of the male genotype $a \cdot$, and let $q_2, q_3, \dots$ be the corresponding frequencies of subsequent generations. Then

$$q_n = 0.40 + 0.20(-\tfrac{1}{2})^{n-1} \quad \text{for} \quad n = 1, 2, 3, \dots . \qquad (8.1.13)$$

For this particular example we obtain

$$q_1 = 0.40 + 0.20 \qquad\quad = 0.60 \;,$$
$$q_2 = 0.40 + 0.20(-\tfrac{1}{2}) = 0.30 \;,$$
$$q_3 = 0.40 + 0.20 \quad (\tfrac{1}{4}) = 0.45 \;,$$
$$q_4 = 0.40 + 0.20(-\tfrac{1}{8}) = 0.375 \;, \quad \text{etc.}$$

As n tends to infinity, the sequence $\{(-\tfrac{1}{2})^{n-1}\}$ tends to zero and, therefore,

$$\lim_{n \to \infty} q_n = 0.40 \;.$$

The behavior of the sequence $\{q_n\}$ is depicted in Fig. 8.3.

In calculus, convergence often appears in a somewhat different form. Instead of the independent variable $n = 1, 2, 3, \dots$, a quantity h occurs which tends to zero. For simplicity we let

$$h = 1/n \qquad (8.1.14)$$

and then consider sequences such as

$$\left\{ \frac{2h - h^2}{h} \right\} .$$

When $h \to 0$, both the numerator and the denominator tend to zero. But $0/0$ has no meaning. Hence, we divide numerator and denominator by h. This is permitted since $h \neq 0$ because of (8.1.14). Thus we get

$$\lim_{h \to 0} \frac{2h - h^2}{h} = \lim_{h \to 0} (2 - h) = 2 \;.$$

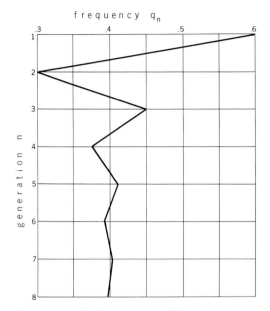

Fig. 8.3. For sex chromosomes the frequency q_n of the male genotype $a \cdot$ oscillates from generation to generation. The oscillation is damped so that q_n tends to a limit (adapted from Li, 1958, p. 62)

Similarly we obtain

$$\lim_{h \to 0} \frac{(h-1)^2 - 1}{(h-2)(h+1)+2} = \lim_{h \to 0} \frac{h^2 - 2h}{h^2 - h} = \lim_{h \to 0} \frac{h-2}{h-1} = +2.$$

8.2. Some Special Limits

In this section we deal with a few limits which will be used later. We begin with the sequence $\{a_n\}$ whose general term is

$$a_n = \left(1 + \frac{1}{n}\right)^{100}.$$

For every natural number n, the base $1 + 1/n$ is greater than one. Raising to the hundredth power will increase the value so that $a_n > 1$. On the other hand, as n tends to infinity, $1 + 1/n \to 1$. Since $1^{100} = 1$, we guess $\lim_{n \to \infty} a_n = 1$. To verify the result we apply logarithms:

$$\log a_n = 100 \log \left(1 + \frac{1}{n}\right).$$

While the first factor remains constant, the second factor tends to $\log 1$ which is zero. Hence $\log a_n \to 0$ and $a_n \to 1$.

Now consider

$$a_n = \left(1 + \frac{1}{100}\right)^n = 1.01^n.$$

This is a special case of an exponential term (8.1.11). Since the base is greater than one, $\{a_n\}$ tends to infinity.

Combining a variable base with a variable exponent we are led to the particular term

$$a_n = \left(1 + \frac{1}{n}\right)^n.$$

a_n is a product consisting of n identical factors $1 + \frac{1}{n}$. As n tends to infinity, the base $1 + \frac{1}{n}$ tends to one. Thus we may think that the power also tends to one. On the other hand, the exponent tends to infinity so that we may argue that the power itself tends to infinity. There exists a curious balance between a tendency of augmenting the power and a tendency of diminishing it. By means of logarithms one can calculate some members of the sequence. In the following table a_n is rounded-off to five significant figures:

n	a_n	n	a_n
1	2	1 000	2.71692
2	2.25	10 000	2.71815
5	2.48832	100 000	2.71825
10	2.59374	1 000 000	2.71828
100	2.70481	10 000 000	2.71828

a_n tends to a limit which has the standard notation e:

$$\lim_{n \to \infty} \left(1 + \frac{1}{n}\right)^n = e = 2.718281828459\ldots[4].\tag{8.2.1}$$

For a proof of convergence we refer the reader to textbooks on calculus. For an application of this limit in biophysics see Rashevsky (1960, vol. 1, p. 435).

---

[4] The letter e was chosen by Leonhard Euler (1707–1783). Born in Switzerland, he became a mathematician in Germany and later in Russia.

Another limit of importance concerns the sine function. Measuring angles in radians, we study the behavior of the quotient

$$\frac{\sin h}{h}$$

as $h = 1/n$ tends to zero. In Fig. 8.4 a part of the unit circle is depicted (cf. Section 5.4). h is the length of the arc from Q to R. $\cos h$ and $\sin h$ are

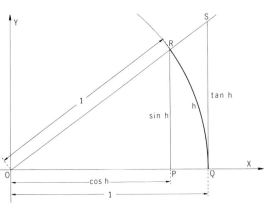

Fig. 8.4. Proof of formula (8.2.3)

the rectangular coordinates of R, and $SQ = \tan h$ because of formula (5.6.2). Now we compare the areas of triangle OPR, of sector OQR, and of triangle OQS with each other. Using an obvious notation we obtain

$$A_{OPR} < A_{OQR} < A_{OQS}.\qquad(8.2.2)$$

For a right triangle, the area is one-half the product of the two legs. Thus

$$A_{OPR} = \frac{1}{2}\cos h \sin h,\quad A_{OQS} = \frac{1}{2}\tan h = \frac{1}{2}\frac{\sin h}{\cos h}.$$

The area of a sector is proportional to the arc. Hence comparing the sector with the full circle of radius $r = 1$ we get

$$A_{OQR} : \pi r^2 = h : 2\pi r.$$

This yields $A_{OQR} = h/2$. From (8.2.2) it follows that

$$\frac{1}{2}\cos h \sin h < \frac{h}{2} < \frac{1}{2}\frac{\sin h}{\cos h}$$

or, on multiplication by $2/\sin h$,

$$\cos h < \frac{h}{\sin h} < \frac{1}{\cos h}. \tag{8.2.3}$$

As h tends to zero, the lower bound $\cos h$ as well as the upper bound $1/\cos h$ tend to one. Hence $h/\sin h$, which remains between the two bounds, must also tend to one. The same holds for the reciprocal $\sin h/h$:

$$\lim_{h \to 0} \frac{\sin h}{h} = 1, \tag{8.2.4}$$

whereas both numerator and denominator tend to zero.

8.3. Series

Does

$$0.99999\ldots = 1? \tag{8.3.1}$$

We obtain this peculiar equality by multiplying both sides of $1/3 = 0.33333\ldots$ by 3. Clearly $0.9 < 1$, $0.99 < 1$, $0.999 < 1$, etc. No matter how many digits we add, the decimal fraction $0.9999\ldots 9$ is always less than one. Yet (8.3.1) need not be false. The sign of equality is correct if we understand the right side as the limit of the left side:

$$\lim_{n \to \infty} \underbrace{0.999\ldots 9}_{n \text{ figures}} = 1.$$

Avoiding decimal fractions we may rewrite the sequence as follows:

$$s_1 = 9/10, \quad s_2 = 9/10 + 9/100, \quad s_3 = 9/10 + 9/100 + 9/1000, \ldots.$$

The general term,

$$s_n = 9/10 + 9/10^2 + \cdots + 9/10^n,$$

is called a *partial sum*. Formally, as n increases, the partial sum tends to a "sum with an infinite number of terms".

In general, let $\{a_n\}$ be any sequence. Then the partial sums

$$s_n = a_1 + a_2 + \cdots + a_n \tag{8.3.2}$$

form a new sequence $\{s_n\}$. As n tends to infinity, the sequence of partial sums is called a *series*. We are interested in the limiting behavior of a series.

A famous example is the *geometric series*. It is generated by a geometric sequence with terms $a_n = aq^n$. Here, the partial sums usually include the term $a_0 = aq^0 = a$. Hence,

$$s_n = a + aq + aq^2 + aq^3 + \cdots + aq^n. \tag{8.3.3}$$

The partial sum can be rewritten in a "closed" form. For this purpose we multiply both sides of (8.3.3) by q:

$$qs_n = aq + aq^2 + aq^3 + \cdots + aq^n + aq^{n+1}. \tag{8.3.4}$$

Subtraction of (8.3.4) term by term from (8.3.3) yields

$$s_n - qs_n = a - aq^{n+1}.$$

Hence,

$$s_n = a\frac{1 - q^{n+1}}{1 - q} \tag{8.3.5}$$

provided that $q \neq 1$.

With this "closed" form we can easily study the limiting behavior of s_n. According to (8.1.12) we get *convergence*, that is, a *finite limit* by assuming that

$$|q| < 1. \tag{8.3.6}$$

Under this condition we get $q^{n+1} \to 0$ and, therefore,

$$\lim_{n \to \infty} s_n = a\frac{1}{1 - q}. \tag{8.3.7}$$

The case $q = 1$ needs special attention: In (8.3.3) the $n + 1$ terms are all equal to a. Hence $s_n = (n + 1)a$, and this does not tend to a finite limit (except in the trivial case where $a = 0$).

The result contained in (8.3.7) is customarily written without the symbol "lim":

$$a + aq + aq^2 + aq^3 + \cdots = \frac{a}{1 - q} \quad \text{for} \quad |q| < 1. \tag{8.3.8}$$

Some special geometric series are:

a) $a = 1, q = \frac{1}{2}$: $1 + \frac{1}{2} + \frac{1}{4} + \frac{1}{8} + \cdots = 2$.

   The partial sums are 1, 1.5, 1.75, 1.875, etc. They are monotone increasing.

b) $a = 1, q = -\frac{1}{2}$: $1 - \frac{1}{2} + \frac{1}{4} - \frac{1}{8} + \cdots = \frac{2}{3}$.

   The partial sums are 1, 0.5, 0.75, 0.625, etc. They are oscillating around the limiting value $2/3 = 0.666 \ldots$.

c) $a = 0.9, q = 0.1$: $0.9 + 0.09 + 0.009 + \cdots = 0.999 \ldots = 1$

   in agreement with (8.3.1).

Example 8.3.1. The known reserves of natural gas in the earth's mantle are 5×10^{13} m^3 as of 1971. In 1972 the world consumption was 1.3×10^{12} m^3 with a yearly increase of 8.7%. Assuming that the growth rate remains constant,

   (a) when will the known reserves be exhausted,
   (b) when will the total reserves, estimated at ten times the known reserves, be exhausted?

Solution: $p = 8.7\%$, $q = 1 + \dfrac{p}{100} = 1.087$.

Consumption in 1972: $a = 1.3 \times 10^{12}$ m^3.
Total consumption in n years:

$$A = a + aq + aq^2 + \cdots + aq^n = a\,\frac{q^{n+1} - 1}{q - 1}.$$

(a) Exhaustion of known reserves:

$$a\,\frac{q^{n+1} - 1}{q - 1} = 5 \times 10^{13} \text{ m}^3, \qquad n = 17.6 \text{ years}.$$

(b) Exhaustion of 10 times known reserves:

$$a\,\frac{q^{n+1} - 1}{q - 1} = 50 \times 10^{13} \text{ m}^3, \qquad n = 42 \text{ years}.$$

(n calculated by logarithms, see Section 6.4.)

Example 8.3.2. Neel and Schull (1958, p. 333) studied the occurrence of *retinoblastoma* under the influence of *mutation, inheritance,* and *selection.* Retinoblastoma is a kind of cancer of the eye in children. Apparently the disease depends upon a single dominant gene, say A. Let a be the normal allele. It is believed that mutation of a into A occurs at a rate of $m = 2 \times 10^{-5}$ in each generation. We exclude the possibility of back-mutations (A into a). With medical care, approximately 70% of the affected persons survive. We further assume that survivors reproduce at half of the normal frequency. The net productive proportion of affected persons will thus be $r = 0.35$. Since A is extremely rare, practically all affected persons are of genotype Aa. Thus we may neglect individuals of genotype AA. Only half of the children of affected individuals are then expected to receive the pathogen allele A, but this reduction is compensated by the fact that each affected person shares reproduction with his unaffected mate. Starting with zero inherited cases in an early generation we obtain for the n-th consecutive generation a rate of

m due to mutation in the n-th generation

mr due to mutation in the $(n-1)st$ generation

mr^2 due to mutation in the $(n-2)nd$ generation

....

mr^n due to mutation in the original generation
                        (which is numbered 0).

For the total rate p_n of occurrence in the n-th generation we get

$$p_n = m + mr + mr^2 + \cdots + mr^n = m\,\frac{1 - r^{n+1}}{1 - r}$$

according to (8.3.5).

As $n \to \infty$, we obtain

$$p = \lim_{n \to \infty} p_n = \frac{m}{1 - r} = \frac{2 \times 10^{-5}}{1 - (0.35)} = 3.08 \times 10^{-5}. \qquad (8.3.9)$$

Therefore, the frequency of persons affected with retinoblastoma will finally be 3.08×10^{-5}, that is, about 50 % higher than the mutation rate. It also follows from formula (8.3.9) that

$$m = (1 - r)p. \qquad (8.3.10)$$

For this result we find a simple intuitive interpretation. In a steady state, the number of A alleles produced by mutation must be compensated by an equal number of losses. Thus, per generation, the production rate is m, whereas the loss rate is $1 - r$ (nonproductive proportion) multiplied by p (frequency of affected persons).

For another instructive example of a geometric series in population genetics, see Sved and Mayo (1970, p. 294).

Example 8.3.3. The particular series

$$1 + \tfrac{1}{2} + \tfrac{1}{3} + \tfrac{1}{4} + \tfrac{1}{5} + \tfrac{1}{6} + \cdots, \qquad (8.3.11)$$

is known as *harmonic series*. The terms are the reciprocals of all natural numbers. The terms tend to zero. Does this guarantee convergence of the series? No, on the contrary, we will prove that the series diverges to ∞. For this purpose we group the terms in the following way:

$$1 + \tfrac{1}{2} + \underbrace{\left(\tfrac{1}{3} + \tfrac{1}{4}\right)}_{g_2} + \underbrace{\left(\tfrac{1}{5} + \tfrac{1}{6} + \tfrac{1}{7} + \tfrac{1}{8}\right)}_{g_4} + \underbrace{\left(\tfrac{1}{9} + \cdots + \tfrac{1}{16}\right)}_{g_8} + \cdots.$$

For each sum $g_2, g_4, g_8, \ldots$ we seek a lower bound:

$$g_2 = \tfrac{1}{3} + \tfrac{1}{4} > \tfrac{1}{4} + \tfrac{1}{4} = \tfrac{1}{2},$$

$$g_4 = \tfrac{1}{5} + \tfrac{1}{6} + \tfrac{1}{7} + \tfrac{1}{8} > \tfrac{1}{8} + \tfrac{1}{8} + \tfrac{1}{8} + \tfrac{1}{8} = \tfrac{1}{2},$$

$$g_8 = \tfrac{1}{9} + \cdots + \tfrac{1}{16} > 8 \cdot \tfrac{1}{16} = \tfrac{1}{2}, \quad \text{etc.}$$

Therefore,

$$1 + \tfrac{1}{2} + g_2 + g_4 + g_8 + \cdots > 1 + \tfrac{1}{2} + \tfrac{1}{2} + \tfrac{1}{2} + \tfrac{1}{2} + \cdots .$$

The right member clearly diverges to infinity. Even more so must the left member diverge.

The example may serve as a warning to the reader not to trust any process where smaller and smaller steps accumulate. It is especially important to ensure convergence when working with a computer.

Example 8.3.4. The terms of the series

$$1 + \frac{1}{1!} + \frac{1}{2!} + \frac{1}{3!} + \frac{1}{4!} + \cdots \tag{8.3.12}$$

are reciprocals of factorials (see Section 1.5). Does the series converge? To answer the question, we compare the series with the geometric series

$$1 + \frac{1}{1} + \frac{1}{2} + \frac{1}{2^2} + \frac{1}{2^3} + \frac{1}{2^4} + \cdots = 1 + 2 = 3.$$

We get

$$\frac{1}{3!} = \frac{1}{2 \cdot 3} < \frac{1}{2^2},$$

$$\frac{1}{4!} = \frac{1}{2 \cdot 3 \cdot 4} < \frac{1}{2^3}, \quad \text{etc.}$$

Each term of the series (8.3.12) is less than (or at most equal to) the corresponding term of the geometric series. Since the geometric series converges to 3, the series (8.3.12) has 3 as an *upper bound*. Later, in Section 10.10 we will learn that the series converges to Euler's number $e = 2.718\ldots$ which we already met in Eq. (8.2.1).

In general, the geometric series is often used for a *test of convergence*. Let

$$a_0 + a_1 + a_2 + \cdots + a_k + a_{k+1} + \cdots \tag{8.3.13}$$

be a series of *positive* terms. We may find an upper bound, if all terms are at most equal to the corresponding terms of a converging geometric series

$$a + aq + aq^2 + \cdots + aq^k + aq^{k+1} + \cdots \qquad (a > 0, 0 < q < 1). \qquad (8.3.14)$$

Characteristic for a geometric series is the equation

$$a_{k+1}/a_k = q. \qquad (8.3.15)$$

If for the terms of (8.3.13)

$$a_{k+1}/a_k \leqq q \qquad (8.3.16)$$

holds, then a_k converges faster to zero than the terms of the geometric series. Therefore, we obtain an upper bound for (8.3.13). This idea is behind the so-called

Ratio test: A series (8.3.13) of positive terms converges, if the inequality (8.3.16) holds for all its terms whereby q is a fixed number less than 1.

Example 8.3.5. We apply the ratio test to

$$\frac{1}{1 \cdot 2} + \frac{1}{2 \cdot 2^2} + \frac{1}{3 \cdot 2^3} + \cdots + \frac{1}{k \cdot 2^k} + \cdots . \qquad (8.3.17)$$

We get

$$a_k = \frac{1}{k \cdot 2^k}, \ a_{k+1} = \frac{1}{(k+1) \, 2^{k+1}}, \ a_{k+1}/a_k = \frac{k}{k+1} \cdot \frac{1}{2}.$$

For all $k = 1, 2, 3, \ldots$ the inequality

$$a_{k+1}/a_k < \tfrac{1}{2}$$

is satisfied. Hence, the series converges according to the ratio test.

Notice that the harmonic series (8.3.11) does not satisfy (8.3.16). Indeed, $a_k = 1/k$, $a_{k+1} = 1/(k+1)$, $a_{k+1}/a_k = k/(k+1)$. As k increases, a_{k+1}/a_k tends to 1. Hence, there exists no fixed number $q < 1$ such that (8.3.16) would hold.

8.4. Limits of Functions

When the domain of a function $y = f(x)$ is not finite, the limiting behavior of y as $x \to \infty$ is of interest. For instance, it may be important to know whether y reaches some finite level, or whether y increases infinitely.

So far in our study of sequences the independent variable n was discrete; more precisely, it was restricted to the natural numbers $1, 2, 3, \ldots$. Now x is a *continuous* variable, that is, x takes on all real values from a certain x_0 on. Fortunately, the theory of limits with continuously changing x differs only slightly from the theory with n increasing discontinuously. The following definition resembles the first definition in Section 8.1:

A function $y = f(x)$ is said to tend or to diverge to infinity as $x \to \infty$, if for any given $C > 0$ it is possible to find a number x_C such that

$$f(x) > C \tag{8.4.1}$$

holds whenever $x > x_C$. In symbols

$$f(x) \to \infty \quad \text{as} \quad x \to \infty . \tag{8.4.2}$$

For instance, consider the common logarithm $y = \log x$. We choose $C = 1000$. Then we try to find x_C such that $\log x > 1000$ for all values $x > x_C$. A suitable bound is $x_C = 10^{1000}$. For all $x > x_C$ we have indeed that $\log x > 1000$. Similarly, for any other choice of C we can find a suitable x_C. Hence, $y = \log x$ tends to infinity as $x \to \infty$.

Other functions that tend to infinity for $x \to \infty$ are:
a) the linear function $y = ax + b$ with $a > 0$.
b) The power functions $y = x^2$, $y = x^3$, ..., $y = x^{1/2}$, $y = x^{3/2}$, ..., $y = x^{1/n}$ for any natural number n.
c) The exponential function $y = q^x$ for $q > 1$. Cf. (8.1.12).

The above definition can easily be modified for functions that tend or diverge to $-\infty$ as $x \to \infty$. Examples of such functions are $y = -x$, $y = -x^2$, $y = -x^{1/2}$, $y = -q^x$ for $q > 1$, and $y = -\log x$.

There are functions that tend to *zero* as $x \to \infty$. For instance, $y = 1/x$, $y = 1/x^2$, $y = q^x$ for $0 < q < 1$. The definition is as follows:

A function $y = f(x)$ is said to tend or to converge to zero, if for any given $\varepsilon > 0$ it is possible to find a number x_ε such that

$$|f(x)| < \varepsilon \tag{8.4.3}$$

holds whenever $x > x_\varepsilon$. In symbols

$$\lim_{x \to \infty} f(x) = 0 . \tag{8.4.4}$$

As an illustration we consider the exponential function $y = (\tfrac{1}{2})^x$. We choose $\varepsilon = 10^{-2}$ and try to find a number x_ε such that $y < \varepsilon$ for $x > x_\varepsilon$. The inequality $y < \varepsilon$ is equivalent to $x \log \tfrac{1}{2} < \log \varepsilon$ or to $x(-0.30103) < (-2)$ which implies $x > 6.65$. Hence, if we put $x_\varepsilon = 7$, say, (8.4.3) is satisfied for all values $x > x_\varepsilon$. For any other choice of $\varepsilon > 0$, we could also find a suitable x_ε. This proves that the function tends to zero.

As $x \to \infty$, the graph of a function such as $y = 1/x$, $y = 1/x^2$, or $y = q^x$ with $0 < q < 1$ approaches the x axis from one side. The "distance" between the graph and the axis converges to zero, but the graph never exactly reaches the x axis. We call the x axis an *asymptote* of the graph. We also say that the graph *reaches* the x *axis asymptotically*. In Fig. 8.5 several functions reaching the x axis asymptotically are plotted.

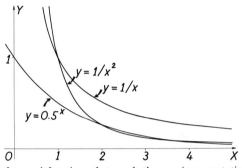

Fig. 8.5. Graphs of several functions that reach the x axis asymptotically. The graphs of $y = 1/x$ and $y = 1/x^2$ are called *hyperbolas*

Example 8.4.1. An application to biophysics may illustrate the theory. Let S be a source of *radiation* such as electromagnetic waves, sound waves, or nuclear radiation. Assume for simplicity that S takes a little space (a "point") and sends its energy uniformly into all directions of the three-dimensional space. Let E be the energy transmitted by the source per second. Then we ask: *What is the intensity I of radiation received at a distance r from the source S?* The intensity is defined to be energy received per second and per unit area. Points of equal intensity are on the surface of a sphere with center in S. Therefore, I is the energy that passes through the surface per second, divided by the area of the surface (cf. formula (4.2.2)):

$$I = \frac{E}{4\pi r^2}. \tag{8.4.5}$$

The greater r is, the smaller is the intensity I. More precisely: I is inversely proportional to the square of r. Doubling of r, for instance, reduces I to one fourth of its original value. With the distance r tending to infinity, we get $\lim I = 0$.

A third definition is also required:

A function $y = f(x)$ is said to tend or to converge to a finite value A as $x \to \infty$, if the modified function

$$f(x) - A \tag{8.4.6}$$

tends to zero. In symbols

$$\lim_{x \to \infty} f(x) = A. \tag{8.4.7}$$

Examples of functions tending to finite limits are:

a) $y = a - b/x$. As $x \to \infty$, the second term tends to zero. Hence $\lim y = a$. The line with equation $y = a$ is an asymptote.

b) $y = \dfrac{a + bx}{c + dx}$ with $d \neq 0$. To find the limiting behavior we divide the numerator and the denominator by x. Thus $y = \dfrac{a/x + b}{c/x + d}$.

For $x \to \infty$, a/x and c/x tend to zero. Hence $\lim y = b/d$.

c) $y = a - bq^x$ with $0 < q < 1$. As $x \to \infty$, we have that $q^x \to 0$ and $\lim y = a$. The graph reaches asymptotically the line with equation $y = a$.

Notice that $y = \sin \alpha$ oscillates between the values $+1$ and -1 as α approaches infinity. Therefore, $\sin \alpha$ does not tend to a limit. We say that $y = \sin \alpha$ *diverges* as $\alpha \to \infty$.

Convergence and divergence of functions may also be studied when $x \to -\infty$ or when $x \to x_0$ where x_0 is any fixed value of the independent variable. There is no difficulty in extending the preceding concept for $x \to -\infty$. We leave it to the reader to find the corresponding definitions.

The case $x \to x_0$, however, needs special attention because it generally makes a difference whether x approaches x_0 from the left or from the right. When x increases and tends to x_0 from the left, we write

$$x \uparrow x_0.$$

Conversely, when x decreases and tends to x_0 from the right, we use the symbol

$$x \downarrow x_0.$$

An example may illustrate the difference. Let $y = 1/x$ and $x_0 = 0$. Let $x \downarrow 0$. Then $1/x \to +\infty$, since the denominator tends to zero over positive values. Conversely, if we begin on the negative x axis, $1/x$ is always negative and, as x approaches zero, $1/x$ tends to minus infinity. Summing up.

$$\frac{1}{x} \to +\infty \quad \text{as} \quad x \downarrow 0,$$

$$\tag{8.4.8}$$

$$\frac{1}{x} \to -\infty \quad \text{as} \quad x \uparrow 0.$$

The result is depicted in Fig. 8.6.

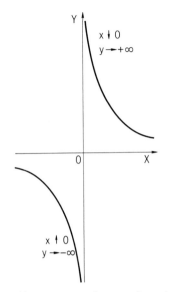

Fig. 8.6. The behavior of $y = 1/x$ as x approaches zero from the right and from the left. The y axis is an asymptote to the hyperbola

This example shows that it is not correct to write $1/0 = +\infty$. First, the relation should be written with the symbol $\rightarrow$ instead of $=$, that is, $1/x \rightarrow \infty$. Second, the result is correct, only if $x \downarrow 0$. In the opposite case, $x \uparrow 0$, we would obtain $1/x \rightarrow -\infty$.

Similarly, it is not correct to write $\tan 90° = +\infty$. It is true that $\tan \alpha = \sin \alpha / \cos \alpha \rightarrow +\infty$ for $\alpha \uparrow 90°$. But for $\alpha \downarrow 90°$ we get $\tan \alpha \rightarrow -\infty$. For a graph the reader is referred to Fig. 5.13.

Another important example is the common logarithm $y = \log x$. This function is only defined for $x > 0$. But we may ask how $\log x$ behaves for $x \downarrow 0$. Let $x = 1/u$. Then

$$\log x = \log 1/u = -\log u$$

according to formula (6.4.3). Now $x \downarrow 0$ is equivalent to $u \rightarrow \infty$. We already know that $\log u \rightarrow \infty$. Hence

$$\log x \rightarrow -\infty \quad \text{as} \quad x \downarrow 0. \qquad (8.4.9)$$

The result is in agreement with the graph of the logarithmic function in Fig. 6.9. The graph reaches the negative y axis asymptotically.

In terms of limits it is possible to define *continuity* of a function. Let x_0 belong to the domain of a function $y = f(x)$. Then *the function is said* *to be continuous at x_0, if both limits*

$$\lim_{x \uparrow x_0} f(x) \quad \text{and} \quad \lim_{x \downarrow x_0} f(x)$$

exist and if they are equal to $f(x_0)$.

If one of the conditions is not satisfied, the function is called *discontinuous* at x_0. In Fig. 8.7 the graph of a rather unusual function is plotted. At x_0 the value of $f(x)$ approaches a certain value $f(x_0)$ from both sides. Thus $f(x)$ is continuous at x_0. The same is true at x_1 despite the different angles under which the lines join at the point with coordinates

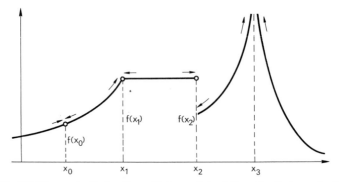

Fig. 8.7. The graph of a function with two points of discontinuity (x_2 and x_3)

x_1 and $f(x_1)$. At x_2 the value of $f(x)$ is $f(x_2)$. But this value cannot be the common limit when x_2 is approached from the left and from the right. In Fig. 8.7 it is assumed that

$$\lim_{x \uparrow x_2} f(x) = f(x_2), \quad \text{but} \quad \lim_{x \downarrow x_2} f(x) < f(x_2).$$

Hence, $f(x)$ is discontinuous at x_2. We say that the value of $f(x)$ *jumps* or has a *saltus* at x_2. A different kind of discontinuity occurs at x_3. Here $f(x)$ has no finite limit as x tends to x_3.

Examples of continuous and discontinuous functions:

a) The power functions $y = ax$, $y = ax^2$, $y = ax^3$, ... are continuous everywhere, that is, for all real values of x.

b) The power functions $y = x^{1/2}$, $y = x^{3/2}$, ... are continuous in their full domain, that is, for $x > 0$.

c) The power function $y = ax^{-1} = a/x$ is continuous everywhere except for $x = 0$. There no finite limit exists, neither from the left nor from the right (see Fig. 8.6).

d) The trigonometric functions $y = \sin \alpha$, $y = \cos \alpha$ are continuous everywhere, that is, for all real values of the angle α (see Fig. 5.8).

e) The trigonometric function $y = \tan \alpha$ is continuous for all real values α except for $\alpha = \pm 90°$, $\pm 270°$, ... (see Fig. 5.13).

f) The exponential function $y = aq^x$ with $q > 0$ is continuous everywhere, that is, for all real values of x. (See Fig. 6.9).

g) The common logarithm $y = \log x$ is continuous for its full domain $x > 0$ (see Fig. 6.9).

h) Let $N = f(t)$ be the number of cats in a household as a function of time. Whenever birth or death occur, $f(t)$ jumps by one or several units. Thus $f(t)$ is discontinuous at those instants (see Fig. 8.8). A function that remains constant in an interval, then jumps to another value, remains constant in an adjacent interval, then jumps, etc., is called a *step function*. Thus $N = f(t)$ is a step function. Similarly, the number of pulses or of scintillations is a step function. In statistics, step functions are especially useful.

i) The water intake of an animal as a function of time is an other example of step function.

k) The sun's azimuth (angle in the horizontal plane beginning North in the clockwise sense) increases as a continuous function of time with one exception, however. At the time of an equinox and at a point of the earth's equator the rising sun keeps the azimuth 90° until it reaches the zenith. Then suddenly the azimuth changes into 270°. Thus in this particular situation, the azimuth is a *discontinuous* function of time.

l) A function $y = f(x)$ whose graph is a "broken line" is continuous in its full domain, since at the "joints" the value of $f(x)$ tends to the same limit from both sides. Cf. x_1 in Fig. 8.7.

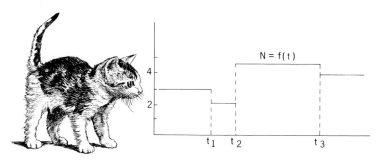

Fig. 8.8. Example of a *step function:* The number N of cats in a household. The function has a saltus at each instant when birth or death occurs

*8.5. The Fibonacci Sequence

In 1202 Fibonacci[5] raised and solved the following problem: Rabbits breed rapidly. It is assumed that a pair of adult rabbits produces a pair of young rabbits every month and that newborn rabbits become adults in two months and produce, at this time, another pair of rabbits. Starting with an adult pair, how big will a rabbit colony be after the first, second, third, etc. month?

During the first month a pair is born so that there are two couples present. During the second month the original pair has produced another pair. One month later both the original pair and the first born pair has produced new pairs so that two adults and three young pairs are present, etc. The figures are shown in Table 8.1.

Table 8.1. *Growth of a rabbit colony*

months	adult pairs	young pairs	total
1	1	1	2
2	1	2	3
3	2	3	5
4	3	5	8
5	5	8	13
6	8	13	21
7	13	21	34
8	21	34	55
9	34	55	89
10	55	89	144

Let a_n denote the number of adult pairs at the end of the n-th month. Thus we get the following sequence:

$$a_1 = 1, a_2 = 1, a_3 = 2, a_4 = 3, a_5 = 5, a_6 = 8, \ldots . \qquad (8.5.1)$$

This is the famous *Fibonacci sequence*. It has the following remarkable property:

$$2 = 1 + 1 \quad \text{or} \quad a_3 = a_1 + a_2,$$
$$3 = 1 + 2 \quad \text{or} \quad a_4 = a_2 + a_3,$$
$$5 = 2 + 3 \quad \text{or} \quad a_5 = a_3 + a_4, \quad \text{etc.}$$

---

[5] Leonardo Pisano, often called Fibonacci (possibly 1170–1230), Italian mathematician. In his book "Liber abaci" he introduced arabic numerals into Europe.

Each number of the Fibonacci sequence is the sum of the two preceding numbers. In concise mathematical language this property may be stated by a so-called *recursion formula:*

$$a_{n+1} = a_{n-1} + a_n \tag{8.5.2}$$

where $n = 2, 3, 4, \ldots$.

In order to study the *growth rate* of successive members, we form the following sequence of ratios

$$b_n = a_{n+1}/a_n \, . \tag{8.5.3}$$

Thus we get

$$
\begin{aligned}
&b_1 = 1 &&b_5 = 8/5 = 1.60 \\
&b_2 = 2/1 = 2 &&b_6 = 13/8 = 1.625 \\
&b_3 = 3/2 = 1.5 &&b_7 = 21/13 = 1.615 \ldots \\
&b_4 = 5/3 = 1.66 \ldots &&b_8 = 34/21 = 1.619 \ldots \quad \text{etc.}
\end{aligned}
\tag{8.5.4}
$$

This sequence seems to converge to a certain limit which falls between 1.60 and 1.62. It would mean that the rabbit colony grows from one month to the next by a percentage that tends to a value slightly higher than sixty percent.

The proof of convergence is left to the reader and may be discovered by working through problems 8.5.4 to 8.5.6 at the end of this chapter. Here we confine ourselves to calculating the exact limit. For this purpose we assume that

$$b = \lim_{n \to \infty} b_n \tag{8.5.5}$$

exists. This implies that the modified sequence

$$c_n = b_n - b \tag{8.5.6}$$

tends to zero. From (8.5.2) it follows that

$$a_{n+1}/a_n = a_{n-1}/a_n + 1 \, .$$

By virtue of (8.5.3) and (8.5.6) this becomes

$$b_n = \frac{1}{b_{n-1}} + 1 \tag{8.5.7}$$

and

$$b + c_n = \frac{1}{b + c_{n-1}} + 1 \, . \tag{8.5.8}$$

Since $c_n \to 0$ for $n \to \infty$ we get

$$b = \frac{1}{b} + 1 \tag{8.5.9}$$

which is equivalent to the quadratic equation

$$b^2 - b - 1 = 0. \tag{8.5.10}$$

The only positive root is

$$b = \tfrac{1}{2}(1 + \sqrt{5}) = 1.618034\ \ldots\ . \tag{8.5.11}$$

This is the desired limit of the sequence $(8.5.4)^6$.

    Fibonacci numbers are found when *arrangement of leaves* (phyllotaxis) is studied. We consider the case where leaves around a stem follow a helical pattern (Fig. 8.9). Proceeding upwards we mark consecutive leaves by L_1, L_2, L_3, etc. Leaf L_2 will be found standing at a certain angle away from L_1 around the stem and at a certain distance along the stem. In Fig. 8.9 a special case is depicted where the angle is 144°. Leaf L_3 is displaced from L_2 in much the same way as L_2 from L_1. We assume the same pattern for all following leaves. In our example it takes five angles of 144° to arrive at a leaf that has the same orientation as L_1 since $5 \times 144° = 720° = 2 \times 360°$. Thus we find a periodicity with a "*period*" consisting of two windings and five leaves.

    In general we may introduce two numbers:
a) m = number of complete turns or windings in a period,
b) n = number of leaves in a period.
    In Fig. 8.9 the special case $m = 2$, $n = 5$ is shown. The numbers m and n can also be defined for whorls of leaves.
    Actual countings on numerous plants have proven that m as well as n take most frequently values such as $1, 2, 3, 5, 8, 13, 21, 34, \ldots$, that is, numbers from the Fibonacci sequence. According to Schips (1922) we find the cases

$m = 1$, $n = 2$	in the two row leaves of several bulbous plants as well as the horizontal twigs of the elm,
$m = 1$, $n = 3$	in sedges, the alder, and the birch,
$m = 2$, $n = 5$	very frequent, in willows, roses, and stone fruit trees,
$m = 3$, $n = 8$	in cabbage, asters, and hawkweed,
$m = 8$, $n = 21$	scales of spruce and fir cones,
$m = 13$, $n = 34$	scales of cones of *Pinus laricio*.

There are also exceptions, but Fibonacci numbers occur so frequently that they cannot be explained by chance.

---

    [6] The number b plays a major role in geometry and aesthetics. Divide a line AB at a point C such that $AB : AC = AC : CB$. Then the division is called the *golden section* or *divine proportion*. The ratio AB/AC is equal to $b = \tfrac{1}{2}(1 + \sqrt{5})$.

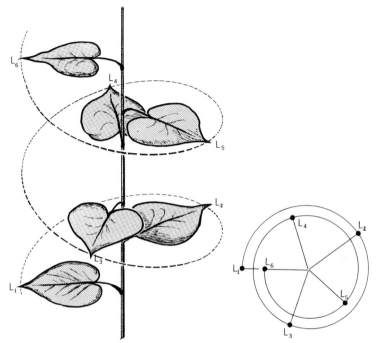

Fig. 8.9. Helical arrangement of leaves on a stem. In the figure it is assumed that the same pattern with five leaves is repeated after two full windings of the helix. This is the case in roses, some willows and cherries. Left: view from the side. Right: View from top

We find Fibonacci numbers also in florets of composite flowers. The following table is taken from a report by Wagner (1957) who counted the number of ray florets in composite flowers of alpine plants:

Species:	number of ray florets	
	mean:	range:
Achillea macrophylla	5	4— 7
Achillea atrata	8	6—13
Rudbeckia nitida	8	6—10
Arnica montana (side flowers)	11	7—17
Rudbeckia speciosa 1	13	13—30
Centaurea montana	13	10—17
Senecio doronicum 1	13	11—21
Senecio uniflorus	13	5—24
Senecio doronicum 2	17—18	11—22
Arnica montana (top flowers)	18—19	10—29
Chrysanthemum leucanthemum	21	13—33
Rudbeckia speciosa 2	21	13—30
Senecio doronicum 3	21	11—22
Doronicum Clusii	34	26—47

As we see from the table, the number of ray florets is subject to random fluctuation. However, the mean values fall frequently on Fibonacci numbers. Exceptions are *Arnica montana* and *Senecio doronicum 2*.

Biologists have tried to explain the peculiar prevalence of Fibonacci numbers in phyllotaxis. Symmetry may play a major role because symmetry maintains the mechanical equilibrium of a stem, gives the leaves the best exposure to light, and supports a regular flow of nutrients. However, science is still far away from a satisfactory explanation. Major works on phyllotaxis were written by Thompson (1917, p. 912–933) and Nelson (1954, p. 48–60). For remarks see also Dormer (1972), Weyl (1952, p. 72).

Occasionally Fibonacci numbers occur in population genetics. It would go well beyond the scope of this book to deal with this section of biology. We refer the reader to Li (1958, p. 68–70, 118).

For a mathematical treatment of the Fibonacci sequence we recommend Vorobyov (1961). A most amusing book on the divine proportion and on Fibonacci numbers was written by Huntley (1970).

Problems for Solution

8.1.1. Find the limits of the following terms as n tends to infinity:

a) $\left(2 + \dfrac{3}{n}\right)\left(4 - \dfrac{100}{n}\right)$,

b) $\left(1 - \dfrac{3}{n}\right) \Big/ \left(\dfrac{1}{n} - 3\right)$,

c) $\dfrac{2n + 5}{7n - 5}$,

d) $\dfrac{an^2 + 400\,n}{bn^2 - 400}$ $(b \neq 0)$.

8.1.2. To what numbers converge the following terms as m tends to infinity?

a) $\left(1 - \dfrac{2}{m} - \dfrac{A}{m^2}\right)\left(1 + \dfrac{B}{m}\right)$,

b) $\left(\dfrac{4}{m} - s\right) \Big/ \left(\dfrac{5}{m^2} - t + 1\right)$,

c) $\dfrac{21 - 100\,m}{28 + 10\,m}$,

d) $\dfrac{p - m - m^2}{q + m - m^2}$ $(q \neq 0)$.

8.1.3. Let $h = 1/n$ and $n \to \infty$. Find

a) $\displaystyle\lim_{h \to 0} \frac{4-h}{2+7h}$,

b) $\displaystyle\lim_{h \to 0} \frac{4h-h^2}{2h}$,

c) $\displaystyle\lim_{h \to 0} \frac{(h-1)^2 - 1}{(h+3)^2 - 9}$,

d) $\displaystyle\lim_{h \to 0} \frac{(x+h)^2 - x^2}{h}$.

8.1.4. Find

a) $\displaystyle\lim_{h \to 0} \frac{2 - 2(1+h^2)}{-3 + 3(1-h^2)}$,

b) $\displaystyle\lim_{h \to 0} \frac{4 - (2+h)^2}{1 - (1-h)^2}$,

c) $\displaystyle\lim_{h \to 0} \frac{(3h-1)(3h+1)+1}{(3h-1)(h-2)-2}$,

d) $\displaystyle\lim_{h \to 0} \frac{(x+h)^3 - x^3}{h}$.

8.1.5. Find the limit of $a_n = n/(n^2 + 1)^{\frac{1}{2}}$ as $n \to \infty$.
(Hint: Divide numerator and denominator by n.)

8.1.6. Find $\displaystyle\lim_{n \to \infty} \frac{\sqrt{4n^2 - 1}}{3n}$.

8.1.7. Find $\displaystyle\lim_{h \to 0} \frac{\sqrt{5+h} - \sqrt{5}}{h}$.

(Hint: Multiply numerator and denominator by $\sqrt{5+h} + \sqrt{5}$.)

8.1.8. Let $x > 0$. Find the limit of $\dfrac{1}{h}(\sqrt{x+h} - \sqrt{x})$ as $h \to 0$.

8.1.9. How do the following terms behave as $n \to \infty$?

a) $100(\frac{1}{2})^n$,

b) $(-\frac{1}{2})^n$,

c) $(5/4)^n$,

d) $(1 + 10^{-3})^n$,

e) $(1 - 10^{-3})^n$.

8.1.10. To study a problem of inbreeding, Kempthorne (1957, p. 86) applies the sequence

$$F_n = \frac{s}{2-s}[1 - (s/2)^n] + (s/2)^n F_0, \qquad n = 1, 2, 3, \ldots$$

whereby $0 < s < 1$. Find the limiting behavior of F_n as n tends to infinity.

8.2.1. Find a) $\displaystyle\lim_{n \to \infty} \left(1 + \frac{1}{n}\right)^{3n}$,   b) $\displaystyle\lim_{h \to 0} \frac{\sin h}{a^2 h}$ $(a \neq 0)$.

8.2.2. Find $\lim\limits_{n \to \infty} \left(1 + \dfrac{1}{3n}\right)^n$.

(Hint: Replace n by $N/3$ and observe that $N \to \infty$.)

8.2.3. Find $\lim\limits_{h \to 0} \dfrac{\sin 5h}{h}$.

(Hint: Replace h by $k/5$ and observe that $k \to 0$.)

8.2.4. What is $\lim\limits_{N \to \infty} \left(1 - \dfrac{3}{N}\right)^N$?

8.3.1. Calculate the following partial sums:

a) $1 + 1/3 + 1/9 + 1/27 + 1/81 + 1/243$

b) $2 + 2/11 + 2/11^2 + \cdots + 2/11^5$

c) $1 - 1/2 + 1/4 - 1/8 + 1/16 - 1/32 + 1/64$.

8.3.2. Find a closed expression for the following partial sums:

a) $1 + r + r^2 + \cdots + r^{10}$

b) $1 - r + r^2 - r^3 + r^4 - \cdots + r^{10}$

c) $1/s + 1/s^2 + 1/s^3 + \cdots + 1/s^n$

d) $e + e/5 + e/5^2 + \cdots + e/5^k$.

8.3.3. Calculate the infinite sums:

a) $1 + r + r^2 + r^3 + \cdots$ assuming that $|r| < 1$

b) $c + c/2 + c/2^2 + c/2^3 + \cdots$

c) $1 + 1/s + 1/s^2 + 1/s^3 + \cdots$ assuming that $|s| > 1$.

8.3.4. Calculate the infinite sums

a) $1 - r + r^2 - r^3 + r^4 - \cdots$ $(-1 < r < +1)$,

b) $a + a/p + a/p^2 + a/p^3 + \cdots$ $(p > 1)$,

c) $1 + n^3 + n^6 + n^9 + n^{12} + \cdots$ $(-1 < n < 1)$.

8.3.5. In finding the average life span of a harmful gene, Li (1961, p. 150) is led to the following infinite sum:

$$U = 1 + 2w + 3w^2 + 4w^3 + \cdots$$

where $0 < w < 1$. Find a closed expression for U.
(Hint: Determine first the difference $U - wU$.)

8.3.6. Find the sum of the infinite series

$$S = A - 2A^2 + 3A^3 - 4A^4 + \cdots \qquad (|A| < 1)$$

by first calculating $T = S + AS$.

8.3.7. The cesium isotope ^{137}Cs loses 2.3% of its mass annually by radioactive decay. ^{137}Cs is a dangerous pollutant contained in radioactive fallout. Assume that each year the same mass M of ^{137}Cs is released to the environment. What is the total mass that will accumulate (a) after n years, (b) when equilibrium is reached ($n \to \infty$)? (Adapted from Woodwell, 1969.)

8.3.8. In 1972 world consumption of mineral oil was 2.7×10^9 t (metric tons). The annual increase was 5.1%. The total reserves of crude oil in the earth (including known and undiscovered) are estimated to be 700×10^9 t. If the annual increase remains constant, when would the reserve be exhausted?

8.3.9. In Iraq an epidemic of methylmercury poisoning killed 459 people during 1972. Human beings became exposed to the poison when they ate homemade bread accidently prepared from seed-wheat treated with a methylmercurial fungicide. Symptoms appeared only after weeks of exposure. Assume for simplicity that a person takes a constant daily dose d of the poison and that a certain percentage p of the accumulated poison is excreted each day. Find a formula which relates the amount of poison stored in the body to the number of days. (Adapted from Bakir et al., 1973.)

8.3.10. Using the ratio test prove that the series

$$\frac{1}{(a+1)\,3} + \frac{1}{(a+2)\,3^2} + \frac{1}{(a+3)\,3^3} + \frac{1}{(a+4)\,3^4} + \cdots$$

converges.

8.3.11. Show that the series

$$1 + \frac{a}{1!} + \frac{a^2}{2!} + \frac{a^3}{3!} + \frac{a^4}{4!} + \cdots .$$

converges for all values $a \in \mathbb{R}$.

8.4.1. Find the limiting behavior of $3/(2 - x)$

a)   as $x \uparrow 2$,   b)   as $x \downarrow 2$.

8.4.2. Find the limiting behavior of $1/x^2$

  a) as $x \uparrow 0$,   b) as $x \downarrow 0$.

8.4.3. Living tissue can only be excited by an electric current if the current reaches or exceeds a certain threshold which we denote by i. The threshold i depends on the duration t of current flow. Weiss' law states that $i = a/t + b$ with positive constants a and b (cf. Defares et al., 1973, p. 73). Describe the behavior of the threshold i when t approaches zero and when t tends to infinity.

8.4.4. Discuss the limiting behavior of $f(x) = 2x/(x^2 - 4)$ for $x \uparrow 2$, $x \downarrow 2$, $x \uparrow (-2)$, and $x \downarrow (-2)$. Plot a graph of $f(x)$.

8.4.5. A bat emits ultrasonic waves which are reflected from a moth and the echo is detected by the bat. Prove that the intensity of the echo heard by the bat is inversely proportional to the fourth power of the distance r between bat and insect. [Hint: Use formula (8.4.5).] [Adapted from Jarman (1970, p. 28).]

*8.5.1. Show that the partial sum of the Fibonacci numbers a_n as given by (8.5.1) can be written in the form

$$a_1 + a_2 + a_3 + \cdots + a_n = a_{n+2} - 1 .$$

(Hint: Add the following equalities $a_1 = a_3 - a_2, a_2 = a_4 - a_3$, etc.)

*8.5.2. Again for Fibonacci numbers show that

  a) $a_1 + a_3 + a_5 + \cdots + a_{2n-1} = a_{2n}$

  b) $a_2 + a_4 + a_6 + \cdots + a_{2n} = a_{2n+1} - 1 .$

(Hint: Add the equalities $a_1 = a_2, a_3 = a_4 - a_2, a_5 = a_6 - a_4$, etc. For the proof of b) use the result of Problem 8.5.1.).

*8.5.3. Consider the sequence of the following fractions:

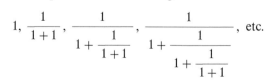

Show that it is equal to the sequence $1/b_1, 1/b_2, 1/b_3, \ldots$ where $b_1, b_2, b_3, \ldots$ are given by (8.5.4).

*8.5.4. By means of $b_1 = 1$, $b_2 = 2$ and formula (8.5.7) show that $1 \leq b_n \leq 2$ for all natural numbers n.

*8.5.5. Let c_n be defined by (8.5.6). Assume that $b = \frac{1}{2}(1 + \sqrt{5})$, that is, satisfies the equation (8.5.10). Verify that

$$c_n = 1 - b + \frac{1}{b + c_{n-1}} = -c_{n-1} \frac{b - 1}{b + c_{n-1}}.$$

*8.5.6. From the result of the preceding problem conclude that

a) $|c_n| < 0.7$ for every $n = 1, 2, 3, \ldots$ and

b) $c_n \to 0$ for $n \to \infty$.

*8.5.7. Without proof we present a formula for the Fibonacci numbers a_n. Let $b = \frac{1}{2}(1 + \sqrt{5})$ and $c = \frac{1}{2}(1 - \sqrt{5})$. Then

$$a_n = \frac{1}{\sqrt{5}} (b^n - c^n).$$

Check this formula for $n = 1$ and $n = 2$. Use the formula to prove that $b_n = a_{n+1}/a_n$ tends to b as $n \to \infty$.

Chapter 9

Differential and Integral Calculus

9.1. Growth Rates

Differential calculus is based on the notion of *rate of change*. The notion appears implicitly in words such as growth rate, relative growth, velocity, acceleration, rate of reaction, density, and slope of a curve. We begin with some introductory examples:

Example 9.1.1. Let N be the number of individuals in an animal or plant population. N changes with time. Thus we may consider N as a function of time t:

$$N = f(t).$$

Let t_1 and t_2 be two time instants, and assume that $t_2 > t_1$. Then $f(t_1)$ and $f(t_2)$ are the corresponding numbers of individuals. The difference

$$\Delta N = f(t_2) - f(t_1) \tag{9.1.1}$$

is the total change of population size in the time interval from t_1 to t_2. For $\Delta N > 0$ we have an increase, and for $\Delta N < 0$ a decrease in size. To judge how fast the population size has changed we have also to consider the length of the time interval

$$\Delta t = t_2 - t_1. \tag{9.1.2}$$

The ratio

$$\frac{\Delta N}{\Delta t} = \frac{f(t_2) - f(t_1)}{t_2 - t_1} \tag{9.1.3}$$

informs us how much the population size changed per unit time. Strictly speaking, it is the average change per unit time within the time interval from t_1 to t_2. We call the quantity (9.1.3) the *average rate of change* or, if misunderstanding is not expected, simply rate of change.

A numerical example may be useful. In a zoo, a colony of ducks consisted of 52 birds on October 1, 1967 ($= t_1$) and of 78 birds on October 1, 1968 ($= t_2$). Thus $f(t_1) = 52$, $f(t_2) = 78$ and $\Delta N = 26$. Let us measure the time in months. Then $\Delta t = 12$ months. Hence, the average rate of change from October 1, 1967 till October 1, 1968 is

$$\Delta N/\Delta t = 26/12 \text{ months}^{-1} = 2.17 \text{ months}^{-1},$$

that is, on the average the colony grew by 2.17 "individuals" each month. Naturally, births were numerous in spring, and deaths were quite irregular over the year. Thus the true growth differs from month to month. The number 2.17 looks quite artifical. Yet it gives a good indication of the average speed of growth.

The ratio (9.1.3) is often called a *growth rate*. Actual growth occurs only for $\Delta N > 0$. If, instead, $\Delta N < 0$, the population size is decreasing. Nevertheless, this property can be interpreted as *negative growth*. This justifies the general use of the expression "growth rate".

Example 9.1.2. In metabolism we are interested in the speed of a chemical reaction. Let $M = f(t)$ be the mass of some nutrient as a function of time. We assume for instance that the nutrient disintegrates chemically and, consequently, that M decreases. Let t_1, t_2 be two consecutive time instants. Let $\Delta t = t_2 - t_1$ be the length of the time interval and $\Delta M = f(t_2) - f(t_1)$ the decrease in mass. We calculate the increase in mass per time unit:

$$\frac{\Delta M}{\Delta t} = \frac{f(t_2) - f(t_1)}{t_2 - t_1}. \tag{9.1.4}$$

This quotient is called the rate of reaction. Under our assumptions, $\Delta M/\Delta t$ is negative. Strictly speaking, $\Delta M/\Delta t$ is the *average rate of reaction* over the time interval from t_1 to t_2. The chemical reaction need not have a constant rate.

Example 9.1.3. Quite similar is the *rate of decay* in nuclear physics. Let $N = N(t)$ be the number of radioactive atoms in a sample at time instant t. Then

$$\frac{\Delta N}{\Delta t} = \frac{N(t_2) - N(t_1)}{t_2 - t_1} \tag{9.1.5}$$

is the average rate of decay over the time interval from t_1 to t_2.

Example 9.1.4. What is ordinarily called *velocity* or *speed* is also a rate of change. Assume that a particle moves along a straight line. Suppose further that at each instant of time over the period of motion we mark its position as a displacement from a fixed point of reference on the line. Let $s = s(t)$ denote the position at instant t. In a time interval from t_1 to t_2 we calculate the increase of displacement, $\Delta s = s(t_2) - s(t_1)$. To get the increase per time unit, we divide by $\Delta t = t_2 - t_1$:

$$\frac{\Delta s}{\Delta t} = \frac{s(t_2) - s(t_1)}{t_2 - t_1}. \tag{9.1.6}$$

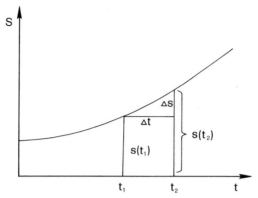

Fig. 9.1. The average velocity $\Delta s/\Delta t$ is represented by the average slope of a curve in a Cartesian coordinate system

This quotient is called the *average velocity* of the particle over the time interval from t_1 to t_2. Thus velocity is a rate of change (Fig. 9.1).

Example 9.1.5. Consider again the motion of a particle along a line. We may ask how fast the velocity increases or decreases. To answer this question we introduce the function $v = v(t)$, that is, the velocity as a function of time. Then we choose two consecutive time instants t_1 and t_2. The ratio

$$\frac{\Delta v}{\Delta t} = \frac{v(t_2) - v(t_1)}{t_2 - t_1} \tag{9.1.7}$$

is the average change of velocity per time unit. This quantity is usually called the *average acceleration*. For $\Delta v > 0$ the acceleration is positive. For $\Delta v < 0$ the velocity decreases, and the acceleration is negative.

Example 9.1.6. To define a rate of change, the independent variable need not be the time. Consider a spherical cell. Its volume V is a function of the radius r. Thus we may write $V = f(r)$. As r increases, V also increases and we may ask: How does V change relative to a change in r? To answer this question consider two radii, say r_1 and r_2, where $r_2 > r_1$. Then $\Delta r = r_2 - r_1$ is the increase of r, and $\Delta V = f(r_2) - f(r_1)$ is the corresponding increase of V. The ratio

$$\frac{\Delta V}{\Delta r} = \frac{f(r_2) - f(r_1)}{.\ \ r_2 - r_1} \tag{9.1.8}$$

is the increase of the volume per unit length of the radius. $\Delta V/\Delta r$ may again be called the rate of change. Strictly speaking, $\Delta V/\Delta r$ is the average rate of change when the radius r increases from r_1 to r_2.

It is worth studying a numerical example. If r is measured in micrometers (μm)[1], then

$$V = f(r) = \frac{4}{3}\pi r^3$$

is the volume of the cell measured in cubic micrometers (μm^3). Let $r_1 = 5.00\,\mu m$ and $r_2 = 8.00\,\mu m$. Then $f(r_1) = 4/3 \times \pi \times 5^3\,\mu m^3 = 0.524 \times 10^3\,\mu m^3$ and $f(r_2) = 4/3 \times \pi \times 8^3\,\mu m^3 = 2.144 \times 10^3\,\mu m^3$. Hence $\Delta V = f(r_2) - f(r_1) = 1.62 \times 10^3\,\mu m^3$, and the rate of change

$$\frac{\Delta V}{\Delta r} = \frac{1.62 \times 10^3\,\mu m^3}{3\,\mu m} = 0.54 \times 10^3\,\mu m^2 .$$

This means that, on the average, the volume has grown by $0.54 \times 10^3\,\mu m^3$ for each μm increase of the radius.

Example 9.1.7. Let particles be spread over the x axis, regularly or irregularly. Let M be the total mass of particles that fall into an interval between the origin O and a point P with positive abscissa x. With each x there is associated a unique value of total mass[2]. Therefore, $M = M(x)$ is a function of x. Let P_1 and P_2 be two points on the x axis with coordinates x_1 and x_2 where $x_2 > x_1$ (Fig. 9.2). $\Delta M = M(x_2) - M(x_1)$ is

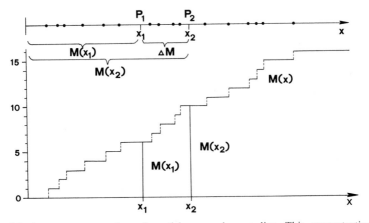

Fig. 9.2. Average concentration of particles spread on a line. This concentration is measured by a difference quotient $\Delta M/\Delta x$ where $M = M(x)$ is the total mass in the interval from 0 to x

---

[1] 1 micrometer $= 1\mu m = 10^{-6}\,m$. A former notation for the same unit is 1 micron (μ). The new notation is more consistent with the metric system and is becoming standard.

[2] For the sake of rigor it should also be stated whether a particle that falls exactly on O or on P is counted or not.

the mass of all particles that fall into the interval between P_1 and P_2. To get a measure of how densely the particles lie between the points P_1 and P_2, we calculate the mass per unit length. Thus we divide ΔM by $\Delta x = x_2 - x_1$:

$$\frac{\Delta M}{\Delta x} = \frac{M(x_2) - M(x_1)}{x_2 - x_1}. \qquad (9.1.9)$$

This quotient is called the *density* or *concentration* of the particles in the interval P_1 and P_2. However, when the particles are not uniformly distributed, it is more accurate to call $\Delta M/\Delta x$ an *average density* or *average concentration*.

Having studied several applications of the notion "rate of change", we are now prepared to deal with this notion from a purely mathematical point of view. Let $y = f(x)$ be any function and let the interval between x_1 and x_2 belong to the domain of this function (see Section 3.4 for the terms "domain" and "range"). The total change of y on the interval between x_1 and x_2 is $\Delta y = f(x_2) - f(x_1)$. We relate this change to the *increment* $\Delta x = x_2 - x_1$ by forming the fraction

$$\frac{\Delta y}{\Delta x} = \frac{f(x_2) - f(x_1)}{x_2 - x_1}. \qquad (9.1.10)$$

We call this fraction an *average rate of change* or a *difference quotient*. In a graph of the function, the average rate of change has an intuitive meaning. In Fig. 9.3 the straight line l is drawn joining two points (x_1, y_1) and (x_2, y_2) of the graph of $y = f(x)$. According to the definition given in Section 3.6, the *average rate of change is identical with the slope of l*, that is,

$$a = \frac{\Delta y}{\Delta x}. \qquad (9.1.11)$$

The line l is the extension of a *chord* of the graph.

Tacitly we have assumed that x and y are measured on an *interval scale* (Section 1.2). A ratio scale is not required.

It is not always satisfactory to consider the average of a rate of change. Intuitively, we are looking for a term which means something like an "actual" or "instantaneous" rate of change which would be opposed to an average. *The transition from an average rate of change to an "instantaneous" rate is the basic idea in differential calculus.* It is the creation of a new concept. For this purpose we must reduce the interval to a point. Hence, keeping x_1 fixed we let x_2 tend toward x_1, that is, in

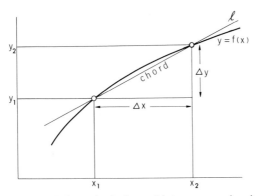

Fig. 9.3. The average rate of change in the interval between x_1 and x_2 is the slope of the straight line joining the points (x_1, y_1) and (x_2, y_2)

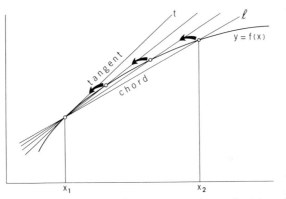

Fig. 9.4. As $x_2 \to x_1$, the line l tends to the tangent t to the graph at (x_1, y_1). The slope $a = \Delta y/\Delta x$ of l tends to the slope a_t of the tangent

the notation of the preceding chapter,

$$x_2 \to x_1, \quad \Delta x \to 0. \tag{9.1.12}$$

Now, Δx is the denominator of $\Delta y/\Delta x$. In order that the quantity $\Delta y/\Delta x$ will tend to a finite limit, we have to assume that not only Δx but also Δy tends to zero. This implies that $y = f(x)$ must be a *continuous function* at $x = x_1$ (see Section 8.4) and that the graph of the function must have a *tangent* at the point (x_1, y_1). Indeed, when $x_2 \to x_1$, the line l tends to the tangent t, if such a tangent exists at all. This process is shown in Fig. 9.4.

When Δy and Δx both tend to zero, their quotient $\Delta y/\Delta x$, that is, the slope of the chord, tends to the slope of the tangent to the curve at the point (x_1, y_1). Denoting this slope by a_t we may summarize the limiting process as follows:

$$x_2 \to x_1, \qquad\qquad \Delta x \to 0,$$
$$y_2 \to y_1, \qquad\qquad \Delta y \to 0,$$
$$a = \frac{\Delta y}{\Delta x} \to a_t \quad \text{or} \quad \lim_{x_2 \to x_1} \frac{\Delta y}{\Delta x} = a_t. \tag{9.1.13}$$

In differential calculus the phrase "slope of a tangent" is sometimes replaced by the word *gradient*.

In many applications the fraction $\Delta y/\Delta x$ and its limit are not interpreted as a slope of a tangent or as a gradient. Instead, the meaning is a rate of change. Thus we return to our original question of how to define an "instantaneous" rate of change. Here is the definition based on the process in (9.1.13):

Let $y = f(x)$ be a function defined in an interval which contains the point x_1. We assume that the limit

$$\lim_{x_2 \to x_1} \frac{\Delta y}{\Delta x} \tag{9.1.14}$$

exists. Then this limit is called the instantaneous rate of change of y at x_1.

Notice that x_2 may tend to x_1 either from the right, that is, $x_2 \downarrow x_1$, or from the left, that is, $x_2 \uparrow x_1$.

When the limit (9.1.14) exists, we call the function $y = f(x)$ *differentiable* at x_1.

There is a variety of different notations for the limit of $\Delta y/\Delta x$. All of them are frequently used. One notation is y' (read: y prime), a similar one $f'(x_1)$; another notation is $\frac{dy}{dx}$ or $\frac{df}{dx}$, the same in print dy/dx or df/dx (for dy/dx we read: dy by dx); and still another notation is $Df(x)$. Thus we have[3]

$$\lim_{x_2 \to x_1} \frac{\Delta y}{\Delta x} = y' = f'(x_1) = \frac{dy}{dx} = \frac{df}{dx} = Df(x_1). \tag{9.1.15}$$

---

[3] One of the inventors of the calculus, Isaac Newton (1642—1727), used the symbol $\dot{y}$ which is still popular in dynamics. Our y' is similar to $\dot{y}$. The other inventor, Gottfried Wilhelm Leibniz (1646—1716), introduced the symbol $\frac{dy}{dx}$. He did not say that Δy and Δx tend to zero, but declared these increments as "infinitesimal" quantities and wrote for them dy and dx. The term "infinitesimal" or "infinitely small" caused confusion for about two centuries. Later mathematicians abandoned this term. In mathematics "infinitesimal" is not defined. When used, it has only intuitive meaning. In formulas, do not confuse dx with $d \cdot x$, that is, with the multiplication of d by x.

The limit is called the *derivative* of the function $y = f(x)$ at $x = x_1$. dy/dx is also called the derivative of y with respect to x.

Example 9.1.8. If $N = f(t)$ is the number of individuals in an *animal or plant population*, we cannot immediately apply the new concept of instantaneous rate of change since $N = f(t)$ is a *discontinuous* function of time (cf. Example (h) at the end of Section 8.4). When the population size

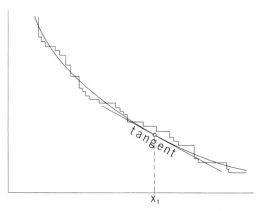

Fig. 9.5. A smooth curve replaces the graph of a discontinuous function. At each point the smooth curve has a tangent. Its slope is called the instantaneous rate of change

remains constant, the instantaneous rate of change is zero. When a birth or a death occurs, it would be infinite.

It also appears to be impossible to consider an instantaneous rate of decay in nuclear physics because fission is spontaneous and not a continuous process. However, the number of radioactive atoms in a sample is usually so high that we may replace the graph of a discontinuous decay by a *smooth curve* without committing a relevant numerical error. Fig. 9.5 illustrates the procedure.

For a smooth curve there exists a tangent at each point. Thus we may define the *instantaneous rate of decay* at x_1 as the slope of the tangent to the point with abscissa x_1. In the notation of differential calculus we get

$$\frac{dN}{dt} = N'(t_1) = \lim_{t_2 \to t_1} \frac{\Delta N}{\Delta t}. \tag{9.1.16}$$

For a very *large animal or plant population* we may employ the same procedure. This allows us not only to calculate an average rate of change, but also an *instantaneous rate of change*. Examples of population growth will be studied in Chapter 11.

Example 9.1.9. A *chemical reaction* is, strictly speaking, a discontinuous process because of the molecular structure of matter. However, the number of participating molecules is usually so high that the process appears to be continuous. Thus it is possible to introduce the notion of an *instantaneous reaction rate*. Following the notation in formula (9.1.4), we may write $dM/dt = M'(t_1)$ for this rate.

Example 9.1.10. According to the law of inertia, *motion* of a body is a continuous process. A body can neither accelerate nor decelerate in zero time. Hence, there is no difficulty proceeding from an average velocity to the notion of *instantaneous velocity* at time t_1. Applying formula (9.1.6) we define it by

$$\frac{ds}{dt} = s'(t_1) = \lim_{t_2 \to t_1} \frac{\Delta s}{\Delta t}. \tag{9.1.17}$$

Similarly the *instantaneous acceleration* at time t_1 is derived from formula (9.1.7) as follows:

$$\frac{dv}{dt} = v'(t_1) = \lim_{t_2 \to t_1} \frac{\Delta v}{\Delta t}. \tag{9.1.18}$$

Notice that in this limiting process $v(t)$ is the instantaneous velocity $\dfrac{ds}{dt}$ as defined by formula (9.1.17).

Example 9.1.11. When forming an instantaneous rate of change, time need not be involved. Assume that a large number of particles is spread on the x axis. Let $M = M(x)$ be the total mass of particles that lie between the origin O and a point with abscissa x (Fig. 9.2). Strictly speaking, $M(x)$ is a discontinuous function. But as in Example 9.1.8 we may replace it by a *smooth* function. Then, based on Eq. (9.1.9), we get

$$\frac{dM}{dx} = M'(x_1) = \lim_{x_2 \to x_1} \frac{\Delta M}{\Delta x}. \tag{9.1.19}$$

This limit means the *density* or the *concentration* of the particles at the point x_1 on the x axis. Notice that such density is "instantaneous" and not an average.

9.2. Differentiation

The operation of finding the derivative of a function is called *differentiation*. We say that we *differentiate* $y = f(x)$ with respect to x.

Differentiation is a special type of limiting process. It is somewhat easier to understand if we replace x_1 by x and x_2 by $x+h$. Thus $\Delta x = h$. The increment h may be positive or negative, but not zero. The value of $y = f(x)$ at the point $x+h$ is denoted by $f(x+h)$. Thus, $\Delta y = f(x+h) - f(x)$ and

$$\frac{\Delta y}{\Delta x} = \frac{f(x+h) - f(x)}{h}. \qquad (9.2.1)$$

Fig. 9.6 illustrates the new notations.

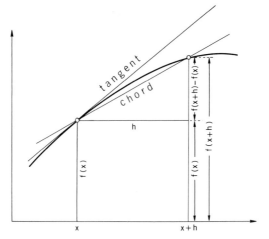

Fig. 9.6. An alternative approach for finding the derivative

In the limiting process, x remains a constant, but h tends to zero. Notice that we cannot simply put $h=0$, since the fraction (9.2.1) would turn into $\frac{0}{0}$ which is meaningless.

The derivative of $f(x)$ at x is now

$$\frac{dy}{dx} = f'(x) = \lim_{h \to 0} \frac{f(x+h) - f(x)}{h}, \qquad (9.2.2)$$

provided that $y = f(x)$ is *differentiable* at x.

An example may illustrate how the limiting process is performed. Let

$$y = f(x) = x^2$$

be a quadratic power function. The domain consists of all real numbers. Then, according to formula (1.14.5), we get

$$f(x+h) = (x+h)^2 = x^2 + 2xh + h^2.$$

Hence

$$\frac{f(x+h)-f(x)}{h} = \frac{2xh+h^2}{h}.$$

Here x is a fixed abscissa, whereas h is a variable quantity tending to zero. Since $h \neq 0$, we may simplify the fraction by dividing numerator and denominator by h. The fraction reduces to $2x+h$. For $h \to 0$, we obtain the derivative

$$y' = f'(x) = (x^2)' = \frac{dy}{dx} = 2x. \qquad (9.2.3)$$

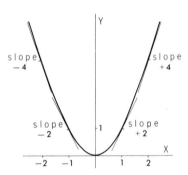

Fig. 9.7. Tangents to the parabola with equation $y = x^2$. The slope of the tangents is $y' = 2x$

If x is given numerically, we have simply to double this number in order to get the rate of change or, geometrically speaking, the slope of the tangent.

During the limiting process, x was a fixed abscissa. Now we learn from (9.2.3) that the result has been found for an arbitrary value of x. To each given x, there exists a unique value of the derivative, namely $2x$. Therefore, the derivative is a function of x. Fig. 9.7 depicts the graph of $y = x^2$, a quadratic parabola, and shows some tangents to it.

For $x = 0$ we get $y' = 0$. Indeed, the tangent to the parabola at $x = 0$ is the x axis with slope 0. For $x = 1$ we find $y' = 2 \times 1 = 2$. For $x > 0$ we obtain $y' > 0$, that is, all tangents on the positive side of the x axis are ascending. For $x = -1$ we obtain $y' = 2 \times (-1) = -2$. For $x < 0$, we get $y' < 0$, that is, all tangents on the negative side of the x axis are descending.

In general, assume that $y = f(x)$ is differentiable at each point of its domain. Then $f'(x)$ *is a function of x on the same domain.* From this viewpoint it would be more appropriate to call the derivative a *derived function.* In our example, the derived function is $y' = 2x$.

It is easy to treat the cubic function $y = x^3$ in the same way as $y = x^2$. We leave it to the reader to prove that $y' = (x^3)' = 3x^2$. However, it is difficult to find the derivative of a power function $y = x^n$, when n is a negative or a fractional number. Fortunately there exists a unified method using logarithms. We will learn this method at the end of Section 10.7. Here we confine ourselves to state the result:

$$\boxed{(x^n)' = nx^{n-1}} \,. \qquad (9.2.4)$$

The formula is correct for all real values of n and all values of x that belong to the domain of $y = x^n$. It is worth learning the result by heart: *The derivative of x^n with respect to x is n times the $(n-1)$st power of x.* The reader may check the special cases $(x^2)' = 2x$ and $(x^3)' = 3x^2$. Included also is the trivial case of a constant $y = x^0 = 1$ which must have derivative $y' = 0$. Generally, for $y = c$, c being a *constant*, we get

$$y' = \frac{dy}{dx} = 0 \,. \qquad (9.2.5)$$

Other special cases of (9.2.4) are

$$(1/x)' = (x^{-1})' = -x^{-2} \,,$$
$$(1/x^2)' = (x^{-2})' = (-2) x^{-3} \,, \qquad (9.2.6)$$
$$(\sqrt{x})' = (x^{\frac{1}{2}})' = \tfrac{1}{2} x^{-\frac{1}{2}} \,.$$

We will postpone the derivative of exponential and logarithmic functions to Chapter 10. But we are prepared to treat trigonometric functions. Let $y = f(x) = \sin x$ and assume that x is measured in *radians* (see Section 5.2). Then

$$\Delta y = f(x + h) - f(x) = \sin(x + h) - \sin x \,.$$

By means of formula (5.7.6) we get

$$\sin(x + h) - \sin x = 2 \cos \frac{2x + h}{2} \sin \frac{h}{2} \,.$$

Thus

$$\frac{\Delta y}{\Delta x} = \frac{2}{h} \cos \frac{2x + h}{2} \sin \frac{h}{2} = \cos\left(x + \frac{h}{2}\right) \frac{\sin h/2}{h/2} \,.$$

In this formula we replace $h/2$ by k and notice that $k \to 0$ as $h \to 0$. Thus

$$\frac{\Delta y}{\Delta x} = \cos(x + k) \frac{\sin k}{k} \,.$$

Because of Eq. (8.2.4) we obtain

$$\frac{\sin k}{k} \to 1 .$$

and therefore,

$$\boxed{(\sin x)' = \cos x} .\qquad\qquad (9.2.7)$$

The assumption that x is measured in radians was implicitly used with (8.2.4).

Special values of this derivative are

x	$(\sin x)'$
0	1
$\pi/2$	0
π	-1

The reader may check these values by using Fig. 5.8.

Also left to the reader (see Problem 9.2.11) is the proof of

$$\boxed{(\cos x)' = - \sin x} .\qquad\qquad (9.2.8)$$

Knowing the derivative of a function $y = f(x)$, we are able to find the derivative of another function $y = c \cdot f(x)$ obtained by multiplying $f(x)$ by a *constant factor* c:

$$\frac{\Delta y}{\Delta x} = \frac{c f(x+h) - c f(x)}{h} = c \frac{f(x+h) - f(x)}{h} \to c f'(x) .$$

Therefore,

$$\boxed{(c \cdot f(x))' = c \cdot f'(x)} .\qquad\qquad (9.2.9)$$

We have simply to carry the same factor c. Thus for $y = 5x^2$ we get $y' = 5 \times 2x = 10x$. A trivial case is the special linear function $y = ax$. Here, $y' = a$ in agreement with the fact that the factor a is the slope of the straight line with equation $y = ax$.

Assume further that the derivatives of two functions $y = f(x)$ and $y = g(x)$ are known. Then the new function $y = f(x) + g(x)$, that is, the *sum* of $f(x)$ and $g(x)$, has derivative

$$\boxed{(f(x) + g(x))' = f'(x) + g'(x)}\ .\qquad(9.2.10)$$

Indeed, $\Delta y = [f(x+h) + g(x+h)] - [f(x) + g(x)] = [f(x+h) - f(x)] + [g(x+h) - g(x)]$. Hence

$$\frac{\Delta y}{\Delta x} = \frac{f(x+h) - f(x)}{h} + \frac{g(x+h) - g(x)}{h} \to f'(x) + g'(x)\ .$$

Example 9.2.1. $y = x^2 + x^3,\ y' = 2x + 3x^2$;

Example 9.2.2. $y = ax + b,\ y' = a + 0 = a$ (slope of a straight line);

Example 9.2.3. $u = a \sin t + b \cos t,\ \dfrac{du}{dt} = a \cos t - b \sin t$.

To differentiate a function such as $y = x^2 \cdot \sin x$, we have to apply the rule for differentiating the *product* $y = f(x) \cdot g(x)$ of two functions $y = f(x)$ and $y = g(x)$. We get

$$\begin{aligned}
\Delta y &= f(x+h)\,g(x+h) - f(x)\,g(x) \\
&= f(x+h)\,g(x+h) - f(x)\,g(x+h) + f(x)\,g(x+h) - f(x)\,g(x) \\
&= [f(x+h) - f(x)]\,g(x+h) + f(x)\,[g(x+h) - g(x)]
\end{aligned}$$

and, therefore,

$$\frac{\Delta y}{\Delta x} = \frac{f(x+h) - f(x)}{h} \cdot g(x+h) + f(x) \cdot \frac{g(x+h) - g(x)}{h}\ .$$

As $h \to 0$, $\Delta y / \Delta x$ tends to

$$\boxed{(f(x) \cdot g(x))' = f'(x) \cdot g(x) + f(x) \cdot g'(x)}\ .\qquad(9.2.11)$$

Hence, to get the derivative of $f(x) \cdot g(x)$ we have first to differentiate $f(x)$ and to multiply by $g(x)$. Second, we have to differentiate $g(x)$ and to multiply by $f(x)$. Third, we have to add the two expressions.

Example 9.2.4. $y = x^2 \cdot \sin x$. We identify $f(x) = x^2$ and $g(x) = \sin x$. We know that $f'(x) = 2x$ and $g'(x) = \cos x$. Hence,

$$(x^2 \cdot \sin x)' = 2x \cdot \sin x + x^2 \cdot \cos x\ .$$

Before we study the quotient of two functions, we derive the important rule for differentiating a "function of a function". We begin with an introductory example: $y = \sin^2 x$. This is the square of sin x. We may disentangle this function by introducing the auxiliary function $u = u(x) = \sin x$. Then $y = u^2$. Hence, y is a quadratic function of u, whereas u is the sine of x. In general we may write

$$y = f(u(x)) \tag{9.2.12}$$

which means that $u = u(x)$ and $y = f(u)$. y is directly a function of u and indirectly a function of x.

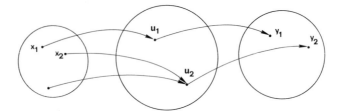

Fig. 9.8. Composite function: A variable x is mapped into u, and u mapped into y

The independent variable x is first mapped into the variable u, and the u into the variable y. Both, u and y, are dependent variables (Fig. 9.8). $u(x)$ is sometimes called the *inner function* and $f(u)$ the *outer function*. $f(u(x))$ is most frequently called a *composite function*.

Now, when x is increased by h, $u = u(x)$ also changes in value, say

$$\Delta u = u(x + h) - u(x) = k \neq 0,$$

provided that $u(x)$ does not remain constant. As a consequence, y changes by

$$\Delta y = f(u + k) - f(u).$$

Hence,

$$\frac{\Delta y}{\Delta x} = \frac{f(u+k) - f(u)}{h}$$

$$= \frac{f(u+k) - f(u)}{k} \cdot \frac{k}{h} = \frac{f(u+k) - f(u)}{k} \cdot \frac{u(x+h) - u(x)}{h}.$$

$f(u)$ and $u(x)$ are assumed to be differentiable. Hence, for $h \to 0$ we also get $k \to 0$, and

$$\frac{dy}{dx} = (f(u(x)))' = f'(u) \cdot u'(x). \tag{9.2.13}$$

This is the famous *chain rule*. It states that we have first to derive $f(u)$ with respect to u, second, to derive $u(x)$ with respect to x and third, to multiply the two expressions.

Formula (9.2.13) may be rewritten in a form which can be easily memorized[4]

$$\frac{df}{dx} = \frac{df}{du} \cdot \frac{du}{dx}. \qquad (9.2.14)$$

Chain rule: The derivative of a composite function is equal to the derivative of the outer function times the derivative of the inner function.

Example 9.2.5. $y = \sin^2 x$. Here, $u(x) = \sin x$ and $f(u) = u^2$. Hence, $du/dx = \cos x$, $df/du = 2u$, and

$$(\sin^2 x)' = 2u \cdot \cos x = 2 \sin x \cos x.$$

Example 9.2.6. $y = \sqrt{1 + 5x}$. Here, $u(x) = 1 + 5x$ and $f(u) = u^{\frac{1}{2}}$. Hence, $du/dx = 5$, $df/du = \frac{1}{2}u^{-\frac{1}{2}}$, and

$$(\sqrt{1 + 5x})' = \frac{1}{2}u^{-\frac{1}{2}} \cdot 5 = \frac{5}{2}(1 + 5x)^{-\frac{1}{2}}.$$

Example 9.2.7. $K = \sin(at + b)$. Here, $u = u(t) = at + b$ and $K = f(u) = \sin u$. Hence, $du/dt = a$, $dK/du = \cos u$, and

$$(\sin(at + b))' = a \cdot \cos(at + b).$$

An especially important application of the chain rule is the differentiation of $1/u(x)$. This is the reciprocal of a function $u(x)$. We have to differentiate first $1/u$ with respect to u and second, u with respect to x. Formula (9.2.4) with $n = -1$ yields $(1/u)' = -1/u^2$. Therefore,

$$(1/u(x))' = -\frac{1}{u(x)^2} \cdot u'(x). \qquad (9.2.15)$$

Further, to derive the quotient of two functions $y = f(x)$ and $y = g(x)$, that is, the new function $y = f(x)/g(x)$, we may combine the formulas (9.2.11) and (9.2.15) as follows:

$$\left(\frac{f(x)}{g(x)}\right)' = \left(f(x) \cdot \frac{1}{g(x)}\right)' = f'(x) \cdot \frac{1}{g(x)} + f(x) \cdot \left(-\frac{1}{g(x)^2} \cdot g'(x)\right).$$

---

[4] The result resembles the rule for reducing a fraction to the lowest terms. However, it should be kept in mind that we introduced $\dfrac{df}{du}$ and $\dfrac{du}{dx}$ as symbols and not as fractions of df and du, or of du and dx, respectively.

The result may be rewritten in the form

$$\left(\frac{f(x)}{g(x)}\right)' = \frac{g(x)f'(x) - f(x)g'(x)}{g(x)^2}. \tag{9.2.16}$$

When learning it by heart, we may use the obvious formulation:

$$\left(\frac{\text{numerator}}{\text{denominator}}\right)'$$

$$= \frac{\text{denominator} \times (\text{numerator})' - \text{numerator} \times (\text{denominator})'}{\text{denominator squared}}.$$

Example 9.2.8.

$$y = \frac{3x - 5}{2x + 7}, \quad y' = \frac{(2x + 7) \cdot 3 - (3x - 5) \cdot 2}{(2x + 7)^2} = \frac{31}{(2x + 7)^2}.$$

Example 9.2.9. $y = \tan x$. We assume that x is measured in radians. Then from (5.5.6) we get $y = \sin x / \cos x$ and from (9.2.7) and (9.2.8) $(\sin x)' = \cos x$, $(\cos x)' = -\sin x$. Hence,

$$(\tan x)' = \left(\frac{\sin x}{\cos x}\right)' = \frac{\cos x(\cos x) - \sin x(-\sin x)}{\cos^2 x}.$$

This becomes, by means of formula (5.7.1),

$$(\tan x)' = 1/\cos^2 x. \tag{9.2.17}$$

Finally, we determine the derivative of an *inverse function*. We introduced the notion of an inverse function in Section 6.3. Given $y = f(x)$, we know that an inverse function $x = g(y)$ exists when $y = f(x)$ is *monotone* over the domain. Now we assume in addition that $f(x)$ is differentiable and ask: Can we express $g'(y)$ in terms of $f'(x)$? This is indeed possible. Let h be an increment of x. Then we denote the corresponding increment of y by k. This means

$$\Delta x = g(y + k) - g(y) = h, \quad \Delta y = f(x + h) - f(x) = k.$$

For $k \neq 0$, it follows that

$$\Delta x / \Delta y = h/k = \frac{1}{k/h} = \frac{1}{\Delta y / \Delta x}.$$

When $h \to 0$, we also have that $k \to 0$, and

$$\frac{\Delta x}{\Delta y} = \frac{g(y + k) - g(y)}{k} \to g'(y) = \frac{dx}{dy}, \quad \frac{\Delta y}{\Delta x} \to f'(x) = \frac{dy}{dx}.$$

Hence,

$$g'(y) = \frac{1}{f'(x)}. \tag{9.2.18}$$

The result is easy to memorize when we change notation:

$$\boxed{dx/dy = \frac{1}{dy/dx}.} \tag{9.2.19}$$

Example 9.2.10. $y = 3x - 4$, $dy/dx = 3$. The inverse function is $x = \frac{1}{3}(y + 4)$. Here, $dx/dy = 1/3$ which is the reciprocal of $dy/dx = 3$.

Example 9.2.11. $y = \sqrt{x}$ with domain $x \geq 0$. According to (9.2.4) we get for $n = 1/2$:

$$\frac{dy}{dx} = (x^{\frac{1}{2}})' = \frac{1}{2}x^{-\frac{1}{2}} = \frac{1}{2\sqrt{x}}.$$

The inverse function is $x = y^2$. Here $dx/dy = 2y$. Using $y = \sqrt{x}$ this becomes $dx/dy = 2\sqrt{x}$ which is indeed the reciprocal of dy/dx.

9.3. The Antiderivative

It frequently happens that we know the derivative of a function, that is, $y' = f'(x)$, and that we want to find $y = f(x)$ itself. $f(x)$ may be called the *antiderivative* of $f'(x)$. For instance, if we are given $y' = 2x$, then we know that $y = x^2$ is an antiderivative.

The question then arises whether $f'(x)$ has a unique antiderivative. The answer is "no". An example may serve to explain the reason. Let $y' = f(x) = 2x$. Then not only $y = x^2$, but also $y = x^2 + c$ with an arbitrary constant c is a correct antiderivative. This follows from formulas (9.2.5) and (9.2.10). The same fact can be seen geometrically: $y' = 2x$ associates with each x a slope of a tangent, but the equation does not specify the point of tangency. Fig. 9.9 shows that every parabola with equation $y = x^2 + c$ satisfies the requirements stated by $y' = 2x$.

Similarly, given $y' = a$, a being a constant, we get $y = ax + c$. The geometric interpretation is left to the reader.

In general, *given a derivative $f'(x)$, there exists an infinite set of anti-derivatives $f(x) + c$. These antiderivatives differ from each other by an additive constant.*

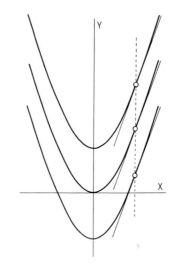

Fig. 9.9. Every function $y = x^2 + c$ is an antiderivative of $y' = 2x$

From formulas derived or quoted in Section 9.2 we get the following table:

derivative	antiderivative
a (constant)	$ax + c$
$n \cdot x^{n-1}$	$x^n + c$
$\cos x$	$\sin x + c$
$-\sin x$	$\cos x + c$
$1/\cos^2 x$	$\tan x + c.$

Applying formulas (9.2.9) and (9.2.10) to these results, we can easily find a host of additional antiderivatives. Examples:

a) $y' = a \cdot \cos x$. The antiderivative is $y = a \cdot \sin x + c$

b) $y' = a \cdot \sin x$. The antiderivative is $y = -a \cdot \cos x + c$

c) $du/dt = 2pt + 3qt^2$. Here we can take the antiderivative of each term separately. We obtain $u = pt^2 + qt^3 + c$.

d) $y' = x^m$. We know that $y = x^n$ implies $y' = nx^{n-1}$. We identify $n - 1$ by m, that is, $n = m + 1$. We also have to compensate for the missing factor n by dividing x^{m+1} by $m + 1$. Thus we obtain for the antiderivative of $y' = x^m$

$$\frac{x^{m+1}}{m+1} + c.$$ (9.3.1)

It is worth checking the result by differentiation. Notice that the division by $m+1$ is only possible if $m \neq -1$. The exceptional case $m = -1$ is not covered by formula (9.3.1). We will postpone this case to Section 10.7.

9.4. Integrals

As we have seen, the major motivations for introducing the differential calculus are problems of growth rate, reaction rate, concentration, velocity, and acceleration. There exists another group of problems, equally important for life scientists, which leads to the *integral calculus*. We invite the reader to study some introductory *examples*:

Example 9.4.1. Let $N = N(t)$ be the number of individuals in an *animal or plant population* as a function of time. In the beginning of Section 9.1 we introduced the average growth rate $\Delta N/\Delta t$. We denote this growth rate now by g and assume for simplicity that $g > 0$. We consider a fixed time interval, say from t_0 to t_z [5]. We subdivide this interval into a number, say n, of smaller intervals:

interval No.	from	to	length of interval
1	t_0	t_1	$\Delta t_1 = t_1 - t_0$
2	t_1	t_2	$\Delta t_2 = t_2 - t_1$
3	t_2	t_3	$\Delta t_3 = t_3 - t_2$
....	....	....	....
n	t_{n-1}	$t_n = t_z$	$\Delta t_n = t_n - t_{n-1}$.

Let $g_1, g_2, ..., g_n$ be the average growth rates in these subintervals, that is, $g_1 = \Delta N_1/\Delta t_1, g_2 = \Delta N_2/\Delta t_2, ...$. Then we may ask: *How big is the total increase in population size expressed in terms of g?* In the first of the subintervals the increase is $\Delta N_1 = g_1 \cdot \Delta t_1$, in the second $\Delta N_2 = g_2 \cdot \Delta t_2$, etc. Hence, the total increase is $\Sigma \Delta N_i$ or, more precisely[6],

$$\sum_{i=1}^{n} g_i \cdot \Delta t_i . \qquad (9.4.1)$$

---

[5] The notations t_0 and t_z were chosen to indicate that t_0 and t_z are prefixed time instants. z is the last letter of the alphabet and may thus help us to remember that t_z is the final time.

[6] See Section 1.9 for the summation sign Σ.

Here we used the average growth rate. In case the growth rate changes continuously, that is, if $g = g(t)$ is a continuous function of time, the solution of the problem is somewhat different. In formula (9.4.1) we may replace each g_i by $g(t_i)$, which has not quite the same value. Thus we introduce an error. However, it is possible to reduce the error and to get a higher precision if we increase the number n of subintervals. The smaller the intervals are, the less g_i differs from $g(t_i)$. We may even try to obtain an exact result by letting n tend to infinity and the length of all subintervals to zero. In Section 9.5 we will show that the sum tends to a limit

$$\lim_{n \to \infty} \sum_{i=1}^{n} g(t_i) \cdot \Delta t_i \qquad (9.4.2)$$

and that this limit is the exact increase in population size.

Notice that we cannot simply put $\Delta t_i = 0$, for no matter how large the number of terms would be, the sum would be zero. Instead, we have to study a limiting process with $\Delta t_i \to 0$.

The limit (9.4.2) is always written in the following standard notation:

$$\int_{t_0}^{t_z} g(t)\,dt . \qquad (9.4.3)$$

We read "integral from t_0 to t_z, $g(t)\,dt$". The word "integral" originated from the idea of making the "whole" out of parts[7].

Example 9.4.2. Assume that particles are spread on the x axis. Let N denote the number of particles that lie between the origin O and a point P with abscissa x_z. As in Section 9.1 we may introduce the notion of *average density* or *average concentration*. We may ask: Can we express N in terms of density or concentration? To answer the question we subdivide the segment from O to P into a number, say n, of subsegments. Let Δx_i be the length of the i-th subsegment and ΔN_i the number of particles in this subsegment ($i = 1, 2, ..., n$). Then, by definition, $C_i = \Delta N_i / \Delta x_i$ is the average concentration of particles in the *i-th* subsegment. Hence,

$$\Delta N_i = C_i \cdot \Delta x_i .$$

The total number of particles, spread over the segment from O to P, is $\Sigma \Delta N_i$, or, more precisely,

$$N = \sum_{i=1}^{n} C_i \cdot \Delta x_i . \qquad (9.4.4)$$

---

[7] The notation in (9.4.3) is due to Gottfried Wilhelm Leibniz (1646—1716), a German philosopher, mathematician, and diplomat. The similarity of (9.4.2) and (9.4.3) is obvious. The integral sign $\int$ is a tall letter S signifying summation. dx was chosen to indicate that Δx tends to zero.

In case $C = C(x)$ is a continuously changing concentration, we replace C_i by a slightly different value $C(x_i)$. To reduce the error we increase the number n of subsegments. As we will show later, an exact result is obtained by a limiting process. We let n tend to infinity and the length of all subsegments tend to zero. Then

$$N = \lim_{n \to \infty} \sum_{i=1}^{n} C(x_i) \cdot \Delta x_i \qquad (9.4.5)$$

or, in standard notation,

$$N = \int_0^{x_z} C(x)\, dx. \qquad (9.4.6)$$

Example 9.4.3. We get still another integral when we determine the *total distance* traveled by a particle in a time interval from t_0 to t_z. Let v_i be the *average velocity* of the particle in the i-th of n subintervals $\Delta t_1, \Delta t_2, \ldots, \Delta t_n$ which subdivide the original interval from t_0 to t_z. Then the distances traveled are $\Delta s_1 = v_1 \cdot \Delta t_1$, $\Delta s_2 = v_2 \cdot \Delta t_2$, etc. Total distance is

$$s = \sum_{i=1}^{n} v_i \cdot \Delta t_i. \qquad (9.4.7)$$

For a continuously changing velocity $v = v(t)$, we adopt the same procedure as in Examples 9.4.1 and 9.4.2. We obtain

$$s = \lim_{n \to \infty} \sum_{i=1}^{n} v(t_i) \cdot \Delta t_i \qquad (9.4.8)$$

or, in standard notation,

$$s = \int_{t_0}^{t_z} v(t)\, dt. \qquad (9.4.9)$$

Example 9.4.4. Assume that a force acts upon a helical spring. The more the spring is extended, the greater is the force F (see Fig. 9.10). To extend a spring by a first segment, say Δs_1, a variable force is required which increases to a value that we denote by F_1. The *mechanical work* done by the force in extending the spring is calculated by

mechanical work = segment × force.

However, in this expression we have to use some *average force*, that is, a suitable value between the forces zero and F_1, say $\bar{F}_1$. Hence, the mechanical work is $\bar{F}_1 \cdot \Delta s_1$.

To extend the spring by an additional segment Δs_2 we have to apply a force which increases continuously from F_1 to F_2 (say). Let $\bar{F}_2$ be a suitable average between F_1 and F_2. Then the mechanical work is $\bar{F}_2 \cdot \Delta s_2$.

Let us repeat this process n times such that the total extension is

$$\Delta s_1 + \Delta s_2 + \cdots + \Delta s_n = s_z .$$

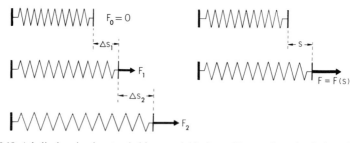

Fig. 9.10. A helical spring is extended by a variable force. The total mechanical work may be expressed by means of an integral

Then we ask: What is the total mechanical work W? Physics shows that mechanical work is some form of energy and that it is additive. Hence,

$$W = \sum_{i=1}^{n} \bar{F}_i \cdot \Delta s_i . \qquad (9.4.10)$$

It is quite inconvenient to operate with average forces $\bar{F}_i$. It is more natural to introduce a variable force $F = F(s)$ where s denotes the segment by which the spring is extended (Fig. 9.10).

$F(s)$ is a continuous function of s. Replacing $\bar{F}_i$ by $F(s_i)$ we commit a small error, but we can reduce the error by subdividing the total segment from 0 to s_z into smaller subsegments. When n tends to infinity and all Δs_i tend to zero, we get the exact mechanical work

$$W = \lim_{n \to \infty} \sum_{i=1}^{n} F(s_i) \cdot \Delta s_i , \qquad (9.4.11)$$

or, in standard notation,

$$W = \int_{0}^{s_z} F(s)\, ds . \qquad (9.4.12)$$

Example 9.4.5. Morphologists have frequently to determine the *area of plane figures*. There is no difficulty in calculating the area of squares, rectangles, triangles, etc. However, when regions are bounded by curves, difficulties arise. We may subdivide the region into squares of equal size, but near the boundary there are squares which overlap the region only partly (see Fig. 9.11). A practical method for estimating an area is *dot counting* (see Hennig, 1967, and Fischmeister, 1967). The region is covered by a network of squares of equal size. When the *center of a square* is inside the boundary, it is marked by a dot. Then all dots are counted. Thus all squares that fall entirely into the region are counted, but only about half

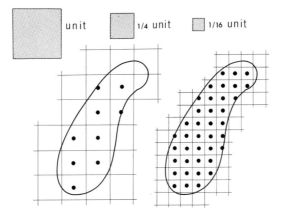

Fig. 9.11. Estimating an area by dot counting. The finer the mesh size is, the smaller the error. In our drawing we get $9 \times 1/4 = 2.3$ units with a coarse network and $39 \times 1/16 = 2.4$ units with a finer network

of the squares intersected by the boundary are counted. Finally the number of dots is multiplied by the area of a single square. This estimates the area of the region.

To get a better approximation we may use a finer network, for instance with a mesh size that is only half of the former one. The process of refining the network may be continued infinitely such that the mesh size tends to zero. The limit is then expected to be the exact area.

In principle, the idea of Example 9.4.5 leads again to the concept of an integral. In order to perform the process analytically, the boundary of the region under consideration or parts of it must be given by an equation. We concentrate on a region R which is enclosed by a curve, by the x axis, and by two lines parallel to the y axis (see Fig. 9.12).

We introduce a segment or an interval on the x axis whose endpoint to the left has abscissa $x = a$ and whose endpoint to the right abscissa

$x = b$. The interval itself is usually denoted by $[a, b]$[8]. The curve is the graph of a continuous function $y = f(x)$. In Fig. 9.12 it is also assumed that $f(x) > 0$ and that the function is monotone decreasing over the interval $[a, b]$.

The region R is a point set and is bounded by the graph of $y = f(x)$, the x axis and lines parallel to the y axis at the endpoints of $[a, b]$.

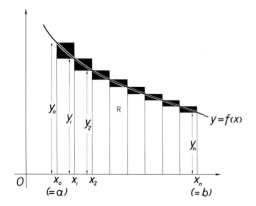

Fig. 9.12. Region R between a curve and the segment $[a, b]$ on the x axis. The area is found by a limiting process

We may approximate the region R by a set of rectangles. For this purpose we subdivide the interval $[a, b]$ into n equal sub-intervals by introducing the abscissas

$$x_0 = a, x_1, x_2, \ldots, x_i, \ldots, x_{n-1}, x_n = b .$$

We put

$$x_i - x_{i-1} = \frac{b - a}{n} = \varDelta x \quad (i = 1, 2, \ldots, n) .$$

The corresponding ordinates are

$$y_0, y_1, y_2, \ldots, y_i, \ldots, y_{n-1}, y_n .$$

Let A be the unknown area of the region R. A can be approximated by the areas of n rectangles that are inside the region R or by the areas

---

[8] $[a, b]$ is a point set which can also be written by the general notation of Chapter 2 as $\{x \mid a \leq x \leq b\}$. Brackets are used to indicate that the two endpoints belong to the interval. In this case the interval is said to be *closed*. If one or both endpoints do not belong to the interval, parentheses are used. Thus $(a, b]$ means that only the upper endpoint, $[a, b)$ that only the lower endpoint belongs to the interval. Finally, (a, b) denotes an *open* interval where both endpoints are missing.

of n rectangles that cover R completely. In the first case the total area is too small, in the second case too large. We denote these approximations by A_l (lower bound) and by A_u (upper bound). From Fig. 9.12 we deduce[9]

$$A_l = y_1 \cdot \Delta x + y_2 \cdot \Delta x + \cdots + y_n \cdot \Delta x,$$
$$A_u = y_0 \cdot \Delta x + y_1 \cdot \Delta x + \cdots + y_{n-1} \cdot \Delta x,$$

(9.4.13)

$$A_l \leqq A \leqq A_u.$$

(9.4.14)

To get an idea of the error of approximation we subtract A_l from A_u term by term:

$$A_u - A_l = y_0 \cdot \Delta x - y_n \cdot \Delta x.$$

Since $y_0 = f(x_0) = f(a)$ and $y_n = f(x_n) = f(b)$, the difference becomes

$$A_u - A_l = (f(a) - f(b)) \Delta x.$$

(9.4.15)

In Fig. 9.12 this difference is the total area of the small black rectangles. Now we proceed to finer and finer subdivisions of the interval $[a, b]$. We let n tend to infinity. Then $\Delta x \rightarrow 0$, and from (9.4.15) it follows that $A_u - A_l \rightarrow 0$. Hence, A_u and A_l tend to the same limit, and (9.4.14) yields

$$A = \lim_{n \to \infty} A_l = \lim_{n \to \infty} A_u.$$

This common limit is written with the integral sign in the form

$$A = \int_a^b f(x)\, dx.$$

(9.4.16)

It is left to the reader to discuss the case where $y = f(x)$ is monotone increasing instead of decreasing and to show that the result is the same.

9.5. Integration

So far, we have introduced integrals as limits of a sum. Now we want to learn how to evaluate an integral. The operation of finding an integral is called *integration*. We will show that integration is, in some sense, the converse of differentiation and involves the antiderivative which we studied in Section 9.3.

To approach the main result we use the geometric interpretation of an integral as it was introduced in the previous section for a function $y = f(x)$ with $f(x) > 0$. There, the integral was the area of a region between a curve and an interval $[a, b]$ on the x axis. Now we slightly

---

[9] For calculating the area of such a rectangle by the formula "length × height" we have to assume that x and y are measured in the same unit of length.

change notation. We let $b = x$ and consider x as variable. The abscissa x may take on any value from the domain of $y = f(x)$. Let A_a^x be the area of the region between the graph of $y = f(x)$ and the interval $[a, x]$ (see Fig. 9.13). For $x = a$ the interval is reduced to a point, hence $A_a^x = A_a^a = 0$. For increasing x, A_a^x also increases. With each x there is uniquely associated an area A_a^x. Hence, A_a^x is a function of x which is called *area function*. We may write

$$A_a^x = F(x). \tag{9.5.1}$$

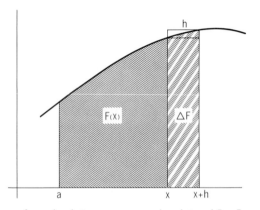

Fig. 9.13. The area of a region between a curve and an interval $[a, x]$ as a function of x

We already know that

$$A_a^a = F(a) = 0. \tag{9.5.2}$$

The next step in our search for a technique of integration is rather surprising: *We try to differentiate $F(x)$ with respect to x.* For this purpose let $\Delta x = h$ be an increase of x, and let $A_a^{x+h} = F(x + h)$ be the new value of the area function. Then $\Delta F = F(x + h) - F(x)$ is the area of the shaded region in Fig. 9.13. ΔF can be approximated by the area of a rectangle. We consider two rectangles, one with sides h and $f(x)$, the other one with sides h and $f(x + h)$. Since, as in Fig. 9.13, $f(x)$ is monotone increasing[10], we get

$$h \cdot f(x) < \Delta F < h \cdot f(x + h). \tag{9.5.3}$$

It follows for the rate of change that

$$f(x) < \frac{\Delta F}{\Delta x} < f(x + h). \tag{9.5.4}$$

---

[10] The reader is invited to study the case where $f(x)$ is monotone decreasing and to show that the limiting process leads to the same result.

Since $f(x)$ is continuous, $f(x+h)$ tends to $f(x)$ as $h \to 0$. Hence, the limit of $\Delta F/\Delta x$, that is, the derivative $dF/dx = F'(x)$ exists and

$$\frac{dF}{dx} = F'(x) = f(x). \qquad (9.5.5)$$

In words: *The derivative of the area function $F(x)$ is the given function $f(x)$*. Or in other words: *The area function $F(x)$ is a certain antiderivative of $f(x)$*.

Let $I(x)$ be an *arbitrary antiderivative of $f(x)$*. Then we know from Section 9.3 that $F(x)$ differs from $I(x)$ only by a certain constant, that is,

$$F(x) = I(x) + c. \qquad (9.5.6)$$

For the particular value $x = a$ we get from (9.5.6) and (9.5.2) $F(a) = I(a) + c = 0$. Hence, $c = -I(a)$ and

$$F(x) = I(x) - I(a). \qquad (9.5.7)$$

Finally we return to our original interval $[a, b]$. In (9.5.7) we have simply to replace x by the fixed value b. Therefore,

$$F(b) = A_a^b = I(b) - I(a).$$

The standard notation for the area A_a^b is the integral (9.4.16). Thus

$$\boxed{\int_a^b f(x)\,dx = I(b) - I(a)} \;. \qquad (9.5.8)$$

In this integral the number a is called the *lower limit* and b the *upper limit*[11].

The function to be integrated is called the *integrand*. Thus in formula (9.5.8) $f(x)$ is the integrand. x is called the *variable of integration* and $[a, b]$ the *interval of integration*.

To integrate a continuous function $f(x)$ from a to b we proceed in the following three steps:
1. Find an antiderivative $I(x)$ of $f(x)$.
2. Take the special values $I(a)$ and $I(b)$ for the upper limit a and the lower limit b, respectively.
3. Subtract $I(a)$ from $I(b)$.

Example 9.5.1. Let $y = f(x) = \sin x$ where x is measured in radians. We want to calculate the area between an arc of the sine curve and the x axis

---

[11] The word "limit" is used in this connection quite differently from the usual sense. It is not the result of a limiting process.

(cf. Fig. 5.8). More precisely, we choose $a = 0$ and $b = \pi$. The interval $[0, \pi]$ corresponds to $[0°, 180°]$. Thus we have to evaluate the integral

$$\int_0^\pi \sin x \, dx \, .$$

We know that $-\cos x$ is an antiderivative of $\sin x$. Thus $I(x) = -\cos x$. Then, $I(a) = I(0) = -\cos 0 = -1$, $I(b) = I(\pi) = -\cos \pi = -(-1) = +1$. Finally, $I(b) - I(a) = +1 - (-1) = 2$. Therefore the area of the proposed region is 2.

The procedure is usually written in a rather stenographic way which is self-explanatory:

$$\int_0^\pi \sin x \, dx = (-\cos x)|_0^\pi = (-\cos \pi) - (-\cos 0) = -(-1) - (-1) = 1 + 1 = 2 \, .$$

In general, the antiderivative $I(x)$ is more often called an *indefinite integral*, since the constant c need not be determined. The indefinite integral is written without lower and upper limits:

$$I(x) = \int f(x) \, dx \, . \tag{9.5.9}$$

Notice that the indefinite integral is a function of x with an arbitrary constant. On the other hand, the integral (9.5.8) is uniquely determined by a and b and is therefore called a *definite integral*. If a and b are fixed numbers, the definite integral is also a number. Sometimes the upper limit is considered to be a variable, say $b = x$. Then the integral is a function of its upper limit (see formula (9.5.7)). We called this function an area function:

$$F(x) = \int_a^x f(x) \, dx \, .$$

In this formula the variable x plays a double role. First, x is the variable of integration. Second, x also denotes the upper limit. To avoid confusion it is customary to use two different letters. Therefore, we rewrite $F(x)$ as follows:

$$F(x) = \int_a^x f(t) \, dt \, . \tag{9.5.10}$$

In a definite integral it makes no difference what letter is used for the variable of integration:

$$\int_a^b f(x) \, dx = \int_a^b f(u) \, du = \int_a^b f(\alpha) \, d\alpha \, . \tag{9.5.11}$$

Equation (9.5.5) applied to formula (9.5.10) yields

$$\boxed{\frac{d}{dx} \int_a^x f(t)\, dt = f(x)}\ . \tag{9.5.12}$$

This result is known as *fundamental theorem of integral calculus*. It states:

The derivative of $F(x)$ as defined by Eq.(9.5.10) is $f(x)$.

Replacing $f(x)$ by $F'(x)$ in Eq. (9.5.10) we also get

$$\int_a^x F'(t)\, dt = F(x)\,. \tag{9.5.13}$$

Formulas (9.5.12) and (9.5.13) show that *differentiation and integration are inverse operations*.

Example 9.5.2. From Eq.(9.2.17) we know the derivative of $\tan x$. The inverse operation, is given by Eqs. (9.5.8) and (9.5.10), leads us to

$$\int_a^x \frac{1}{\cos^2 t}\, dt = \tan x - \tan a\,.$$

Here, a and x have to be restricted to an interval in which $\cos t$ does not vanish. Let $a = \pi/4$. Then $\tan \pi/4 = 1$ and

$$\int_{\pi/4}^x \frac{1}{\cos^2 t}\, dt = \tan x - 1\,.$$

According to the fundamental theorem of integral calculus, as stated in Eq.(9.5.12), the derivative of the integral with respect to x is the integrand written with the variable x, that is, $1/\cos^2 x$.

Example 9.5.3. We are not yet able to integrate a function such as $y = u \cdot \sin u$. Yet it is possible to differentiate the integral

$$\int_a^x u \cdot \sin u\, du\,,$$

with respect to the upper limit. Formula (9.5.12) yields

$$\frac{d}{dx} \int_a^x u \cdot \sin u\, du = x \cdot \sin x\,.$$

We derived a main result, formula (9.5.8), by interpreting the integral as an area. The pedagogical advantage of such an interpretation is obvious. However, to cover various applications it is essential to have a more "abstract" concept of integrals in mind. The reader may review the beginning of this section and verify that *all steps can be performed without referring to areas*. He may also use the Examples 9.4.1 through 9.4.4 for finding other interpretations of the same mathematical procedure.

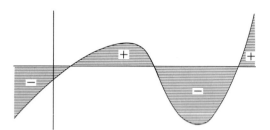

Fig. 9.14. Areas of regions above the *x* axis are positive, of regions below the *x* axis negative

In general, the integrand may take *negative values*. Thus, if $f(x) < 0$ in an interval $[a, b]$, the integral as a limit of $\Sigma f(x_i) \cdot \Delta x$ will also be negative. The geometric interpretation by areas requires that we assign a negative area to a region below the *x* axis (see Fig. 9.14).

Now we learn a few *rules* which facilitate integration:

1. *The interval of integration may be broken up into a number of subintervals*, and the integration performed over each interval separately. Thus, breaking $[a, b]$ up into $[a, c]$ and $[c, b]$ we obtain

$$\int_a^b f(x)\, dx = \int_a^c f(x)\, dx + \int_c^b f(x)\, dx. \tag{9.5.14}$$

The result follows from the definition of an integral as the limit of a sum.

Example 9.5.4. $\int_0^{2\pi} \sin x\, dx = (-\cos x)\big|_0^{2\pi} = (-\cos 2\pi) - (-\cos 0) = (-1)$

$-(-1) = -1 + 1 = 0$. It may be surprising that the region between the sine curve and the *x* axis has area zero. But when we break up the interval of integration into $[0, \pi]$ and $[\pi, 2\pi]$ the reason becomes clear. Indeed, for angles between 0 and π, we find $\int_0^{\pi} \sin x\, dx = +2$ and, for angles between π and 2π, we find $\int_\pi^{2\pi} \sin x\, dx = -2$. The second integral is negative since the region is below the x axis. Hence,

$$\int_0^{2\pi} \sin x \, dx = \int_0^{\pi} \sin x \, dx + \int_{\pi}^{2\pi} \sin x \, dx = 2 + (-2) = 0 .$$

The two areas cancel each other.

2. *Interchanging the limits changes the sign of the integral*, that is,

$$\int_a^b f(x)\, dx = - \int_b^a f(x)\, dx . \qquad (9.5.15)$$

This is an extension of the integral to the case where the upper limit is smaller than the lower limit. Equation (9.5.8) remains valid.

3. *A sum of functions is integrated term by term.* Thus

$$\int [f(x) + g(x)]\, dx = \int f(x)\, dx + \int g(x)\, dx . \qquad (9.5.16)$$

Here the integrals may be definite or indefinite. This rule can be traced back to (9.2.10).

Example 9.5.5. $\int (x^2 + x + 1)\, dx = \int x^2 dx + \int x dx + \int dx = \dfrac{x^3}{3} +$
$+ \dfrac{x^2}{2} + x + c.$ Notice that $\int dx$ stands for $\int 1 \cdot dx.$

4. *A constant factor of the integrand can be put in front of the integral sign:*

$$\int k \cdot f(x)\, dx = k \int f(x)\, dx . \qquad (9.5.17)$$

Here again, the integral may be definite or indefinite. The rule is based on formula (9.2.9).

Example 9.5.6. $\int R \cos t \, dt = R \int \cos t \, dt = R \sin t + c.$

Special techniques of integration, "integration by parts" and "integration by substitution" will be studied in Section 9.10.

9.6. The Second Derivative

Discussion of graphs is a frequent task in all sciences. It is greatly facilitated by introducing the second derivative of the original function.

Let $y = f(x)$ be a differentiable function. The graph of the function is a certain curve. When the *curve rises*, its slope is positive, that is,

$$f'(x) > 0 . \qquad (9.6.1)$$

Conversely, when the *curve falls*, the slope is negative, that is,

$$f'(x) < 0. \tag{9.6.2}$$

We may now ask what occurs at a point where

$$f'(x) = 0. \tag{9.6.3}$$

The curve neither rises nor falls at such a point. We say that the *function* remains *stationary*. Fig. 9.15 shows several points with this property. In case (*a*), *y* reaches a *maximum value* at the point with $f'(x) = 0$. In the

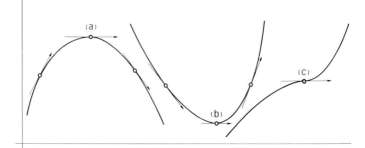

Fig. 9.15. A function remains stationary when $f'(x) = 0$. In case (*a*) there is a maximum point, in case (*b*) a minimum point, and in case (*c*) a point of inflection

neighboring points *y* takes on smaller values. The point itself is called a *maximum point*. In case (*b*), *y* reaches a *minimum value* at a point with $f'(x) = 0$. In the neighboring points *y* takes on greater values. The point itself is called a *minimum point*. In case (*c*), we have neither a maximum nor a minimum point but a *point of inflection* with horizontal tangent.

Thus, a point where the function remains stationary can have quite different properties. We try now to make these distinctions by a more detailed study of the slope. In case (*a*), the slope is continuously decreasing when we move from the left to the right. To the left of the maximum point the slope is positive, to the right it becomes negative. If we consider the slope as a function of *x*, this function is decreasing in the neighborhood of a maximum point. We may reformulate this fact by using the *rate of change of the slope*. In our case, this rate of change is negative. Analytically, the rate of change is the derivative of the slope, that is, the derivative of $f'(x)$. This derivative is called the *second derivative* of $y = f(x)$. We denote it by $f''(x)$ (read: *f* double prime of *x*) or by

$$\frac{d^2 y}{dx^2} \qquad \text{(read: } d \text{ second } y \text{ by } dx \text{ second).}$$

In the neighborhood of a maximum point, and especially at the point itself, the second derivative is usually negative.

For practical applications the following *rule* is useful:

A function $y = f(x)$ reaches a maximum value at x_0 if $f'(x_0) = 0$ and $f''(x_0) < 0$.

In case (*b*), when we move from the left to the right, the slope is an increasing function of x. To the left of the minimum point the slope is negative, to the right it becomes positive. Hence, the slope has a positive rate of change. In other words: The derivative of $f'(x)$ or the second derivative of $f(x)$ is positive. Thus $f''(x) > 0$ in the neighborhood of a minimum point, and usually at the point itself.

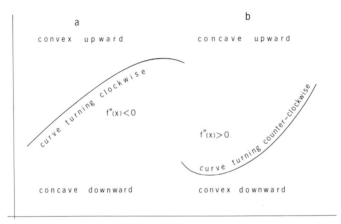

Fig. 9.16. Concavity and convexity of a curve can be judged by the sign of the second derivative $f''(x)$

Again, for practical applications we have the *rule*:

A function $y = f(x)$ reaches a minimum value at x_0 if $f'(x_0) = 0$ and $f''(x_0) > 0$.

The second derivative is also helpful in deciding whether a curve turns clockwise or counter-clockwise when we move from the left to the right. In Fig. 9.16 two cases are distinguished. In case (*a*) the curve turns *clockwise*, the slope is decreasing, and therefore, $f''(x) < 0$ for all values of x. The same property may be formulated by saying that the curve is *convex upward* and *concave downward*.

In case (*b*) the curve turns *counter-clockwise*, the slope is increasing, and therefore, $f''(x) > 0$ for all values of x. The same property may be formulated by saying that the curve is *concave upward* and *convex downward*.

At a *point of inflection* the curve changes from a clockwise turn into a counter-clockwise turn or conversely. It follows that by passing through a point of inflection, $f''(x)$ changes its sign. Hence, at a point of inflection we must have

$$f''(x) = 0. \tag{9.6.4}$$

Fig. 9.17 depicts two points of inflection on a bell-shaped curve.

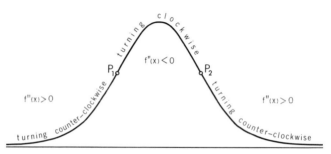

Fig. 9.17. At points of inflection P_1 and P_2, the second derivative is zero

Example 9.6.1. $y = f(x) = x^2$, $f'(x) = 2x$, $f''(x) = 2$. Since the second derivative is positive, the first derivative is an increasing function of x. The graph (see Fig. 9.18) turns counter-clockwise. In other words, the graph is concave upward and convex downward. A minimum point is reached at $x = 0$.

Example 9.6.2. $y = f(x) = 2 - x^2$, $f'(x) = -2x$, $f''(x) = -2$.
Since the second derivative is negative, the first derivative is a decreasing function of x. The graph (see Fig. 9.18) turns clockwise. In other words, the graph is convex upward and concave downward. A maximum point is reached at $x = 0$.

Example 9.6.3. $y = f(x) = \frac{1}{6}x^3 - x$, $f'(x) = \frac{1}{2}x^2 - 1$, $f''(x) = x$.
The second derivative is negative for $x < 0$. Hence, over the negative x axis the slope is decreasing and the curve turns clockwise (see Fig. 9.19). Over the positive x axis the second derivative is positive. Hence, the slope is increasing and the curve is turning counter-clockwise. At $x = 0$ the second derivative is zero and, therefore, the curve has a point of inflection. Finally, $f'(x) = 0$ for $x^2 = 2$. At $x = \sqrt{2}$ the curve has a minimum and at $x = -\sqrt{2}$ a maximum point.

The application of derivatives raises a *logical problem*, namely the distinction between different types of conditions. We have seen

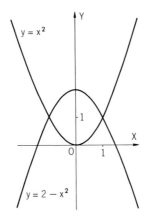

Fig. 9.18. Concavity and convexity of two quadratic parabolas

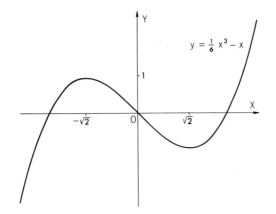

Fig. 9.19. Graph of a cubic function. The curve is called a cubic parabola

that a maximum point can occur at x only if $f'(x)=0$. Such a *condition* is said to be *necessary*, since it has to be satisfied. However, the condition $f'(x)=0$ does not guarantee a maximum point, because it could also indicate a minimum or a point of inflection. Therefore, the condition is said to be *not sufficient*. A sufficient condition for a maximum point at x would be: $f'(x)=0$ and $f''(x)<0$.

The two notions "necessary condition" and "sufficient condition" are very convenient. They have been standard in mathematics for several decades. Now they are entering the sciences slowly. A few more *examples* may help the reader to catch the idea:

a) To get a contagious disease it is necessary to have the infectious agents. But the condition is not sufficient, because an infection need not result in a disease.

b) To become poisoned it is sufficient to swallow ten grams of strychnine. However, the condition is not necessary. A dose of one gram would still suffice. Moreover, it need not be strychnine. There are other chemicals which may also cause poisoning.

c) A triangle is equilateral if every angle has sixty degrees. The condition "all angles measure sixty degrees" is sufficient. If one angle differs from sixty degrees, the triangle cannot be equilateral. Hence, the condition is also necessary. Thus the condition "all angles measure sixty degrees" is necessary and sufficient for a triangle to be equilateral. An alternative phrase to express the same statement is as follows: A triangle is equilateral if, and only if, all three angles are of size sixty degrees.

The expression

if, and only if,

is abbreviated in hand-writing and in print by *iff.*

The reader will find more examples in Problems 9.6.9 through 9.6.14 at the end of this chapter. In addition he may enjoy exploiting Problem 2.10.1 on networks (end of Chapter 2) for the establishment of numerous necessary and/or sufficient conditions.

In many applications, calculation and/or discussion of the second derivative is no easy matter. In order to find out what type of extreme occurs at a point x_0, it is possible to operate with $f'(x)$ alone. In Fig. 9.15a, we get $f'(x) > 0$ to the left of the maximum and $f'(x) < 0$ to the right. Together with $f'(x_0) = 0$, the condition is sufficient for a maximum. For a minimum we find a corresponding statement from Fig. 9.15b. Thus we obtain the

Rule: Let $f(x)$ be differentiable in an interval $[a, b]$. The function reaches a maximum at x_0 in the interior if

$$f'(x_0) = 0 \begin{cases} f'(x) > 0 & when \quad x < x_0 \\ f'(x) < 0 & when \quad x > x_0. \end{cases} \tag{9.6.5}$$

For a minimum the corresponding condition is

$$f'(x_0) = 0 \begin{cases} f'(x) < 0 & when \quad x < x_0 \\ f'(x) > 0 & when \quad x > x_0. \end{cases} \tag{9.6.6}$$

Example 9.6.4. Let

$$f(x) = \frac{2x^2 - 1}{x^2 + 1}.$$

This function is defined for all real values of x. To get extreme values of $f(x)$ we differentiate:

$$f'(x) = \frac{6x}{(x^2 + 1)^2}.$$

The only possibility for $f'(x) = 0$ is $x = x_0 = 0$. It would be laborious to calculate the second derivative and to decide its sign at x_0. It is more economic to apply the preceding rule. Since $f'(x) < 0$ for $x < 0$ and $f'(x) > 0$ for $x > 0$, we conclude by (9.6.6) that $f(x)$ reaches a *minimum* at $x_0 = 0$. The minimum itself is $f(x_0) = -1$.

We end the section with another application of the second derivative:
Let a particle move along a straight line and let $s = s(t)$ be its distance from a fixed point on the line. Then we know from Section 9.1 that the derivative $s'(t)$ is the (instantaneous) *velocity* of the particle:

$$v = v(t) = \frac{ds}{dt} = s'(t). \tag{9.6.7}$$

If the velocity is not a constant, we may ask for the (instantaneous) acceleration, that is, the rate of change of velocity. Hence, we have to differentiate $v(t)$ with respect to t. Then $v'(t)$ is the second derivative of $s(t)$ and the first derivative of $s'(t)$. Thus the *acceleration* of the particle is:

$$a = a(t) = \frac{dv}{dt} = v'(t) = \frac{d^2 s}{dt^2} = s''(t). \tag{9.6.8}$$

Example 9.6.5. For a falling body that is not subject to air resistance, Galilei[12] found the formula

$$s = \frac{g}{2} t^2,$$

t being the time and s the vertical distance traveled. When t is measured in seconds and s in meters, then $g = 9.81$ m/sec^2 at the surface of the earth. From (9.6.7) and (9.6.8) it follows that $v = gt$ and $a = g$. Hence, for a falling body the velocity increases linearly with time and the acceleration remains constant during the motion.

Example 9.6.6. A body oscillates in such a way that its center of gravity remains on the x axis and has abscissa

$$x = 1 + \sin\left(\frac{2\pi}{T} t\right)$$

---

[12] Galileo Galilei (1564—1642), Italian physicist and astronomer.

where $t =$ time (sec), $T = 3$ sec, and x is measured in meters. To discuss the motion we first notice that x is restricted to the interval $[0\,\text{m}, 2\,\text{m}]$. When time t increases from 0 sec to 3 sec, $\dfrac{2\pi}{T}\,t$ grows from 0 to 2π.

Hence, in this time interval the sine completes a full period. This happens again from $t = 3$ sec to $t = 6$ sec, from $t = 6$ sec to $t = 9$ sec, etc. Therefore, x is a periodic function of *period* $T = 3$ sec.

We obtain the instantaneous velocity by differentiation:

$$ v = \frac{dx}{dt} = (2\pi/T) \cos\left(\frac{2\pi}{T} t\right). $$

For example, when $t = 0$ sec, the velocity reaches the value $v = 2\pi/3$ (m/sec). The acceleration is given by the second derivative:

$$ a = \frac{d^2 x}{dt^2} = -(2\pi/T)^2 \sin\left(\frac{2\pi}{T} t\right). $$

At time instant $t = 0$ sec, we obtain $a = 0$ (m/sec^2).

9.7. Extremes

During evolution all those functions of an organism that are essential for surviving under severe conditions have to be maximized. Conversely, "cost" such as requirement for food and protection have to be minimized. Thus in plants the leaves should receive a maximum amount of sunlight and the roots should have access to as many minerals as possible. On the other hand, the time required to adapt to changing conditions should be minimal. In animals the ability to obtain food and to protect against enemies should be as high as possible, whereas susceptibility to disease must be minimized.

From these few examples we see that nature pursues economy in that some quantities are maximized and others minimized. If such a quantity can be expressed as a function of other variables, differential calculus provides a method for finding extreme values, that is, maxima and minima.

Let $y = f(x)$ be a function which is differentiable over an interval, say from a lower boundary $x = a$ to an upper boundary $x = b$ (see Fig. 9.20). As we have seen in Section 9.6, a sufficient condition for a *maximum point* at $x = x_0$ is $f'(x_0) = 0$ and $f''(x_0) < 0$. An example is the point Q in Fig. 9.20.

For a *minimum point* the corresponding condition is $f'(x_0) = 0$ and $f''(x_0) > 0$. In Fig. 9.20 the point P is such a minimum point.

We have also to consider a different type of extreme, namely *extremes taken at boundary points*. At the point R in Fig. 9.20 the function reaches a maximum, since all neighboring points of the curve are lower. Similarly, at S the function reaches a minimum.

The property of a point being an extreme point is purely local. Thus we speak of *local maxima and minima*. Often, however, we are asked to find the highest maximum and the lowest minimum. Then we are concerned with the *absolute maximum* and the *absolute minimum*. In Fig. 9.20 the absolute maximum is taken at R and the absolute

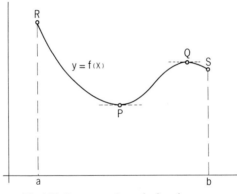

Fig. 9.20. Extreme values of a function

minimum at P. The two absolute extremes determine the range of the function.

Example 9.7.1. $y = 1/x$ defined over the interval from $x = 1$ to $x = 3$. The function is monotone decreasing in the interval. Hence, the maximum value is $y = 1$ for $x = 1$ and the minimum value is $y = 1/3$ for $x = 3$. The two extremes are taken at the boundary. There is no local extreme within the interval from $x = 1$ to $x = 3$. This is also revealed by the first derivative $y' = -1/x^2$ which cannot vanish since equating $y' = 0$ yields no solution.

Example 9.7.2. $f(u) = u^3 - 6u^2 + 32$ over the interval from $u = -1$ to $u = 10$. At the boundary we have the values $f(-1) = 25$ and $f(10) = 432$. The first derivative, $f'(u) = 3u^2 - 12u$, is positive for $u = -1$ as well as for $u = 10$, that is, at both boundary points the function is increasing. Hence, $f(-1) = 25$ is a local minimum and $f(10) = 432$ a local maximum. The first derivative, $f'(u)$, vanishes twice within the interval, namely for $u = 0$ and $u = 4$. The second derivative, $f''(u) = 6u - 12$, is negative for $u = 0$ and positive for $u = 4$. Hence $f(0) = 32$ is a local maximum and

$f(4) = 0$ a local minimum. Finally, the absolute maximum is $f(10) = 432$ and the absolute minimum $f(4) = 0$.

Example 9.7.3. The *bee's cell* is a regular hexagonal prism with one open end and one trihedral apex (see Fig. 9.21 a). We may construct the surface by starting with a regular hexagonal base *abcdef* with side s (Fig. 9.21 b). Over the base we raise a right prism of a certain height h and

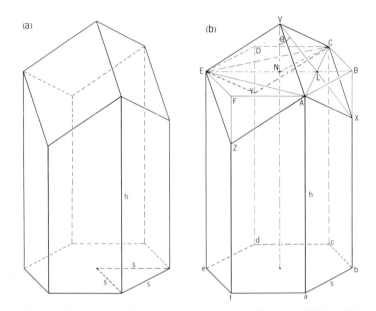

Fig. 9.21. The bee's cell. The figure is redrawn from Thompson (1917, p. 330)

with top $ABCDEF$. The corners B, D, F are cut off by planes through the lines AC, CE, EA, meeting at a point V on the axis VN of the prism, and intersecting Bb, Dd, Ff at X, Y, Z. The three cut-off pieces are the tetrahedrons $ABCX$, $CDEY$, $EFAZ$. We put these pieces on top of the remaining solid such that X, Y, and Z coincide with V. Hereby, the lines AC, CE, EA act as "hinges". The faces $AXCV$, $CYEV$, $EZAV$ are rhombuses, that is, quadrilaterals with equal sides. The new body is the bee's cell and has the same volume as the original prism. The hexagonal base *abcdef* is the open end.

The bees form the faces by using wax. When the volume is given, it is economic to spare wax and, therefore, to choose the angle of inclination, $\theta = \measuredangle NVX$[13], in such a way that the *surface of the bee's cell is minimized.*

The problem can be solved mathematically as follows. Let L be the intersection of CA and VX. Then L bisects the segment NB and, hence, $NL = s/2$. The segment CL is the height of the equilateral triangle BCN. Therefore,

$$CL = \frac{s}{2}\sqrt{3}. \tag{9.7.1}$$

In the triangle NLV we have the relationship

$$VL = \frac{s}{2\sin\theta}. \tag{9.7.2}$$

The rhombus $AXCV$ has its center in L and consists of four congruent right triangles with legs equal to CL and VL. Therefore, from (9.7.1) and (9.7.2) we get

$$\text{area } AXCV = 4 \cdot \frac{1}{2} \cdot \frac{s}{2}\sqrt{3} \cdot \frac{s}{2\sin\theta} = \frac{s^2\sqrt{3}}{2\sin\theta}. \tag{9.7.3}$$

The surface of the bee's cell contains three such areas.

The six lateral faces of the bee's cell, such as $abXA$, are congruent trapezoids. Since $BX = VN$, we obtain from triangle VNL

$$BX = \frac{s}{2}\cot\theta. \tag{9.7.4}$$

Hence,

$$\text{area } abXA = \frac{s}{2}(aA + bX) = \frac{s}{2}(h + h - BX) = hs - \frac{s^2}{4}\cot\theta. \tag{9.7.5}$$

The total area made of wax amounts to

$$6hs - \frac{3}{2}s^2\cot\theta + \frac{3}{2}\frac{s^2\sqrt{3}}{\sin\theta}. \tag{9.7.6}$$

This area is a function of the variable angle θ and, thus, we denote it by $f(\theta)$. We may rewrite $f(\theta)$ in the form

$$f(\theta) = 6hs + \frac{3}{2}s^2\left(-\cot\theta + \frac{\sqrt{3}}{\sin\theta}\right). \tag{9.7.7}$$

---

[13] θ is the lower case Greek letter theta. This letter is frequently used to denote angles.

Only the expression in the parentheses contains the variable θ. Some numerical values rounded-off to two decimals are given in the following table:

θ	$-\cot\theta + \dfrac{\sqrt{3}}{\sin\theta}$
10°	4.30
20°	2.32
30°	1.73
40°	1.50
50°	1.42
60°	1.42
70°	1.48
80°	1.58
90°	1.73

The minimum of $f(\theta)$ is reached somewhere between $\theta = 50°$ and $\theta = 60°$. To get the optimal angle, say θ_0, we differentiate $f(\theta)$ (cf. Problem 9.2.20):

$$f'(\theta) = \frac{3}{2}s^2\left(\frac{1}{\sin^2\theta} - \frac{\sqrt{3}\cos\theta}{\sin^2\theta}\right). \tag{9.7.8}$$

The derivative vanishes if, and only if,

$$1 = \sqrt{3}\cos\theta. \tag{9.7.9}$$

Hence, $\cos\theta_0 = 1/\sqrt{3} = 0.57735$ and $\theta_0 = 54.7°$. Notice that the optimal angle θ_0 is independent of the choice of s and h.

It is worth comparing the result with the actual angle chosen by the bees. It is difficult to measure this angle. However, the average of all measurements does not differ significantly from the theoretical value $\theta_0 = 54.7°$. Therefore, the bees prefer strongly the optimal angle. It is unlikely that the result is due to chance. We may rather suppose that selection pressure had an effect on the angle θ.

For more details and for the amazing history of this problem see Thompson (1917, p. 323 ff.). For comments see also Bailey (1967, p. 8), Tóth (1964) and Weyl (1952, p. 90).

Example 9.7.4. It is well known that *homing pigeons* avoid flying over large areas of water unless they are forced to do so. The reason for this behavior is not known at the present time. In our example we suppose that pigeons prefer a detour around a lake since at daytime the air is falling over the cool water, a phenomenon which increases the energy required for maintaining altitude in flight.

In Fig. 9.22 we assume that a pigeon is released from a boat (point B) floating on the west side of a lake, whereas the loft (point L) is located on the south-east bank. The shortest route from B to L is indicated by a dashed line. However, the pigeon makes a detour. First it heads to a certain point P on the southern bank, not too far away from B, then it follows the bank eastward to L. For simplicity, we assume that the bank is straight in the east-west direction. The question arises: Where

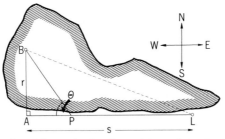

Fig. 9.22. A pigeon released from a boat makes a detour when returning to its loft. It is assumed that the total energy required for homing be minimized

should the point P be chosen in order to minimize the energy required for the flight from B to L? In other words, what is the optimal angle BPL?

Let A be the point on the bank exactly south of B and let $\theta = \measuredangle APB$. We put $AB = r$ and $AL = s$. Then

$$BP = r/\sin\theta, \qquad AP = r \cdot \cot\theta.$$
$$PL = AL - AP = s - r \cdot \cot\theta. \tag{9.7.10}$$

We denote the energy required for flying one unit of length over the lake by e_1 and along the bank by e_2. We assume that no horizontal wind interferes. For reasons mentioned above, we have $e_1 > e_2$ or $e_1 = ce_2$ with a certain constant $c > 1$. The energy required for flying from B to P is $e_1 \cdot BP$ and for continuing from P to L is $e_2 \cdot PL$. Hence, the total energy E turns out to be

$$\begin{aligned} E &= e_1 \cdot BP + e_2 \cdot PL \\ &= e_1 r/\sin\theta + e_2(s - r \cdot \cot\theta) \\ &= e_2 s + e_2 r(c/\sin\theta - \cot\theta). \end{aligned} \tag{9.7.11}$$

Only the expression in the last parentheses depends on θ. Hence it suffices to minimize the function

$$y = c/\sin\theta - \cot\theta. \tag{9.7.12}$$

The angle θ ranges from its highest value $90°$ to its lowest value $\measuredangle PLB$. If L is sufficiently far away, the optimal angle, say θ_0, is expected to fall in between. To determine θ_0, we differentiate y with respect to θ and equate the result to zero. By a calculation similar to that leading to formula (9.7.8) we get

$$\frac{dy}{d\theta} = \frac{1 - c \cdot \cos\theta}{\sin^2\theta}. \tag{9.7.13}$$

This expression vanishes for $1 - c \cdot \cos\theta = 0$. Hence,

$$\cos\theta_0 = 1/c. \tag{9.7.14}$$

(Numerical example: $c = 2$, $\cos\theta_0 = \frac{1}{2}$, $\theta_0 = 60°$).

One may check that the energy E increases whenever θ deviates from θ_0. Notice that the optimal angle θ_0 does not depend on r and s.

Example 9.7.5. The *blood vascular system* consists of arteries, arterioles, capillaries, and veins. The transport of blood from the heart through all organs of the body and back to the heart should be as effective as possible. With a minimal energy expenditure, the body should be

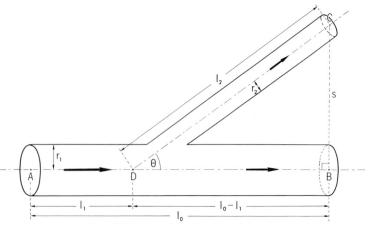

Fig. 9.23. Branching of blood vessels and the search for an optimal angle θ

fed quickly by the constituents of blood. Optimality has to be reached in several ways. For instance, each vessel should be wide enough to avoid turbulence, and erythrocytes should be kept at a size as to minimize viscosity.

In our example, we restrict ourselves to a special optimization problem, that of *vascular branching*. We assume that a main vessel of radius r_1 runs along the horizontal line from A to B in Fig. 9.23. A point C should be reached by a branch of a given radius r_2. For simplicity, we have chosen B such that CB is perpendicular to AB. Let $CB = s$ and let D be the point where the axis of the branching vessel intersects the axis of the main vessel. We denote the angle BDC by θ. The problem that we consider is then: *Find the particular angle $\theta = \theta_0$ which minimizes the total resistance of the blood along the path ADC.* The smaller the resistance is, the less energy is spent by the pumping heart.

To solve the problem, we need a law due to Poiseuille (cf. Section 4.5). In laminar flow, the resistance R is proportional to the length l of the vessel and inversely proportional to the fourth power of its radius r, that is,

$$R = k \frac{l}{r^4}, \qquad (9.7.15)$$

k being a constant factor determined by the viscosity of blood.

Let $AB = l_0, AD = l_1, DC = l_2$. Then it follows by inspection of the right triangle BDC that

$$l_2 = \frac{s}{\sin\theta}, \qquad l_0 - l_1 = s \cdot \cot\theta. \qquad (9.7.16)$$

This permits us to express l_1 and l_2 in terms of l_0, s and θ. The total resistance R along the path ADC is the sum of resistance R_1 along AD and resistance R_2 along DC. It follows from formula (9.7.15) that

$$R = R_1 + R_2 = k \frac{l_1}{r_1^4} + k \frac{l_2}{r_2^4} \qquad (9.7.17)$$

and from (9.7.16) that

$$R = k \left(\frac{l_0 - s \cdot \cot\theta}{r_1^4} + \frac{s}{r_2^4 \sin\theta} \right). \qquad (9.7.18)$$

R is a function of the variable angle θ so that we may write $R = R(\theta)$. To get the minimum of R we differentiate $R(\theta)$. The calculation is similar to that leading to formulas (9.7.8) or (9.7.13). We obtain

$$R'(\theta) = k \left(\frac{s}{r_1^4 \sin^2\theta} - \frac{s \cdot \cos\theta}{r_2^4 \sin^2\theta} \right). \qquad (9.7.19)$$

We may factor out $s/\sin^2\theta$. The derivative vanishes if the expression in the parentheses is zero, that is, if

$$\frac{1}{r_1^4} - \frac{\cos\theta}{r_2^4} = 0.$$

Hence, the optimal angle θ_0 is determined by

$$\cos\theta_0 = r_2^4/r_1^4. \qquad (9.7.20)$$

To show that θ_0 actually minimizes the total resistance, we confine ourselves to a numerical example: Let $r_2/r_1 = 3/4$. Then $\cos\theta_0 = (3/4)^4 = 0.316$ and $\theta_0 = 72°$. For this angle, the resistance takes on the value (14.00) k/r_1^4 if $l_0 = 5$ cm and $s = 3$ cm are chosen. For neighboring angles, say $\theta_0 + 6°$ and $\theta_0 - 6°$ we get $R = (14.06)\ k/r_1^4$, that is, a value only slightly above the minimum.

The problem of vascular branching may be raised in different ways. For more elaborate work we refer the reader to Rosen (1967, p. 42–55).

Example 9.7.6. A fish swims upstream at a constant speed v relative to the water. The water itself has a velocity v_1 relative to ground. The fish intends to reach a point at distance s upstream. The energy required is essentially determined by friction in water and by the time t necessary to reach the goal. Experiments have proved that this energy is $E = c v^k t$ where $c > 0$ and $k > 2$ are certain constants (k depending on the shape of the fish). Given v_1, what speed v minimizes the energy?

Fig. 9.24. The most economic speed of a fish swimming upstream depends on the velocity v_1 of the water and on a parameter k connected with the shape of the fish

As Fig. 9.24 shows, the speed of the fish relative to the ground is $v - v_1$. On the other hand, this traveling speed is also equal to s/t. Thus,

$$v - v_1 = \frac{s}{t}, \qquad t = \frac{s}{v - v_1}. \tag{9.7.21}$$

Elimination of t in the formula for the energy E yields

$$E = c \cdot v^k \frac{s}{v - v_1}. \tag{9.7.22}$$

Since c and k are constants, and since s and v_1 are given values, E is a function of v. The domain is $v > v_1$. To obtain the minimum energy expenditure we differentiate E with respect to v:

$$\frac{dE}{dv} = cs\,\frac{v^{k-1}[(k-1)\,v - kv_1]}{(v-v_1)^2}. \tag{9.7.23}$$

$dE/dv = 0$ implies

$$v = v_0 = \frac{k}{k-1}\,v_1. \tag{9.7.24}$$

That E reaches a minimum for this particular value v_0 is seen from the numerator of (9.7.23). For $v = v_0$ the term in brackets vanishes, whereas for $v < v_0$ it is negative and for $v > v_0$ positive.

The most economic speed relative to water is given by Eq. (9.7.24). For example, if $k = 3$, the fish should try to keep a velocity which is $3/2$ of the river's velocity relative to the ground.

9.8. Mean of a Continuous Function

In a laboratory, provision was made to keep the temperature constant. A recording of temperature showed slight fluctuations. The problem arose: How to determine the mean temperature? Similar situations occur frequently in research. Humidity, light intensity, reaction rates, etc. change continuously. Yet, usually only mean values are of interest. In the study of biological rhythms all bodily activities are observed to vary in intensity. To get rid of the ups and downs one may ask for average intensities. For instance, the question may be: What is the average output of the kidneys per hour?

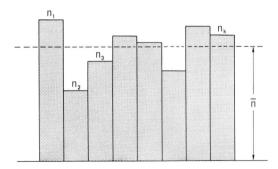

Fig. 9.25. The arithmetic mean $\bar{n}$ is interpreted as the total area of the histogram divided by the number of bars

We know how to determine an average in the case of discrete data. If we are given countings $n_1, n_2, \ldots, n_k$ at k different times or k locations, we simply calculate the arithmetic mean

$$\bar{n} = \frac{n_1 + n_2 + \cdots + n_k}{k} = \frac{1}{k} \sum_{i=1}^{k} n_i. \tag{9.8.1}$$

We may plot the countings using a histogram in which the area of each bar is denoted by the corresponding n (Fig. 9.25). The numerator in formula (9.8.1) is then the total area of the histogram, and the average $\bar{n}$ is simply the total area divided by the number of bars.

This interpretation provides us with a clue for defining the mean of a continuous function. Assume that a quantity u is a function of time, say $u = f(t)$, over an interval from t_0 to t_1 (Fig. 9.26). Let T be the length of the time interval, that is, $T = t_1 - t_0$. Then we determine the area over the interval which is measured in units $[t] \cdot [u]$ where $[t]$ and $[u]$

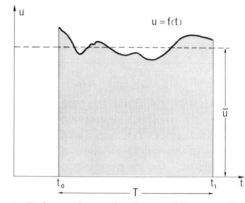

Fig. 9.26. The mean $\bar{u}$ of a continuous function $u = f(t)$ over an interval from t_0 to t_1

denote the units of t and u, respectively. Finally we divide the area by T and call the result the *time average* or the *mean* of u over the interval from t_0 to t_1. We denote this mean by $\bar{u}$. Thus we obtain

$$\bar{u} = \frac{1}{T} \int_{t_0}^{t_1} f(t)\, dt. \tag{9.8.2}$$

An even stronger geometric interpretation would proceed as follows: The figure with a curvilinear edge is replaced by a rectangle with the same base T and the same area. The height $\bar{u}$ of this rectangle is the desired mean of u. With this view in mind, it is not difficult to estimate $\bar{u}$ graphically.

The independent variable need not be the time. In the following two applications, time is actually not involved.

Example 9.8.1. The investigation of the *internal structure of tissue* calls sometimes for estimating the volume of a component which is "randomly" included in the tissue. The term "random" is not well defined. In this connection it means "irregular, but not in clusters". Fig. 9.27 illustrates the situation.

A tissue can be investigated by sections. A tissue section is a slice not thicker than a few micrometers ($10^{-6}m$). Such a section may be considered as a two-dimensional sample from a three-dimensional body. The question then arises: Can the composition of tissue be estimated from a thin slice?

To answer this question we partly follow a presentation given by Weibel (1963, p. 12): Suppose that a cube with volume $V = L^3$ contains

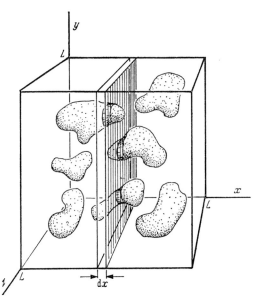

Fig. 9.27. Estimation of the amount of granules in a tissue by means of a two-dimensional section. The figure is reproduced from Weibel (1963, p. 13)

granules of any shape and size evenly, that is, without forming clusters. All granules together are of volume $v < V$, say

$$v = \varrho \cdot V \quad (\varrho < 1). \tag{9.8.3}$$

It is our purpose to estimate the rate ϱ by observing a thin slice. As shown in Fig. 9.27, we let one face of the cube coincide with the x, y-plane, another face with the x, z-plane, and a third face with the y, z-plane. Consider a slice of thickness dx parallel to the y, z-plane[14]. Its volume is

$$dV = L^2 dx.$$

In this slice, let

$$dv = \eta(x) \cdot dV = \eta(x) L^2 dx \qquad (9.8.4)$$

denote the volume occupied by segments of granules[15]. The rate $\eta(x)$ depends on the position x where the slice is cut. Following formula (9.8.2) we introduce the *average rate*

$$\bar{\eta} = \frac{1}{L} \int_0^L \eta(x)\, dx. \qquad (9.8.5)$$

Now, from formula (9.8.4) we get

$$v = \int dv = \int_0^L \eta(x)\, L^2 dx = L^2 \cdot \int_0^L \eta(x)\, dx.$$

It follows from (9.8.5) that

$$v = L^2 \cdot L \cdot \bar{\eta} = V \cdot \bar{\eta}.$$

Comparing the result with (9.8.3) we finally obtain

$$\varrho = \bar{\eta}. \qquad (9.8.6)$$

In words: *The three-dimensional rate ϱ of a tissue component is equal to the two-dimensional average rate $\bar{\eta}$.* This rule was discovered by Delesse in 1842[16].

The experimenter measures η for a particular x and uses η as an estimate for $\bar{\eta}$ and indirectly for ϱ. For more details see Weibel (1963, p. 11 ff.) and Hennig (1967, p. 100).

Example 9.8.2. For another application of the mean of a continuous function, we return to the laminar flow of blood as considered in Section 4.5. A velocity diagram is depicted in Fig. 4.7b. We may ask: *What is the mean velocity of blood at a particular cross section of a blood vessel?*

Formula (4.5.1) states that the velocity $v = v(r)$ is given by

$$v(r) = c(R^2 - r^2) \qquad (0 \le r \le R) \qquad (9.8.7)$$

---

[14] In Section 9.4 we used $\varDelta x$ to denote a quantity which tends to zero, whereas dx was a mere symbol. In applied mathematics, however, it is customary to use dx both with the meaning of $\varDelta x$ and as a symbol.

[15] ϱ (= rho) and η (= eta) are lower case Greek letters.

[16] Achille Ernest Delesse (1817—1881), French geologist.

where $c = P/4\eta l$ is a constant. Now we subdivide R into a number, say n, of equal parts of length Δr. Thus $R = n \cdot \Delta r$. Then we introduce the sequence of radii

$$r_0 = 0, \quad r_1 = \Delta r, \quad r_2 = 2\Delta r, \ldots, r_i = i \cdot \Delta r, \ldots, r_n = R.$$

For a fixed value of i we consider the ring formed by the two circles of radii r_i and $r_{i+1} = r_i + \Delta r$, respectively. The area of this ring is approximately $2\pi r_i \cdot \Delta r$. Over the ring as part of the cross section, the velocity changes only slightly and may be approximated by $v(r_i)$. Hence, the volume of blood flowing through the ring per time unit is nearly

$$2\pi r_i \cdot \Delta r \cdot v(r_i).$$

The total blood volume passing the cross section per time unit is approximated by

$$\sum_{i=0}^{n-1} 2\pi r_i \cdot v(r_i) \cdot \Delta r.$$

We can reduce the error by making a finer subdivision of R, that is, by increasing n. As n tends to infinity, the sum tends to a limit which is the exact volume of blood per time unit. We discussed similar procedures in Section 9.4. Thus we are led to the integral

$$\int_0^R 2\pi r \cdot v(r) \cdot dr. \tag{9.8.8}$$

To get the average velocity of blood, we simply divide the integral by the total area of the cross section, that is, by πR^2. Thus we obtain

$$\bar{v} = \frac{1}{\pi R^2} \int_0^R 2\pi r \cdot v(r) \cdot dr.$$

Formula (9.8.7) yields

$$\bar{v} = \frac{1}{\pi R^2} 2\pi \cdot c \int_0^R (R^2 r - r^3)\, dr$$

$$= \frac{2c}{R^2} \left(R^2 \cdot \frac{r^2}{2} - \frac{r^4}{4} \right) \Big|_0^R = \frac{2c}{R^2} \left(\frac{R^4}{2} - \frac{R^4}{4} \right)$$

and finally

$$\bar{v} = \frac{c}{2} R^2. \tag{9.8.9}$$

As (9.8.7) shows, the maximum velocity is $v_{max} = cR^2$. Therefore, our average speed $\bar{v}$ is just one-half of it, that is,

$$\bar{v} = \tfrac{1}{2} v_{max} \, . \tag{9.8.10}$$

9.9. Small Changes

In systems analysis we study the interdependence of several parts of an organism. If the system is stable, a small change in one part will generate only small changes in other parts. The problem of determining small changes is solved by differential calculus provided that the variables are continuous.

Example 9.9.1. Consider the iris of our eye which acts as a diaphragm. How much does a small change of its width affect the intensity of the entering light? Let r be the radius of the pupil. Then its area (aperture) is proportional to r^2, and the same is true for the light intensity, say I. Hence,

$$I = cr^2 \tag{9.9.1}$$

for a suitable constant c. When r increases by a certain increment Δr, the corresponding increment of I is

$$\Delta I = c(r + \Delta r)^2 - cr^2 = 2cr \cdot \Delta r + c(\Delta r)^2 \, . \tag{9.9.2}$$

Now we assume that Δr is small compared with r, say at most one-tenth of r. Then $(\Delta r)^2$ is so small compared with $r \cdot \Delta r$ that the last term in

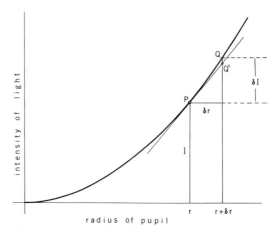

Fig. 9.28. A small increase δr causes a small change δI. The ratio $\delta I / \delta r$ is approximately equal to the slope of the tangent, that is, the derivative dI/dr

(9.9.2) can be neglected. To indicate that Δr should be sufficiently small, it is customary to write δr instead of Δr[17]. Therefore, (9.9.2) turns into the approximate formula[18]

$$\delta I \approx 2cr \cdot \delta r. \qquad (9.9.3)$$

There is, however, a faster way to obtain formula (9.9.3). To explain the basic idea, we look at Fig. 9.28. The graph of the function (9.9.1) is a piece of a parabola. From a given point P with coordinates r, I we proceed to a new point Q with coordinates $r + \delta r, I + \delta I$. However, we would reach nearly the same point by following the tangent to the parabola up to the point Q' with abscissa $r + \delta r$. Since the slope of the tangent is

$$\frac{dI}{dr} = 2cr, \qquad (9.9.4)$$

the ordinate of Q' is $I + 2cr \cdot \delta r$. Hence, $\delta I \approx 2cr \cdot \delta r$ in agreement with formula (9.9.3). Notice that the approximation would not be satisfactory if δr were of the same order of magnitude as r (cf. Section 1.10 for the notion "order of magnitude").

In general, let $y = f(x)$ be a differentiable function of x, and let δx be an increment which is sufficiently close to zero. The increment δx may be positive or negative. Using the absolute value we can reformulate the condition imposed on δx by saying that $|\delta x|$ *has to be sufficiently small*. With δy we denote the corresponding increment of y by δy. Then the ratio $\delta y/\delta x$ is approximately equal to the slope $dy/dx = f'(x)$. Therefore, we get

$$\delta y \approx f'(x) \cdot \delta x. \qquad (9.9.5)$$

In each particular application, a special examination is required to determine the meaning of "sufficiently small". This depends on the properties of the function as well as the desired accuracy.

Example 9.9.2. Consider a blood vessel of length l and of radius r. Then one of *Poiseuille's laws* states that the resistance R of the blood vessel is given by formula (9.7.15). How does a small change δr of r affect the resistance R? We differentiate R with respect to r and get

$$\frac{dR}{dr} = R'(r) = kl(r^{-4})' = kl(-4)r^{-5} = -\frac{4kl}{r^5}. \qquad (9.9.6)$$

---

[17] δ is the lower and Δ the upper case Greek letter delta.
[18] The symbol $\approx$ is read "is approximately equal to". Some authors prefer the symbols $\sim$, $\cong$ or $\doteq$.

It follows from formula (9.9.5) that

$$\delta R \approx \left(- \frac{4kl}{r^5}\right) \cdot \delta r .\tag{9.9.7}$$

The result states that the resistance R diminishes as the radius r increases. The rate of change is $-4kl/r^5$.

Example 9.9.3. We know that $\sqrt{a^2} = a$ for $a \geq 0$. If we add a small quantity b to the radicand, we may ask: How far does $\sqrt{a^2 + b}$ deviate from a? To answer the question, we consider the function $y = \sqrt{x} = x^{1/2}$. We differentiate y with respect to x and deduce from formula (9.9.5)

$$\delta y \approx \frac{1}{2\sqrt{x}} \cdot \delta x .$$

With the new notations $x = a^2$, $\delta x = b$, $y = a$, it follows that $\delta y \approx b/2a$ and, hence,

$$\sqrt{a^2 + b} \approx a + \frac{b}{2a} .\tag{9.9.8}$$

This formula is convenient for numerical calculations. Example: $a = 50$, $b = 9$. Then $\sqrt{2509} \approx 50 + 9/100$ or 50.09. The exact value is $\sqrt{2509} = 50.0899 \ldots$. The error of approximation is negligible.

Example 9.9.4. Formula (9.9.5) is also used in problems of *error propagation*. Suppose that an experimenter is unable to keep the electric resistance R of a wire constant. He observes fluctuations of the electric current within an interval $I + \delta I$. The question is: Can we estimate the corresponding interval $R \pm \delta R$ for the resistance ? In other words: How can an error of R be related to a deviation of I? Ohm's law states[19]

$$\frac{V}{I} = R .\tag{9.9.9}$$

We assume that the voltage V is constant. Therefore, R is a function of I. We differentiate R with respect to I and obtain from (9.9.5)

$$\delta R \approx - \frac{V}{I^2} \cdot \delta I .\tag{9.9.10}$$

Hence, the desired interval ranges approximately from $R - VI^{-2} \cdot \delta I$ to $R + VI^{-2} \cdot \delta I$.

---

[19] Georg Simon Ohm (1789—1854), German mathematician and physicist.

*9.10. Techniques of Integration

For natural scientists it is more important to understand the notion of an integral than to be a master of integration. To concentrate mainly on techniques of calculus may distract him from the major problems of his area of research. Nevertheless, the more methods one knows, the safer one feels in studying mathematical models.

It frequently occurs that an indefinite integral

$$F(x) = \int f(x)\,dx \qquad\qquad (9.10.1)$$

can only be worked out by introducing a new variable of integration. For this purpose we let

$$x = x(t)$$

and apply the fundamental theorem of integral calculus (Eq. (9.5.5)) and the chain rule (Eq. 9.2.14) to $F(x)$. Thus we obtain

$$\frac{dF}{dt} = \frac{dF}{dx} \cdot \frac{dx}{dt} = f(x)\frac{dx}{dt} = f(x(t))\frac{dx}{dt}\,.$$

Upon integration with respect to t we get

$$\boxed{F(x(t)) = \int f(x(t))\frac{dx}{dt} \cdot dt\,.} \qquad\qquad (9.10.2)$$

This formula contains the rule of the so-called *integration by substitution*. We may interpret the rule as the *inverse of the chain rule*.

Problem 9.10.1. Find

$$F(x) = \int \sin(ax + b)\,dx \qquad (a \neq 0)\,.$$

Solution: Let $ax + b = t$. Then

$$x = \frac{1}{a}(t - b)\,, \qquad \frac{dx}{dt} = \frac{1}{a}\,.$$

From Eq. (9.10.2) we get

$$F(x(t)) = \int \sin t \cdot \frac{1}{a} \cdot dt\,.$$

Hence,

$$F(x(t)) = -\frac{1}{a}\cos t + C$$

or

$$F(x) = -\frac{1}{a}\cos(ax + b) + C\,.$$

Problem 9.10.2. Find

$$F(x) = \int \sqrt{1 - 2x}\, dx. \qquad (x \le \tfrac{1}{2})$$

Solution: Let $1 - 2x = t$. Then

$$x = \frac{1}{2}(1 - t), \qquad \frac{dx}{dt} = -\frac{1}{2}.$$

From Eq. (9.10.2) we obtain

$$F(x(t)) = \int \sqrt{t} \cdot (-\tfrac{1}{2})\, dt = (-\tfrac{1}{2}) \int t^{1/2}\, dt.$$

Hence,

$$F(x(t)) = (-\tfrac{1}{2}) \frac{t^{3/2}}{3/2} + C = (-\tfrac{1}{3}) t^{3/2} + C$$

or

$$F(x) = (-\tfrac{1}{3})(1 - 2x)^{3/2} + C.$$

A second, frequently used technique is called *integration by parts*. We know already from Eq. (9.2.11) how to differentiate a product of two functions. Now we use the same formula to integrate some (but by no means all) products of two functions. Rewriting Eq. (9.2.11) we get

$$f(x)\, g'(x) = (f(x) \cdot g(x))' - f'(x)\, g(x)$$

or, upon integration term by term,

$$\boxed{\int f(x)\, g'(x)\, dx = f(x)\, g(x) - \int f'(x)\, g(x)\, dx} \qquad (9.10.3)$$

Under favorable conditions, this formula can reduce an integration problem to one that is easier to solve.

Problem 9.10.3. Find

$$\int x \cos x\, dx.$$

Solution: We identify in Eq. (9.10.3)

$$f(x) = x, \qquad g'(x) = \cos x$$

with

$$f'(x) = 1, \qquad g(x) = \sin x.$$

Hence,

$$\int x \cos x\, dx = x \sin x - \int 1 \cdot \sin x\, dx.$$

Now the integration is much easier to perform on the right side than on the left side. We obtain immediately

$$\int x \cos x\, dx = x \sin x + \cos x + C.$$

Problem 9.10.4. Find

$$I = \int \sin^2 \alpha\, d\alpha.$$

Solution: We identify

$$f(\alpha) = \sin\alpha, \qquad g'(\alpha) = \sin\alpha,$$

with

$$f'(\alpha) = \cos\alpha, \qquad g(\alpha) = -\cos\alpha.$$

Eq. (9.10.3) yields

$$I = \sin\alpha(-\cos\alpha) - \int \cos\alpha(-\cos\alpha)\, d\alpha$$
$$= -\sin\alpha\cos\alpha + \int \cos^2\alpha\, d\alpha.$$

Integration on the right side is as much of a problem as on the left side. However, using $\cos^2\alpha = 1 - \sin^2\alpha$, we get

$$\int \cos^2\alpha\, d\alpha = \int (1 - \sin^2\alpha)\, d\alpha = \alpha - I + C.$$

Hence,

$$I = -\sin\alpha\cos\alpha + \alpha - I + C,$$
$$2I = -\sin\alpha\cos\alpha + \alpha + C,$$
$$I = \tfrac{1}{2}(-\sin\alpha\cos\alpha + \alpha + C).$$

There are numerous other techniques of integration which the reader may find in books on calculus. We question whether a natural scientist should spend much time on such techniques just to learn that there are many easy looking integrals which cannot be worked out in a closed form. An example of such an integral is

$$\int \frac{\sin x}{x}\, dx.$$

There are also *tables* available in which the reader can find a wealth of more sophisticated integrals and their solutions. See Dwight (1961), Gradštejn and Ryžik (1965), Gröbner and Hofreiter (1965, 1966), Meredith (1967), Peirce (1929).

Recommended for further reading: Defares *et al.* (1973), Gelbaum and March (1969), Guelfi (1966), Lefort (1967), McBrien (1961), C. A. B. Smith (1966), Stibitz (1966).

Problems for Solution

9.1.1. A group of tourists started for a 45 km hike at 10 : 00. The group reached a shelter 31 km away from the starting point at 17 : 30. There they stayed overnight. Next morning at 8 : 00 the group continued to hike and arrived at their goal at 11 : 00. Is the average velocity of the second day greater or smaller than that of the first day?

9.1.2. Mass is spread along the x axis (unit 1 cm). Let $M = M(x)$ be the total mass between the origin and the point with abscissa x. For $x_1 = 4$ cm, $x_2 = 7$ cm, $x_3 = 9$ cm, the following values are available: $M(x_1) = 83$ g, $M(x_2) = 141$ g, $M(x_3) = 179$ g. Does the average density in the interval $[0, x]$ increase or decrease as x increases?

9.1.3. Assume that a population of size 25000 (at time $t = 0$) grows according to the formula $N = 25000 + 45\,t^2$ where the time t is measured in days. Find the average growth rate in the time intervals a) from $t = 0$ to $t = 2$, b) from $t = 2$ to $t = 10$, c) from $t = 0$ to $t = 10$.

9.1.4. Assume that a protein (mass M in grams) disintegrates into amino acids according to the formula $M = 28/(t + 2)$ where the time t is measured in hours. Find the average reaction rate for the time intervals a) from $t = 0$ to $t = 2$, b) from $t = 0$ to $t = 1$, c) from $t = 0$ to $t = \frac{1}{2}$.

9.1.5. Mass is spread along the x axis continuously. Units of length is 1 cm and of mass 1 g. $M(x)$ denotes the mass that falls in the interval $[0, x]$. Given $M(x) = x^2/3$. Find the average density in the intervals $[10, 10 + h]$ where h takes on the values 1 cm, 0.1 cm, 0.01 cm, respectively.

9.1.6. Assume that a particle moves from the point $s = 2$ (meters) at time $t = 1$ (sec) to points with $s > 2$ along an s axis. The segment s (meters) is the following function of time t (sec): $s = 2\sqrt{t}$. Find the average velocity of the particle in the time interval from $t = 1$ to $t = 1 + h$ where h takes on the decreasing values 1, 0.1, 0.01, 0.001 (sec).

9.2.1. Find the difference quotient and the derivative of the function $y = ax^2 + b$.

9.2.2. Find the difference quotient and the derivative of the function $y = x^3$.

9.2.3. Using the general formula for differentiating the function $y = x^n$, find the derivatives of the following power functions:

a) $y = x^{-2}$,   b) $u = w^5$,   c) $V = r^{2/3}$,
d) $M = 1/t^3$,   e) $A = \sqrt[3]{Q}$,   f) $Z = 1/\sqrt{P}$.

9.2.4. Use the formula for $(x^n)'$ to find the derivations of

a) $y = x^{-3}$,   b) $Q = u^{10}$,   c) $Z = 1/v^4$,
d) $s = \sqrt[4]{Q}$,   e) $\lambda = \sqrt{t^3}$,   f) $x = 1/\sqrt[3]{t}$.

9.2.5. Find the derivatives of

a) $v = at + b/t + c$,   b) $U = az^2 + b\sqrt{z} + c/\sqrt{z}$ (a, b, c are constants).

9.2.6. Find the derivatives of

a) $U(t) = pt^{3/4} - qt^{-2}$,

b) $S(t) = \dfrac{3p}{t} - \dfrac{q}{t^3}$,

c) $T(u) = au - b/u^2$,

d) $h(x) = \dfrac{ax+b}{x^2}$.

9.2.7. The size of a slowly growing bacteria culture is approximately given by

$$N = N_0 + 52\,t + 2t^2 \quad \text{(time } t \text{ in hours)}.$$

Find the growth rate at $t = 5\text{h}$.

9.2.8. When protein was synthesized in a cell, the mass M of protein as a function of time t increased according to the formula

$$M = p + qt + rt^2 \quad (p, q, r \text{ are constants}).$$

Find the reaction rate as a function of t.

9.2.9. In a metabolic experiment the mass M of glucose decreased according to the formula

$$M = 4.5 - (0.03)\,t^2 \quad (t \text{ in hours}).$$

Find the rate of reaction a) at $t = 0$, b) at $t = 2$, c) in the interval from $t = 0$ to $t = 2$ (average).

9.2.10. The altitude of a right circular cone measures $h = 20\,\text{cm}$. The radius r of the base (in cm) is increasing. The formula for the volume is $V = \frac{1}{3}\pi r^2 h$. Find the growth rate of the volume.

9.2.11. Find the derivative of $y = \cos x$ by employing the method described in Section 9.2 for the function $y = \sin x$.

9.2.12. Find the derivatives of

a) $y = 3\cos x - 2\sin x$,

b) $Q(t) = a\cos t + 2\sqrt{t}$,

c) $P(s) = 1 - s - \frac{1}{2}\sin s$,

d) $r(\alpha) = 1 - \cos\alpha$.

9.2.13. Using the rule for differentiating products find the derivatives of

a) $y = (x+7)(x-3)$,

b) $y = x \cdot \sin x$,

c) $z(t) = (1-t)\cos t$,

d) $Q(\alpha) = \sin\alpha\,\cos\alpha$,

e) $n(x) = x^{1/3}(1-2x)$,

f) $f(y) = a \cdot \sqrt{y}\,\sin y$.

9.2.14. Find the derivative of $y = (x^2 + 1)(x - 2)$ in two ways, a) by removing the parentheses, b) by the rule for differentiating a product.

9.2.15. Find

a) $\dfrac{d}{dx}(\sin x \cdot \cos x)$,

b) $\dfrac{d}{dt}(2t - 3)(t^2 + 5)$,

c) $\dfrac{d}{du}(u \cdot \cos u)$,

d) $\dfrac{d}{dw}(1 - w)\sqrt{w}$.

9.2.16. Find the derivative of $y = (2x + 3)^2$ in three ways, a) by working out the parenthesis, b) by the rule for differentiating a product, c) by the chain rule.

9.2.17. Using the chain rule, find the derivatives of the following functions:

a) $y = (x + 5)^2$,
b) $Y = (u^2 - 3)^2$,
c) $s = 1/(t - 2)$,

d) $K = 2/(1 - v)$,
e) $v = (4 - 3t)^{\frac{1}{3}}$,
f) $p = \sin(4\alpha - 5)$.

9.2.18. Find the derivatives of

a) $y = \sin(2x - 4)$,
b) $U(\beta) = p\cos 3\beta$,

c) $R(s) = 1/(as + b)$,
d) $z = \sqrt{3x + 1}$.

9.2.19. Find the derivatives of

a) $u(x) = (x - 1)^{-4}$,
b) $E(w) = \dfrac{1}{w} + \dfrac{1}{w - 1}$,

c) $z = (t - 1)^{1/2} + (t + 1)^{-1/2}$,
d) $f(\alpha) = (\alpha \cos \alpha)^2$.

9.2.20. Differentiate

a) $y = \dfrac{x}{x - 3}$,
b) $K = \dfrac{1 + r}{1 - r}$,

c) $y = 1/\sin\theta$,
d) $y = \cot\theta = \cos\theta/\sin\theta$.

The results of c) and d) are used in examples at the end of Section 9.7.

9.2.21. Differentiate

a) $y = \dfrac{x^3}{1 - x}$,
b) $S(t) = \dfrac{at + b}{ct + d}$,

c) $Q(\alpha) = \dfrac{1 - \sin\alpha}{1 + \cos\alpha}$,
d) $P(t) = \dfrac{t^3}{1 + t^2}$,

e) $f(u) = (pu^3 + q)^{1/3}$,
f) $h(\phi) = \sin 2\phi/\cos 3\phi$.

9.2.22. Find

a) $\dfrac{d}{dt}(2t - 3)\sqrt{at}$,
b) $\dfrac{d}{dz}\dfrac{1 - uz}{1 - z}$,

c) $\dfrac{d}{du}\dfrac{1 - uz}{1 - z}$,
d) $\dfrac{d}{d\lambda}(\lambda \sin a\lambda)^3$.

9.2.23. Let $y = x^2 + 1$. Calculate dy/dx and, from the inverse function, dx/dy. Verify $dx/dy = (dy/dx)^{-1}$.

9.2.24. Let $y = x^{3/2}$ for $x > 0$. By using the inverse function verify the formula $dx/dy = (dy/dx)^{-1}$.

9.3.1. Find the antiderivatives of the following functions:

a) $y' = 6x^2$, b) $y' = 8x - 7$, c) $u' = at + b$
$\qquad\qquad\qquad\qquad\qquad\qquad\qquad\qquad$ (a, b are constants),

d) $\dfrac{dy}{dx} = 5x^3$, e) $\dfrac{dW}{dt} = 2t - 8$, f) $\dfrac{dU}{dx} = U_0 + \cos x$,

g) $y' = \frac{1}{3}\cos t$, h) $U' = \cos 2x$, i) $K' = 1/u^2$.

9.3.2. Find the antiderivatives of the following functions:

a) $y' = 3x^4 - x^2$, b) $y' = 1 - x^{2/3}$,

c) $\dfrac{dy}{dx} = ax^3 - bx^5$, d) $\dfrac{ds}{dt} = a\sqrt{t}$,

e) $\dfrac{du}{ds} = \dfrac{k}{s^3}$, f) $\dfrac{dQ}{d\alpha} = a\cos\alpha + b\sin\alpha$,

g) $\dfrac{dy}{dt} = \dfrac{pt^2 + q}{t^2}$, h) $\dfrac{dP}{dv} = av^2 + bv + c + \dfrac{d}{v^2}$.

9.5.1. Find an approximate value for the integral $\int\limits_1^2 \dfrac{4}{x}\, dx$ by dot counting.

(Use 5 equidistant intervals.)

9.5.2. Plot a graph of the function $y = 5 + 2x - \frac{1}{2}x^2$ over the interval from $x = 0$ to $x = 3$. Find

$$\int\limits_0^3 y\, dx$$

a) approximately by the method of dot counting, b) exactly by integration.

9.5.3. Evaluate the definite integrals

a) $\int\limits_1^2 \dfrac{1}{r^2}\, dr$, b) $\int\limits_{-1}^1 (5 - w)\, dw$, c) $\int\limits_0^t (at^2 + bt + c)\, dt$,

d) $\int\limits_1^3 t\, dt$, e) $\int\limits_{-1}^1 dx$, f) $\int\limits_{-\pi/2}^{\pi/2} \cos t\, dt$.

9.5.4. Evaluate

a) $\int_0^2 (x^2 + 1)\, dx$,
b) $\int_1^2 \frac{1}{r^2}\, dr$,
c) $\int_{-\beta}^{\beta} \sin\alpha\, d\alpha$,
d) $\int_p^q x^2 dx$,
e) $\int_a^{a+1} \frac{du}{u^2}$,
f) $\int_0^s t^{2/3} dt$.

9.5.5. Find the following indefinite integrals:

a) $\int x^{-6} dx$,
b) $\int t^{-1/3} dt$,
c) $\int (u + 2u^2 + 3u^3)\, du$,
d) $\int s^{0.1} ds$,
e) $\int (A\sin\theta + B\cos\theta)\, d\theta$,
f) $\int \frac{dQ}{\sqrt{8Q^3}}$.

9.5.6. Find

a) $\int_0^{10} \frac{du}{(u + 2)^2}$,
b) $\int_{-p}^p A \cdot dr$,
c) $\int_1^{3/2} t\, dt$,
d) $\int_{-1}^1 dx$.

9.5.7. Applying the fundamental theorem of integral calculus find the following derivatives:

a) $\dfrac{d}{dx} \int_0^x \frac{1}{t + 2}\, dt$,
b) $\dfrac{d}{dx} \int_1^x \frac{au - 1}{u + 1}\, du$,
c) $\dfrac{d}{dt} \int_2^t \frac{u - 3}{\sin u}\, du$,
d) $\dfrac{d}{ds} \int_0^s \sqrt{t^3 + 1}\, dt$.

9.5.8. If a helical spring is extended moderately, Hooke's law is valid. It states that the amount of extension, s, is proportional to the extending force, F, that is, $F = ks$ ($k > 0$, constant). Let s be measured in meters and F in Newtons. Then the energy (work) required for the extension is measured in Joules. Find the energy, W, to extend the spring from $s = 0$ to $s = s_0$. (Hint: Study Example 9.4.4.)

9.6.1. Find the second derivatives of the following functions:

a) $y = 1 - x^3$, b) $u = 2z^5 - 3z^3$, c) $W = 3/t$, d) $p = 2\sqrt{s}$.

9.6.2. Find

a) $\dfrac{d^2}{dx^2}(ax^3 + bx + c)$,
b) $\dfrac{d^2}{dt^2} t^{2/3}$,
c) $\dfrac{d^2}{du^2}\sqrt{u + 2}$,
d) $\dfrac{d^2}{d\alpha^2}\cos(2\alpha + 3)$,
e) $\dfrac{d^2}{d\theta^2}(\cos\theta - \sin\theta)$,
f) $\dfrac{d^2}{dr^2}(r - 1)^{1/3}$.

9.6.3. Given that a particle is at rest at time $t=0$ and, from then on, moves along a straight line with constant acceleration a, find the "law of motion", that is, find velocity v and distance s traveled as a function of time.

9.6.4. Show that the first derivative of a quadratic function is linear, and that the second derivative is constant. Compare the results with the first and second difference introduced in Section 4.4.

9.6.5. By means of the second derivative, determine whether the graphs of the following functions are convex upward or downward:
a) $y = \frac{1}{2}x^2 + 3x - 5$, b) $F = 4 - 2t - t^2$, c) $V = (-1/3)u^2 + u$,
d) $y = 2x - 4$, e) $y = x^3 - x$.

9.6.6. Determine whether the graphs of the following functions turn clockwise or counter-clockwise:
a) $y = 1/x$ $(x > 0)$, b) $y = 1/x^2$ $(x > 0)$,
c) $y = 1/(1 - x)$ $(0 < x < 1)$, d) $y = \dfrac{x+2}{x-3}$ $(x > 3)$.

9.6.7. Find maximum and minimum points of the following functions:
a) $y = x^3 - x$, b) $v = 1 + 2t + \frac{1}{2}t^2$, c) $u = p(1 - p)$.

9.6.8. Find extremum values of
a) $Z = t^5 - 5t$, b) $U = 1 + s^2 - s^3$, c) $R = \sin 2\alpha$.

9.6.9. Is the condition "traffic light is green", a) necessary, b) sufficient for the right to cross an intersection?

9.6.10. What are necessary and sufficient conditions for a telephone call to reach its destination?

9.6.11. Denote the sides of a triangle by a, b, c. Is the condition $a^2 + b^2 = c^2$ necessary and/or sufficient for the triangle to be a right triangle?

9.6.12. Is the condition "diagonal bisects a quadrilateral into congruent triangles" necessary and/or sufficient for a rectangle?

9.6.13. Is the condition "all four sides are equal" a) necessary, b) sufficient for a quadrilateral to be a square?

9.6.14. Verify the statement "An integer is divisible by 6 if, and only if, it is divisible by 2 and by 3".

9.6.15. Given $y = x^4$. For $x = 0$, the second derivative is zero. Yet the graph has no point of inflection. Is the condition $f''(x) = 0$ necessary or sufficient for a point of inflection?

9.6.16. Assume that the graph of a function $y = f(x)$ consists of two arcs which join at a point P. Thus P is a vertex. Is the function at P a) continuous, b) differentiable, c) neither, d) both? Is the condition "$f(x)$ is continuous" necessary or sufficient for differentiability?

9.7.1. Find local and absolute maxima and minima of the following functions:

a) $y = x^2 - 3x$ for $0 \leq x \leq 5$,

b) $v = 1 + 2t + \frac{1}{2} t^2$ for $-3 \leq t \leq 3$,

c) $U = 1/(2v + 3)$ for $1 \leq v \leq 3$,

d) $y = x^3 - 3x$ for $-3 \leq x \leq 3$.

9.7.2. Find local and absolute extrema for

a) $y = 3x^2 - 6x$ for $0 \leq x \leq 2$,

b) $z = 1 + u - u^2/2$ for $-2 \leq u \leq 2$,

c) $S = 5 - t^{2/3}$ for $0 \leq t \leq 8$.

9.7.3. Let

$$Q(s) = \frac{s^2 - 10s + 5}{s + 1}$$

be defined for the domain $s \geq 0$. Find local and absolute maxima and minima.

9.7.4. Find local and absolute extrema for $y = 2x/(x^2 + 3)$.

9.7.5. A substance is distributed continuously over the interval [0 cm, 10 cm] of the x axis. The concentration is given by the function $C = 10x - x^2$ (C measured in mg/cm).
(a) Where is the maximum concentration located?
(b) What is the total mass M?

9.7.6. In water and in solution the product of the concentrations of hydronium ions, $[H_3O^+]$, and of hydroxyl ions, $[OH^-]$, is very close to 10^{-14} (measurements are made in moles). Let

$$S = [H_3O^+] + [OH^-].$$

Determine the value of $[H_3O^+]$ which minimizes S. (From Thrall et al., 1967, ex. MA 7.2) Hint: Let $[H_3O^+] = x$.

9.7.7. In an auto-catalytic reaction one substance is converted into a new substance, the product, in such a way that the product catalyzes its own formation. We assume that the reaction rate is proportional to the amount x of the product at time t and also proportional to the still available amount of the original substance. If a denotes the original amount of the substance, it decreases to $a - x$ at time t. Therefore,

$$\frac{dx}{dt} = kx(a - x) \quad (k \text{ is a positive constant}).$$

Find the particular value of x which maximizes the reaction rate (from Thrall et al., 1967, ex. MA 8.2).

9.7.8. Let $Q(a) = \sum_{i=1}^{n} (x_i - a)^2$ where $x_1, x_2, \ldots, x_n$ are given measurements. Determine a in such a way that $Q(a)$ reaches a minimum. Is the result plausible?

9.7.9. Let v be the speed of a bird versus air. Let W be its weight and ϱ the air density. Pennycuick (1969) finds the following formula for the power P which the bird has to maintain during flight:

$$P = \frac{W^2}{2\varrho Sv} + \frac{1}{2} \varrho A v^3.$$

In this formula, S and A are certain quantities connected with the bird's shape and size. Find the particular speed v_0 which minimizes the power P.

9.7.10. The energy expenditure of some flying birds could be measured. For the Budgerigar (Melopsittacus undulatus, an Australian parakeet) the energy expenditure in $J g^{-1} km^{-1}$ (Joule per gram mass per kilometer) can be well described by the formula

$$E = \frac{1}{v} \{0.31(v - 35)^2 + 92\}$$

where v is velocity of the bird in $km\,h^{-1}$ (wind speed is not considered). What velocity is most economic? (Adapted from Tucker and Schmidt-Koenig, 1971.)

9.7.11. In certain tissues, cells are found having the form of an upright circular cylinder of altitude h and radius r. If the volume is fixed, find the particular radius r that minimizes the total surface area. Also determine the corresponding ratio h to r.

9.7.12. A flea leaping in a vertical direction reached the following height h (in m) as a function of the time t (in sec):

$$h = (4.4)\,t - (4.9)\,t^2.$$

Find the velocity at time $t = 0$, the maximum height reached, and the acceleration caused by gravitation.

9.8.1. Find the mean values of the following functions:

a) $y = \frac{1}{2}x + 3$ for $0 \leq x \leq 6$,

b) $s = at^2$ for $2 \leq t \leq 3$,

c) $F = \cos \alpha$ for $-\frac{\pi}{2} \leq \alpha \leq \frac{\pi}{2}$.

9.8.2. Find the mean values of the following functions:

a) $y = 4 - \frac{1}{3}x$ for $-2 \leq x \leq 2$,

b) $U = 5 + \frac{1}{4}w - \frac{1}{2}w^2$ for $0 \leq w \leq 3$,

c) $K = r^{-2}$ for $1 \leq r \leq 3$.

9.9.1. The surface area of a spherical cell is $S = 4\pi r^2$ and the volume $V = \frac{4}{3}\pi r^3$. How are S and V affected by a small increase δr of r?

9.9.2. When a muscle contracts against a force F (e.g. a weight), the speed v of shortening decreases with increasing force. A. V. Hill discovered the following equation in 1938:

$$(F + a)(v + b) = c$$

with suitable positive constants a, b, c. Express v in terms of F. How is v affected by a small change δF of F? (For Hill's law see Abbott and Brady, 1964, p. 349).

9.9.3. By inspection of Figure 9.28 prove that for $\delta x > 0$

$$(\text{Min } f'(x)) \cdot \delta x \leq \delta y \leq (\text{Max } f'(x)) \cdot \delta x.$$

The minimum and maximum of $f'(x)$ are taken in the interval from x to $x + \delta x$.

*9.10.5. Integrate by substitution

a) $\int \cos \omega x \, dx$,

b) $\int (1 - 8u)^3 \, du$,

c) $\int \sqrt{p + qs} \, ds$ $(q \neq 0)$,

d) $\int_0^2 (4t + 3)^4 \, dt$,

e) $\int_0^{\pi/2} \cos 3\theta \, d\theta$,

f) $\int_1^3 \frac{dx}{(5x - 1)^2}$.

*9.10.6. Integrate by substitution

a) $\int \left(\frac{x}{3} - 1\right)^5 dx$,

b) $\int (a + bv)^{1/3} \, dv$,

c) $\int 1/(3 - u)^{1/2} du$,

d) $\int_3^5 \frac{dw}{(w - 2)^{1/2}}$,

e) $\int_0^3 \frac{ds}{(4 - s)^3}$,

f) $\int_{-\pi/2}^{\pi/2} \sin 2\phi \, d\phi$.

*9.10.7. Integrate by parts

a) $\int x \sin x \, dx$,

b) $\int t \cos \omega t \, dt$,

c) $\int u(u + 1)^{1/2} \, du$.

*9.10.8. Integrate by parts

a) $\int_0^\pi x \cos 2x \, dx$,

b) $\int_0^{\pi/2} x^2 \cos x \, dx$,

c) $\int_0^{1/2} t(1 - t)^{-1/2} \, dt$.

Chapter 10

Exponential and Logarithmic Functions II

10.1. Introduction

In the preceding chapter we carefully avoided applying calculus to exponential and logarithmic functions although these functions are of fundamental importance for all kinds of mathematical and statistical treatment in the life sciences. The functions

$$y = q^x \quad \text{and} \quad y = \log_{10} x \qquad (10.1.1)$$

which we considered in Chapter 6 are not easy to differentiate or integrate unless we introduce a *special base*, namely the number

$$e = 2.718281828459\ldots \qquad (10.1.2)$$

which we defined in formula (8.2.1) as the limit of a special sequence. The particular exponential function which we have in mind is $y = e^x$, and the particular logarithmic function is $y = \log_e x$.

At a glance it seems strange to use the complicated, irrational number e as a base. However, this is a purely subjective view. In fact, e^x and $\log_e x$ have very simple derivatives. As we shall prove in the following sections, these derivatives are

$$(e^x)' = e^x, \quad (\log_e x)' = \frac{1}{x}, \qquad (10.1.3)$$

that is, the derivative of e^x is identical with e^x itself, and the derivative of $\log_e x$ is simply the reciprocal of x.

For this reason we consider e as a *natural base* for exponential and logarithmic functions. It is customary, indeed, to call $\log_e x$ the *natural logarithm* of x and to denote it by log nat x or briefly by ln x. Consequently, we could call e^x the natural exponential function, but this term is not frequently used.

We conclude this introduction with the rule: *Whenever exponential and logarithmic functions have to be differentiated or integrated, we rewrite the functions with the number e as a base.*

In Section 10.2 we will first define $\ln x$ quite independently of the concept of a logarithm. In the following section we will study the properties of this function which reveal some relationship to logarithms. In Section 10.4 we will introduce the exponential function e^x as the inverse of $\ln x$. This will include the proof that $\ln x$ is indeed a logarithm and identical with $\log_e x$.

There is no straightforward way to define these functions. The procedure rather resembles a detective story. In spite of the complicated logical pattern, we hope that the reader will be able to follow each single step without undue effort. A real understanding of exponential and logarithmic functions opens the door for a host of applications.

Before studying this chapter, the reader should be familiar with the major concepts of Chapters 6, 8, and 9.

10.2. Integral of $1/x$

In formula (9.3.1) we learned that the antiderivative of x^m is $x^{m+1}/(m+1)+C$ except for $m=-1$. Now we are going to investigate this exceptional case. For $m=-1$, x^m reduces to $1/x$. In Fig. 10.1 the function $y=1/t$ is plotted for positive values of t. The reason why we

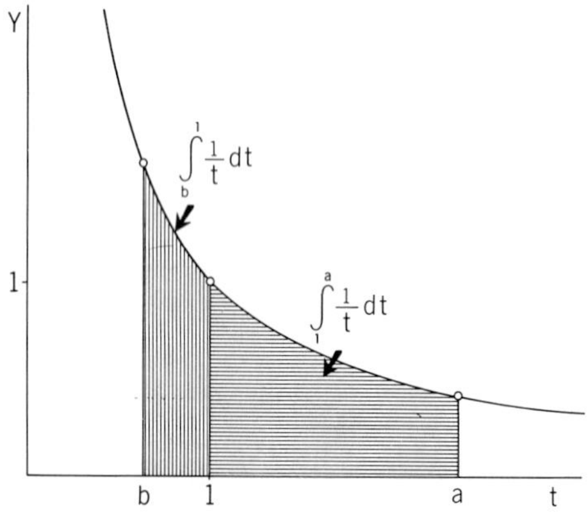

Fig. 10.1. The definition of $\ln a$ for $a>1$ and $\ln b$ for $0<b<1$

denote the abscissa by t and not by x will be clear when we look at formula (10.2.1). The same unit of length is chosen on both axes. Thus integration of the function is equivalent to finding the area of a "curvilinear quadrilateral" between an interval on the t axis and the graph of the function. We will consider two such intervals. First, let $a>1$ be a fixed number. On the horizontal axis the point $t=a$ is to the right of the point $t = 1$. The area over the interval $[1, a]$ is

$$\int_1^a \frac{1}{t}\,dt\,.$$

We will denote this quantity by $\ln a$ and read "*ln* of a". For the moment we forget that $\ln a$ has anything to do with a logarithm. Second, let $0 < b < 1$ be another fixed number. The point $t = b$ is to the left of the point $t = 1$. The area over $[b, 1]$ is

$$\int_b^1 \frac{1}{t}\,dt\,.$$

By means of formula (9.5.14) we can interchange the two limits of the integral which yields for the area

$$-\int_1^b \frac{1}{t}\,dt\,.$$

The integral is of the same form as the integral over $[1, a]$. Hence, it is quite convenient to define

$$\ln b = \int_1^b \frac{1}{t}\,dt = -\int_b^1 \frac{1}{t}\,dt\,.$$

Thus, when $0 < b < 1$, $\ln b$ is a negative number.

Now we perform the same logical step as in Section 9.5 when we introduced the area function $F(x)$. We consider the upper limit of the integral as a variable and denote it by x rather than by a or b. To each $x > 0$ there is assigned a unique value of the area. Hence, the area is a function of x. Therefore, $\ln x$ is also a function of x:

$$\ln x = \int_1^x \frac{1}{t}\,dt \qquad (x > 0)[1]. \tag{10.2.1}$$

If $x = 1$, the interval collapses into one point so that the area is zero. Hence, $\ln 1 = 0$. For $x > 1$, the function $\ln x$ takes on positive values. For $x < 1$, however, we have to interpret x as b in Fig. 10.1. We know already that $\ln b < 0$. Therefore, $\ln x$ takes on negative values for $0 < x < 1$. Numerical values may be approximately determined by the method of dot counting (see Fig. 9.11 and Problem 10.2.1). A table of some values

---

[1] The function $\ln x$ can only be defined for positive values of x. If x were negative, the interval of integration would contain the point $t = 0$, but it follows from formula (10.3.10) that the integral would not be limited to finite values.

rounded off to five decimals follows:

Table 10.1

x	$\ln x$	x	$\ln x$
0.5	—0.69315	2.70	0.99325
1.0	0	2.71	0.99695
1.5	+0.40547	2.72	1.00063
2.0	+0.69315	2.73	1.00430
2.5	+0.91629		
3.0	+1.09861		
3.5	+1.25276		

In Fig. 10.2 the function $y = \ln x$ is plotted. The curve reminds us of the graph of $\log_2 x$ represented in Fig. 6.9. Four-place values of $\ln x$ are given in Table F of the Appendix.

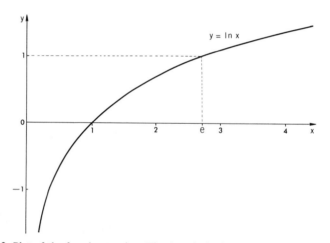

Fig. 10.2. Plot of the function $y = \ln x$. The domain is the positive x axis. The function is monotone increasing

10.3. Properties of ln x

We defined $\ln x$ as the area of a "curvilinear quadrilateral" between the interval $[1, x]$ and the graph of $1/t$. This area changes continuously as x changes. Hence, $\ln x$ is a continuous function of x. Moreover, we can differentiate $\ln x$ by applying formula (9.5.12) to the integral in (10.2.1). Thus we obtain

$$\frac{d}{dx}(\ln x) = (\ln x)' = \frac{1}{x}. \tag{10.3.1}$$

The derivative of $\ln x$ is the integrand in (10.2.1) written with the variable x. Conversely, $\ln x$ is an antiderivative of $1/x$. Hence, the indefinite integral of $1/x$ is

$$\int \frac{1}{x}\,dx = \ln x + C \quad (x > 0) \tag{10.3.2}$$

where C is an arbitrary constant.

Now we study the composite function $\ln(kx)$ where k is a positive number. The chain rule, stated in formula (9.2.14), allows us to differentiate the function. Letting $u = kx$ we get

$$\frac{d}{dx}(\ln kx) = \frac{d}{du}(\ln u) \cdot \frac{du}{dx} = \frac{1}{u} \cdot k = \frac{1}{kx} \cdot k \tag{10.3.3}$$

or

$$\frac{d}{dx}(\ln kx) = \frac{1}{x}. \tag{10.3.4}$$

Surprisingly, the result does not depend on k, that is, $\ln kx$ has the same derivative as $\ln x$. Therefore, the antiderivatives differ only by a certain constant:

$$\ln kx = \ln x + C.$$

To determine C, we let $x = 1$ and get $\ln k = \ln 1 + C = 0 + C$ or $C = \ln k$. It follows that $\ln kx = \ln x + \ln k$. For convenience we rewrite this formula with letters a and b:

$$\ln ab = \ln a + \ln b \quad (a > 0, b > 0). \tag{10.3.5}$$

In words: The function "ln" applied to a product of two positive numbers is equal to the sum of the "ln" taken for each single number. We have already encountered such a property in connection with logarithms (cf. formula (6.4.2)).

For the particular value $b = 1/a$ we conclude from (10.3.5) that $\ln\left(a \cdot \dfrac{1}{a}\right) = \ln a + \ln 1/a$ or, by virtue of $\ln\left(a \cdot \dfrac{1}{a}\right) = \ln 1 = 0$, that

$$\ln 1/a = -\ln a \quad (a > 0). \tag{10.3.6}$$

For example, if $a = 2$, we get $\ln 0.5 = -\ln 2$ (cf. Table 10.1).

Furthermore, by replacing b with $1/c$ in formula (10.3.5) and by observing (10.3.6) for $a = c$ we get

$$\ln a/c = \ln a - \ln c \quad (a > 0, c > 0). \tag{10.3.7}$$

We may apply formula (10.3.5) repeatedly and derive

$$\ln a^2 = \ln a + \ln a = 2 \cdot \ln a,$$
$$\ln a^3 = \ln a^2 + \ln a = 2 \cdot \ln a + \ln a = 3 \cdot \ln a, \quad \text{etc.}$$

In general, for any natural number n it follows that

$$\ln a^n = n \cdot \ln a \quad (a > 0, n = 1, 2, 3, 4, \ldots). \tag{10.3.8}$$

Now we assume that $a > 1$. This implies $\ln a > 0$. Then, as n tends to infinity, $n \cdot \ln a$ also tends to infinity. Therefore, letting $a^n = x$, we obtain from (10.3.8)

$$\ln x \to \infty \quad \text{as} \quad x \to \infty. \tag{10.3.9}$$

If instead $0 < a < 1$, we know that $\ln a < 0$. Hence, as n tends to ∞, the product $n \cdot \ln a$ tends to $-\infty$. At the same time, $x = a^n$ tends to zero (cf. formula (8.1.12)). Thus, (10.3.8) yields

$$\ln x \to -\infty \quad \text{as} \quad x \to 0. \tag{10.3.10}$$

Whereas the domain of the function is the positive x axis, the results (10.3.9) and (10.3.10) indicate that the range is the whole y axis. Cf. Fig. 10.2.

In Section 10.5 we will extend Eq. (10.3.8) for n being any real number. Thus, if x denotes an arbitrary real number, the formula will read

$$\ln a^x = x \cdot \ln a \quad (a > 0, x \in \mathbb{R}). \tag{10.3.11}$$

In Section 10.4 we will already use the result in order to get a definition of e^x.

10.4. The Inverse Function of $\ln x$

Since the derivative of $\ln x$ is $1/x$, and since $1/x$ is positive for every $x > 0$, the function $\ln x$ increases whenever x increases. Hence, $y = \ln x$ is a *monotone* function for all $x > 0$. We introduced this notion in Section 6.3. There we learned that a monotone function has always an *inverse* function. This means: With each value of y there is associated a unique value of x. For instance, if we choose $y = 0$, there exists only one corresponding value of x, namely $x = 1$.

To write the inverse function of $y = \ln x$ in a convenient form, we consider first the special value $y = 1$. The equation $\ln x = 1$ has, as we know, a unique solution. From Fig. 10.2 we see that $x \approx 2.7$ and, with higher accuracy, from Table 10.1 that x must be between 2.71 and 2.72. We denote this particular value of x by e. The reader will hardly be surprised to learn that e is identical with the number (10.1.2). The proof for the identity will be given later in Section 10.8. Now we claim that the exponential function

$$x = e^y$$

is the inverse function to $y = \ln x$. To prove this statement we take the "ln" value on both sides and apply Eq. (10.3.11):

$$\ln x = \ln e^y,$$
$$\ln x = y \ln e,$$
$$\ln x = y.$$

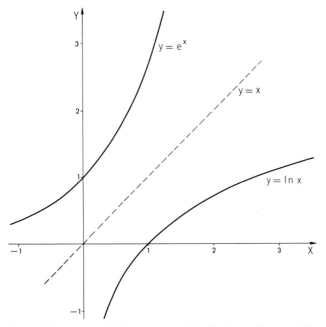

Fig. 10.3. Graph of the exponential function $y = e^x$ obtained by reflection of the graph of $y = \ln x$ about the line $y = x$

When using inverse functions it is customary to exchange x for y (cf. Section 6.3). Thus, we obtain as the inverse function of $y = \ln x$

$$y = e^x \quad \text{(read: } e \text{ to the } x\text{)}, \tag{10.4.1}$$

that is, we get an *exponential function with the special base e*. To plot a graph of this function, we have only to interchange x and y in Fig. 10.2 or, equivalently, to reflect the curve about the line $y = x$ (see Fig. 10.3).

The domain of $y = e^x$ is the whole x axis, and the range is the positive y axis.

Four-place values of e^x are given in Table E of the Appendix.

For e^x there exists another notation which is especially suitable for typing and printing:

$$e^x = \exp x .\qquad (10.4.2)$$

In the newer literature, this notation is frequently used. For instance, instead of $e^{-\frac{1}{2}u^2}$ one writes $\exp(-u^2/2)$.

The inverse function of $y = e^x$ is our original function $y = \ln x$. In Chapter 6 we called the inverse function of an exponential function a *logarithmic function*. Hence, $y = \ln x$ is a special logarithmic function, namely the one with base e, that is,

$$\ln x = \log_e x .\qquad (10.4.3)$$

$\ln x$ is called the *natural logarithm*[2] of x.

Since taking "ln" and "exp" are inverse operations, they cancel each other. Therefore we may write

$$\ln(\exp x) = \ln e^x = x \qquad (10.4.4)$$

and

$$\exp(\ln x) = e^{\ln x} = x .\qquad (10.4.5)$$

10.5. The General Definition of a Power

We already know from Eq. (10.3.8) that

$$\ln a^n = n \cdot \ln a \qquad (a > 0)$$

is valid for $n = 1, 2, 3, \dots$. Now we will prove the formula to be correct for arbitrary real values of n.

First, we assume that n is a negative integer, say $n = -3$. Applying Eqs. (10.3.6) and (10.3.8) we get

$$\ln a^{-3} = \ln 1/a^3 = -\ln a^3 = (-3)\ln a .$$

The exponent may also be $1/n$ where $n = 1, 2, 3, \dots$. Consider for instance

$$\ln a^{1/5} .$$

We multiply this expression by 5. It then follows from formula (10.3.8) that

$$5 \cdot \ln a^{1/5} = \ln(a^{1/5})^5 .$$

---

[2] "ln" was originally introduced as an abbreviation of the Latin logarithmus naturalis. In books on pure mathematics the symbol "ln" is not common. Instead "log" is used since a confusion with other logarithms is not likely.

This reduces to

$$\ln a^{\frac{1}{5} \cdot 5} = \ln a.$$

Hence, on division by 5,

$$\ln a^{1/5} = \frac{1}{5} \cdot \ln a.$$

We may even go a step further and consider

$$\ln a^{3/5}.$$

Here formula (10.3.8) leads to

$$\ln a^{3/5} = \ln (a^{1/5})^3 = 3 \ln a^{1/5},$$

and our previous result yields

$$\ln a^{3/5} = 3 \cdot \frac{1}{5} \ln a = \frac{3}{5} \ln a.$$

In all these cases we obtained

$$\ln a^x = x \cdot \ln a \quad (a > 0). \tag{10.5.1}$$

This formula is true for all fractional numbers x whether positive or negative, that is, *for all rational numbers x* (see the number system in Section 1.14).

Formula (10.5.1) states that $\ln a^x$ is proportional to x for a fixed base a and for all rational numbers x. It is now a minor step to extend this formula for *all real numbers x*. Indeed, a real number can be considered as a limit of a sequence of rational numbers, as for instance $\sqrt{3}$ is the limit of the sequence

$$1.7, \ 1.73, \ 1.732, \ 1.7320, \ \dots .$$

Eq. (10.5.1) makes it possible to define powers which are as strange as

$$3^{\sqrt{2}}, \quad 2^\pi, \quad (\tfrac{1}{2})^{-\sqrt{5}}, \quad \sqrt{3^{\sqrt{2}}}.$$

We only have to apply the exponential function to Eq. (10.5.1):

$$\boxed{a^x = e^{x \cdot \ln a}} \quad (a > 0). \tag{10.5.2}$$

Every power can be written with the special base e.

In Chapter 11 we will have to integrate $1/x$ in the case where x is negative. The antiderivative cannot be $\ln x$ since the logarithm of a negative number does not exist. We try with $\ln(-x)$. According to the chain rule the derivative is $-1/(-x) = 1/x$. This is exactly what we wanted. It is convenient to replace $-x$ by the absolute value of x, that is, by $|x|$ (see Section 1.6). Thus, a result can be stated which is valid for both positive and negative values of x:

$$\int \frac{1}{x} dx = \ln |x| + C \qquad (x \neq 0). \tag{10.7.3}$$

We may also be interested in differentiating $\log x$ where "log" means the common logarithm. By means of formulas (10.6.3) and (10.7.2) we obtain

$$(\log x)' = \left(\frac{\ln x}{\ln 10} \right)' = \frac{1}{\ln 10} \cdot \frac{1}{x}. \tag{10.7.4}$$

This formula is more complicated than (10.7.2). Therefore, we try to avoid it.

The derivative $(\ln x)' = 1/x$ will also serve us in finding the derivative of e^x. For this purpose we use the formula for the derivative of an inverse function

$$dx/dy = \frac{1}{dy/dx}$$

established at the end of Section 9.2. Since $x = e^y$ is equivalent to $y = \ln x$, it follows that

$$\frac{d}{dy} e^y = \frac{dx}{dy} = \frac{1}{dy/dx} = \frac{1}{d(\ln x)/dx} = \frac{1}{1/x} = x = e^y.$$

Replacing y by x we obtain

$$\frac{d}{dx} e^x = (e^x)' = (\exp x)' = e^x \tag{10.7.5}$$

as predicted in (10.1.3).

More frequently we have to differentiate composite functions such as e^{mx} with an arbitrary constant $m \neq 0$. By the chain rule as stated in formula (9.2.13) we get

$$(e^{mx})' = e^{mx} \cdot m. \tag{10.7.6}$$

The antiderivative is therefore

$$\int e^{mx} dx = \frac{1}{m} e^{mx} + C \qquad (m \neq 0) \tag{10.7.7}$$

where C is an arbitrary constant.

Sometimes we are faced with the problem of differentiating the general exponential function $a^x (a > 0)$. By using formula (10.5.2) we introduce the special base e. Then we apply the chain rule as shown in formula (10.7.6). Thus we get

$$(a^x)' = (e^{x \cdot \ln a})' = e^{x \cdot \ln a} \cdot \ln a = a^x \cdot \ln a . \qquad (10.7.8)$$

To avoid the clumsy factor $\ln a$, it is preferable to express all powers in terms of the special base e.

Finally, we exploit the exponential function e^{mx} for solving a problem which we left open in differential calculus, namely the problem of *differentiating the power function* $y = x^n$ where n is any real number. In formula (9.2.4) we merely anticipated the result. The proof is based on formula (10.5.2) and on the chain rule:

$$x^n = e^{n \cdot \ln x} \quad \text{for} \quad x > 0 ,$$

$$\frac{d}{dx}(x^n) = \frac{d}{dx}(e^{n \cdot \ln x}) = e^{n \cdot \ln x} \cdot n \cdot \frac{1}{x} = x^n \cdot \frac{n}{x} = nx^{n-1}$$

for $x > 0$. The formula may be extended to negative values of x if one assumes that n is an integer. The result is the same, but we skip the proof.

10.8. Some Limits

According to formula (9.2.2) the derivative of $\ln x$ is defined by

$$(\ln x)' = \lim_{h \to 0} \frac{\ln(x + h) - \ln x}{h} . \qquad (10.8.1)$$

Since we know already the derivative of $\ln x$, we obtain the following limit of a difference quotient

$$\lim_{h \to 0} \frac{\ln(x + h) - \ln x}{h} = \frac{1}{x} \qquad (x > 0) . \qquad (10.8.2)$$

A numerical example may illustrate this result. Let $x = 2$. Then we get from a table of natural logarithms:

h	$\ln(2 + h)$	$\dfrac{\ln(2 + h) - \ln 2}{h}$
1	1.09861	0.405
0.1	0.74194	0.488
0.01	0.69813	0.498
0.001	0.69365	0.499

We see that the difference quotient approaches the value $1/x = 0.500$ as predicted by formula (10.8.2). For the particular value $x = 1$, we conclude that

$$\frac{\ln(1+h)}{h} \to 1 \quad \text{as} \quad h \to 0$$

since $\ln 1 = 0$. Here we replace h by a/n where a is any positive or negative constant and n is a natural number tending to infinity. Hence

$$\ln\left(1 + \frac{a}{n}\right) \Big/ \frac{a}{n} = \frac{n}{a} \cdot \ln\left(1 + \frac{a}{n}\right) \to 1 \,.$$

We multiply by a and apply formula (10.3.8). Thus we get

$$\ln\left(1 + \frac{a}{n}\right)^n \to a \tag{10.8.3}$$

or by employing the inverse function

$$\left(1 + \frac{a}{n}\right)^n \to e^a \quad \text{as} \quad n \to \infty \quad (a \text{ real})\,. \tag{10.8.4}$$

This is an important limit which occurs frequently both in theoretical and applied work. We already met the special case $a = 1$ in Section 8.2. With this result we are able to identify the base e with the number $2.71828\ldots$ which was established in Section 8.2.

10.9. Applications

No other functions have found such a diversity of applications in the life sciences as have the exponential and logarithmic functions. We have already studied various biological examples in Sections 6.5 and 6.6. Now we are going to deal with many more applications in this section as well as in problems at the end of the chapter.

Example 10.9.1. A chicken egg was incubated for three days at a temperature of $t_0 = 37°$ C. Subsequently during a period of 40 minutes, the temperature t was reduced and the number N of heart beats per minute measured[3]:

$t(°\,C)$	N	$t(°\,C)$	N
36.3	154	31.1	82
35.0	133	30.4	75
33.9	110	24.7	38
32.4	94	24.2	36
31.8	83		

---

[3] Unpublished data. The author is indebted to Dr. G. Wagner, Bern, Switzerland for the permission to use his data.

Show graphically that N can be approximately represented by an exponential function of t. Find the parameters for this exponential function.

In Section 7.2 we introduced the semilogarithmic plot and we learned that the graph of an exponential function is a straight line. Thus we have to plot $\log N$ versus t. A slight change, however, is advisable. Instead of the temperature t it is more revealing if we plot the difference between t and $37°$, that is, the quantity τ (lower case Greek letter tau) defined by

$$\tau = 37° - t. \qquad (10.9.1)$$

For numerical purposes we operate with common logarithms. Thus the original table of data turns into the new table:

$\tau(°\,C)$	$\log N$	$\tau(°\,C)$	$\log N$
0.7	2.188	5.9	1.914
2.0	2.124	6.6	1.875
3.1	2.041	12.3	1.580
4.6	1.973	12.8	1.556
5.2	1.919		

These data are plotted in Fig. 10.4. The dots with coordinates $(\tau, \log N)$ are located very close to a straight line. It is easy to adjust such a line by eye. It follows that $\log N$ is a linear function of τ, that is,

$$\log N = a + b\tau \qquad (10.9.2)$$

with certain positive constants a and b. For $\tau = 0$ we get $a = \log N$. From Fig. 10.4 we read $a = 2.21$. The coefficient of τ is the slope b. In a right triangle with $\Delta\tau = +10$, we read $\Delta \log N = -0.496$. Hence, $b = (-0.496)/10 = -0.0496$. By the slope b we mean the decrease of $\log N$ per degree Celsius. The inverse function of (10.9.2) is

$$N = 10^{a+b\tau} = 10^a \cdot 10^{b\tau} = (\text{antilog } a) \cdot 10^{b\tau} = 162 \cdot 10^{b\tau}.$$

However, it is customary to use e as a base for the exponential function. This is possible by applying formula (10.5.2) for $a = 10$ and $x = b\tau$. We get

$$N = 162 \cdot e^{b(\ln 10)\tau}, \qquad (b = -0.0496)$$

or with the numerical value of $\ln 10$ listed in (10.6.4)

$$N = 162\, e^{-0.114\tau}. \qquad (10.9.3)$$

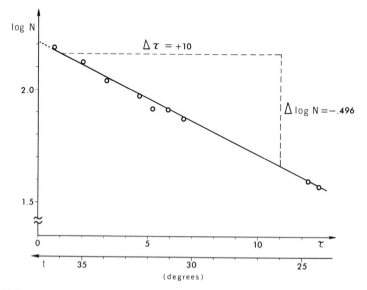

Fig. 10.4. Decrease of the pulse rate N in chicken eggs with decreasing temperature. The slope of the straight line means the decrease of $\log N$ per degree Celsius

The function is of the form $N = N_0 e^{-k\tau}$ with parameters N_0 and k. Here $N_0 = 162$ means the number of heart beats per minute for $\tau = 0°$ or $t = 37°$. The second parameter is $k = 0.114$.

When the rate of decrease is of interest, it follows from formula (10.9.3) that

$$\frac{dN}{d\tau} = 162(-0.114)\, e^{-0.114\tau} = -(18.5)\, e^{-0.114\tau}. \qquad (10.9.4)$$

This rate is negative and depends on τ. For $\tau = 0$ it amounts to -18.5. This means: At the very beginning of the experiment the number of heart beats per minute decreased by 18.5 per $1°$ C.

Example 10.9.2. We consider a substance containing radioactive atoms and assume that only one sort of radioactive isotope occurs[4]. Let N denote the number of radioactive atoms present in the substance at time t. Then experiments show that the radioactive decay follows the law

$$N = N_0 e^{-\lambda t} = N_0 \cdot \exp(-\lambda t). \qquad (10.9.5)$$

---

[4] In many texts this assumption is not explicitly mentioned. However, the exponential law (10.9.5) does not hold if two or more different radioactive isotopes are contained in the substance.

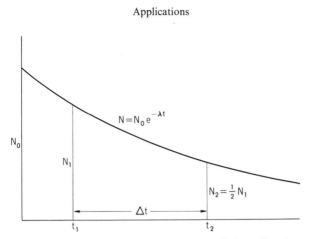

Fig. 10.5. During the time interval of length Δt, half of the radioactive atoms decay

N_0 is the number of radioactive atoms at time $t = 0$. The number λ (lower case Greek letter lambda) is positive and is called the *decay constant*. Strictly speaking, N is an integer, but for reasons explained in Fig. 9.5 we may approximate the integer by a continuous variable.

The decay constant λ has no immediate intuitive meaning. Therefore, it is desirable to introduce the so-called *half-life*. This is the length Δt of a time interval in which 50% of the radioactive atoms decay. To relate Δt with λ, we consider two time instances, say t_1 and $t_2 = t_1 + \Delta t$ (see Fig. 10.5), and let N_1 and N_2 denote the number of radioactive atoms at these time instances. From formula (10.9.5) we get

$$N_1 = N_0 e^{-\lambda t_1}, \quad N_2 = N_0 e^{-\lambda t_2} = N_0 e^{-\lambda(t_1 + \Delta t)} = N_0 e^{-\lambda t_1} \cdot e^{-\lambda \cdot \Delta t}$$

or

$$N_2 = N_1 \cdot e^{-\lambda \cdot \Delta t}. \qquad (10.9.6)$$

Since we assume that half of the radioactive atoms decay during the time interval, we get $N_2 = \frac{1}{2} N_1$. Then formula (10.9.6) yields

$$e^{-\lambda \cdot \Delta t} = \frac{1}{2}, \quad -\lambda \cdot \Delta t = \ln \frac{1}{2} = -0.69315.$$

Hence, the half-life Δt is

$$\Delta t = \frac{-\ln 0.5}{\lambda} = \frac{0.69315}{\lambda}. \qquad (10.9.7)$$

The half-life increases when the decay constant decreases and conversely. Both parameters λ and Δt are typical for the isotope under consideration. Notice that Δt neither depends on N_0 nor on the time instant t_1. This means: *Whenever a time interval of length Δt elapses, the number of remaining radioactive atoms is reduced by one half.*

From formula (10.9.5) and from the chain rule it follows that the *rate of decay* becomes

$$\frac{dN}{dt} = N_0 e^{-\lambda t}(-\lambda) = -\lambda N . \tag{10.9.8}$$

Hence, the rate of decay is proportional to N, that is, to the number of remaining radioactive atoms.

Example 10.9.3. Several strains of tobacco virus, such as Aucuba, on leaves of *Nicotiana sylvestris* and other tobacco plants were exposed to X-rays of different wave lengths. By radiation, part of the virus particles were inactivated so that reproduction ceased. Gowen (1964) reports that the *number y of surviving particles decreased exponentially with the roentgen dosage r applied.* Hence, with satisfactory approximation

$$y = y_0 e^{-ar} = y_0 \cdot \exp(-ar) \tag{10.9.9}$$

where a is a certain positive constant depending on the biological material.

We may ask: What is the proper dosage to inactivate 90% of the virus? To answer this question we let $y = y_0/10$ and denote the required dosage by r_{90}. Then formula (10.9.9) yields

$$e^{-a \cdot r_{90}} = 1/10 .$$

Hence,

$$r_{90} = \frac{-\ln 1/10}{a} = \frac{\ln 10}{a} = \frac{2.303}{a} . \tag{10.9.10}$$

Example 10.9.4. In lakes and in the sea, plant life can only exist in the top layer which is roughly 10 meters deep since *daylight is gradually absorbed by the water.* We may ask: How does light intensity decrease with increasing thickness of the layer? The answer is the *Bouguer-Lambert law*[5]. Consider a vertical beam entering the water with original intensity I_0. Let I be the reduced intensity in a depth of x meters. Then the law states that

$$I = I_0 e^{-\mu x} . \tag{10.9.11}$$

The parameter $\mu > 0$ is called the *absorption coefficient* (μ is the lower case Greek letter mu). It depends on the purity of water and the wave length of the beam. Strictly speaking, the intensity I will never be

---

[5] Pierre Bouguer (1698—1758), French scientist and explorer, studied the absorption of light in the atmosphere. Johann Heinrich Lambert (1728—1777), Alsatian mathematician, astronomer and physicist, studied the law in general.

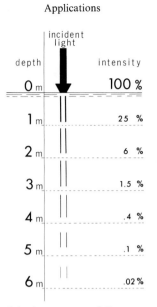

Fig. 10.6. Absorption of daylight in sea water follows an exponential law. With an increase of depth by 1 m, over 75 percent of the light is absorbed

exactly zero. However, for sufficiently large x the remaining light can no longer be perceived. Fig. 10.6 illustrates how the light intensity decreases in fairly clean sea water (for μ we assume the value $1.4\,\mathrm{m}^{-1}$).

The Bouguer-Lambert law is applicable to any homogeneous, transparent substance such as glass, plexiglass, liquids, and thin layers viewed under the microscope. In addition to light waves, other electromagnetic waves, such as X-rays and gamma rays, behave the same way. The Bouguer-Lambert law is basic in photometry.

Example 10.9.5. A more sophisticated application of the exponential function occurs in connection with mortality by natural causes. Fig. 10.7 depicts the number of surviving rats as a function of time (measured in months). The data were collected from 144 laboratory rats of the same strain. All these rats had reached an age of seven months. From then on their ages at the time of natural death were registered. The survival function is a step function, but it can well be approximated by a smooth curve according to a formula due to Gompertz[6]:

$$N = a \cdot e^{-b \cdot e^{kt}} = a\,\exp(-b \cdot \exp kt). \qquad (10.9.12)$$

The formula contains three parameters a, b, k. They are all positive numbers. The number N of surviving animals is a composite function

[6] Benjamin Gompertz (1779—1865), English mathematician and actuary.

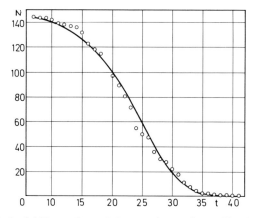

Fig. 10.7. Survival of 144 rats from their seventh month on. The dots represent the observed number of surviving rats as a function of time (months). The fitted line is a Gompertz curve. The Figure is reproduced from Miescher (1955, p. 34)

of the time t. It is essentially "an exponential function of an exponential function". The graph is S-shaped and therefore called a *sigmoid curve*.

In (10.9.12), $t = 0$ does not refer to birth, but to some fixed age (in our example, $t = 0$ means the age of 7 months when the observations began). The initial number of animals is $N_0 = a \exp(-b)$.

The Gompertz formula is used by actuaries who have to estimate the risk of death in life insurance. Attempts have been made to justify the formula for biological reasons. See Strehler (1963, Chap. 4) for a thorough account. With a negative value of k, formula (10.9.12) is sometimes used in the study of growth (see Thrall *et al.*, 1967, CA 13.4.) and in ecology (Wilbur and Collins, 1973).

10.10. Approximations and Series Expansions

For many functions $y = f(x)$ it is a practical problem to find numerical values of y given corresponding values of x. Sometimes accurate figures are required; sometimes very rough approximations will do. There are many possible ways of calculating functional values with reasonable accuracy. Here we introduce a method which has proved to be especially successful in the area of exponential and logarithmic functions, the *approximation of $f(x)$ by polynomials in x*.

We begin with a polynomial of the first degree

$$p_1(x) = a_0 + a_1 x$$

and try to determine the coefficients in such a way that

$$f(x) \approx p_1(x)$$

holds in the vicinity of $x = 0$. Assuming that $f(x)$ is differentiable, we get simultaneously

$$f(x) \approx a_0 + a_1 x,$$
$$f'(x) \approx a_1.$$

We seek exact equality at $x = 0$. Hence we put

$$f(0) = a_0, \qquad f'(0) = a_1.$$

Thus $p_1(x)$ turns out to be

$$p_1(x) = f(0) + f'(0) \cdot x, \qquad f(x) \approx p_1(x). \qquad (10.10.1)$$

The accuracy of approximation cannot be judged in general.

We apply formula (10.10.1) first to $f(x) = e^x$. Since $f'(x) = e^x$ and $e^0 = 1$, we get

$$e^x \approx 1 + x \qquad (10.10.2)$$

for values of x that deviate slightly from 0. This approximation is often used. Notice that the graph of the linear function $y = 1 + x$ is the tangent to the graph of $y = e^x$ at the point $x = 0$, $y = 1$ (cf. Fig. 10.3). A numerical example may prove the usefulness of the approximation. For $x = 0.02$ we get from (10.10.2) $e^{0.02} \approx 1.02$, whereas the exact value is $e^{0.02} = 1.0202...$.

Another frequently used approximation deals with the composite function $f(x) = \ln(1 + x)$. Here $f'(x) = 1/(1 + x)$, $f(0) = \ln 1 = 0$, $f'(0) = 1$. Hence, it follows from (10.10.1) that

$$\ln(1 + x) \approx x \qquad (10.10.3)$$

in the vicinity of $x = 0$. For instance, if $x = 0.02$, we obtain $\ln 1.02 \approx 0.02$, whereas the exact value is $\ln 1.02 = 0.01980...$.

If better approximations are required, we replace $p_1(x)$ by a second degree or higher degree polynomial in x:

$$p_2(x) = a_0 + a_1 x + a_2 x^2,$$
$$p_3(x) = a_0 + a_1 x + a_2 x^2 + a_3 x^3,$$
$$\cdots\cdots\cdots\cdots\cdots\cdots\cdots\cdots\cdots \qquad (10.10.4)$$
$$p_n(x) = a_0 + a_1 x + a_2 x^2 + \cdots + a_n x^n$$

where n is a natural number.

We assume that $f(x)$ has higher order derivatives. Then we try to approximate simultaneously

$$f(x) \approx p_n(x), \quad f'(x) \approx p_n'(x), \quad f''(x) \approx p_n''(x), \quad \text{etc.}$$

We seek equality at $x = 0$, that is, we put

$$f(0) = p_n(0), \quad f'(0) = p_n'(0), \quad f''(0) = p_n''(0), \quad \text{etc.} \quad (10.10.5)$$

Higher derivatives are denoted by $f^{(3)}$, $f^{(4)}$, ..., $f^{(n)}$ (read: fn prime). The derivatives of $p_n(x)$ are

$$p_n'(x) = a_1 + 2a_2 x + 3a_3 x^2 + 4a_4 x^3 + \cdots + na_n x^{n-1},$$
$$p_n''(x) = 2a_2 + 2 \cdot 3a_3 x + 3 \cdot 4a_4 x^2 + \cdots + (n-1) na_n x^{n-2},$$
$$p_n^{(3)}(x) = 2 \cdot 3a_3 + 2 \cdot 3 \cdot 4a_4 x + \cdots + (n-2)(n-1) na_n x^{n-3}, \quad (10.10.6)$$
$$\cdots\cdots\cdots\cdots\cdots\cdots\cdots\cdots\cdots\cdots\cdots\cdots\cdots\cdots\cdots\cdots\cdots$$
$$p_n^{(n)}(x) = 2 \cdot 3 \cdot 4 \cdots (n-1) na_n.$$

Here products of the form $2 \cdot 3 \cdot 4 \cdots k$ appear frequently. They are called *factorials*. As a convenient abbreviation one writes

$$1 \cdot 2 \cdot 3 \cdots k = k! \quad \text{(read: } k \text{ factorial)}. \quad (10.10.7)$$

It follows from (10.10.6) that

$$p_n'(0) = a_1, \quad p_n''(0) = 2a_2, \quad p_n^{(3)}(0) = 3! \, a_3,$$
$$p_n^{(4)}(0) = 4! \, a_4, \dots, p_n^{(n)}(0) = n! \, a_n. \quad (10.10.8)$$

Finally, from the equalities (10.10.5) we obtain

$$a_0 = f(0), \quad a_1 = f'(0), \quad a_2 = \frac{1}{2!} f''(0),$$

$$a_3 = \frac{1}{3!} f^{(3)}(0), \dots, a_n = \frac{1}{n!} f^{(n)}(0).$$

Hence

$$f(x) \approx f(0) + f'(0) x + \frac{f''(0)}{2!} x^2 + \cdots + \frac{f^{(n)}(0)}{n!} x^n \quad (10.10.9)$$

in the vicinity of $x = 0$.

Example 10.10.1. Again we apply the result to $f(x) = e^x$. Since all higher derivatives of e^x are e^x, we get $f(0) = f'(0) = f''(0) = \cdots = 1$. Hence, (10.10.9) yields

$$e^x \approx 1 + x + \frac{x^2}{2!} + \frac{x^3}{3!} + \frac{x^4}{4!} + \cdots + \frac{x^n}{n!} \quad (10.10.10)$$

in the vicinity of $x = 0$. This formula is used by computers to calculate values of e^x.

Example 10.10.2. In the case of our composite function $f(x) = \ln(1 + x)$, the higher derivatives are still relatively simple:

$$f'(x) = (1+x)^{-1}, \quad f''(x) = (-1)(1+x)^{-2},$$
$$f^{(3)}(x) = (-1)(-2)(1+x)^{-3} \quad \text{etc.}$$

Hence,

$$f'(0) = 1, \quad f''(0) = -1, \quad f^{(3)}(0) = 2!, \quad f^{(4)}(0) = -3!, \quad \text{etc.}$$

It follows from (10.10.9) that

$$\ln(1+x) \approx x - \frac{x^2}{2} + \frac{x^3}{3} - \frac{x^4}{4} + \frac{x^5}{5} - \cdots \pm \frac{x^n}{n} \quad (10.10.11)$$

in the vicinity of $x = 0$.

Example 10.10.3. Let $f(x) = (1 + x)^s$ where s is any real number. The higher derivatives are:

$$f'(x) = s(1+x)^{s-1} \qquad f'(0) = s$$
$$f''(x) = s(s-1)(1+x)^{s-2} \qquad f''(0) = s(s-1)$$

Hence,

$$(1+x)^s \approx 1 + sx + \frac{s(s-1)}{2!}x^2 + \frac{s(s-1)(s-2)}{3!}x^3 + \cdots +$$

$$(10.10.12)$$

$$+ \frac{s(s-1)\ldots(s-n+1)}{n!}x^n.$$

A special case of (10.10.12) is the *binomial formula* for $s = 1, 2, 3, \ldots$ which will be treated in Section 13.7.

One is tempted to stress the accuracy and to ask: Does the polynomial $p_n(x)$ with coefficients given in formula (10.10.9) converge to $f(x)$ when n tends to infinity? In other words: Does the equality sign hold in

$$f(x) = \lim_{n \to \infty} \left(f(0) + f'(0)x + \frac{f''(0)}{2!}x^2 + \cdots + \frac{f^{(n)}(0)}{n!}x^n \right)? \quad (10.10.13)$$

To answer this question we would need mathematical tools which go far beyond the scope of this book. We will confine ourselves to remarking that convergence can be established for most differentiable functions. In the case of $f(x) = e^x$, we have convergence not only in the vicinity of $x = 0$, but surprisingly *for all values of x*, that is, even for $x = 1000$ or $x = -1000$. (See Problem 8.3.11.)

For most other functions, however, x has to be restricted to a suitable interval $|x| < C$ if (10.10.13) is to converge. For instance, the polynomial (10.10.11) converges to $\ln(1 + x)$ only if $|x| < 1$.

Formula (10.10.13) may be interpreted by saying that we add an infinite number of terms. It is customary to call such a representation of $f(x)$ an *expansion into a series*. One even drops the symbol for the limit although this may cause difficulties of understanding. Thus one simply writes

$$f(x) = f(0) + f'(0) \cdot x + \frac{f''(0)}{2!} x^2 + \cdots \qquad (|x| < C) \quad (10.10.14)$$

and, in particular,

$$e^x = 1 + x + \frac{x^2}{2!} + \frac{x^3}{3!} + \cdots \qquad \text{(all } x), \qquad\qquad (10.10.15)$$

$$\ln(1 + x) = x - \frac{x^2}{2} + \frac{x^3}{3} - \frac{x^4}{4} + - \cdots \qquad (|x| < 1). \quad (10.10.16)$$

Formula (10.10.14) is the famous *Maclaurin series*[7].

*10.11. Hyperbolic Functions

In biological theory there are occasions when the sum or the difference of e^x and e^{-x} enter a formula. At such an occasion it is customary to rewrite the formula by making use of the following functions:

The *hyperbolic sine*, defined by

$$\sinh x = \frac{1}{2}(e^x - e^{-x}), \qquad\qquad (10.11.1)$$

the *hyperbolic cosine*, defined by

$$\cosh x = \frac{1}{2}(e^x + e^{-x}), \qquad\qquad (10.11.2)$$

the *hyperbolic tangent*, defined by

$$\tanh x = \sinh x/\cosh x, \qquad\qquad (10.11.3)$$

the *hyperbolic cotangent*, defined by

$$\coth x = \cosh x/\sinh x. \qquad\qquad (10.11.4)$$

---

[7] Colin Maclaurin (1698—1746), Scottish mathematician.

All four functions are called *hyperbolic functions*. Both the word hyperbolic as well as sine, cosine, etc. need a justification. For this purpose we calculate

$$\cosh^2 x - \sinh^2 x = \frac{1}{4}(e^{2x} + 2 + e^{-2x}) - \frac{1}{4}(e^{2x} - 2 + e^{-2x})$$

which leads to

$$\cosh^2 x - \sinh^2 x = 1 . \tag{10.11.5}$$

This relationship is analogous to $\cos^2 \alpha + \sin^2 \alpha = 1$ which we know from formula (5.7.1). The only formal difference is the minus sign in (10.11.5). Moreover, if α is a variable polar angle, the points (x, y) with

$$x = \cos \alpha , \qquad y = \sin \alpha ,$$

plotted in a rectangular coordinate system, are located on the circumference of the *unit circle* whose equation is $x^2 + y^2 = 1$ (see Section 5.4). In the same way, by plotting the points (x, y) with

$$x = \cosh \alpha , \qquad y = \sinh \alpha , \tag{10.11.6}$$

we find a curve with equation $x^2 - y^2 = 1$ because of formula (10.11.5). This curve is a *hyperbola* with asymptotes given by the equations $y = x$ and $y = -x$. Thus the hyperbolic sine and cosine are related to a hyperbola in much the same way as the ordinary sine and cosine are connected with the circle.

Even though the preceding results would offer enough justification for the terms hyperbolic sine and hyperbolic cosine, there exist more striking analogies. For instance, we may study the derivatives:

$$(\cosh x)' = \frac{d}{dx} \frac{1}{2}(e^x + e^{-x}) = \frac{1}{2}(e^x - e^{-x}) = \sinh x , \tag{10.11.7}$$

$$(\sinh x)' = \frac{d}{dx} \frac{1}{2}(e^x - e^{-x}) = \frac{1}{2}(e^x + e^{-x}) = \cosh x . \tag{10.11.8}$$

These formulas are quite analogous to $(\cos \alpha)' = -\sin \alpha$ and $(\sin \alpha)' = \cos \alpha$.

Despite many analogies there are also major differences. The hyperbolic functions are *not periodic*. This is immediately seen from a graph of these functions (see Fig. 10.8).

Readers interested in biological applications of hyperbolic functions are referred to K. S. Cole (1965, p. 141; 1968, p. 71 ff.), Defares *et al.* (1973), Fisher (1965, p. 124), Moran (1962, p. 27, 137), Rashevsky (1960, Vol. 1, p. 56 ff., 61, 316; Vol. 2, p. 177–178). The graph of $\tanh x$ is closely related to the logistic curve which will be treated in Section 11.5.

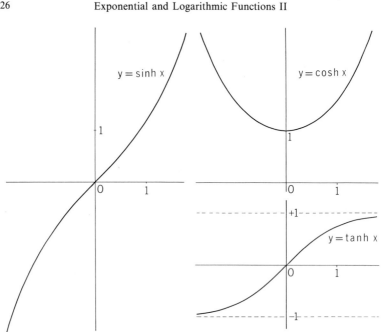

Fig. 10.8. Graphs of three hyperbolic functions

Brief list of formulas. It is worth learning the following formulas by heart:

$$e^u \cdot e^v = e^{u+v}, \qquad e^u/e^v = e^{u-v},$$

$$(e^u)^r = e^{ur}, \qquad q^x = e^{x \cdot \ln q} \quad (q > 0),$$

$$\left. \begin{aligned} \ln a + \ln b &= \ln(ab) \\ \ln a - \ln b &= \ln(a/b) \end{aligned} \right\} \quad (a > 0,\ b > 0),$$

$$\ln a^s = s \cdot \ln a \qquad (a > 0),$$

$$\ln e = 1,$$

$$\frac{d}{dx} e^x = e^x, \qquad \int e^x \, dx = e^x + C \quad (\text{all } x),$$

$$\frac{d}{dx} \ln x = \frac{1}{x}, \qquad \int \frac{1}{x} \, dx = \ln|x| + C \quad (x \neq 0),$$

$$\left. \begin{aligned} e^x &\approx 1 + x \\ \ln(1 + x) &\approx x \end{aligned} \right\} \text{ for } x \text{ sufficiently close to zero.}$$

Recommended tables of e^x and $\ln x$: Up to six decimal places: Allen (1947), Diem *et al.* (1970), Meredith (1967). Up to 18 decimal places:

National Bureau of Standards (1947), contains a table of e^x for x ranging from -2.5 to 10 in steps of 0.0001. National Bureau of Standards (1941), the four volumes contain tables of $\ln x$ for x ranging from 0.0001 to 5 in steps of 0.0001 and from 5 to 100,000 in steps of 1.

Recommended for further reading: Bak and Lichtenberg (1966), Defares *et al.* (1973), Lefort (1967).

Problems for Solution

10.2.1. Find an approximate value of

$$\int_1^3 \frac{1}{x}\,dx = \ln 3$$

by the method of dot counting (see Fig. 9.11) and compare the result with Table 10.1 or Table F of the Appendix.

10.2.2. In the same way as in the previous problem, find an approximate value of $\ln 0.5$.

10.3.1. Using Eq. (10.3.8.) and Table F, find numerical values of

a) $\ln 2^{10}$,   b) $\ln 4^3$,   c) $\ln 10000$.

10.3.2. In the same way find

a) $\ln(2.5)^5$,   b) $\ln 125$,   c) $\ln 1600$.

10.4.1. Using Table E calculate

a) $e^2 + e^{-2}$,   b) $1 - e^{-0.5}$,   c) $1/\exp(0.7)$.

10.4.2. Using Table E find $\exp 1.5$ and $\exp(-1.5)$ and show that the two values are mutually reciprocal.

10.4.3. Plot $y = e^{2x}$ and $y = \frac{1}{2}e^{2x}$.

10.4.4. Plot $y = e^{-x}$ and $y = -e^{-x}$.

10.4.5. Plot $y = e^{0.5x}$ and $y = 2e^{0.5x}$.

10.4.6. Find

a) $\lim_{x \to \infty} (1 - e^{-x})$,   b) $\lim_{s \to \infty} \dfrac{e^{-s} + 2}{2e^{-s} + 3}$,

c) $\lim_{u \to \infty} \dfrac{1}{3 + e^u}$.

10.4.7. Find

a) $\lim\limits_{w\to\infty} \dfrac{e^w + p}{2e^w + 3p}$,

b) $\lim\limits_{u\to\infty} \dfrac{1000}{e^{u/2} - 1}$,

c) $\lim\limits_{t\to\infty} \dfrac{3 + ae^t}{a^2 - e^t}$.

10.5.1. Calculate

a) $e^{\sqrt{3}}$,

b) $(0.4)^{1.5}$,

c) $(1.5)^{-0.4}$.

10.5.2. Calculate

a) $e^{1/3}$,

b) $\exp(-\sqrt{2})$,

c) $1/e^{-3/5}$.

10.5.3. Using formula (10.5.2) rewrite the following exponential functions in the form $e^{\cdots}$ or $\exp(\ldots)$:

a) 2^x

b) 10^u

c) 4.43^s

d) 2.8^{-3t}

e) $0.77^{(1.65)x}$

10.5.4. Rewrite in the form $\exp(\ldots)$

a) e^{s-t},

b) 2^{3n},

c) 5^{1-s},

d) $2.8^{0.35}$,

e) 6^{-3x},

f) 100^{ax^2}.

10.5.5. Rewrite $e^{2x+3.5}$ in the form $A \cdot e^{cx}$.

10.5.6. Rewrite in the form $A \cdot \exp ct$

a) e^{t+3},

b) $\exp(3 + 2t)$,

c) 10^{t-1}.

10.6.1. In Table D find common logarithms and in Table F natural logarithms of 10, 11, 12 and show that corresponding logarithms are proportional.

10.6.2. Using Table D and Eq. (10.6.3.) find $\ln 3.54$ to four decimal places.

10.7.1. Differentiate the following composite functions:

a) $\dfrac{d}{dx} e^{3x}$

b) $\dfrac{d}{du} e^{1-2u}$

c) $\dfrac{d}{dt} \exp\left(-\dfrac{1}{2}t^2\right)$

d) $\dfrac{d}{dx} \ln(5x+4)$

e) $\dfrac{d}{dv} \ln(v^2 - 2)$

f) $\dfrac{d}{ds} \ln\left(1 + \dfrac{1}{s}\right)$

g) $\dfrac{d}{dt} (t \cdot e^{\frac{1}{2}t})$

h) $\dfrac{d}{du} (u \cdot \ln 3u)$

i) $\dfrac{d}{dr} \ln \dfrac{r}{1-r}$

10.7.2. Find the following first and second derivatives:

a) $\dfrac{d^2}{dx^2} e^{3x}$,

b) $\dfrac{d}{ds} e^{as+b}$,

c) $\dfrac{d}{du} e^{1/u}$,

d) $\dfrac{d^2}{dt^2} e^{2t-1}$,

e) $\dfrac{d}{d\alpha} e^{\sin \alpha}$,

f) $\dfrac{d}{dr} \exp(1-rs)$,

g) $\dfrac{d^2}{dv^2} \exp(kv)$,

h) $\dfrac{d^2}{dx^2} \exp(-x^2/2)$,

i) $\dfrac{d}{dt} \ln(ct+1)$,

k) $\dfrac{d^2}{dt^2} \ln(ct+1)$,

l) $\dfrac{d}{d\beta} \ln(\sin \beta)$,

m) $\dfrac{d}{du} \ln(1+u^2)$.

10.7.3. If $f(x)$ is differentiable and $f(x) > 0$, prove that

$$(\ln f(x))' = f'(x)/f(x).$$

10.7.4. Let $g(x)$ be differentiable. Solve

$$\frac{d}{dx} \ln(g(x))^{-1}$$

in two ways, a) by immediate differentiation, b) by first using Eq. (10.5.1.).

10.7.5. Find the indefinite integrals of the following functions:

a) e^u

b) e^{2t}

c) $\exp(-x)$

d) $\dfrac{1}{w}$

e) $\dfrac{1}{x+1}$

f) $\dfrac{1}{2t+5}$

10.7.6. Solve

a) $\int_0^1 e^u du$,

b) $\int_{-1}^1 e^{2t} dt$,

c) $\int_a^b \exp(-x) dx$,

d) $\int_1^x \dfrac{dw}{w}$,

e) $\int_0^1 \dfrac{dv}{v+1}$,

f) $\int_{T_1}^{T_2} \dfrac{dT}{T} \ (T_1 > 0, T_2 > 0)$,

g) $\int_{-1}^{-2} \dfrac{dx}{x}$,

h) $\int_{-1}^{-2} e^{-x} dx$,

i) $\int_{-S_0}^{S_0} \exp(S+1) dS$,

k) $\int \dfrac{pt^2+q}{t} dt$,

l) $\int \left(ax^2 + bx + c + \dfrac{d}{x} + \dfrac{e}{x^2} \right) dx$.

10.7.7. Let $y = \ln x$. How does a small error δx of x affect y? (Hint: use formula (9.9.5)).

10.7.8. Solve the same problem for $y = \exp x$.

10.7.9. The exponential function $y = \exp(-x^2)$ plays a dominant role in probability and statistics in connection with the normal or Gaussian distribution. Show that

a) $y > 0$ for all values of x,
b) $\lim y = 0$ as $x \to \infty$ or $x \to -\infty$,
c) the function reaches a maximum at $x = 0$,
d) points of inflection are located at $x = \pm 1/\sqrt{2}$,
e) the graph is bell-shaped.

10.7.10. In *Drosophila melanogaster* the reproduction rate drops sharply when the population density is increased. If x denotes the number of flies per bottle and y the progeny per female per day, it was found empirically that

$$y = 34.53 \, e^{-0.018x} \cdot x^{-0.658} .$$

Calculate y for $x = 20$. (The example is taken from Strehler, 1963, p. 74.)

10.7.11. Consider the function

$$y = c(e^{-at} - e^{-bt})$$

with positive parameters a, b, c and domain $t \geq 0$. Assume that $b > a$. Show that

a) $y = 0$ for $t = 0$,
b) $y > 0$ for $t > 0$,
c) $\lim y = 0$ as $t \to \infty$,
d) y reaches a maximum at $t = 1/(b-a) \cdot \ln b/a$,
e) the function has only one point of inflection.

This function is used to fit the concentration-time relationship for a drug injected into the blood stream (see Heinz, 1949, p. 482, or Defares *et al.*, 1973, p. 235).

10.8.1. Find the following limits:

a) $\lim\limits_{n \to \infty} (1 + \frac{3}{n})^n$,

b) $\lim\limits_{n \to \infty} (1 - \frac{3}{n})^n$,

c) $\lim\limits_{n \to \infty} \left(1 + \dfrac{1}{2n}\right)^n$,

d) $\lim\limits_{h \to 0} \dfrac{\ln(1 + 3h)}{h}$ (hint: let $3h = k$).

10.8.2. Find

a) $\left(1 + \dfrac{3}{2n}\right)^n$ as $n \to \infty$,

b) $\left(1 - \dfrac{r}{m}\right)^m$ as $m \to \infty$,

c) $\dfrac{1}{h} \ln(1 + ah)$ as $h \to 0$.

10.9.1. When a body is surrounded by a cooling liquid of constant temperature T_0, the temperature of the body decreases according to the formula

$$T = T_0 + a \cdot \exp(-kt)$$

where t is the time and a, k are positive constants. Plot a graph of this function. Find the rate of decrease, that is, dT/dt.

10.9.2. Let x be the amount of fertilizer applied to a certain cultivation. The yield y cannot be raised indefinitely by using more and more fertilizer. Instead, there exists an upper bound B for y. A workable approximation is given by Mitscherlich's formula

$$y = B(1 - e^{-kx})$$

with a positive constant k. Show that y is a monotone increasing function and that the line $y = B$ is an asymptote to the graph of the function (cf. Section 11.4).

10.9.3. Find the half-life of the radioactive substances a) ^{131}I, b) ^{18}F whose decay constants are $0.086\,d^{-1}$ and $0.371\,h^{-1}$, respectively (see formula (10.9.7); notations: $d = day$, $h = hour$). These isotopes are used in medicine for diagnosis as well as for therapy.

10.9.4. Tritium (^3H) has a half-life of 12.3 years. Let M_0 be the mass of ^3H at time instant $t = 0$. Find the mass $M(t)$ at the variable time instant t (measured in years).
(Hint: Use Eq. (10.9.7).)

10.9.5. Radioactive decay follows the law given in formula (10.9.5). What proportion of atoms disintegrate in a time interval from t_1 to $t_1 + \delta t$ where δt is sufficiently small?

10.9.6. A certain RNA molecule consisting of 218 nucleic acids is known to replicate in a test tube (extracellular replication). Under favorable conditions one molecule will generate 10^{12} copies in 20 minutes (autocatalysis).
a) Find an exponential function which describes this process (N = number of molecules, t = time in sec.)
b) What time is needed to generate a second molecule,
c) the first 1000 molecules? (Adapted from Mills *et al.*, 1973.)

10.9.7. Consider the function (10.9.12) of Gompertz for $t > 0$. Show that the function is monotone decreasing. Determine the mortality rate dN/dt. Prove that the graph has a point of inflection at $t = -(\ln b)/k$.

10.10.1. Show that

$$1 - e^{-x} \approx x$$

holds in the vicinity of $x = 0$.

10.10.2. Approximate the following function by a linear polynomial in the vicinity of $x = 0$:

a) $\sqrt{1+x}$,   b) $1/(1+x)$,   c) $1/(1+x)^2$.

(Hint: use Eq. (10.10.12).)

10.10.3. Approximate $\sin x$ and $\cos x$ in the vicinity of $x = 0$ by polynomials of degree ≤ 3.

10.10.4. In the vicinity of $x = 8$ approximate $\sqrt{x+1}$ by a linear function. (Hint: Replace the graph by its tangent at $x = 8$.)

10.10.5. Differentiate the power series (10.10.16) term by term and show that the result is consistent with $\dfrac{d}{dx}\ln(1+x) = 1/(1+x)$.

10.10.6. Differentiate the power series (10.10.15) term by term and compare the result with $(e^x)' = e^x$.

10.10.7. Consider the exponential function

$$y = \left(1 + \frac{p}{100}\right)^t$$

with the constant $p > 0$ and variable time t. a) Find the doubling time for y by an exact formula. b) Show that this doubling time is approximately equal to $70/p$ for sufficiently small values of p. (Hint: Use $\ln(1+x) \approx x$.)

10.10.8. For large values of n, the factorial $n! = 1 \cdot 2 \cdot 3 \cdots n$ is well approximated by Stirling's formula

$$n! \approx \sqrt{2\pi n} \cdot (n/e)^n .$$

The right side is particularly suitable for logarithmic evaluation. Find a rough approximation of $100!$.

*10.11.1. Prove the following formulas:

a) $\sinh(-x) = -\sinh x$,
b) $\cosh(-x) = \cosh x$,
c) $\sinh x + \cosh x = e^x$,
d) $\sinh(x + y) = \sinh x \cosh y + \cosh x \sinh y$.

*10.11.2. Show that

$$\frac{d}{dx} \tanh x = 1/\cosh^2 x .$$

*10.11.3. Using series expansions of e^x and e^{-x}, find series expansions for $\sinh x$ and $\cosh x$. The series converge for all values of x.

*10.11.4. Verify that the function $y = \frac{1}{2}(1 + \tanh x)$ can be written in the form

$$\frac{1}{1 + e^{-2x}} .$$

This function is a special case of the *logistic function* discussed in Section 11.5.

For the following problems familiarity with integration by substitution and by parts is required.

*10.11.5. Work out

a) $\int x e^x dx$,

b) $\int (x + k) e^x dx$,

c) $\int x^2 \ln x \, dx$,

d) $\int_1^2 \frac{dz}{5 - 2z}$,

e) $\int_0^1 a^x dx \quad (a > 0)$,

f) $\int_a^{2a} (at + b)^{-1} dt$.

*10.11.6. Solve

a) $\int x e^{ax} dx$,

b) $\int (ps + q) e^s ds$,

c) $\int t^n \ln t \, dt$,

d) $\int_0^5 (10 - z)^{-1} dz$,

e) $\int_0^5 (10 - z)^{-2} dz$,

f) $\int_0^k (x + k) e^{kx} dx$.

Chapter 11

Ordinary Differential Equations

11.1. Introduction

When the derivative $y' = f'(t)$ of an unknown function $y = f(t)$ is given, we usually have to find the antiderivative. We treated this problem in Sections 9.3 and 9.5. Sometimes the derivative y' is not given as a function of t, but is involved in an equation which contains also the unknown function $y = f(t)$. As an example, consider the equation

$$y' = ay + bt + c$$

with known coefficients a, b, c. Such an equation is called a *differential equation* since it contains not only the unknown function but also its derivative. The problem consists in finding a suitable function which satisfies the differential equation.

Differential equations occur frequently in the *analysis of physiological systems* and of *ecological systems*. We may briefly speak of *systems analysis*. When a quantity varies in one part of a system, its rate of change usually depends on quantities in other parts. In addition, any change of a quantity may indirectly influence the quantity itself, a phenomenon which is called *feedback*. The study of feedback systems originated in engineering, but its application to the life sciences turns out to be most fruitful.

The *independent variable* is usually *time*. Therefore, we denote it by t in most parts of this chapter. There are a few exceptions. Examples where the independent variable is not time are given in Example 11.3.5 and in Section 11.6 on allometry.

Dependent variables are denoted by $x = x(t)$, $y = y(t)$, $m = m(t)$, $N = N(t)$, $Q = Q(t)$, etc.

Chapters 9 and 10 are prerequisites for the understanding of differential equations. As the scope of this book is limited, we will not go deeply into the theory. A comprehensive study of differential equations would be a task covering several years. Fortunately, there are lists of differential equations and their solutions available. We recommend Kamke (1948).

There are differential equations whose solutions cannot be written in a manageable form. They are solved by *computers* either by applying methods of *numerical analysis* or by computer *simulation*. In this introductory book we will not deal with these methods.

11.2. Geometric Interpretation

For a better understanding of differential equations and their solutions we introduce a conceptual device. We interpret $y' = dy/dt$ as a *slope* in a rectangular t, y-coordinate system (see Section 9.2). Then a given differential equation assigns a slope y' to each point (t, y). As an example we consider the equation

$$y' = y - t^2 . \qquad (11.2.1)$$

Here y' is uniquely associated with a point (t, y). Hence we can draw a straight line through each point (t, y) with slope y' determined by (11.2.1). Thus we get the plane full of slopes or directions. We call such a plane a *slope field* or a *direction field*. Fig. 11.1 depicts the slope field of Eq. (11.2.1). We check a few slopes: For the point $(0,0)$ the equation yields $y' = 0$. With the point $(0,2)$ there is associated the slope $y' = 2 - 0 = 2$ (Choose $\Delta t = 1$ in the unit of the t axis. Then plot $\Delta y = 2$ in the unit of the y axis). Finally, for the point $(2,1)$ we get $y' = 1 - 2^2 = -3$ (let $\Delta t = 1$, then $\Delta y = -3$).

Fig. 11.1 seems to depict a moving liquid or gas. Two *streamlines* are also shown in the figure. At each point a streamline follows the slope given by the differential equation. Hence, the straight lines with slopes y' are tangents to the streamlines. Assume that a streamline can be represented by a function $y = f(t)$. Then $y' = df/dt$ is equal to $y' = y - t^2$, that is, $y = f(t)$ is a solution of the differential equation (11.2.1).

Finding solutions of a differential equation is often called *integration of a differential equation*. A solution is also called an *integral*, and the graph of such an integral is said to be an *integral curve*. Hence, streamlines and integral curves mean the same thing.

We see from a slope field that a differential equation has not a single integral curve, but an infinity of them. Fig. 11.1 shows only two special integral curves, one of them passing through the point $(0,1)$, the other through the point $(0, 2.5)$.

Since we found an infinite number of integral curves, there are also *infinitely many solutions or integrals* of a given differential equation. Each single one is called a *particular solution* or a *particular integral*. An expression which contains all particular solutions as special cases is called the *general solution*.

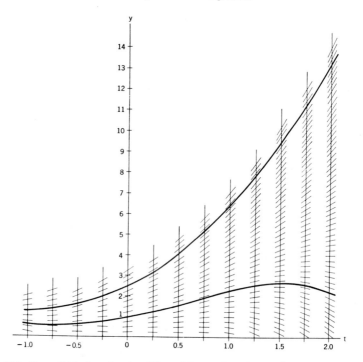

Fig. 11.1. Slope field given by the differential equation $y' = y - t^2$. The slope field resembles the picture of a moving fluid or gas. The streamlines are the integral curves. The
drawing is reproduced from Levens (1968, p. 751)

Slope fields do not only contribute to the intuitive understanding
of differential equations. They also provide an easy graphical method
for finding approximate solutions.

11.3. The Differential Equation $y' = a y$

One of the simplest differential equations is

$$\frac{dy}{dt} = ay \tag{11.3.1}$$

where a is a given constant. The integration is usually performed by a
rather symbolic procedure. We know that dy/dt is the limit of the
difference quotient $\Delta y / \Delta t$ as $\Delta t \to 0$. However, we did not define dy/dt
as a quotient of two quantities dy and dt. We treated dy and dt
merely as symbols. Now, neglecting this fact, we multiply the Eq.

(11.3.1) by dt and get symbolically

$$dy = ay \cdot dt .\qquad(11.3.2)$$

Next, upon division by y, the equation becomes

$$\frac{dy}{y} = a \cdot dt \quad (y \neq 0).\qquad(11.3.3)$$

Here the variable y occurs only on the left side and t only on the right side. We say that we have *separated the variables*. Integration yields

$$\int \frac{dy}{y} = \int a \cdot dt$$

or, according to formula (10.7.3),

$$\ln|y| = at + C \quad (y \neq 0)\qquad(11.3.4)$$

where C is an arbitrary constant. We may remove the natural logarithm by applying the inverse function. Hence, the explicit solution of Eq. (11.3.1) is

$$|y| = e^{at + C} .$$

However, the solution is seldom written in this form. Since $e^{u+v} = e^u \cdot e^v$, and since y can take on positive and negative values, we rewrite the solution in the form

$$y = c \cdot e^{at} \quad \text{or} \quad y = c \cdot \exp(at)\qquad(11.3.5)$$

where c stands for $\pm e^C$. The solution is an *exponential function with given coefficient a and arbitrary constant c*. As long as c is undetermined, we call (11.3.5) the *general solution* of the differential equation $y' = ay$.

Since we derived the solution by a rather symbolic method, the result should be verified. From (11.3.5) it follows by differentiation that

$$\frac{dy}{dt} = cae^{at}$$

which can be rewritten in the form $dy/dt = ay$. Hence, (11.3.5) does indeed satisfy the differential equation (11.3.1).

Fig. 11.2a illustrates the slope field for $a = \frac{1}{2}$. Depending on the choice of c, we get an infinite number of integral curves. A few of them are depicted ($c = 0$, $c = \frac{1}{4}$, $c = \frac{1}{2}$).

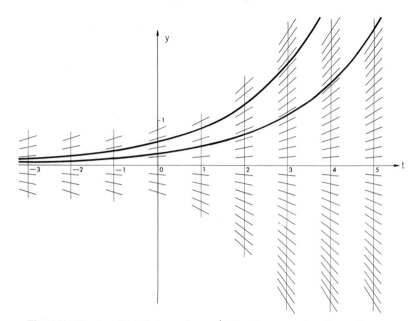

Fig. 11.2a. The slope field of $y' = ay$ for $a = \frac{1}{2}$. Also shown are three integral curves

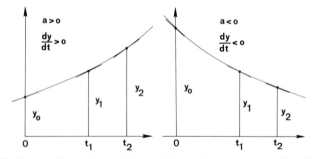

Fig. 11.2b. For positive y the solution of $y' = ay$ is an increasing function if $a > 0$ and a decreasing function if $a < 0$

In Fig. 11.2b the effect of changing from positive to negative values of a is depicted. For $a > 0$ we obtain growth in the normal sense, but for $a < 0$ we obtain *negative growth* (in special cases reduction, decrease, or decay).

Applications

Example 11.3.1. *Growth of a Cell.* Assume a cell is of mass m_0. In an ideal environment the cell grows. Thus its mass is a function of time, and we may write $m = m(t)$ with $m = m_0$ at $t = 0$. Assume that chemicals pass quickly through the cell wall, and that growth is only determined by the speed of metabolism inside the cell. Since the output of metabolism depends on the mass of participating molecules, it is reasonable to expect that *the growth rate is proportional to the mass at each time instant*, that is, $dm/dt \propto m$ or

$$\frac{dm}{dt} = am \tag{11.3.6}$$

with a certain positive constant a.

Of course, there is a limitation: If the mass m of the cell reaches a certain size, the cell will divide rather than continue to grow. Thus we add a restriction, say $m < m_1$.

The differential equation (11.3.6) is of the form (11.3.1). Therefore, the general solution follows from (11.3.5):

$$m = c \cdot e^{at}.$$

By our assumption that $m = m_0$ at time instant $t = 0$, we can determine the constant c. We get $c = m_0$. Hence the particular integral of (11.3.6) is

$$m = m_0 \, e^{at} \tag{11.3.7}$$

with the above mentioned restriction $m < m_1$.

With our assumptions we have gone slightly beyond experience. We have introduced some theoretical arguments. It is customary to say that we are *model-making*. Whether or not our model is biologically meaningful can only be tested by experiments. Here and in subsequent models we share G. F. Gause's view (Gause, 1934, p. 10): "There is no doubt that [growth, etc.] is a biological problem, and that it ought to be solved by experimentation and not at the desk of a mathematician. But in order to penetrate deeper into the nature of these phenomena, we must combine the experimental method with the mathematical theory, a possibility which has been created by [brilliant researchers]. The combination of the experimental method with the quantitative theory is in general one of the most powerful tools in the hands of contemporary science."

It is worth discussing the above growth model under different aspects. Since dm/dt was assumed to be proportional to m, we may introduce the *specific* or *relative growth rate* defined by

$$\frac{1}{m} \cdot \frac{dm}{dt}. \tag{11.3.8}$$

It is the quotient of the *absolute growth rate dm/dt* and the mass *m*. Our differential equation (11.3.6) then states: *At each time instant, the specific growth rate remains constant.*

The notion of a specific growth rate needs some illustration. Assume that a plant which has reached the mass $m = 300$ g, grows 12 g during the next 24 hours. Then the average growth rate is 12 g/24 hours $= 0.5$ g/h. Assuming that the growth rate does not fluctuate, we may consider 0.5 g/h as a good approximation of the instantaneous growth rate dm/dt. We may ask: Is this growth rate large or small? The answer depends very much on the present mass of the plant. For a plant of mass $m = 10$ g only, our growth rate would be tremendous, whereas for a large tree of (living) mass $m = 1000$ kg the same growth must be called tiny. Therefore, we have to relate 0.5 g/h with the present mass, in our case with 300 g. The quotient is

$$\frac{0.5 \text{ g/h}}{300 \text{ g}} = 0.0017 \, \text{h}^{-1}.$$

This quotient is called the specific growth rate. With the same specific growth rate, the tree of (living) mass 1000 kg would gain 1.7 kg per hour.

The specific growth rate is an important concept. There are two steps involved. First, when forming dm/dt, we relate the increase of mass with time which gives us some measure of velocity of growth. Second, we relate the velocity of growth with the mass present.

Let us finally consider another aspect of the differential equation (11.3.6). With increasing *m*, the growth rate dm/dt also increases. This growth rate, in turn, determines future values of *m*. Thus we have a simple example of a *feedback mechanism* with a single *loop*:

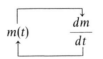

Example 11.3.2. *A Birth Process.* Let *N* stand for the number of individuals in an animal or plant population. This number is time dependent so that we may write $N = N(t)$. Strictly speaking, $N(t)$ takes on only integral values and is a discontinuous function of *t* (cf. Example (h) of Section 8.4). However, as we pointed out in Example 9.1.8, $N(t)$ may be approximated by a continuous and differentiable function as soon as the number of individuals is large enough.

In microorganisms reproduction occurs by simple cell division. In multicellular individuals we distinguish between vegetative and sexual reproduction. We will include all these possibilities in our study.

We assume that the proportion of reproductive individuals remains constant in the growing population. In addition we assume constant fertility. Then the rate of birth is proportional to the number $N(t)$ of individuals. If we finally exclude death, emigration and immigration, the growth rate coincides with the birth rate. Thus

$$\frac{dN}{dt} = \lambda N \tag{11.3.9}$$

where λ (lambda) is a certain constant. Referring to the concept introduced in (11.3.8), we may call λ the *specific birth rate*.

The differential equation (11.3.9) is of type (11.3.1). Hence the solution is

$$N = N_0 e^{\lambda t}$$

where N_0 denotes the population size at $t = 0$.

This birth process turns out to be quite realistic in a large population that grows under ideal conditions, that is, when all factors inhibiting growth are absent.

In a small population we cannot expect that the occurrence of birth is distributed evenly over time. Instead, we face random fluctuations. Then the process has to be modified in the light of probability theory. Such a refined model is called a *stochastic birth process*. For a presentation see Bailey (1964, Chap. 8), Chiang (1968, Chap. 3), Goel and Richter-Dyn (1974), Iosifescu and Tautu (1973), Pielou (1969).

Example 11.3.3. *A Birth-and-Death Process.* Let us consider an animal or plant population under the conditions outlined in the preceding application. Now we will extend the model by allowing for death. The net change in population size may be positive or negative. Within a time interval of length Δt we get

$$\text{net change} = \text{number of births}$$
$$\text{minus number of deaths}$$

or, in a convenient notation,

$$\Delta N = \Delta B - \Delta D .$$

Upon division by Δt we get the average rate of change

$$\frac{\Delta N}{\Delta t} = \frac{\Delta B}{\Delta t} - \frac{\Delta D}{\Delta t} . \tag{11.3.10}$$

As in previous cases we treat $N = N(t)$ as a continuous and differentiable function of time even though this means only an approximation to reality. Similarly we assume a large number of births and deaths so that the number of births $B = B(t)$ and of deaths $D = D(t)$ may also be considered as differentiable functions. As Δt tends to zero, we obtain from (11.3.10)

$$\frac{dN}{dt} = \frac{dB}{dt} - \frac{dD}{dt},$$ (11.3.11)

that is, the *rate of net change is equal to the rate of birth minus the rate of death*. The rate dN/dt may be positive or negative depending on whether occurrences of birth or of death prevail.

In Example 11.3.2 we stated assumptions such that the birth rate becomes proportional to the number of individuals $N(t)$. Under corresponding assumptions on death, the death rate also becomes proportional to $N(t)$. Hence,

$$\frac{dB}{dt} = \lambda N, \qquad \frac{dD}{dt} = \mu N,$$ (11.3.12)

λ denoting the *specific birth rate* and μ (mu) the *specific death rate*. Combining (11.3.11) and (11.3.12) we obtain

$$\frac{dN}{dt} = \lambda N - \mu N = (\lambda - \mu)N.$$ (11.3.13)

If we identify $\lambda - \mu$ with the coefficient a in (11.3.1), we can immediately solve the differential equation (11.3.13) and obtain

$$N = N_0 e^{(\lambda - \mu)t}$$ (11.3.14)

where N_0 stands for the population size at time $t = 0$. When the birth rate prevails, that is, when $\lambda > \mu$, the population size increases exponentially. We have an eruption. When instead $\lambda < \mu$, the population size decreases, and the population will die out. Only for $\lambda = \mu$ will the population remain stable.

This model of a birth-and-death process does not account for random fluctuations. It is therefore called *non-stochastic* or *deterministic*.

Example 11.3.4. *Radioactive Decay.* Let us assume that a substance contains only one sort of radioactive atom. The simplest assumption about decay is that there exists no preferred time for decay and that all atoms have the same chance of disintegration independent of each other. This implies that we expect twice as many scintillations per time unit with a supply of twice as many atoms, three times as many scintillations with a triple amount of atoms, etc. In general, the model requires that the rate of decay is proportional to the number N of radioactive atoms present, that is,

$$\frac{dN}{dt} = -\lambda N \tag{11.3.15}$$

where λ is a certain positive constant called the *decay constant*. Since λ is positive by definition and dN/dt must be negative, the minus sign in (11.3.15) is required. The differential equation is of the form (11.3.1) and its solution is therefore

$$N = N_0 e^{-\lambda t}, \tag{11.3.16}$$

N_0 denoting the original number of radioactive atoms at time $t = 0$. The result coincides with formula (10.9.5). The agreement with experimental facts is excellent.

In Example 10.9.2 we treated the exponential law for radioactive decay as a purely empirical fact. The model-making procedure in our present section offers the additional advantage that we have found an intuitively simple explanation for this law. *The differential equation gives more insight into the process than does the solution.* It is the purpose of all theoretical reasoning to find conceptual relationships which are both logically simple and empirically valid.

Example 11.3.5. *Living Tissue Exposed to Ionizing Radiation.* An ionizing beam of particles consists of either protons, neutrons, deuterons, electrons, γ ray quanta, or the like. If high polymers such as proteins or nucleic acids are hit by an ionizing beam, they may be irreversibly altered. New bonds may be formed between chains or existing bonds may be broken. We simply say that polymers become damaged.

Let n_0 be the original number of undamaged molecules of a specific chemical compound which are present in a cell and which are assumed to be susceptible to radiation. Let D be the number of ionizing particles which cross the unit area of the target. We simply call D the *dose of radiation.* Let n be the number of undamaged molecules after exposure to radiation $(n < n_0)$. The question then arises: How does n depend on the dose?

When n and D are large numbers, we may operate with these quantities as if they were continuous variables. To answer our question we assume that n is a function of D. Then we consider the rate dn/dD after exposure to different doses of radiation. Since a higher dose inflicts more damage, the rate dn/dD must be negative. When building a model it is plausible to assume that dn/dD is proportional to n. Thus we get the differential equation

$$\frac{dn}{dD} = -S \cdot n \tag{11.3.17}$$

where S denotes a certain positive constant. This equation is again of type (11.3.1). We notice that the independent variable is not the time but the dose D. The differential equation leads once more to an exponential law:

$$n = n_0 e^{-SD}. \tag{11.3.18}$$

This example is adapted from Ackerman (1962, p. 305).

Example 11.3.6. *Radioactive Tracer.* Before we treat a specific problem, let us introduce some generalities on a very useful method in biophysics, the *compartment analysis*. Milhorn (1966, p. 36) defines the term "compartment" in the following way:

If a substance is present in a biological system in several distinguishable forms or locations, and if it passes from one form or location to another form or location at a measurable rate, then each form or location constitutes a separate *compartment* for the substance.

Milhorn illustrates the special case of a single compartment with a *tracer dose* of radioactive iodine ^{131}I injected into the blood stream. Let Q_0 be the original mass of iodine at time $t = 0$ and denote the mass remaining in the blood at time instant t by $Q = Q(t)$. The blood stream plays the role of the compartment. We assume that the iodine is distributed evenly in the entire blood stream before any loss occurs. Part of the iodine leaves the blood and enters the urine. It is plausible to assume that the rate of loss is proportional to $Q(t)$ at each time instant t. Hence, we may equate this rate to $k_1 Q$ where k_1 is a certain positive constant. Another part of the iodine enters the thyroid gland at a rate which is also assumed to be proportional to $Q(t)$. For this second rate we may write $k_2 Q$ $(k_2 > 0)$. Finally, there is loss by radioactive decay at a certain rate $k_3 Q$ $(k_3 > 0)$. The total rate of change is therefore

$$\frac{dQ}{dt} = -k_1 Q - k_2 Q - k_3 Q = -(k_1 + k_2 + k_3) Q. \tag{11.3.19}$$

Writing $k_1 + k_2 + k_3 = k$ for simplicity, the solution of the differential equation becomes

$$Q = Q_0 e^{-kt}, \tag{11.3.20}$$

that is, the concentration of iodine in the blood decreases exponentially. This simple law is upheld even though the iodine leaves by multiple pathways at different rates.

Example 11.3.7. *Dilution of a Substance.* We consider a second problem that may be approached via compartment analysis.

In a tube containing 2000 g of water, 50 g of sucrose are dissolved. By stirring, the sucrose will be distributed evenly at all times. Through a pipe, 10 g of water flow into the tube per minute, and through another pipe, 10 g of water leave the tube per minute removing some sucrose at the same time (Fig. 11.3). We may ask: How does the mass of sucrose decrease as a function of time?

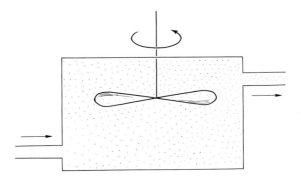

Fig. 11.3. Steady reduction of the concentration of a solute

Let $M = M(t)$ be the mass of sucrose in the tube. By assumption, we have $M_0 = 50$ g at time $t = 0$. In 10 g of water the mass of dissolved sucrose is

$$M(t) \cdot \frac{10}{2000} = (0.005) M .$$

In a time interval of length Δt, the loss of sucrose from the tube amounts to

$$\Delta M = (-0.005) \bar{M} \Delta t$$

where $\bar{M}$ denotes a certain average of $M(t)$ during the time interval. As Δt tends to zero, we get for the rate of decrease

$$\frac{dM}{dt} = (-0.005) M . \tag{11.3.21}$$

This implies an exponential decrease of sucrose. Our question is answered by the function

$$M = M_0 e^{-(0.005)t} \tag{11.3.22}$$

where t is measured in minutes.

Example 11.3.8. *Chemical Kinetics.* Gaseous nitrogen pentoxide decomposes as stated by the equation

$$2\,N_2O_5 \rightarrow 4\,NO_2 + O_2 .$$

We are interested in the speed of this reaction when the temperature is kept constant. Let $C = [N_2O_5]$ be the concentration of nitrogen pentoxide measured in moles per liter. The concentration $C = C(t)$ is a decreasing function of time so that the derivative dC/dt is negative. This derivative is called the *reaction rate*.

The reaction rate depends on the concentration $C = [N_2O_5]$. Intuitively we expect that the higher the concentration is, the more frequently collisions of two N_2O_5 molecules will occur with the possible emergence of the new bonds NO_2 and O_2. One may theorize that under constant temperature the reaction rate is proportional to C, that is,

$$\frac{dC}{dt} = -kC \qquad\qquad (11.3.23)$$

where k denotes a positive constant. The solution of this differential equation is

$$C = C_0 e^{-kt} , \qquad\qquad (11.3.24)$$

C_0 being the concentration of N_2O_5 at time $t = 0$. The experimental facts are in good agreement with this model. As (11.3.24) shows the concentration C will asymptotically tend to zero. It will never reach zero exactly.

11.4. The Differential Equation $y' = ay + b$

For $b = 0$, this equation reduces to $y' = ay$ which we studied in the preceding section. With $b \neq 0$, we are dealing with a slight generalization. The differential equation is solved by the same method. Here are the steps:

$$\frac{dy}{dt} = ay + b \qquad (a \neq 0) ,$$

$$\frac{dy}{dt} = a\left(y + \frac{b}{a}\right). \qquad\qquad (11.4.1)$$

For simplicity, let $b/a = p$. Then we *separate the variables:*

$$dy = a(y + p)dt , \qquad \frac{dy}{y + p} = a \cdot dt .$$

Upon integration we get

$$\int \frac{dy}{y+p} = \int a \cdot dt, \quad \ln|y+p| = at + C,$$

$$y + p = \pm e^{at+C} = c \cdot e^{at}$$

where $c = \pm e^C$. Finally,

$$y = c \cdot e^{at} - \frac{b}{a}. \tag{11.4.2}$$

This is the general solution of the differential equation (11.4.1). It is valid for an arbitrary value of c as shown by verification of the result. Indeed, upon differentiation of (11.4.2) with respect to t we get

$$\frac{dy}{dt} = cae^{at} = a\left[ce^{at} - \frac{b}{a}\right] + b = ay + b$$

which is the given equation (11.4.1).

Applications

Example 11.4.1. *Restricted Growth.* No organism and no population grow indefinitely. There are limitations set by shortage of food supply or shelter, by lack of space, by intolerable physical conditions, or by some control mechanism.

Assume that there exists a fixed upper bound for the size y of an individual, a tissue, a population or a crop. The size may be a volume, a weight, a diameter, a number, etc. We denote the upper bound by B. Then $y = y(t)$ may approach B asymptotically. This implies that the growth rate dy/dt tends to zero as $B - y$ becomes smaller and smaller. A plausible mathematical formulation of such a model is given by the differential equation

$$\frac{dy}{dt} = k(B - y) \tag{11.4.3}$$

where k is a positive constant which determines how fast dy/dt tends to zero. If y is small relative to B, then we have approximately $y' \approx kB = \text{constant}$, that is, the size y increases approximately as a linear function of time. However, if y is close to B, then $B - y$ is a small positive quantity, and so is the growth rate dy/dt.

Eq. (11.4.3) may be rewritten in the form $y' = kB - ky = ay + b$ with $a = -k$ and $b = kB$. Thus, we get the general solution by rewriting (11.4.2):

$$y = ce^{-kt} + B. \tag{11.4.4}$$

Since $y < B$, the constant c of integration has to be negative.

A particular solution is derived by assuming that $y=0$ at time $t=0$. In this case, $c=-B$ and

$$y = B(1 - e^{-kt}).\qquad(11.4.5)$$

This model was proposed by E. A. Mitscherlich in 1939. It fits some experimental data in agriculture quite well. The reader is invited to draw a graph of formula (11.4.5) or to solve Problem 10.9.2.

For more details see von Bertalanffy (1951, p. 359). Cf. also Thrall *et al.* (1967, CA 13). Numerous biological examples are contained in Brody (1945, Chap. 5 and 16).

Another model of restricted growth which is used more frequently will be studied in Example 11.5.1.

Example 11.4.2. *A Birth-and-Immigration Process.* In Example 11.3.2 of the preceding section we studied a birth process. Several assumptions were made about the population. We maintain these assumptions with the exception that we now permit *immigration of individuals at a constant rate.* The rate is measured in number of individuals per time unit and denoted by v (Greek nu). Thus Eq. (11.3.9) turns into

$$\frac{dN}{dt} = \lambda N + v \qquad (\lambda > 0, v > 0).\qquad(11.4.6)$$

The differential equation is obviously of the form (11.4.1). Hence, the general solution is

$$N = ce^{\lambda t} - \frac{v}{\lambda}.$$

Assuming that $N = N_0$ at $t = 0$, we obtain $c = N_0 + \dfrac{v}{\lambda}$. Thus the particular solution of (11.4.6) is

$$N = N_0 e^{\lambda t} + \frac{v}{\lambda}(e^{\lambda t} - 1).\qquad(11.4.7)$$

Here the growth of the population depends on two terms. The first term is an exponential function determined by the specific birth rate alone. The second term is also rapidly increasing but it depends on both rates λ and v.

Example 11.4.3. *Cooling.* Consider a body without internal heating whose temperature is higher than that of the surrounding medium. The body will then cool. We want to know how the temperature of the body drops as a function of time.

Let $T = T(t)$ be the temperature of the body at the time instant t, T_0 its temperature at $t = 0$, and T_s the constant temperature of the surrounding medium.

The derivative dT/dt is called the rate of cooling. Since T decreases, this rate is negative. It is dependent on the difference $T - T_s$. Under favorable conditions the rate of cooling is proportional to $T - T_s$, that is,

$$\frac{dT}{dt} = -k(T - T_s) \tag{11.4.8}$$

where k is a positive constant determined by the physical conditions of heat exchange. Since the right side can be written as $(-k)T + kT_s$, the differential equation is of the form (11.4.1). From (11.4.2) we deduce the general solution and obtain

$$T = ce^{-kt} + T_s. \tag{11.4.9}$$

Finally we satisfy the initial condition $T = T_0$ at time $t = 0$. This leads to $c = T_0 - T_s$ and

$$T = T_s + (T_0 - T_s)e^{-kt}. \tag{11.4.10}$$

As t tends to infinity, the second term tends to zero, and T approaches T_s asymptotically. Eq. (11.4.8) is known as Newton's law of cooling.

This and the following application may be considered as illustrations of compartment analysis (cf. Example 11.3.6).

Example 11.4.4. *A Diffusion Problem.* We assume that a cell of constant volume is suspended in a homogeneous liquid which contains a solute of concentration c_0, constant in space as well as in time. Let $c = c(t)$ be the concentration of the solute inside the cell at the time instant t and assume that the solute is almost evenly distributed over the cell at all times so that $c = c(t)$ depends only on the time.

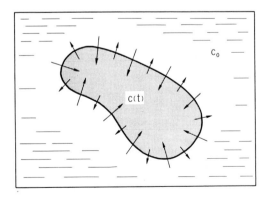

Fig. 11.4. Diffusion of molecules through a cell wall. In the figure it is assumed that $c_0 > c(t)$ and, therefore, that more molecules enter the cell than leave the cell

By *diffusion*, molecules of the solute will enter the cell from the surrounding liquid, but there will be also molecules of the solute which will leave the cell. Thus there is a flow of molecules through the cell membrane in both directions (Fig. 11.4). The *net flow* is from the liquid into the cell if c_0 is higher than $c(t)$ and conversely. We are interested in finding the function $c(t)$.

Let $m = m(t)$ be the mass of solute in the cell, A be the area of the cell membrane, and V be the volume of the cell. Then by definition of concentration

$$m(t) = V \cdot c(t). \qquad (11.4.11)$$

The derivative dm/dt is the rate of increase of m and may be called the *net flow rate* in our problem. Fick's law[1] states that dm/dt is proportional to the area of the membrane and to the difference in concentration on both sides of the membrane. Thus

$$\frac{dm}{dt} = kA(c_0 - c). \qquad (11.4.12)$$

If $c < c_0$, that is, if the solute has a lower concentration inside the cell than outside, m will increase. Hence, k is a positive constant. This constant is determined by the structure and thickness of the membrane. It is called the *permeability of the membrane* for the particular solute.

By means of (11.4.11) we can replace dm/dt by $V \cdot dc/dt$ in our differential equation. Thus we obtain

$$\frac{dc}{dt} = \frac{kA}{V}(c_0 - c). \qquad (11.4.13)$$

We integrate this equation by using the explicit solution (11.4.2) and get

$$c = K \cdot \exp\left(-\frac{kA}{V}t\right) + c_0 \qquad (11.4.14)$$

where K denotes the constant of integration. As t tends to infinity, $c(t)$ approaches c_0 asymptotically. The constant K may be determined by some initial condition, say $c = c^*$ at time $t = 0$. We leave it to the reader to discuss the two cases $c^* > c_0$ and $c^* < c_0$.

This application is adapted from Thrall *et al.* (1967, CA 10).

It should be noted however that this model is a crude approximation to reality. Diffusion through cell membranes is a complicated process which cannot be treated adequately in this connection.

---

[1] Adolf Fick (1829—1901), German physiologist and biophysicist.

Example 11.4.5. *Nerve Excitation.* The cells of a nerve fiber may be conceived as an electric system. The protoplasm contains a large number of different ions, both cations (positive electric charge) and anions (negative electric charge). When an electric current is applied to the nerve fiber, the cations move to the cathode, the anions to the anode, and the electric equilibrium is disturbed. This leads to an excitation of the nerve. The mechanism is not known in detail.

Based on the observation that the excitation originates at the cathode, N. Rashevsky developed a theory which postulates that two different sorts of cation are responsible for the process. One sort is exciting, the other inhibiting. The two sorts are therefore said to be *antagonistic factors.*

Let $\varepsilon = \varepsilon(t)$ (Greek epsilon) be the concentration of the exciting cations and $j = j(t)$ be the concentration of the inhibiting cations near the cathode at time instant t. Then, the theory states that excitation occurs whenever the ratio ε/j exceeds a certain threshold value. Denoting the threshold value by c, the statement is as follows:

$$\frac{\varepsilon}{j} \geq c \qquad \text{excitation},$$

$$\frac{\varepsilon}{j} < c \qquad \text{no excitation}.$$

Let ε_0 and j_0 be the concentrations at rest of exciting and inhibiting cations, respectively. When ε grows and j remains limited, excitation becomes likely; when instead ε does not grow as fast as j, excitation will not occur.

Denote the intensity of the stimulant electric current by I. For simplicity, I is assumed to be constant during a certain time interval. Then Rashevsky assumes that $d\varepsilon/dt$ consists of two terms, one term proportional to I and the second negative term proportional to $\varepsilon - \varepsilon_0$ which accounts for the loss of cations by diffusion, that is,

$$\frac{d\varepsilon}{dt} = KI - k(\varepsilon - \varepsilon_0) \tag{11.4.15}$$

with positive constants K and k. Using formula (11.4.2) it is easy to solve this differential equation and to write an explicit formula for $\varepsilon = \varepsilon(t)$.

Similarly, for $j = j(t)$ the theory postulates a differential equation

$$\frac{dj}{dt} = MI - m(j - j_0) \qquad (M, m \text{ positive constants}) \tag{11.4.16}$$

for which an explicit solution can also be found. The ratio $\varepsilon(t)/j(t)$ finally determines whether excitation occurs and when. The result is in good agreement with experiments.

For a full account of the theory see Rashevsky (1960, Vol. 1, Chap. 32). Competitive models are given in K. S. Cole (1968, p. 121 ff.), Griffith (1971), Hodgkin and Huxley (1952), and Johnson, Eyring, Polissar (1954, Chap. 12). Similar differential equations are used in a theory of the central nervous system (see Rashevsky, 1961, Chap. 5, or 1964, Chap. 24).

11.5. The Differential Equation $y' = a y^2 + b y + c$

In the differential equation $y' = ay + b$ which we have studied in the preceding section, the unknown terms $y = y(t)$ and $y' = dy/dt$ are in the first power. Therefore, the equation is said to be linear. In the equation

$$\frac{dy}{dt} = ay^2 + by + c \tag{11.5.1}$$

the term y^2 occurs. For this reason, (11.5.1) is called a *nonlinear differential equation*.

Before we solve the Eq. (11.5.1), we consider the quadratic polynomial on the right side and the related quadratic equation

$$ay^2 + by + c = 0. \tag{11.5.2}$$

We may solve this equation using formula (4.6.2). We denote the two *roots* by A and B and assume that they are *two different real numbers*. The quadratic polynomial may then be rewritten in the form

$$ay^2 + by + c = a(A - y)(B - y). \tag{11.5.3}$$

Indeed, if $y = A$, the factor $A - y$ vanishes and hence the quadratic equation is satisfied. The same argument is valid if $y = B$.

With (11.5.3) in mind we rewrite our original differential equation in the form

$$\frac{dy}{dt} = a(A - y)(B - y) \qquad (A \neq B, y \neq A, y \neq B). \tag{11.5.4}$$

The method of *separating the variables* leads to

$$\frac{dy}{(A - y)(B - y)} = a \cdot dt. \tag{11.5.5}$$

To simplify integration, we rewrite the fraction on the left side in the form

$$\frac{1}{(A-y)(B-y)} = \frac{1}{B-A}\left(\frac{1}{y-B} - \frac{1}{y-A}\right). \tag{11.5.6}$$

This formula can easily be verified. It is valid only under the assumption $A \neq B$. The terms $1/(y-B)$ and $1/(y-A)$ are called *partial fractions*.

Now we are able to solve (11.5.5). In view of (11.5.6), the differential equation becomes

$$\left(\frac{1}{y-B} - \frac{1}{y-A}\right)dy = a(B-A)dt.$$

Upon integrating term by term using formula (10.7.3) we get

$$\ln|y-B| - \ln|y-A| = a(B-A)t + C$$

where C is an arbitrary constant of integration. Then we rewrite the equation by virtue of the formula $\ln u - \ln v = \ln(u/v)$ and obtain

$$\ln\frac{|y-B|}{|y-A|} = a(B-A)t + C. \tag{11.5.7}$$

Applying the exponential function on both sides we get

$$\frac{|y-B|}{|y-A|} = e^{a(B-A)t+C}.$$

Dropping the vertical bars and replacing $\pm e^C$ by a constant $+k$ or $-k$, the equation becomes

$$\frac{y-B}{y-A} = -k \cdot e^{a(B-A)t}. \tag{11.5.8}$$

The minus sign on the right side is no requirement, but it is convenient in some applications. Finally, we solve (11.5.8) with respect to y and obtain the explicit solution of (11.5.4):

$$y = A + \frac{B-A}{1+ke^{a(B-A)t}}. \tag{11.5.9}$$

It is suggested that the reader verify the result by differentiation.

Applications

Example 11.5.1. *Restricted Growth.* We return to the study of growth of populations. Let $y = y(t)$ be the number of individuals in a population at the time instant t. The differential equation $y' = ay$ with $a > 0$

provides for an unrestricted exponential growth, whereas $y' = a(B - y)$ with $a > 0$, $B > 0$ results in growth which is nearly linear in the beginning and levels off later (cf. applications to growth in the preceding sections).

To get a model of growth which is biologically more meaningful we may combine the two approaches, that is, assume that y' is proportional to y as well as to $B - y$. This idea leads to the differential equation

$$\frac{dy}{dt} = \lambda y(B - y) \tag{11.5.10}$$

where λ is a certain positive constant [2]. This equation is a special case of (11.5.4) with $A = 0$ and $a = -\lambda$. Hence, the general solution follows immediately from (11.5.9):

$$y = \frac{B}{1 + ke^{-\lambda Bt}}. \tag{11.5.11}$$

In our model, y can never exceed B. Therefore, the denominator in (11.5.11) must be greater than 1, and k has to be restricted to positive values. The quantity y increases monotonically since the differential equation (11.5.10) implies that $dy/dt > 0$. For $t \to -\infty$, y tends to zero, and for $t \to +\infty$, y tends to B. The growth starts slowly, then becomes faster and finally tapers off. Growth is fastest in the neighborhood of the *point of inflection*. To get its location we have to equate the second derivative of $y = y(t)$ to zero (see formula (9.6.4)). Differentiating (11.5.10) we obtain

$$\frac{d^2 y}{dt^2} = \lambda \left[\frac{dy}{dt}(B - y) - y\frac{dy}{dt} \right] = \lambda(B - 2y)\frac{dy}{dt}. \tag{11.5.12}$$

This expression can only vanish if $B - 2y = 0$ or

$$y = B/2, \tag{11.5.13}$$

that is, the point of inflection is halfway between the lines $y = 0$ and $y = B$. To get the abscissa, we let $y = B/2$ in Eq. (11.5.11) and solve it with respect to t:

$$t = \frac{\ln k}{B\lambda}. \tag{11.5.14}$$

This particular abscissa is positive or negative depending on whether $k > 1$ or $k < 1$.

Fig. 11.5 shows a graph of the function (11.5.11) in a particular application. The curve is *S-shaped* or *sigmoid*. Formula (11.5.11) is generally known as the *logistic function*. It was introduced into popula-

---

[2] Cf. rule at the end of Section 12.1.

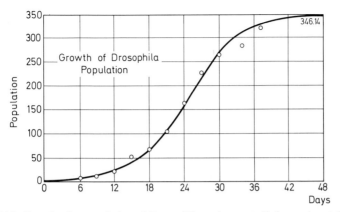

Fig. 11.5. Growth of a population of Drosophila under controlled experimental conditions. The figure is reproduced from Lotka (1956, p. 69). Data are attributed to R. Pearl and S. L. Parker

tion dynamics by Verhulst in 1838[3]. The graph of the logistic function is called the logistic curve[4]. There are numerous experimental growth data, especially for protozoa and bacteria, for which fitting a logistic curve was quite successful. But in some other populations the fit was poor, and prediction was misleading.

For more details and experimental data see D'Ancona (1954, p. 58–77), Gause (1934, p. 35 ff.), Kostitzin (1939, Chap. 4), Lotka (1956, Chap. 7), Pielou (1969, Chap. 2), Schanz (1974). For critical remarks the reader is referred to Feller (1940) and Slobodkin (1961, p. 122).

Example 11.5.2. *Spread of Infection.* How does an infectious disease spread in a community of susceptible individuals? Of course, this depends on many circumstances. For simplicity, we will make a few assumptions. Into a population of equally susceptible individuals we introduce a single infective. By contact between individuals the disease will spread, that is, the number of infectives will increase. In the beginning, the number of infectives will increase slowly, then the process will accelerate and finally level off when most of the individuals have become infectives. Further we assume that an individual that is once infected will remain infective during the process and that no one is removed.

---

[3] P.F. Verhulst (1804—1849), Belgian mathematician. It is not known why he chose the word "logistic".

[4] The logistic function can be derived from the *hyperbolic function* $y = \tanh x$ which we introduced in Section 10.11. Only a translation and a change of scale are required. As is known from $\tanh x$, the sigmoid curve is *symmetric about its point of inflection.* Cf. Problem 10.11.4 at the end of Chapter 10.

Let $x = x(t)$ be the number of susceptibles, $y = y(t)$ the number of infectives at time instant t, and n the total size of the population into which an infective was introduced. Thus, at any time,

$$x + y = n + 1 . \tag{11.5.15}$$

As in previous work on population growth we treat x and y as continuous variables. The rate at which the number of infectives increases is then dy/dt. The more infectives and susceptibles are present, the more frequently will contacts occur that lead to infection. It is therefore plausible to assume that dy/dt is proportional to y as well as to x. Thus we get the differential equation [5]

$$\frac{dy}{dt} = \beta y x$$

where β is a positive constant called the *specific infection rate*. In virtue of (11.5.15) the equation becomes

$$\frac{dy}{dt} = \beta y (n + 1 - y) . \tag{11.5.16}$$

This equation is of the form (11.5.10) and can be integrated immediately by formula (11.5.11):

$$y = \frac{n + 1}{1 + k e^{-\beta(n+1)t}} . \tag{11.5.17}$$

So far the constant k is undetermined. We assume now that the process started at time $t = 0$ with the single infective. Thus from (11.5.17) we have $y(0) = 1 = (n+1)/(1+k)$ from which we deduce $k = n$. Hence, the particular solution is

$$y = \frac{n + 1}{1 + n e^{-\beta(n+1)t}} . \tag{11.5.18}$$

Under our simplifying assumptions the spread of disease follows a *logistic law*.

For more elaborate models on epidemics, both stochastic and deterministic, see Bailey (1957 and 1964), Bartlett (1960), Goel and Richter-Dyn (1974), Iosifescu and Tautu (1973), Waltman (1974).

Example 11.5.3. *Chemical Kinetics.* Let us consider the reaction of n-amyl fluoride and sodium ethoxide:

$$n\text{–}C_5H_{11}F + NaOC_2H_5 \rightarrow NaF + n\text{–}C_5H_{11}OC_2H_5 .$$

---

[5] See the rule at the end of Section 12.1.

Let A be the original concentration of n-amyl fluoride and B be the original concentration of sodium ethoxide, and assume that $A \neq B$. Since during the reaction one molecule of the first compound removes exactly one molecule of the second compound, A and B decrease by the same amount $x = x(t)$ at any time t. Hence, the remaining concentrations of the components are $A - x$ and $B - x$.

We call dx/dt the *reaction rate*. Since the reaction requires collision of molecules of n-amyl fluoride with molecules of sodium ethoxide, it is plausible to assume that dx/dt is proportional to the number of molecules of both components present. This is equivalent to saying that dx/dt is proportional to $A - x$ and to $B - x$. Hence, we obtain the differential equation

$$\frac{dx}{dt} = r(A - x)(B - x) \tag{11.5.19}$$

with a *specific reaction rate* $r > 0$. In order that r remain constant we have to assume that the temperature is kept constant during the reaction. The Eq. (11.5.19) is of the form (11.5.4)[6] and the general solution is stated in (11.5.9). To satisfy the initial condition $t = 0$, $x = 0$, we must equate k with $-B/A$. After rearranging terms the particular solution may be written in the form

$$x = A\left(1 + \frac{B - A}{A - Be^{r(B - A)t}}\right). \tag{11.5.20}$$

When $t \to \infty$, x tends either to A, if $A < B$, or to B, if $A > B$. It should be noted that this function is not a logistic function.

There is good agreement between the model and the experiment. For a full account of this example see Latham (1964, p. 103).

Example 11.5.4. *Autocatalysis.* Another sort of chemical reaction leads to a related differential equation. Consider the process by which trypsinogen is converted into trypsin (an enzyme). The reaction starts only in the presence of some trypsin, that is, the product of reaction acts as a catalyst.

Let y_0 be the initial concentration of trypsin at time $t = 0$, and let $y = y(t)$ be the additional concentration gained by the reaction at time t so that the total concentration is $y_0 + y(t)$. Let B denote the initial concentration of trypsinogen. Since each molecule of trypsinogen yields one molecule of trypsin, B decreases by the same amount as the

---

[6] In chemistry this reaction is called a *second order reaction* and (11.5.19) is said to be a *second order equation*. However, it should be noted that in mathematics the term "second order" in connection with differential equations has quite a different meaning in that it refers to equations containing the second derivative y''.

trypsin concentration increases, that is, by y. Therefore, at time t the concentrations of trypsinogen and of trypsin are $B - y$ and $y_0 + y$, respectively.

It is reasonable to assume that the reaction rate dy/dt is proportional to $y_0 + y$ and to $B - y$. Thus we get the differential equation

$$\frac{dy}{dt} = r(y_0 + y)(B - y) \tag{11.5.21}$$

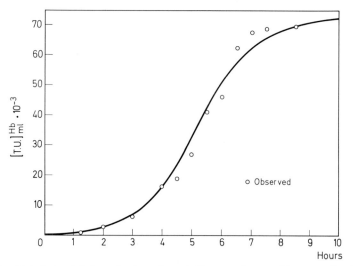

Fig. 11.6. Autocatalytic activation of crystalline trypsinogen. The curve fitted to the observed points is a logistic curve. The figure is reproduced from Northrop, Kunitz, and Herriot (1948, p. 126). See also B. Stevens (1965, p. 83)

where r is a constant. To compare this equation with (11.5.4) we let $r = -a$ and $y_0 = -A$ so that a formal agreement is reached. The general solution, as indicated by (11.5.9), is

$$y = -y_0 + \frac{B + y_0}{1 + ke^{-r(B + y_0)t}} . \tag{11.5.22}$$

With the initial condition $t = 0$, $y = 0$ we are able to determine k. Thus we get a particular solution with $k = B/y_0$. By comparison with (11.5.11) it is easy to see that the graph of (11.5.22) is a *logistic curve*. As shown in Fig. 11.6, the agreement with the experimental facts is satisfactory.

11.6. The Differential Equation $dy/dx = k \cdot y/x$

It is easy to observe that a child's head grows more slowly than his body and that his eyes grow even more slowly than the head. This different pattern of growth is studied in allometry.

Let $x = x(t)$ be the size (length, volume, or weight) of an organ and $y = y(t)$ the size of another organ or part in the same individual at time instant t. The growth rates of these two parts are dx/dt and dy/dt. When we state that one part grows more slowly than another part, we will not simply mean that $dx/dt \neq dy/dt$, since for example a part which is smaller in size than another part will naturally gain less per unit time. Rather we will mean that the *specific or relative growth rates*, introduced in (11.3.8), are different. In other words, our statement implies that

$$\frac{1}{y}\frac{dy}{dt} \neq \frac{1}{x}\frac{dx}{dt}.$$

Numerous empirical data support the claim that *the specific growth rates are approximately proportional*, that is, that the equation

$$\frac{1}{y}\frac{dy}{dt} = k\frac{1}{x}\frac{dx}{dt} \tag{11.6.1}$$

holds with satisfactory precision. The constant k is positive and depends only on the nature of the two organs or parts under consideration. The relationship (11.6.1) is called the *allometric law*.

The differential equation (11.6.1) may be simplified by eliminating time. We first show that dy/dt divided by dx/dt simply yields dy/dx. We have only to recall that dy/dt and dx/dt are limits of the difference quotients $\Delta y/\Delta t$ and $\Delta x/\Delta t$. Since $\Delta t \neq 0$, we get

$$\frac{\Delta y}{\Delta t} \bigg/ \frac{\Delta x}{\Delta t} = \frac{\Delta y}{\Delta x}.$$

As $\Delta t \to 0$, we obtain the general formula

$$\frac{dy}{dt} \bigg/ \frac{dx}{dt} = \frac{dy}{dx}. \tag{11.6.2}$$

With this result, the Eq. (11.6.1) turns into

$$\frac{dy}{dx} = k\frac{y}{x}. \tag{11.6.3}$$

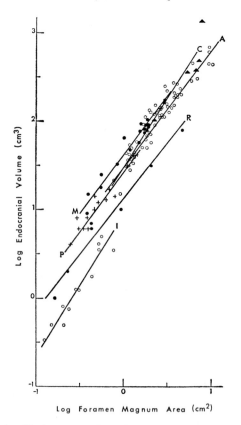

Fig. 11.7. Relationship between endocranial volume and foramen magnum area in five orders of mammals. Letters refer to straight lines fitted to the date of each group. *I* insectivores, *R* rodents, *P* prosimian primates, *M* new and old world monkeys combined, *C* fissiped carnivores, *A* artiodactyls. The figure is reproduced from Radinsky (1967)

This new differential equation is solved by *separating the variables* (cf. Section 11.3):

$$\frac{dy}{y} = k\frac{dx}{x},$$

$$\ln y = k \ln x + C \quad (x > 0, y > 0), \tag{11.6.4}$$

$$y = e^{k \ln x + C} = e^{C}(e^{\ln x})^{k}$$

or

$$y = cx^{k} \quad (x > 0, y > 0) \tag{11.6.5}$$

where c stands for e^{C}. The Eq. (11.6.5) defines a *power* function (for a definition see Chapter 4). The exponent is the constant $k > 0$, and the coefficient of the power is $c > 0$. The function contains two parameters, c and k.

For numerical and graphical purposes, it is practical to take common logarithms. Thus (11.6.5) is equivalent to

$$\log y = k \cdot \log x + \log c \qquad (11.6.6)$$

or, in the notation of Section 7.3,

$$Y = kX + B. \qquad (11.6.7)$$

A double-logarithmic plot of (11.6.5) is therefore a straight line with a slope k and Y-intercept $B = \log c$. A typical example is shown in Fig. 11.7.

The allometric law may be stated either by the differential equation (11.6.1), by the power function (11.6.5), or by the logarithmic equation (11.6.6). The three statements are equivalent.

Notice that the allometric law is not primarily concerned with the speed of growth since time is eliminated. An individual may grow as a function of time following an exponential, a logistic or any other law. This leaves the allometric relationship unaffected.

The allometric law was applied successfully not only to the relative growth of parts of a body, but also to metabolism, to dose-response problems, to racial differences, and to evolutionary history. There exist also attempts to justify the allometric law theoretically. For details see Grande and Taylor (1965), Huxley (1932), Rosen (1967, Chap. 5), Schüepp (1966), Simpson, Roe, Lewontin (1960, p. 396 ff.), Teissier (1960), von Bertalanffy (1951, p. 311–332), Wilbur and Owen (1964).

11.7. A System of Linear Differential Equations

In physiological systems as well as in populations there are usually two or even more functions of time under consideration. Let $x = x(t)$ and $y = y(t)$ be two such functions of time. We assume some kind of interaction between the quantities x and y. Thus the rates of change dx/dt and dy/dt may depend on both quantities x and y. A relatively simple case consists of the two *simultaneous equations*

$$
\begin{aligned}
\frac{dx}{dt} &= a \cdot x + b \cdot y, \\
\frac{dy}{dt} &= c \cdot x + d \cdot y.
\end{aligned}
\qquad (11.7.1)
$$

where a, b, c, d are given constants. The unknown functions $x = x(t)$ and $y = y(t)$ and their derivatives appear to the first power. Therefore, (11.7.1) is said to be a *system of linear differential equations*.

It is worth studying the logical content of such a system. The quantities x and y interact in such a way that they fully determine the rates dx/dt and dy/dt. As time elapses, x and y will increase or decrease according to the rates dx/dt and dy/dt. Thus the rates will determine future values of x and y. The information originally given to dx/dt and to dy/dt is then "fed back" to x and y. We get two feedback loops as indicated in the following graphical presentation:

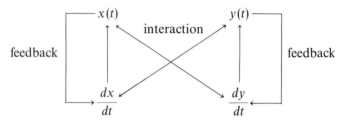

The system (11.7.1) is solved by the *method of trial functions*. We try the following special functions:

$$x = Ae^{\lambda t}, \qquad y = Be^{\lambda t} \tag{11.7.2}$$

where A, B, λ are some constants to be determined later. To avoid trivial solutions we may assume that $A \neq 0$ and $B \neq 0$. Substituting (11.7.2) in (11.7.1) we obtain

$$A\lambda e^{\lambda t} = aAe^{\lambda t} + bBe^{\lambda t},$$
$$B\lambda e^{\lambda t} = cAe^{\lambda t} + dBe^{\lambda t}$$

or, after removing the common factor $e^{\lambda t}$ and rearranging the terms

$$\begin{aligned}(a - \lambda)A + bB &= 0, \\ cA + (d - \lambda)B &= 0.\end{aligned} \tag{11.7.3}$$

From this it follows that

$$\frac{A}{B} = -\frac{b}{a - \lambda} = -\frac{d - \lambda}{c}. \tag{11.7.4}$$

Thus we obtain two different expressions in terms of the unknown λ which both equal A/B. This imposes some limitation on λ. Indeed, from (11.7.4) we deduce the quadratic equation

$$bc = (a - \lambda)(d - \lambda)$$

or, in the standard form

$$\lambda^2 - (a + d)\lambda + (ad - bc) = 0. \tag{11.7.5}$$

This equation can be solved with respect to λ by means of formula (4.6.2). For simplicity, we assume that we get *two different real roots* λ_1 and λ_2.

The Eq. (11.7.5) is called the *characteristic equation* of the system (11.7.1) of differential equations and λ_1, λ_2 the *characteristic roots*. So far we can only claim: If there exists a solution to (11.7.1) of the form (11.7.2), λ has to be one of the roots of the characteristic equation.

With $\lambda = \lambda_1$ we find suitable constants A_1 and B_1 from (11.7.4):

$$\frac{A_1}{B_1} = -\frac{b}{a - \lambda_1} = -\frac{d - \lambda_1}{c}. \tag{11.7.6}$$

The coefficients A_1 and B_1 are not fully determined. One of them can be chosen arbitrarily.

Similarly, with $\lambda = \lambda_2$ we find A_2 and B_2 such that

$$\frac{A_2}{B_2} = -\frac{b}{a - \lambda_2} = -\frac{d - \lambda_2}{c}. \tag{11.7.7}$$

Thus we have obtained two particular solutions to (11.7.1):

$$x = A_1 e^{\lambda_1 t}, \qquad y = B_1 e^{\lambda_1 t},$$

and

$$x = A_2 e^{\lambda_2 t}, \qquad y = B_2 e^{\lambda_2 t}.$$

To get the *general solution* we combine the particular solutions as follows:

$$x = A_1 e^{\lambda_1 t} + A_2 e^{\lambda_2 t},$$
$$\qquad\qquad (\lambda_1 \neq \lambda_2) \tag{11.7.8}$$
$$y = B_1 e^{\lambda_1 t} + B_2 e^{\lambda_2 t}.$$

In each pair of coefficients A_1, B_1 and A_2, B_2, one coefficient can be chosen arbitrarily. Thus the general solution contains two arbitrary constants. To verify the result we have only to substitute x and y in (11.7.1) by the formulas (11.7.8) and to show that we get the same expressions on both sides of the equations. We leave this task to the reader.

The case where λ_1 and λ_2 are not real numbers will be discussed in Chapter 15. There exists also a mathematical theory dealing with the exceptional case $\lambda_1 = \lambda_2$, but we will omit this part.

A numerical example may help the reader to understand the procedure. Let

$$\frac{dx}{dt} = 3x - 2y,$$

$$\frac{dy}{dt} = 2x - 2y$$

be the given system. According to (11.7.5) the characteristic equation is

$$\lambda^2 - \lambda - 2 = 0.$$

Its roots are $\lambda_1 = -1$ and $\lambda_2 = 2$. Hence, particular solutions are of the form

$$x = A_1 e^{-t}, \qquad y = B_1 e^{-t}$$

and

$$x = A_2 e^{2t}, \qquad y = B_2 e^{2t}.$$

From (11.7.6) and (11.7.7) we find two side conditions:

$$\frac{A_1}{B_1} = \frac{1}{2}, \qquad \frac{A_2}{B_2} = 2.$$

Hence we may put $A_1 = k$, $B_1 = 2k$, $A_2 = 2m$, $B_2 = m$, with arbitrary, fixed values for k and m. Thus the general solution becomes

$$x = ke^{-t} + 2me^{2t},$$
$$y = 2ke^{-t} + me^{2t}.$$

As t increases, e^{-t} tends to zero, and e^{2t} tends to infinity. Hence, x and y both tend to $+\infty$ if $m > 0$ or to $-\infty$ if $m < 0$. In either case, the system (x, y) is "exploding".

For the use of *determinants* to establish characteristic equations see Section 14.9.

Applications

Example 11.7.1. *Ecology.* Let us consider a closed area which contains two animal or plant species. Let $N_1 = N_1(t)$ and $N_2 = N_2(t)$ be the number of individuals of the two species as a function of time. We assume that there is some interaction between the two species. The first species may be a food source for the second species. Or, one species consisting of trees may reduce the incident light on another plant species. But there are many other possibilities of interaction: One species may provide shelter for the other species; pollination of a plant species may be performed by an animal species; soil may be poisoned by one species, thus inhibiting the growth of a plant species.

During a given time interval of length Δt we may describe the process by the following equations:

$$\text{population of species 1:} \quad \left\{\begin{matrix} \Delta N_1 = \text{net gain} \\ \text{(or loss)} \end{matrix}\right\} = \left\{\begin{matrix} \text{change in the} \\ \text{absence of} \\ \text{interaction} \end{matrix}\right\} + \left\{\begin{matrix} \text{change due to} \\ \text{interaction} \\ \text{with species 2} \end{matrix}\right\}$$

$$\text{population of species 2:} \quad \left\{\begin{matrix} \Delta N_2 = \text{net gain} \\ \text{(or loss)} \end{matrix}\right\} = \left\{\begin{matrix} \text{change due to} \\ \text{interaction} \\ \text{with species 1} \end{matrix}\right\} + \left\{\begin{matrix} \text{change in the} \\ \text{absence of} \\ \text{interaction} \end{matrix}\right\}$$

We divide both equations by Δt and assume that each term tends to a limit as Δt tends to zero. Thus the limiting process leads to

$$\frac{dN_1}{dt} = \left\{ \begin{array}{l} \text{rate of change} \\ \text{in the absence of} \\ \text{interaction} \end{array} \right\} + \left\{ \begin{array}{l} \text{rate of change} \\ \text{due to interaction} \\ \text{with species 2} \end{array} \right\}$$

$$\frac{dN_2}{dt} = \left\{ \begin{array}{l} \text{rate of change} \\ \text{due to interaction} \\ \text{with species 1} \end{array} \right\} + \left\{ \begin{array}{l} \text{rate of change} \\ \text{in the absence of} \\ \text{interaction} \end{array} \right\}.$$

In the absence of interaction we may assume, for simplicity, that in both populations there is a constant birth and death rate so that the rate of change becomes proportional to the size of the population (see Example 11.3.3). Also for simplicity, we may assume that the rate of change due to interaction is proportional to the size of the interfering population. This leads to differential equations such as

$$\frac{dN_1}{dt} = a \cdot N_1 + b \cdot N_2,$$

$$\frac{dN_2}{dt} = c \cdot N_1 + d \cdot N_2$$

(11.7.9)

where a, b, c, d, are certain constants (positive, negative or zero).

The equations are of the form (11.7.1). The solution (11.7.8) indicates that much depends on the roots λ_1 and λ_2 of the characteristic equation. If one or both λ's are positive, the ecological system will explode since $\exp(\lambda t)$ with $\lambda > 0$ tends to infinity. If, on the other hand, λ_1 and λ_2 are both negative, the population will decrease in size since $\exp(\lambda t)$ with $\lambda < 0$ tends to zero. When the roots are not real, we will have to postpone the discussion to Chapter 15. There we will show that the ecological system oscillates.

For a more detailed discussion see Keyfitz (1968, p. 271–287).

Example 11.7.2. *Passage of Food in Ruminants.* Ruminants, such as deer, sheep, goats, oxen, have a complicated stomach. Newly eaten but unchewed food is passed to a storage compartment called the rumen. Later, when chewed, food passes through the omasum into the abomasum where it is further processed. From there it slowly enters the intestines. To get a mathematical description of the passage of food through the digestive tract, Blaxter, Graham, and Wainman (1956) proposed the following model:

Let $r = r(t)$ be the amount of food in the rumen at time instant t. At $t = 0$ this amount is a known quantity, say r_0. By $u = u(t)$ we denote the amount of food in the abomasum at time instant t. At $t=0$ we have $u=0$. The rate of decrease of $r=r(t)$ is likely to be proportional to r. It is also reasonable to assume that du/dt consists of two terms: a rate of increase equal to the rate of decrease of r, and a rate of decrease proportional to u. Thus we obtain the linear differential equations

$$\frac{dr}{dt} = -k_1 r,$$

$$\frac{du}{dt} = +k_1 r - k_2 u \tag{11.7.10}$$

where k_1 and k_2 are positive constants. They may be called *specific rates of digestion*. We assume that $k_1 \neq k_2$. The first equation contains a minus sign since r decreases.

Further we denote the total amount of food that has entered the duodenum at time t by $v=v(t)$ and the amount of faeces by $w=w(t)$. Since the duodenum receives exactly the same amount that leaves the abomasum, we get the differential equation

$$\frac{dv}{dt} = k_2 u \tag{11.7.11}$$

where $v=0$ at time $t=0$. Finally, we assume that the faeces leave the animal with a constant time delay which we denote by τ. Neglecting the loss of matter that enters the blood vessels, we get

$$w(t) = v \quad \text{at time} \quad t-\tau,$$

that is,

$$w(t) = v(t-\tau) \quad \text{for} \quad t > \tau^7. \tag{11.7.12}$$

The overall process may be summarized graphically by

$$r \xrightarrow{\ \ k_1\ \ } u \xrightarrow{\ \ k_2\ \ } v \xrightarrow{\ \text{time delay}\ } w \ .$$
$$\text{rumen} \qquad \text{abomasum} \qquad \text{duodenum} \qquad \text{faeces}$$

We first solve the Eqs. (11.7.10). By (11.7.5) the characteristic equation turns out to be

$$\lambda^2 + (k_1 + k_2)\lambda + k_1 k_2 = 0$$

so that $\lambda_1 = -k_1$, $\lambda_2 = -k_2$ where by assumption $\lambda_1 \neq \lambda_2$. As (11.7.8) indicates, the general solution may be written in the form

$$r = A_1 e^{-k_1 t} + A_2 e^{-k_2 t},$$

$$u = B_1 e^{-k_1 t} + B_2 e^{-k_2 t} \ .$$

---

[7] The reader should not confuse the functional value $v(t-\tau)$ with a product of two factors v and $t-\tau$.

The coefficients have to satisfy (11.7.6) and (11.7.7). Thus[8]

$$\frac{A_1}{B_1} = \frac{k_2 - k_1}{k_1}, \qquad \frac{A_2}{B_2} = \frac{k_2 - k_2}{k_1} = 0,$$

or

$$B_1 = \frac{k_1}{k_2 - k_1} A_1, \qquad A_2 = 0.$$

We have also imposed the initial conditions $r = r_0$ and $u = 0$ at time $t = 0$. This leads to

$$r_0 = A_1 + A_2, \qquad 0 = B_1 + B_2.$$

Combining the results we get for the coefficients

$$A_1 = r_0, \qquad A_2 = 0,$$

$$B_1 = \frac{k_1}{k_2 - k_1} r_0, \qquad B_2 = -\frac{k_1}{k_2 - k_1} r_0.$$

Thus we obtain the particular solution

$$r = r_0 e^{-k_1 t},$$

$$(11.7.13)$$

$$u = \frac{r_0 k_1}{k_2 - k_1} (e^{-k_1 t} - e^{-k_2 t}).$$

From (11.7.11) it follows by integration that

$$v = k_2 \int_0^t u(t) dt = r_0 - \frac{r_0}{k_2 - k_1} (k_2 e^{-k_1 t} - k_1 e^{-k_2 t}). \quad (11.7.14)$$

Finally, (11.7.12) yields a formula for $w = w(t)$ which originates from (11.7.14) by merely replacing t with $t - \tau$.

Graphs of the functions $r(t)$, $u(t)$, $w(t)$ are plotted in Figure 11.8.

We have treated the digestive tract as a *three-compartment system*. Our application may then be interpreted as an example of compartment analysis (cf. Example 11.3.6). More on this subject may be found in Matis and Hartley (1971).

Example 11.7.3. *Excretion of a Drug.* We consider a drug D which is dissolved in the plasma of blood. By diffusion there is an exchange of molecules of D between the plasma and the tissue that is fed by the blood. Molecules of D are also being excreted by the kidneys into the urine. We may consider the plasma and the tissue as two compartments with one of them "leaking". The situation is schematically shown in Fig. 11.9.

---

[8] Note that the formula $A_1/B_1 = -b/(a - \lambda_1)$ yields the indeterminate form $0/0$, whereas the formula $A_1/B_1 = -(d - \lambda_1)/c$ gives the desired result.

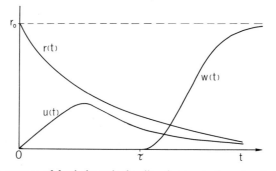

Fig. 11.8. The passage of food through the digestive tract of ruminants. The amount of food in the rumen and in the abomasum are denoted by $r(t)$, $u(t)$, respectively. $w(t)$ is the amount of faeces

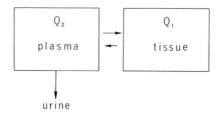

Fig. 11.9. A two-compartment system with excretion

Under simplifying assumptions similar to those in Example 11.3.6 of Section 11.3, we are able to find a system of differential equations for the exchange. Let $Q_1 = Q_1(t)$ and $Q_2 = Q_2(t)$ be the mass of D in the tissue and in the plasma, respectively. The rate of change dQ_1/dt is the sum of a rate of increase (molecules from plasma to tissue) and a rate of decrease (molecules from tissue to plasma). These rates are assumed to be proportional to Q_2 and Q_1. The rate dQ_2/dt consists of three such terms, the third one meaning the rate of excretion. Thus we obtain

$$\frac{dQ_1}{dt} = k_2 Q_2 - k_1 Q_1 ,$$

$$\frac{dQ_2}{dt} = k_1 Q_1 - k_2 Q_2 - k_3 Q_2 \tag{11.7.15}$$

where k_1, k_2, k_3 are positive constants. Upon factoring out Q_2 in the second equation, we see that (11.7.15) is of the form (11.7.1). The behavior of the solution is determined by the roots of the characteristic equation.

The roots λ_1 and λ_2 turn out to be negative. The general solution (11.7.8) then reveals that the process tapers off steadily.

Readers interested in learning more about compartment analysis are referred to Atkins (1969), Defares *et al.* (1973, p. 609 ff.), Jacquez (1972), Metzler (1971), Milhorn (1966), Rubinow (1973), and Solomon (1960).

11.8. A System of Nonlinear Differential Equations

In Section 11.5 we considered a *nonlinear* differential equation, and in Section 11.7 we introduced a *system* of linear differential equations. In this section we will combine the two concepts. To give the reader the proper motivation for doing so, we will begin with a biological example.

D'Ancona reported that in the years 1910–1923 the fish population in the upper Adriatic changed considerably. During the first world war, 1914–1918, fishing was suspended. After the war, sharks and other voracious species became more numerous relative to herbivorous types of fish. D'Ancona concluded that the suspension of fishing allowed the fish population to grow and that this gave the predatory species an advantage over the prey species (see D'Ancona, 1954, p. 209).

This observation stimulated a mathematical model on population dynamics in the case where *one species, called predator, feeds on another species, called prey.* We assume that the prey population finds ample food at all times, but that the food supply of the predator population

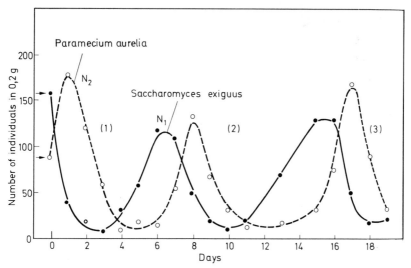

Fig. 11.10. Fluctuation of population size of *Paramecium aurelia* which feed upon *Saccharomyces exiguus*. The experiment was performed by G. F. Gause. The figure is reproduced from D'Ancona (1954, p. 244)

depends entirely on the prey population. We also assume that during the process the environment does not change in favor of one species, and that genetic adaptation is sufficiently slow.

When the prey population increases in size, the predatory species obtains a larger food base. Hence, with a certain time delay it will also become more numerous. As a consequence, the growing pressure for food will reduce the prey population. After awhile food becomes rare for the predator species so that its propagation is inhibited. The size of the predator population will decline. The new phase favors the prey population. Slowly it will grow again, and the pattern in changing population sizes may repeat. When conditions remain the same, the process continues in cycles. Fig. 11.10 illustrates three such cycles.

Let us consider the process during a time interval of length Δt. The changes are given by the following equations:

$$
\left\{\begin{array}{l}\text{change in}\\ \text{the number}\\ \text{of prey}\end{array}\right\} = \left\{\begin{array}{l}\text{natural}\\ \text{increase}\\ \text{in prey}\end{array}\right\} - \left\{\begin{array}{l}\text{destruction}\\ \text{of prey by}\\ \text{predator}\end{array}\right\}
$$

$$
\left\{\begin{array}{l}\text{change in}\\ \text{the number}\\ \text{of predator}\end{array}\right\} = \left\{\begin{array}{l}\text{increase in}\\ \text{predator resulting}\\ \text{from devouring prey}\end{array}\right\} - \left\{\begin{array}{l}\text{natural}\\ \text{loss in}\\ \text{predator}\end{array}\right\} .
$$

(11.8.1)

Let $x = x(t)$ be the number of prey individuals and $y = y(t)$ the number of predator individuals at time instant t. As in previous studies in population dynamics we assume that $x(t)$ and $y(t)$ are differentiable functions. By dividing both Eqs. (11.8.1) by Δt and letting Δt tend to zero, we obtain the rates of change

$$
\frac{dx}{dt} = \left\{\begin{array}{l}\text{birth rate}\\ \text{of prey}\end{array}\right\} - \left\{\begin{array}{l}\text{destruction rate}\\ \text{of prey}\end{array}\right\}
$$

$$
\frac{dy}{dt} = \left\{\begin{array}{l}\text{birth rate}\\ \text{of predator}\end{array}\right\} - \left\{\begin{array}{l}\text{death rate of}\\ \text{predator}\end{array}\right\} .
$$

(11.8.2)

The birth rate of the prey species is likely to be proportional to x (cf. birth process in Section 11.3), that is, equal to $a \cdot x$ with a certain constant $a > 0$. The destruction rate depends on x and on y. The more prey individuals are available, the easier it is to catch them, and the more predator individuals are around, the more stomachs have to be fed. It is reasonable to assume that the destruction rate is proportional to x and to y, that is, equal to $b \cdot xy$ with a certain constant $b > 0$.

The birth rate of the predator population depends on food supply as well as on its present size. We may assume that the birth rate is proportional to x and to y, that is, equal to $c \cdot xy$ with a certain constant $c > 0$. Finally, the death rate of the predator species is likely to be proportional to y, that is, equal to $d \cdot y$ with a certain $d > 0$.

Under these simplifying assumptions the Eqs. (11.8.2) become

$$\frac{dx}{dt} = a \cdot x - b \cdot xy,$$

$$\frac{dy}{dt} = c \cdot xy - d \cdot y .$$

$$(11.8.3)$$

All four constants a, b, c, d are positive. This system is said to be *nonlinear* since the equations contain the product xy of two unknown functions $x = x(t)$ and $y = y(t)$.

The prey-predator model was discovered by Lotka and independently by Volterra around 1925[9].

There exists no explicit solution of the differential equations (11.8.3). When numerical values or graphs of $x(t)$ and $y(t)$ are required, the integration can be performed by a computer employing methods of numerical analysis. Fortunately, it is possible to derive a rather easy relationship between x and y which discloses some properties of the solution. For this purpose we eliminate the time t. This is possible by applying formula (11.6.2). Thus upon dividing the second Eq. (11.8.3) by the first equation we get

$$\frac{dy}{dx} = \frac{c \cdot xy - d \cdot y}{a \cdot x - b \cdot xy} = \frac{y(cx - d)}{x(a - by)} . \qquad (11.8.4)$$

Since at a specific time instant t, the variable y is uniquely associated with x, we may consider y as a function of x. Hence, (11.8.4) is a differential equation for this function. Upon separating the variables we get

$$\left(\frac{a}{y} - b\right) dy = \left(c - \frac{d}{x}\right) dx ,$$

$$(11.8.5)$$

$$\int \left(\frac{a}{y} - b\right) dy = \int \left(c - \frac{d}{x}\right) dx \qquad (x > 0, y > 0) ,$$

$$a \cdot \ln y - by = cx - d \cdot \ln x + C \qquad (11.8.6)$$

---

[9] Alfred James Lotka (1880—1949), born in Austria, American biophysicist. Vito Volterra (1860—1940), Italian mathematician.

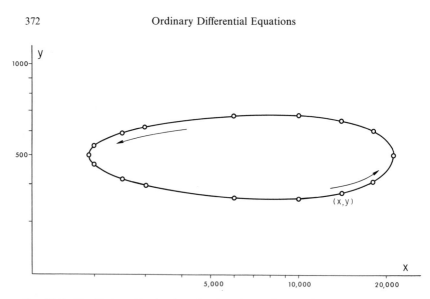

Fig. 11.11. Double-logarithmic plot of points (x, y) that satisfy the equation (11.8.6) with the particular constants $a = 1$, $b = 0.002$, $c = 0.00001$, $d = 0.08$, $C = 5.8$. The solution set is a closed curve. If we consider x and y as functions of time, the point (x, y) moves along the closed curve in the counter-clockwise direction, thus describing cycle by cycle

where C is an integration constant. It is not possible to solve this equation explicitly with respect to y. However, with rather laborious calculations that require some training in numerical analysis, it is possible to find pairs (x, y) which satisfy the Eq. (11.8.6). In a rectangular coordinate system, the points (x, y) form a closed curve. An example is shown in Fig. 11.11.

We may also plot x and y versus time. Then we get a regular up-and-down in the two population sizes as shown in Fig. 11.12.

Some experimental data are in agreement with the outcome of the Lotka-Volterra model, but disagreement in other experiments was also reported. The model may be modified and refined by adding more terms to the Eqs. (11.8.3) or by allowing for chance fluctuations. The reader will find many more details in the main works on the Lotka-Volterra model: D'Ancona (1954), Gause (1934), Kostitzin (1939), Lotka (1956), Volterra (1931). For more modern views and additions see Bailey (1964), Barnett (1962), Bartlett (1960), Chapman (1967), Chiang (1964), Goel et al. (1971), Goel and Richter-Dyn (1974), Kemeny and Snell (1962), Keyfitz (1968), Leigh et al. (1968), Leslie (1957), Leslie and Gower (1960), Levin (1969), Levins (1968), MacArthur and Connell (1966), MacArthur (1972), Montroll (1972), Pielou (1969), Pearce (1970), Slobodkin (1961), J. M. Smith (1968), von Bertalanffy (1951), Watt (1968).

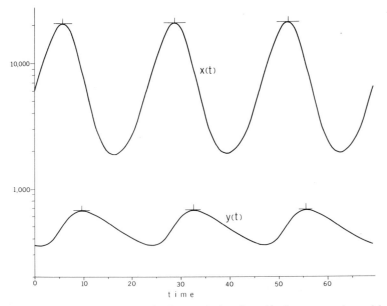

Fig. 11.12. The ups and downs in the population size $x(t)$ of a prey species and in the population size $y(t)$ of a predator species. The graph is based on the same constants as in Fig. 11.11. Notice that a maximum of $x(t)$ is followed by a maximum of $y(t)$, and that the same sequence occurs for the minima

Systems of nonlinear differential equations are also applied in the theory of *epidemics* (see Bailey, 1957, Lotka, 1956, Waltman, 1974, Watt, 1968) and in *enzyme synthesis* [see Ackerman (1962), p. 323, and J. M. Smith (1968), p. 109].

*11.9. Classification of Differential Equations

There are several ways to distinguish between different types of differential equations.

We begin with the *order* of a differential equation. When only the first derivative occurs, an equation is said to be of the *first order*. Thus all differential equations discussed in the previous sections are first order differential equations. When the second derivative of an unknown function occurs, the equation is said to be of the *second order*. For instance,

$$\frac{d^2y}{dt^2} = ay,$$

$$y'' = ay' + by + c$$

are second order equations. In general, by the order of a differential equation we mean the order of the highest order derivative appearing in it. This meaning of "order" should not be confused with the usage of the word "order" in chemical kinetics (cf. footnote in Example 11.5.3).

A further distinction is concerned with the role of the highest order derivative. A differential equation is said to be *explicit* if the derivative of the highest order is given as a function of all the other variables. Otherwise, it is called *implicit*. All differential equations discussed so far are explicit. Examples of implicit equations are

$$(y')^2 = ay + b ,$$

$$y' + \log y' = cy .$$

Here, y' need not be a function of y.

An important class of differential equations is formed by the *linear equations*. Such an equation may be written as a "linear aggregate" equated to zero. As an example we consider

$$\frac{dy}{dt} + py + q = 0.$$

Here p and q are either constants or are functions of t, that is, $p = p(t)$ and $q = q(t)$. The second order equation

$$y'' = ay' + by + c$$

is also linear if a, b, c are constants or known functions of the independent variable.

Examples of *nonlinear equations* have been studied in Sections 11.5 and 11.8.

Within the class of linear differential equations we distinguish between *homogeneous* and *nonhomogeneous* equations. A linear differential equation is homogeneous if that particular term is missing which does not contain the unknown function. Thus

$$y' = ay + c \quad \text{or} \quad y'' = ay' + by + c$$

are homogeneous if, and only if, $c = 0$. A homogeneous equation has an important property: If $y = u(t)$ is a particular solution, then any multiple $k \cdot u(t)$ with an arbitrary constant k is also a solution. In addition, if $u(t)$ and $v(t)$ are particular solutions, all linear combinations $ku(t) + mv(t)$ with constants k, m are also solutions.

In previous sections we dealt only with differential equations having *constant coefficients*. We did not consider a linear equation such as

$$\frac{dy}{dt} = f(t) \cdot y + g(t).$$

It would be more difficult to solve this equation than to solve $dy/dt = ay + b$ with constant coefficients a and b. In linear differential equations, the distinction between equations with *constant* and *variable coefficients* is very important.

Finally, there is a class of differential equations which contain the so-called partial derivatives. This concept will be introduced in Chapter 12. There we will consider a *partial differential equation*. All differential equations discussed so far are called *ordinary differential equations* to distinguish them from partial differential equations.

Recommended for further reading: Atkins (1969), Defares *et al.* (1973), Kynch (1955), Latham (1964), Lefort (1967), Milhorn (1966), Pielou (1969), Rashevsky (1964), Riggs (1963, 1970), Rubinow (1973), Saunders and Fleming (1957), C.A.B. Smith (1966), B. Stevens (1965), von Bertalanffy (1951), Watt (1968).

Problems for Solution

11.1.1. Verify that $y = ax^2 + bx$ is a solution of the differential equation

$$y' = \frac{y}{x} + ax.$$

11.1.2. Verify the following statements:

a) $y = \sin x$ is solution of $y'' = -y$,
b) $y = \cos \omega t$ is solution of $d^2 y/dt^2 = -\omega^2 y$,
c) $y = at + e^{kt}$ is solution of $dy/dt = ky + a(1 - kt)$.

11.2.1. Plot direction fields of the following differential equations:

a) $y' = y/x$, b) $y' = y - x$, c) $y' = 1 + xy$, d) $y' = 1/y$.

11.2.2. Plot direction fields of the following differential equations:

a) $y' = -x/y$, b) $y' = x + y$, c) $y' = (x+y)^2$, d) $y' = \frac{1}{3}(y+1)$.

11.3.1. Solve the following differential equations by separating the variables:

a) $y' = y/x$ (compare with the plot of Problem 11.2.1a)),
b) $y' = ay^2$, c) $y' = kyt$, d) $du/dt = u^2 t$.

11.3.2. By separating variables solve
   a) $dy/dx = -x/y$ (compare with the plot of Problem 11.2.2a)),
   b) $dy/dt = ayt^2$,
   c) $dv/dt = v^{3/2}$ $(v > 0)$,
   d) $dz/dt = z^{1/2}t^{3/2}$ $(t > 0, z > 0)$.

11.3.3. Solve

$$\frac{dv}{dt} = k\frac{v}{2-v} \qquad (0 < v < 2)$$

with the initial condition $v = 1$ at $t = 0$. (This equation occurs in genetics, see Sved and Mayo, 1970, p. 296).

11.3.4. Solve

$$\frac{dy}{dx} = k\left(1 - \frac{y-1}{K}\right)^m \qquad (k > 0, K > 0)$$

for $m = 1, 2, 3$ and initial condition $y = 1$ at $x = 0$. (This differential equation occurs in genetics, see Robertson, 1970, p. 267.)

11.3.5. Assume that a population grows in such a way that the specific growth rate $\dfrac{1}{N}\dfrac{dN}{dt}$ remains constant. Let N_1 be the number of individuals at the time instant t_1. Find $N = N(t)$.

11.3.6. For the worker bees of the species *Bombus humilis Ill.* the specific death rate is approximately $0.04\,d^{-1}$ where d denotes day. Let $y(t)$ be the number of worker bees at time t. Find the differential equation for $y(t)$ and the half-life of the worker bees. (Adapted from Brian, 1965, p. 35).

11.3.7. Some 435 striped bass (*Roccus saxatilis*) from the Atlantic Ocean were planted in San Francisco Bay in 1879 and 1881. In 1899 the commercial net catch alone was 1,234,000 pounds (MacArthur and Connell, 1966, p. 122). Assume that the average weight of a bass fish is one pound and that in 1899 every tenth bass fish was caught. The growth of the population was so fast that it is reasonable to operate with the differential equation $dN/dt = \lambda N$. Find a lower bound for λ.

11.3.8. Given are the following growth rates:
   a) $dy/dt = (3.5)\,y$,   b) $dy/dt = \sqrt{y}$,   c) $du/dt = u^2 + 1$.
   Find the *specific* growth rates.

11.3.9. Given the following *specific* growth rates of $y = y(t)$:
   a) 0.76,   b) $2/\sqrt{y}$,   c) $1 - 1/y$.
   Find dy/dt.

11.3.10. In the birth-and-death process, assume that the coefficient $\lambda - \mu$ in Eq. (11.3.13) is not constant but proportional to $N^{1/k}$ with a certain positive constant k. Find the modified differential equation and integrate it. The result seems to fit the population explosion in developing countries very well (see Watt, 1968, p. 10).

11.3.11. In formula (10.9.11) we stated an equation for the absorption of light in a transparent medium. What differential equation does the light intensity $I = I(x)$ satisfy?

11.4.1. Plot a slope field for the differential equation $y' = (y - 1)/2$, and plot the solution with initial condition $y = 2$ at $t = 0$.

11.4.2. Plot a slope field for $y' = 1 - y$ and the solution with initial condition $y = 0$ at $t = 0$.

11.4.3. Combining the birth-and-death process of Example 11.3.3 with the birth-and-immigration process of Example 11.4.2, find the differential equation for a birth-death-immigration process. Integrate the equation assuming that $N = N_0$ at $t = 0$.

11.4.4. By infusion the glucose concentration of blood is increased at a constant rate R (in $mg\,l^{-1}\,min^{-1}$). At the same time, glucose is converted and excreted at a rate which is proportional to the present concentration of glucose. Hence, if $C = C(t)$ denotes this concentration, we obtain the differential equation $dC/dt = R - kC$ with a constant $k > 0$. Solve the differential equation.

11.4.5. Find the particular integral of

$$1 + \frac{dz}{dx} = z$$

for $z(0) = 2$.

11.4.6. Find the particular integral of

$$dy/dt = \tfrac{1}{3}(y - A) \quad (A > 0)$$

for initial value $y = 2A$ at $t = 1$.

11.4.7. Find an implicit function which satisfies

$$\frac{dQ}{dx} = \frac{x - 3}{Q}$$

subject to the condition $Q(2) = \sqrt{2}$.

11.5.1. By means of Eq. (11.5.9) find the solution of the following differential equations:

a) $dy/dt = (2 - y)(5 - y)$,
c) $dz/dt = 3 - 4z + z^2$,

b) $du/dt = \frac{1}{2}(3 - u)(-u)$,
d) $dy/dx = 4 - y^2$.

11.5.2. Use Eq. (11.5.9) to integrate

a) $dy/dt = (1 - y)(6 - y)$,
c) $dv/dt = v^2 - 9$,

b) $du/dt = u^2 - 3u$,
d) $dy/dx = y^2 - y - 12$.

11.5.3. Adjust Eq. (11.5.11) of logistic growth to the following conditions:

a) $k = 1$, $\lambda = 0.5$, $y(0) = 1$,
b) $k = \frac{1}{3}$, $\lambda = 1$, $y(0) = 3$,
c) $B = 1$, $\lambda = \frac{1}{3}$, $y(0) = 0.5$,
d) $B = 4$, $y(0) = 2$, $y(1) = 3$.

11.5.4. In the bimolecular reaction $2\,NO_2 \rightarrow N_2O_4$ the concentration $C = [NO_2] = C(t)$ satisfies the differential equation $dC/dt = -kC^2$ where k is a positive constant. Let $C(0) = C_0$. a) Find the solution of the differential equation. b) Find the limit of $C(t)$ for $t \rightarrow \infty$.

11.6.1. A mushroom was observed while growing. All linear dimensions were growing proportionally. Therefore, the volume increased according to the formula $V = cL^3$ where L is any linear dimension and c is a suitable constant. Show that

$$\frac{1}{V} \cdot \frac{dV}{dt} = 3 \frac{1}{L} \cdot \frac{dL}{dt},$$

that is, that the specific growth rate of the volume is three times the specific growth rate of a linear dimension.

11.6.2. The growth of a cell depends on the flow of nutrients through the surface. Let $W = W(t)$ be the weight of the cell. Assume that for a limited time the growth rate dW/dt is proportional to the area of the surface. If the form of a growing cell does not change, the area of the surface is proportional to the square of a linear dimension (such as a diameter, as defined in Section 4.2) and therefore proportional to $W^{2/3}$. Hence, $dW/dt = kW^{2/3}$ with a positive constant k. Solve this differential equation and interpret the result (cf. von Bertalanffy, 1951).

11.6.3. Solve $dy/dx = my/x$ for $y(2) = -1$. Assume $x > 0$.

11.6.4. "Growth of a function" may have several different meanings. Let $y = f(t)$ be a monotone increasing function. What do we mean

a) by growth in an interval $[t_1, t_2]$,
b) by percent increase in an interval $[t_1, t_2]$,
c) by average growth rate in an interval $[t_1, t_2]$,
d) by instantaneous growth rate at a time instant t_1,
e) by specific growth rate at a time instant t_1?

11.6.5. Apply the five concepts of the previous problem to the function $y = A e^{kt}$ $(A > 0, k > 0)$.

11.7.1. Find the general solution of the following system of differential equations:

$$\frac{dx}{dt} = 7x - 4y,$$

$$\frac{dy}{dt} = -9x + 7y.$$

11.7.2. Adjust the general solution of the previous problem to the special conditions $x(0) = 4$, $y(0) = 1$.

11.7.3. Prove that the characteristic equation (11.7.5.) has real valued roots whenever b and c are of equal sign. [Hint: Rewrite the discriminant in the form $(a - d)^2 + \cdots$]

11.7.4. To study the concept of feedback, consider the following feedback systems in human life: a) behavior of criminals – effort of police, b) adjustment of youths in view of the generation gap, c) quality of a performance under the influence of applause.

11.7.5. Discuss qualitatively the following feedback systems:

a) Population growth in developing countries – poverty,
b) Energy production by fossil fuels – damage by SO_2 pollution – public concern – pollution abatement – energy production,
c) Standard of living – diseases – production,
d) Production of cars – density of traffic – road construction,
e) Administration of a (healing) drug – habituation – disease,
f) Slashing and burning forests for agriculture – change of climate – land loss.

11.7.6. Discuss qualitatively the following control mechanisms:

a) change of light intensity – opening or closing pupil of eye – feedback by retina.

b) Change of ambient temperature – insulation of warmblooded body and/or heat production or sweating – feedback by sensory nerves of skin.

*11.9.1. Some biological rhythms are described by the second order differential equation

$$\frac{d^2 x}{dt^2} + kx = 0 \quad (k > 0).$$

Find solutions in the form $x = A\cos\omega t + B\sin\omega t$.

*11.9.2. Classify the following differential equations with respect to order, explicit-implicit, linear-nonlinear, homogeneous-non-homogeneous, constant-variable coefficients:

a) $u' = 5 - u$ b) $dN/dt = at\,N$

c) $z'' = az$ d) $z'' = az^2$

e) $dy/dx = y \cdot \sin x$ f) $dv/dt = v + f(t)$

g) $y = 1/\sin y'$ h) $yy' = 1 - y$.

Chapter 12

Functions of Two or More Independent Variables

12.1. Introduction

We recall the formula

$$z = (xy)^{\frac{1}{2}} \qquad (x \geq 0, \ y \geq 0) \qquad (12.1.1)$$

for the geometric mean of two numbers x and y. Consider x and y as variables whose values can be chosen independently of each other. Then with each pair (x, y) there is uniquely associated a number z, the geometric mean. In Chapter 3 we called such an association a function. We say that z is *a function of the pair* (x, y), or the pair (x, y) is *mapped into z*. It is also customary to call z *a function of two variables x and y*.

The numbers x and y are known as *independent variables*, whereas z is called the *dependent variable*. The *domain D* of the function (12.1.1) is the set of all pairs (x, y) with $x \geq 0$ and $y \geq 0$. In set theory (Chapter 2) we would denote this domain by

$$D = \{(x, y) | x \geq 0, y \geq 0\} .$$

The *range R* of the function (12.1.1) is the set of all numbers $z \geq 0$ or

$$R = \{z | z \geq 0\} .$$

For a *graph* of a function of two independent variables we may use a *three-dimensional coordinate system* with pair-wise perpendicular x, y, z axes. A pair (x, y) is represented by a dot in the xy-plane. With each such dot there is associated a coordinate z plotted on a line through the dot and plotted perpendicular to the xy-plane. This coordinate determines a point. We denote this point simply by

$$(x, y, z) .$$

If z is a function of (x, y), the points (x, y, z) form a *surface* which may be continuous or not. Usually a perspective view of the coordinate system and of the surface is drawn. Fig. 12.1 depicts the surface defined by the function (12.1.1).

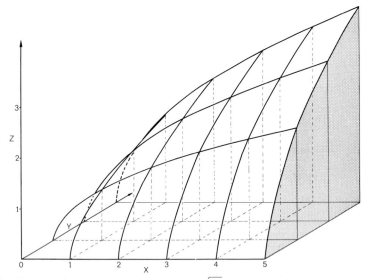

Fig. 12.1. Graph of the geometric mean $z = \sqrt{xy}$ in a rectangular coordinate system

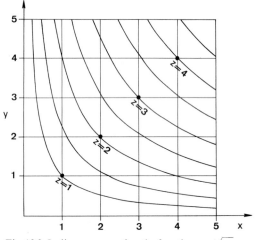

Fig. 12.2. Isolines representing the function $z = \sqrt{xy}$

It is not always convenient to plot or to use a perspective view. For continuous functions, an *isoline chart* may be easier to prepare as well as to interpret. *Isolines* are lines along which the functional value remains constant, much in the same way as *contour lines* on a geographical map connect points of constant elevation. An isoline chart of the function (12.1.1) is presented in Fig. 12.2.

Under certain conditions a mapping of (x, y) into z can also be depicted by a *nomogram* (Section 7.5). A nomogram of the geometric mean is shown in Fig. 7.16b.

A function of two or more variables is denoted in a way similar to that for a function of one variable. Thus, if z is a function of x and y, we write

$$f : (x, y) \mapsto z$$

or

$$z = f(x, y).$$

In Section 11.5 and later we were tacitly concerned with functions of two variables when we studied rates of change which are products of two variables. Consider, for instance, the destruction rate of a prey population inflicted by a predator population. We assumed that the destruction rate is proportional to the number x of prey individuals and also to the number y of predator individuals. Then we concluded that the destruction rate is equal to

$$b \cdot xy$$

where b is a constant.

A similar conclusion occurs frequently in the natural sciences. We state the following *rule:*

When a quantity $z = f(x, y)$ is proportional to x as long as y remains constant, and when z is proportional to y as long as x remains constant, then z is proportional to the product xy, that is,

$$z = c \cdot xy \tag{12.1.2}$$

where c is a positive or negative constant.

With the symbol $\propto$ introduced in Section 3.5 we may briefly write: From $z \propto x$ and $z \propto y$ it follows that $z \propto xy$.

To prove the statement we proceed in two steps. First, we keep y at a fixed value, say $y = y_0$. Since z is assumed to be proportional to x, we get $z = A \cdot x$ where A is a constant as long as $y = y_0$. Second, if we keep x at a fixed value, say $x = x_0$, and vary y, we conclude from $z = Ax$ that $z = A \cdot x_0$. However, A is no longer a constant, but a function of y. Since z is proportional to y, we conclude that A must be of the form $A = cy$ with a constant c. Replacing A in $z = Ax$ with cy we finally get (12.1.2).

As an example consider the voltage V of an electric circuit. It is proportional to the resistance R when the intensity I of current is kept constant. The voltage V is also proportional to I when R is kept constant. Hence $V = cRI$.

In model making, the assumption that we can keep one of the variables x and y constant is not always satisfied. Rather x and y are dependent on each other. Even then z is frequently expressed in the form (12.1.2). For examples see formulas (11.4.12), (11.5.10), (11.5.16), (11.5.19), (11.5.21), and (11.8.3).

12.2. Partial Derivatives

Rates of change for $z = f(x, y)$ can be defined in much the same way as for functions of a single variable. First, we keep y fixed, say $y = y_0$, and study a change of x only. Let x increase (or decrease) from a value $x = x_0$ by a certain amount $\Delta x = h$ to $x_0 + \Delta x$ or $x_0 + h$. Then the corresponding change of z is

$$\Delta z = f(x_0 + h, y_0) - f(x_0, y_0). \tag{12.2.1}$$

The *average rate of change* is defined by the *difference quotient*

$$\frac{\Delta z}{\Delta x} = \frac{f(x_0 + h, y_0) - f(x_0, y_0)}{h}. \tag{12.2.2}$$

If this quotient tends to a limit as Δx tends to zero, we call the result the *instantaneous rate of change* or simply *rate of change*:

$$\lim_{\Delta x \to 0} \frac{\Delta z}{\Delta x} = \lim_{h \to 0} \frac{f(x_0 + h, y_0) - f(x_0, y_0)}{h}. \tag{12.2.3}$$

This limit is also known as the *partial derivative of z with respect to x* at (x_0, y_0). If the limit exists, we say that $z = f(x, y)$ is *differentiable* with respect to x at (x_0, y_0). The word "partial" indicates that we are only concerned with the rate of change in the x direction.

Several notations are used for partial derivatives. The most frequent notations for the limit (12.2.3) are

$$\frac{\partial z}{\partial x}\bigg|_{x_0, y_0} \quad \text{and} \quad f_x(x_0, y_0). \tag{12.2.4}$$

The differentials dz and dx are no longer written with an ordinary letter d. Instead, a *rounded* ∂ is used for a partial differential symbol. We read $\partial z / \partial x$ as follows: "partial dz by dx".

If z is differentiable for any pair (x, y) belonging to a certain domain D, we drop the subscript zero and simply write:

$$\frac{\partial z}{\partial x} = f_x(x, y), \quad (x, y) \in D. \tag{12.2.5}$$

Thus, the partial derivative $\partial z / \partial x$ is a *function* of x and y.

In the same way we define the partial derivative of z with respect to y. We keep x fixed, say $x = x_0$, and study a change of y only. Let y increase (or decrease) from a value y_0 by a certain amount $\Delta y = k$ to $y_0 + \Delta y$ or $y_0 + k$. Then we get

$$\Delta z = f(x_0, y_0 + k) - f(x_0, y_0) ,$$

$$\frac{\Delta z}{\Delta y} = \frac{f(x_0, y_0 + k) - f(x_0, y_0)}{k} ,$$

and, if the limit exists,

$$\frac{\partial z}{\partial y}\bigg|_{x_0, y_0} = f_y(x_0, y_0) = \lim_{k \to 0} \frac{f(x_0, y_0 + k) - f(x_0, y_0)}{k} . \quad (12.2.6)$$

We call the result the *partial derivative of z with respect* to y at (x_0, y_0). We also say that $z = f(x, y)$ is *differentiable* with respect to y at (x_0, y_0). Assuming that z is differentiable for any pair (x, y) belonging to a certain domain D, we drop the subscript zero and write

$$\frac{\partial z}{\partial y} = f_y(x, y), \quad (x, y) \in D . \quad (12.2.7)$$

Thus, the partial derivative $\partial z / \partial y$ is a function of x and y.

Higher partial derivatives are also common. If we differentiate twice with respect to x, we obtain

$$f_{xx}(x, y) = \frac{\partial}{\partial x}\left(\frac{\partial z}{\partial x}\right) = \frac{\partial^2 z}{\partial x^2} .$$

(Read: Partial d second z by dx second.)

If we first differentiate a function with respect to x and then to y, we get

$$f_{xy}(x, y) = \frac{\partial}{\partial y}\left(\frac{\partial z}{\partial x}\right) = \frac{\partial^2 z}{\partial y \, \partial x} .$$

Such a derivative is called a mixed derivative.

The order in which we differentiate does not matter, that is,

$$\frac{\partial}{\partial y}\left(\frac{\partial z}{\partial x}\right) = \frac{\partial}{\partial x}\left(\frac{\partial z}{\partial y}\right) \quad \text{or} \quad f_{xy}(x, y) = f_{yx}(x, y) \quad (12.2.8)$$

provided that the mixed derivatives are continuous functions of x and y. We shall omit the proof.

Example 12.2.1. We consider again the geometric mean $z = (xy)^{\frac{1}{2}}$. With y being kept constant, z is proportional to $x^{\frac{1}{2}}$. Hence, using formula (9.2.4), the partial derivative turns out to be

$$\frac{\partial z}{\partial x} = y^{\frac{1}{2}} \cdot (x^{\frac{1}{2}})' = y^{\frac{1}{2}} \cdot \frac{1}{2} x^{-\frac{1}{2}} = \frac{1}{2} x^{-\frac{1}{2}} y^{\frac{1}{2}}.$$

Similarly, when x is being kept constant,

$$\frac{\partial z}{\partial y} = x^{\frac{1}{2}} \cdot (y^{\frac{1}{2}})' = \frac{1}{2} x^{\frac{1}{2}} y^{-\frac{1}{2}}.$$

Example 12.2.2. We consider the electric current I which flows through a resistor that is kept under constant temperature. According to Ohm's law,

$$I = V/R \tag{12.2.9}$$

where V denotes the voltage and R the resistance. The current I is a function of two independent variables V and R. If the voltage V changes while R remains constant, we get for the rate of change of I

$$\frac{\partial I}{\partial V} = \frac{1}{R} = R^{-1}. \tag{12.2.10}$$

If, however, the resistance R changes while V remains constant, the rate of change becomes

$$\frac{\partial I}{\partial R} = -\frac{V}{R^2} = -VR^{-2}. \tag{12.2.11}$$

Further differentiation leads to

$$\frac{\partial^2 I}{\partial V^2} = 0, \quad \frac{\partial}{\partial R}\left(\frac{\partial I}{\partial V}\right) = -R^{-2}, \quad \frac{\partial^2 I}{\partial R^2} = 2VR^{-3}. \tag{12.2.12}$$

To get the mixed derivative we first differentiated with respect to V and then with respect to R. Interchanging the order of differentiation we obtain

$$\frac{\partial^2 I}{\partial V \partial R} = \frac{\partial}{\partial V}\left(\frac{\partial I}{\partial R}\right) = \frac{\partial}{\partial V}(-VR^{-2}) = -R^{-2},$$

that is, the same result as in (12.2.12).

12.3. **Maxima and Minima**

Partial derivatives are frequently applied in problems of *optimization*. We consider first an introductory example:

$$z = 2x + y - \frac{1}{6}(x^2 + y^2). \qquad (12.3.1)$$

As a domain of this function we assume the whole xy-plane. Our aim is to find maxima and/or minima of z. In a rectangular coordinate system the graph of (12.3.1) is a surface with a single peak (Fig. 12.3). Since z is a differentiable function of x and y, we can apply differential calculus.

In Section 9.7 we studied maxima and minima for a function $f(x)$ of one independent variable. A *sufficient condition for a maximum* was

$$f'(x) = 0, \qquad f''(x) < 0.$$

To apply these results to our function (12.3.1), we first keep y constant, say $y = y_0$, and ask for possible extrema when moving in the x direction. In terms of partial derivatives we find a *maximum point* if

$$\frac{\partial z}{\partial x} = 0, \qquad \frac{\partial^2 z}{\partial x^2} < 0$$

and a *minimum point* if

$$\frac{\partial z}{\partial x} = 0, \qquad \frac{\partial^2 z}{\partial x^2} > 0.$$

In our example

$$\frac{\partial z}{\partial x} = 2 - \frac{1}{3}x, \qquad \frac{\partial^2 z}{\partial x^2} = -\frac{1}{3}.$$

The first derivative $\partial z/\partial x$ vanishes for $x = 6$, and the second derivative is always negative. Hence, for any fixed y we get a maximum point at $x = 6$. In Fig. 12.3 these points are located on a line which is denoted by L_1.

Second, we keep x fixed, say $x = x_0$, and search for extrema when moving in the y direction. A *maximum point* is present if

$$\frac{\partial z}{\partial y} = 0, \qquad \frac{\partial^2 z}{\partial y^2} < 0$$

and a *minimum point* if

$$\frac{\partial z}{\partial y} = 0, \qquad \frac{\partial^2 z}{\partial y^2} > 0.$$

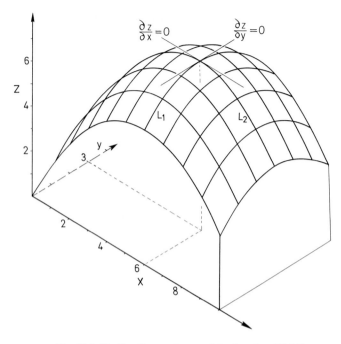

Fig. 12.3. Finding the maximum of the function (12.3.1)

In our example

$$\frac{\partial z}{\partial y} = 1 - \frac{1}{3} y, \qquad \frac{\partial^2 z}{\partial y^2} = -\frac{1}{3}.$$

The first derivative vanishes for $y = 3$, and the second derivative is always negative. Hence, for any fixed x we get a maximum point at $y = 3$. In Fig. 12.3 such points are located on a line which is denoted by L_2.

In order that a point is a maximum or minimum point with respect to the x and the y directions, we have to satisfy the combined condition

$$\frac{\partial z}{\partial x} = 0, \qquad \frac{\partial z}{\partial y} = 0. \qquad (12.3.2)$$

In our example there is only one pair which satisfies (12.3.2), namely $x = 6$, $y = 3$. It determines the intersection of L_1 and L_2. Fig. 12.3 shows that the point is a maximum point. At this point, z reaches its highest value, $z = 7.5$. For all other pairs (x, y) the ordinate z takes on smaller values.

The condition (12.3.2) does not guarantee a maximum or a minimum point. For a counterexample the reader may solve Problem 12.3.5 at the end of this chapter. To get a sufficient condition for an extremum, we would have to add a condition for the second derivatives. Since this condition is rather complicated for functions of more than one variable, we will skip it.

In general, assume that we are given a differentiable function of n independent variables. If $t_1, t_2, \ldots, t_n$ denote these variables and u the dependent variable, we may write

$$u = f(t_1, t_2, \ldots, t_n). \tag{12.3.3}$$

A necessary but not sufficient condition for a maximum or a minimum is that all n equations

$$\frac{\partial u}{\partial t_1} = 0, \quad \frac{\partial u}{\partial t_2} = 0, \ldots, \quad \frac{\partial u}{\partial t_n} = 0 \tag{12.3.4}$$

must be satisfied at such a point.

We add an application which also introduces the reader to the famous *method of least squares*. In Fig. 12.4 some data (x_1, y_1), (x_2, y_2), ..., (x_m, y_m) are plotted in a rectangular coordinate system. The dots are close to a straight line. It is our aim to find a straight line with equation

$$y = ax + b \tag{12.3.5}$$

which fits the data "best". Of course, optimality could be defined in many different ways. For instance, one may postulate that no dot is farther away from the straight line than a given amount. However, it is customary to proceed as follows:

Consider the ordinates of the linear function (12.3.5) at $x_1, x_2, \ldots, x_m$. They are

$$ax_1 + b, \quad ax_2 + b, \quad \ldots, \quad ax_m + b. \tag{12.3.6}$$

They need not coincide with the given ordinates $y_1, y_2, \ldots, y_m$. Instead we will have some positive or negative differences called *deviations:*

$$ax_1 + b - y_1, \quad ax_2 + b - y_2, \quad \text{etc.}$$

Let $i = 1, 2, \ldots, m$ denote a variable subscript. Then we introduce the general deviation δ_i (delta sub i) by

$$\delta_i = ax_i + b - y_i \quad (i = 1, 2, \ldots, m). \tag{12.3.7}$$

Which of the deviations are positive, negative or exactly zero depends on the choice of the parameters a and b. As a condition of optimality

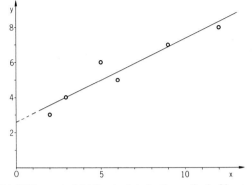

Fig. 12.4. Fitting a straight line to data by the method of least squares

we minimize the sum of squares of deviations ("least squares"), that is, we determine a and b in such a way that

$$\sum_{i=1}^{m} \delta_i^2 \tag{12.3.8}$$

takes on the smallest possible value. In doing so we proceed along the lines which we already indicated in Section 1.9.

We denote the expression (12.3.8) by S. In view of (12.3.7) we may write S in the form

$$S = \sum_{i=1}^{m} (ax_i + b - y_i)^2 . \tag{12.3.9}$$

Since x_i and y_i are given numbers and only a and b are not yet determined, S is a function of the two variables a and b. Thus we may also write

$$S = S(a, b) .$$

If a minimum exists at all, the condition (12.3.2) must be satisfied. In our notation this is

$$\frac{\partial S}{\partial a} = 0, \qquad \frac{\partial S}{\partial b} = 0 . \tag{12.3.10}$$

In order to differentiate (12.3.9) we could work out all the squares in the formula, but this would be rather tedious. It is easier to apply the chain rule. Thus, the derivatives of the general term become

$$\frac{\partial}{\partial a} (ax_i + b - y_i)^2 = 2(ax_i + b - y_i) \cdot x_i ,$$

$$\frac{\partial}{\partial b} (ax_i + b - y_i)^2 = 2(ax_i + b - y_i) \cdot 1 .$$

It follows from (12.3.10) that

$$2\Sigma(ax_i + b - y_i) x_i = 0,$$
$$2\Sigma(ax_i + b - y_i) \quad = 0$$

or, by removing the factor 2 and by working out the sums, that

$$a \cdot \Sigma x_i^2 + b \cdot \Sigma x_i - \Sigma x_i y_i = 0,$$
$$a \cdot \Sigma x_i + b \cdot m - \Sigma y_i = 0. \qquad (12.3.11)$$

These are two linear equations in the unknowns a and b. We will solve them with respect to the slope a by multiplying the first equation by m, the second equation by $-\Sigma x_i$, and by adding the equations:

$$a = \frac{m\Sigma x_i y_i - \Sigma x_i \Sigma y_i}{m\Sigma x_i^2 - (\Sigma x_i)^2}. \qquad (12.3.12)$$

It is hardly worthwhile to find an explicit solution for the y-intercept b. It is easier to express b by a using the second Eq. (12.3.11):

$$b = \tfrac{1}{m}(\Sigma y_i - a\Sigma x_i). \qquad (12.3.13)$$

Example 12.3.1. Fig. 12.4 presents the following data graphically:

x_i	y_i	$x_i y_i$	x_i^2	
2	3	6	4	
3	4	12	9	
5	6	30	25	$(m = 6)$
6	5	30	36	
9	7	63	81	
12	8	96	144	
Total 37	33	237	299	

With these data the Eqs. (12.3.11) become

$$299a + 37b = 237,$$
$$37a + 6b = 33.$$

The solution is

$$a = 201/425 = 0.473,$$
$$b = 1098/425 = 2.58.$$

The best line in the sense of "least squares" is given by the equation

$$y = (0.473)\, x + 2.58$$

with slope 0.473 and y-intercept 2.58.

In general, that the solution really minimizes the function (12.3.9) may be seen by the following argument: A sum of squares, such as S, cannot be negative. For suitable values of a and b, the sum S takes on values as large as we want. Thus the function cannot have a maximum. On the other hand, S must reach a lowest value since the function is continuous and cannot take on negative values. There is only one such point where $\partial S/\partial a$ and $\partial S/\partial b$ vanish simultaneously. Hence, it must be a minimum point. This is also indicated by the fact that $\partial^2 S/\partial a^2 = 2\Sigma x_i^2 > 0$ and that $\partial^2 S/\partial b^2 = 2m > 0$.

For instructive applications of maxima and minima to population genetics see Wright (1969, Vol. 2, Chapters 3, 4).

*12.4. Partial Differential Equations

Whereas the whole of Chapter 11 was devoted to ordinary differential equations, we will look only briefly into the large area of partial differential equations. These equations contain partial derivatives of an unknown function.

We will derive one of the biologically relevant differential equations, the *diffusion equation*. We begin with the physical circumstances. Assume that we inject a substance into a tube filled with a solvent liquid (Fig. 12.5). The molecules of the substance are in random motion. They will spread in all directions. We assume that the solvent liquid is not in motion.

At locations with high concentration, the molecules will tend to decrease in number and, conversely, at locations with low concentration they will tend to increase in number. This random migration process, called diffusion, will finally end with the molecules at equal density throughout the tube.

In Fig. 12.5 we place an x axis parallel to the axis of the tube. For simplicity, we assume that the concentration of the substance varies only in the x direction. At each x the concentration depends also on the time instant t. Thus we may denote the *concentration* by

$$C = C(x, t). \tag{12.4.1}$$

As a unit of measurement we take for instance g/cm^3. The function $C(x, t)$ is plotted in Fig. 12.5 for three different time instants t_1, t_2, t_3. Our aim is to determine the unknown function $C(x, t)$ or, at least, to find an equation which $C(x, t)$ must satisfy.

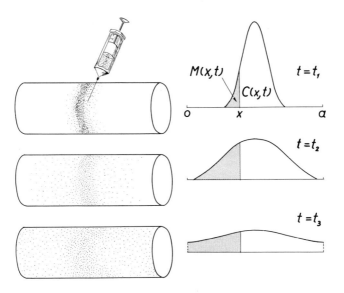

Fig. 12.5. The molecules of a substance dissolved in a liquid are in random motion. This results in a slow "migration", called diffusion. The density $C(x, t)$ of the dissolved substance tends to uniformity in space

For this purpose we assume that $C(x, t)$ is differentiable with respect to x and to t. Strictly speaking, matter is discrete. However, the number of molecules is so large that the error in using a smooth function is negligible (cf. Fig. 9.5).

Let $x = 0$ coincide with the left end of the tube and $x = a$ with the right end. We denote the (constant) area of the cross section by A. Let $M = M(x, t)$ be the mass of the substance which is contained between the left end of the tube and the cross section at x at time instant t (Fig. 12.5).

For a first result we consider $M(x, t)$ as a function of x only and keep the time t constant. Let x be a fixed value and $\Delta x = h$ be an increase of x. The mass $M(x + h, t)$ consists of the mass $M(x, t)$ and the additional mass ΔM located between x and $x + h$ (Fig. 12.6). Thus

$$\Delta M = M(x + h, t) - M(x, t).$$

The volume occupied by this mass is $A \cdot h$. Its *average density* is therefore

$$\frac{\Delta M}{A \cdot h} = \frac{1}{A} \cdot \frac{M(x + h, t) - M(x, t)}{h}.$$

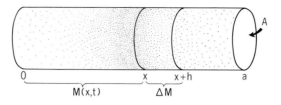

Fig. 12.6. Explanation of ΔM

As h tends to zero, the left side tends to $C(x, t)$ and the second factor on the right side to $\partial M/\partial x$ according to formulas (12.2.3) and (12.2.4). Hence,

$$\frac{\partial M}{\partial x} = A \cdot C(x, t). \tag{12.4.2}$$

To obtain a second result, we consider $M(x, t)$ as a function of t only and keep x constant. The mass $M(x, t)$ increases in time if the net flow of molecules at x is directed to the left. This case occurs if $C(x, t)$ increases with increasing x, that is, if $\partial C/\partial x > 0$ (see Fig. 12.5). We call $\partial C/\partial x$ the *gradient of the concentration*. It is intuitively clear that $M(x, t)$ increases faster as this gradient becomes larger. It is reasonable to expect that the rate of increase $\partial M/\partial t$ is proportional to $\partial C/\partial x$ as well as to the area A of cross section. Indeed, this assumption is well supported by experimental facts. Hence, in virtue of the rule stated in formula (12.1.2),

$$\frac{\partial M}{\partial t} = D \cdot A \cdot \frac{\partial C}{\partial x} \tag{12.4.3}$$

where D denotes a positive constant known as the *diffusion constant*. A similar argument holds when $\partial C/\partial x < 0$. Then the net flow is to the right, and $M(x, t)$ decreases as a function of time. The same relationship (12.4.3) results.

Formula (12.4.3) contains two unknown functions, $M(x, t)$ and $C(x, t)$. As we are interested in $C(x, t)$, we will eliminate $M(x, t)$. For this purpose we differentiate (12.4.3) with respect to x and obtain

$$\frac{\partial}{\partial x} \left(\frac{\partial M}{\partial t} \right) = D \cdot A \cdot \frac{\partial^2 C}{\partial x^2}.$$

In view of formula (12.2.8) we can interchange the order of differentiation on the left side. Then we replace $\partial M/\partial x$ by $A \cdot C(x, t)$ according to formula (12.4.2). Finally, after removing the factor A we get

$$\frac{\partial C}{\partial t} = D \cdot \frac{\partial^2 C}{\partial x^2}.$$ (12.4.4)

This is the famous *diffusion equation in one dimension*. It can be generalized for all three spacial dimensions x, y, z.

We will not learn how to solve the second order partial differential equation (12.4.4) but will confine ourselves to verifying two particular solutions.

Example 12.4.1. Find coefficients a and b such that

$$C(x, t) = \exp(ax + bt)$$

satisfies the diffusion equation (12.4.4). Applying the chain rule (9.2.14) we get

$$\frac{\partial C}{\partial t} = b e^{ax + bt}$$

and

$$\frac{\partial^2 C}{\partial x^2} = a^2 e^{ax + bt}.$$

Substituting these derivatives in (12.4.4) we obtain

$$b e^{ax + bt} = D \cdot a^2 e^{ax + bt}.$$

This equation is satisfied for all values of x and t if we choose the coefficient a arbitrarily and if we let

$$b = Da^2.$$

Example 12.4.2. Verify that

$$C(x, t) = t^{-\frac{1}{2}} \exp(-x^2/4 Dt)$$ (12.4.5)

is a particular solution of the diffusion equation (12.4.4). $C(x, t)$ is a product of two functions of t. Hence, we apply the rule (9.2.11):

$$\frac{\partial C}{\partial t} = (t^{-\frac{1}{2}})' \exp(-x^2/4 Dt) + t^{-\frac{1}{2}} [\exp(-x^2/4 Dt)]'.$$

Then by means of formulas (9.2.4) and (9.2.14) we get

$$\frac{\partial C}{\partial t} = \left(-\frac{1}{2}\right) t^{-\frac{3}{2}} \exp\left(\frac{-x^2}{4 Dt}\right) + t^{-\frac{1}{2}} \left(+\frac{x^2}{4 Dt^2}\right) \exp\left(\frac{-x^2}{4 Dt}\right)$$

$$= \left(-\frac{1}{2} t^{-\frac{3}{2}} + \frac{x^2}{4 D} t^{-\frac{5}{2}}\right) \exp\left(\frac{-x^2}{4 Dt}\right).$$

Similarly we obtain

$$\frac{\partial C}{\partial x} = t^{-\frac{1}{2}}\left(\frac{-x}{2Dt}\right)\exp\left(\frac{-x^2}{4Dt}\right),$$

$$\frac{\partial^2 C}{\partial x^2} = t^{-\frac{1}{2}}\left(-\frac{1}{2Dt}\right)\exp\left(\frac{-x^2}{4Dt}\right) + t^{-\frac{1}{2}}\left(\frac{-x}{2Dt}\right)^2\exp\left(\frac{-x^2}{4Dt}\right).$$

Now, substitution of the derivatives in formula (12.4.4) reduces the verification to showing that

$$-\frac{1}{2}t^{-\frac{3}{2}} + \frac{x^2}{4D}t^{-\frac{5}{2}} = D\left(t^{-\frac{1}{2}}\left(-\frac{1}{2Dt}\right) + t^{-\frac{1}{2}}\frac{x^2}{4D^2t^2}\right).$$

By simple algebra we see that this equation holds for all values of t and x.

As formula (12.4.5) indicates, $C(x, t)$ is bell-shaped for every fixed t (cf. the formula for the normal distribution in Section 13.11). An approximate graph of $C(x, t)$ is shown in Fig. 12.5.

Even without an explicit solution, the Eq. (12.4.4) reveals some major properties of the solution. Consider Fig. 12.7a. There we assume that $C(x, t)$ is a linear function of x between two points with abscissas x_1 and x_2. This means that $\partial^2 C/\partial x^2 = 0$. The differential equation (12.4.4) then implies $\partial C/\partial t = 0$, that is, $C(x, t)$ does not change as a function of t. Hence, we have a *steady state flow* of molecules for all x in $[x_1, x_2]$.

If, however, $C(x, t)$ is convex downward as in Fig. 12.7b, we know from Section 9.6 that the second derivative $\partial^2 C/\partial x^2$ is positive. By virtue of Eq. (12.4.4) the rate of increase $\partial C/\partial t$ is also positive. Hence, $C(x, t)$ will increase as a function of t for all x in $[x_1, x_2]$. The result is quite plausible since Fig. 12.7b indicates a "trough" of concentration between x_1 and x_2.

Quite the opposite is true in Fig. 12.7c where we have a "peak" of concentration. The function $C(x, t)$ is convex upward. Hence, $\partial^2 C/\partial x^2$ and $\partial C/\partial t$ are both negative. The density $C(x, t)$ will decrease as a function of time for all x in $[x_1, x_2]$.

Within the dimensions of a living cell, mass transport by diffusion is quite efficient. It acts in seconds or at most within minutes. In a larger body, molecules cannot move fast enough by diffusion. Distribution of nutrients would be much too slow. In man it would take a lifetime to get sugar which was fed into the stomach to diffuse into the feet and hands (Went, 1968). Therefore, a faster system of distribution quite different from diffusion is required. This is *convection* by the blood stream.

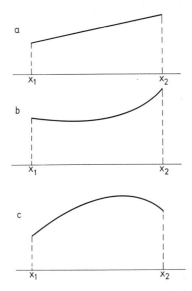

Fig. 12.7. Three different types of density function $C(x, t)$. In (a) the density is a linear function of x, in (b) the density is convex downward, and in (c) convex upward. The flow of molecules depends essentially on this shape

For more information on the mathematical aspect of diffusion in the life sciences see Bailey (1964), Beier (1962), Rashevsky (1960, Vol. 1), Rubinow (1973), Setlow and Pollard (1962), J. M. Smith (1968).

The importance of the differential equation (12.4.4) goes far beyond diffusion. In *ecology*, the *invasion of a large area by a new species* may be mathematically treated by the same method as the random motion of molecules (see Pielou, 1969, Chap. 11). Similarly, in *epidemics* the spread of an infectious disease over a large area follows the same pattern. An exchange of kinetic energy among neighboring molecules known as *heat conduction* is treated mathematically in the same way as diffusion.

More sophisticated applications of the diffusion equation are made in *population genetics* and *evolution* (see Crow and Kimura, 1970; Ewens, 1969; Feller, 1951; Goel and Richter-Dyn, 1974; Iosifescu and Tautu, 1973; Kimura, 1964, 1970; Kimura and Ohta, 1971; Moran, 1962).

Another biologically important partial differential equation is of the form

$$\frac{\partial^2 V}{\partial x^2} - a^2 \frac{\partial^2 V}{\partial t^2} + b \frac{\partial V}{\partial t} + CV = d. \tag{12.4.6}$$

This equation is known as the *cable equation* since it was used to solve a problem related to the first transatlantic telegraph cable. Here, $V = V(x, t)$ denotes the voltage at abscissa x of the cable and at time instant t. Today the cable equation is applied in the theory of *nerve conduction* (see Beier, 1962, and K. S. Cole, 1968).

Whenever random fluctuations are being considered in the treatment of biological processes, probability theory enters the scene. The proper mathematical method is the use of *stochastic processes*. They occur in theoretical work on population growth, on competition between species, on the spread of infection, etc. Among the mathematical tools required for the study of stochastic processes is a great variety of partial differential equations. Here we have to skip details. The reader is referred to Bailey (1964) and to Chiang (1968).

Recommended for further reading: Defares et al. (1973), Morowitz (1970), Rubinow (1973), C. A. B. Smith (1966).

Problems for Solution

12.1.1. Draw an isoline chart of the function

$$z = x^2 + y^2 + 1$$

for the interval $-3 \leq x \leq 3$, $-3 \leq y \leq 3$ and levels $z = 2, 4, 6, 8, 10$.

12.1.2. Draw an isoline chart of the function

$$z = 2x + y + 1$$

for the interval $0 \leq x \leq 8$, $0 \leq y \leq 5$ and levels $z = 2, 4, 6, ..., 20$.

12.1.3. Draw an isoline chart as well as a perspective view of the function

$$z = y/x$$

for the interval $1 \leq x \leq 4$, $0 \leq y \leq 3$.

12.1.4. Draw an isoline chart of the function

$$z = 10 - (x - 5)^2 - (y - 3)^2$$

for the levels $z = 0, 2, 4, 6, 8, 10$.
(Hint: The isolines are circles with common center)

12.2.1. Find the first order partial derivatives of the following functions:

a) $f(x, y) = ax - by$
b) $g(u, v) = u^2 - 2uv + v^2 + c$
c) $\Phi(s, t) = e^{s+t}$
d) $h(x, y, z) = \exp(ax + by + cz)$
e) $\Psi(\alpha, \beta) = \sin(3\alpha + \beta)$
f) $Q(v, w) = w \cdot \ln v$
g) $F(r, s) = (r - 2s)^{-1}$
h) $G(s, t) = at/(bs - ct)$

12.2.2. Find

a) $\dfrac{\partial}{\partial x}(x^3y - y^2)$, b) $\dfrac{\partial}{\partial y}(x^3y - y^2)$, c) $\dfrac{\partial}{\partial u}(u^2/v)$,

d) $\dfrac{\partial}{\partial v}(u^2/v)$, e) $\dfrac{\partial}{\partial \alpha}\sin(2\alpha - \beta)$, f) $\dfrac{\partial}{\partial \beta}\sin(2\alpha - \beta)$,

g) $\dfrac{\partial}{\partial s}\exp(as^2 - bt^2)$, h) $\dfrac{\partial}{\partial t}(as^2 - bt^2)$, i) $\dfrac{\partial}{\partial r}\ln(pr + qr^{-1})$.

12.2.3. Find the first and second order partial derivatives of the following functions:

a) $f(x,y) = ax^2 + bxy + cy^2$ b) $h(u,v) = (uv)^n$
c) $Q(v,w) = w \cdot \ln v$ d) $S(x,y,z) = x^2 + y^2 + z^2$
e) $\Phi(s,t) = se^{at}$ f) $\Psi(\alpha, \beta) = A\sin\alpha + B\sin\beta$.

12.2.4. Verify that for $f(x,y) = ax^n y^m$ the equality

$$\frac{\partial}{\partial x}\left(\frac{\partial f}{\partial y}\right) = \frac{\partial}{\partial y}\left(\frac{\partial f}{\partial x}\right)$$

holds.

12.2.5. For $Q = (x^2 + y^2)^{\frac{1}{2}}$ verify that

$$\frac{\partial^2 Q}{\partial x^2} + \frac{\partial^2 Q}{\partial y^2} = \frac{1}{Q}.$$

12.2.6. Given $Q = (x^2 + y^2 + z^2)^{\frac{1}{2}}$. Find

$$\frac{\partial^2 Q}{\partial x^2} + \frac{\partial^2 Q}{\partial y^2} + \frac{\partial^2 Q}{\partial z^2}.$$

12.3.1. Find the minimum of the function

$$z = 9x^2 - 6xy + 2y^2 - 6y + 11.$$

12.3.2. Find the minimum of the function

$$S(u,v) = u^2 - uv + v^2 - au.$$

12.3.3. Find an extremum of

$$z = \exp(15 + 4x - 3y + 5xy - x^2 - 2y^2).$$

12.3.4. Find an extremum of

$$Q(r,s) = \ln(2r^2 + 2s^2 - 3rs - 7s).$$

12.3.5. Consider the function $z = xy$. At $x = 0$, $y = 0$ the partial derivatives $\partial z/\partial x$ and $\partial z/\partial y$ vanish. Show that despite this fact the point $(0, 0, 0)$ is neither a maximum nor a minimum point of the function. Sketch a pictorial view of the surface. At $x = 0$, $y = 0$ the surface has a so-called *saddle point*.

12.3.6. Given the three points

$$x_1 = 1, \qquad y_1 = 6,$$
$$x_2 = 3, \qquad y_2 = 4,$$
$$x_3 = 5, \qquad y_3 = 3,$$

and using least squares, fit the best straight line.

12.3.7. In five cities of the U.S.A. a relation was found between the sulfur dust content of the air (in $\mu g/m^3$) and the number of female absentees in industry. Only absences of at least seven days were counted.

City	$\mu g/m^3$	Number of absences per 1000 employees
Cincinnati	7	19
Indianapolis	13	44
Woodbridge	14	53
Camden	17	61
Harrison	20	88

a) Plot the number of absentees versus the sulfur dust content.
b) Fit a linear regression line by least square technique. (adapted from Dohan, 1961).

12.3.8. In Section 12.3 the method of least squares was applied to fit a straight line. Let $\bar{x} = \Sigma x_i/m$ and $\bar{y} = \Sigma y_i/m$ be the means of x_i and y_i, respectively. The point $(\bar{x}, \bar{y})$ is known as *center of gravity*. Using formulas (12.3.11), show that the optimal straight line passes through the center of gravity.

12.3.9. When explaining the method of least squares, we have used deviations δ_i in the y-direction. In statistics it is customary to also employ deviations ε_i in the x-direction. Adjust a "best" straight line to the data of Problem 12.3.7. minimizing $\Sigma \varepsilon_i^2$. (Hint: Write the equation of the straight line in the form $x = py + q$).

Chapter 13

Probability

13.1. Introduction

Consider the process of sexual reproduction. Among the sperm of a male the cells differ in their genetic message, and the same is true for the reproductive cells in an ovary. As a consequence, the traits of descendants differ in many ways. As it depends on chance which of the reproductive cells combine to become a fertilized cell, the outcome cannot be predicted. Yet there is no chaos. If we count particular traits in a large number of descendants, we find some rules. For instance, we see that a trait known as "heterozygous" occurs in the descendants at a predictable ratio. The well-known rules by Mendel[1] can best be formulated by using probability theory.

The laws of inheritance were the first major application of probability in the life sciences. Today we know many more applications: occurrence of mutations, risk of disease, chance of survival, distribution and interaction of species, etc.

The most important application, however, is made in *statistics*. No observation and no experiment can be accurately planned and analyzed without some statistical method. Even if we keep the experimental conditions as constant as possible, repetition of an observation or an experiment hardly ever results in exactly the same outcome. There are always fluctuations. Therefore, all conclusions based on empirical data are necessarily inflicted with uncertainty. We try to express the degree of uncertainty in terms of probability. Thus, if an experimenter reports "significance at a five percent level", he admits the possibility of an erroneous statement, but at the same time he claims that the "probability" of an error is at most five percent. A basic knowledge of probability is required before statistical methods can be understood.

The only prerequisites for the study of Sections 13.2–13.8 are Chapters 1 and 2. Later sections, however, require some knowledge of functions and of calculus.

---

[1] Gregor Johann Mendel (1822-1884), Moravian botanist.

13.2. Events

The result of a single observation or the outcome of a measurement is generally called an *event*. When a migrating bird is seen to select a direction, say northeast, or when we read the length of a cell as 1.2 µm, we are in both cases observing an event.

Some events can be decomposed into "simple events". If we carefully watch a bird vanishing at the horizon, we are able to distinguish between the azimuths 0°, 10°, 15°, etc. These angles may then be called *simple events*, whereas "northeast" is said to be a *compound event*. It comprises the simple event 35°, 40°, 45°, 50°, 55°. Similarly, if in a health study all persons with systolic blood pressure above 160 mm are called sick, this is a compound event as compared with the individual measurements of blood pressure which are called simple events.

Two different simple events cannot occur at the same time. They exclude one another. Therefore, we call such events *mutually exclusive*. On the other hand, compound events can occur simultaneously. For instance, the events $x > 5$ and $x < 8$ are not exclusive if x can take on values such as 6 and 7. Similarly, in a study of animal behavior a dog may be observed as being awake and barking. The two events "awake" and "barking" are not mutually exclusive.

Now we consider a particular experiment and list all possible simple events. We call the set of these events the *outcome space*[2].

In terms of set theory, a simple event is a member of the outcome space. However, it is customary to identify a simple event also with a subset containing only one element. Thus in the outcome space $\{A, B, C, D\}$ we may either talk of the event A or of the event $\{A\}$[3]. Similarly, since a compound event comprises some simple events, we may *identify the compound event with a subset of the outcome space*. We may either say "the event A or B" or "the event $\{A, B\}$".

Example 13.2.1. In coin tossing we consider only two possible outcomes: head (H) and tail (T). They are exclusive events. Hence, the outcome space is the set $\{H, T\}$.

If we toss two distinct coins at the same time, we observe *ordered pairs* (H, H), (H, T), (T, H), (T, T) for which we shall also write HH, HT, TH, TT. The four events are mutually exclusive. Thus the out-

---

[2] The terminology "sample space" is more frequently used. However it is hoped that "outcome space" causes less confusion. With the word "space" we associate usually the concept of Euclidean space in two or three dimensions. But in mathematics the word "space" is often used in the general sense of "universe set".

[3] In Section 2.3 we emphasized that A and $\{A\}$ are logically not the same thing. However, when we call both an event, this will hardly cause any confusion.

come space is
$$\{HH, HT, TH, TT\}.$$

Example 13.2.2. In genetics, assume there are two alleles A and a to fill a certain gene locus. Then we know of only three possible outcomes: AA, Aa, aa. The pair Aa is not ordered, since it is not possible to distinguish between two different genotypes Aa and aA. In other words, it does not matter whether a or A came from father or mother. Hence, the outcome space is the set
$$\{AA, Aa, aa\}.$$

Example 13.2.3. The height H of adult human males may range from 120 cm to 250 cm. We are not sure, however, whether 120 cm is the smallest and 250 cm the largest value. On the other hand, to define a set it should be clear whether an element belongs to a set or not. To overcome the difficulty, we use the word "possible event" in a liberal way and include events that are conceptually, but not practically possible. We allow for H all positive values. Hence, the outcome space for the height of adult males is defined by
$$\{H \mid H > 0\}.$$

This space is an *infinite set* for two reasons. First, there is no upper bound for H. Second, we assume quite artificially that measurements could be performed with any degree of accuracy. This implies that every finite interval such as [150 cm, 200 cm] contains an infinite number of values. The example also serves to indicate that an outcome space can often be defined in different ways for the same type of observation or experiment.

Example 13.2.4. If $\{AA, Aa, aa\}$ is the outcome space for the two-allele genetic model, the event "homozygous" consists of the simple events AA and aa. Thus the event is a compound event and may be identified with the subset $\{AA, aa\}$. Using the symbol for "contained in" introduced in Section 2.3 we may write:
$$\{AA, aa\} \subset \{AA, Aa, aa\}.$$

When a die is thrown, the number of dots on the upper face is a simple event. There are six possible outcomes. Thus the outcome space is $\{1, 2, 3, 4, 5, 6\}$. The event "even number" is the subset $\{2, 4, 6\}$. Another subset $\{3, 4, 5, 6\}$ means the event of obtaining either 3, 4, 5 or 6, that is, a number which is at least 3.

To sum up, *simple and compound events are subsets of the outcome space.*

To get further insight we apply the *algebra of sets* introduced in Sections 2.5 through 2.7. As an illustrative example we use again a die and the compound events

$$E_1 = \{2, 4, 6\}, \quad E_2 = \{3, 4, 5, 6\}.$$

The *intersection* of the two sets is

$$E_1 \cap E_2 = \{4, 6\}.$$

It contains the outcomes 4 and 6. These two outcomes are common members to both sets. When the event 4 occurs, then E_1 and E_2 happen simultaneously. The same is true when 6 occurs. Conversely, when E_1 and E_2 happen simultaneously, the outcome must be either 4 or 6, that is, the event $\{4, 6\}$.

In general, *the intersection of two events E_1 and E_2 means another event which occurs when E_1 and E_2 happen simultaneously.* Briefly,

$$E_1 \cap E_2 \tag{13.2.1}$$

is the event "E_1 *and* E_2".

Notice the special case when E_1 and E_2 have no common member. Here

$$E_1 \cap E_2 = \emptyset \tag{13.2.2}$$

where $\emptyset$ denotes the empty set. The events E_1 and E_2 cannot occur simultaneously, and $\emptyset$ may be interpreted as the *impossible event.* We call E_1 and E_2 *mutually exclusive* events. For example, in the die throwing problem, $\{2, 4, 6\}$ and $\{1, 3\}$ are mutually exclusive events.

Consider now the *union* of $E_1 = \{2, 4, 6\}$ and $E_2 = \{3, 4, 5, 6\}$. We get

$$E_1 \cup E_2 = \{2, 3, 4, 5, 6\}.$$

The union contains all simple events except the number 1. When one of the events 2 through 6 occurs, either E_1 happens or E_2 happens or both of them happen. Let us use the word "or" in the weak sense of "and/or". Then, in general, *the union of E_1 and E_2 means that E_1 or E_2 occurs.* Briefly,

$$E_1 \cup E_2 \tag{13.2.3}$$

is the event "E_1 *or* E_2".

Readers who have become acquainted with Section 2.8 know already the correspondence of $\cap$ with "and" and of $\cup$ with "or".

Of some interest is the special case where the union of two events is equal to the outcome space. With the notation Ω (upper case Greek omega) for the outcome space, the relationship is

$$E_1 \cup E_2 = \Omega. \qquad (13.2.4)$$

This means that either E_1 or E_2 must occur at each trial of the experiment. Hence, Ω may be interpreted as the *certain event*. For instance, $E_1 = \{1, 3, 5\}$ and $E_2 = \{1, 2, 4, 6\}$ are two sets satisfying (13.2.4) if $\Omega = \{1, 2, 3, 4, 5, 6\}$ is the outcome space of a die. Two events that satisfy Eq. (13.2.4) are called *exhaustive*. They "exhaust" all possible outcomes. An even more special case is most important. Let Ω be again the outcome space of a die. We choose the event $E = \{1, 5\}$ and ask: Which event occurs whenever E does not occur? We call this event the *complementary* event and denote it by $\bar{E}$. In our example we find $\bar{E} = \{2, 3, 4, 6\}$. This concept corresponds to the complementary set introduced in Section 2.5. In general, a set E and its complementary set $\bar{E}$ satisfy the following conditions

$$E \cup \bar{E} = \Omega, \qquad E \cap \bar{E} = \emptyset, \qquad (13.2.5)$$

that is, E and $\bar{E}$ are *exhaustive and mutually exclusive*.

13.3. The Concept of Probability

There are experiments that can be repeated a great many times under fairly constant conditions. Such experiments are coin tossing, most physical experiments, and many experiments in genetics. They have a common property which we will describe by using an illustrative example.

Let us consider the occurrence of *male and female births*. If we disregard hermaphrodites, the outcome space is simply $\{\male, \female\}$. Let n be the total number of descendants under consideration and k be the number of male descendants. We call k the *absolute frequency* or simply the *frequency* of the event "male". Due to chance fluctuations, the frequency can be any number $0, 1, 2, \ldots, n$ that is,

$$0 \leq k \leq n. \qquad (13.3.1)$$

Particularly, when n is small, say $n = 3$, the frequency k could easily take on the extreme values 0 and 3. We only have to think of families with three children, all of them being girls or all of them boys.

To come closer to the concept of probability we introduce the *relative frequency* h by the formula

$$h = k/n. \qquad (13.3.2)$$

We may memorize this formula in the form

$$\text{relative frequency of an event} = \frac{\text{observed frequency of the event}}{\text{number of replications}}$$

Whereas k ranges from zero to n, the relative frequency h ranges from 0 to 1:

$$0 \leqq h \leqq 1. \tag{13.3.3}$$

The relative frequency is often expressed as percentage. It then ranges from 0% to 100%.

In our example, h means the relative frequency of male births. In a steadily growing survey the following numbers were obtained:

$n =$	10	100	1,000	10,000	100,000
$k =$	7	57	512	5,293	52,587
$h =$	0.7	0.57	0.51	0.529	0.526

The figures in the last row are suitably rounded off. As n tends to infinity, the relative frequency seems to approach a certain limit. We call this empirical property the *stability* of the relative frequency. This property, however, is not well defined. Notice that the way in which h approaches a limit is quite irregular.

We may theorize and assume that there exists a fixed number, say p, which is approached by h in a long run of observations. The hypothetical number p is known as the *probability* of the event under consideration, in our case the probability of a male birth. We do not know the exact numerical value of p, but our observations indicate that p is close to 0.53 or 53%. From other surveys it is well-known that the sex ratio is not exactly $1:1$, but that male births are slightly more frequent. It is also known that the ratio changes slightly from one region to another.

In statistics the relative frequency of an event is used to estimate the probability of this event. There are rules which demonstrate how reliable such an estimate is. The reader is referred to books on statistics.

Let E be an event from an outcome space. Then the probability of E is a number associated with E and denoted by

$$P(E).$$

Notice that $P(E)$ is a function on the events of an outcome space Ω.

Example 13.3.1. Living tissue is exposed to X-rays in order to produce mutation. With tissue of the same sort and a constant dose of radiation, the experiment can be repeated many times under almost constant conditions. The relative frequency of observed mutations is called the *mutation rate*. This rate varies as the experiment continues, but it remains stable in that it approaches a certain value. We assume that there is some kind of limit which we then call the *probability of a mutation*.

Example 13.3.2. In some cases we are able to predict the value of a probability. In coin tossing, for instance, we see no reason why either head (H) or tail (T) should occur more frequently. Hence, we suppose that both events, head and tail, have probability 1/2. There is no mathematical proof possible, but experience indicates that our assumption is correct. In mathematical symbols,

$$P(H) = \frac{1}{2}, \quad P(T) = \frac{1}{2}.$$

Example 13.3.3. There are numerous examples in *genetics* where probabilities can be predicted. If we mate individuals of genotype AA and Aa, the gene A of an AA individual meets either the gene A or the gene a of an Aa individual. The fertilized cells are of genotype AA and Aa, that is, the outcome space is $\{AA, Aa\}$. Since we have no reason to assume that one of the two events occurs more frequently than the other one, we predict the probability of either event to be 1/2. We may write

$$P(AA) = \frac{1}{2}, \quad P(Aa) = \frac{1}{2}.$$

The assumption is supported by experimental facts.

Example 13.3.4. Let the outcome space consist of all integers from 0 to 99, that is,

$$\Omega = \{0, 1, 2, ..., 99\}.$$

Assume that we are given a mechanism which produces a practically infinite sequence of events from Ω. Assume further that each of the hundred numbers is given the same chance to occur at each stage. Such a sequence may run like

$$37, 95, 11, 18, 48, 07, 22, 65, 23, 99, 50, 11, 80, ...$$

In a long sequence each event has approximately the same relative frequency of 1/100. We may idealize and postulate that all of the hundred events are *equally probable*, that is,

$$P(0) = P(1) = P(2) = \cdots = P(99) = \frac{1}{100}.$$

A list of numbers based on the assumption of equal probability is called a *table of random numbers*. A small table of two-digit random numbers is given in the Appendix (Table K). Random numbers are very useful in a variety of applications.

Example 13.3.5. We consider a population consisting of a finite number of individuals, such as a human population, an animal or plant population, or a set of individual objects. For many purposes we have often to make a "random selection" of one or more out of N individuals. Each individual should have the same probability of being chosen. This requirement can hardly be met without employing the following procedure:

All N individuals are labeled $1, 2, 3, ..., N$ in an arbitrary order. Then a table of random numbers in consulted. If N has k digits, we combine columns so as to obtain random k-digit numbers. We simply discard numbers which do not occur in the sequence $1, ..., N$. Then the very first admissible number (or any such number chosen without inspection of its value) gives us a random selection of an individual. For instance, let $N = 38$. In Table K, the very first number satisfying the requirement is 12 (first column, fifth row). Hence, our random selection is individual labeled 12.

If one out of N individuals is chosen this way, we say that it is *randomly selected*. The procedure is called *random selection* or *random sampling*. Each individual has the same probability $1/N$ of being selected.

If a second individual is being selected, we have to take in account that the population has decreased to $N - 1$ individuals by the first selection. Therefore, the probability of a randomly selected second individual is $1/(N - 1)$. A corresponding remark is appropriate for a third, fourth, ... individual. This procedure is called *selection without replacement*. Less frequently, selection is performed by replacing each selected individual. In this case of *selection with replacement* the probability of an individual being selected remains $1/N$.

To what extent probability is applicable in the life sciences can be decided neither by mathematical methods nor by experimental facts. It is today and will perhaps remain a matter of controversy. There are scientists who go far beyond the frequency interpretation of probability. They apply probability for the degree of belief. To an event which is unlikely, but not impossible, they assign a probability close to zero, say 0.01, and to a nearly certain event a value close to 1, say 0.99. There are also rules for assigning probabilities. These scientists think that questions such as the following are meaningful: "What is the probability of life on Mars?" or "What is the probability that the cancer problem can be solved within the next decade?"

13.4. The Axioms of Probability Theory

In Section 13.2 we introduced the concept of an outcome space. It is the finite or infinite set of all possible outcomes or simple events. Let Ω denote such an outcome space, and let E_i $(i = 1, 2, ...)$ be some events belonging to Ω. Each event can be interpreted as a subset of Ω. Thus

$$E_i \subset \Omega \quad (i = 1, 2, ...). \tag{13.4.1}$$

Following Section 13.3 we associate a probability p_i with each event that belongs to Ω. Hence we may write

$$P(E_i) = p_i \quad (i = 1, 2, ...). \tag{13.4.2}$$

Each probability is an idealization of a relative frequency. Therefore, with regard to formula (13.3.3) we have to postulate that

$$0 \leq p_i \leq 1 \quad (i = 1, 2, ...). \tag{13.4.3}$$

This is the content of our first axiom.

Axiom 1. With each event belonging to an outcome space there is associated a number, called the probability of the event. This number is restricted to the interval from 0 to 1[4].

Special cases are the impossible event which never occurs and the certain event which always occurs. The relative frequencies for the two cases are $h = 0$ and $h = 1$, respectively. We postulate that the corresponding probabilities are also 0 and 1. The impossible event is characterized by the empty set $\emptyset$ and the certain event by the outcome space Ω. Therefore, we get the following axiom:

Axiom 2. With the impossible event there is associated the probability 0 and with the certain event the probability 1. In symbols

$$P(\emptyset) = 0, \quad P(\Omega) = 1. \tag{13.4.4}$$

Let E_1 and E_2 be two events belonging to an outcome space Ω and assume that they are mutually exclusive, that is, events which cannot happen simultaneously. In set theoretical notation the assumption is stated in formula (13.2.2). In n performances of the experiment we observe the frequency with which E_1 and E_2 occur. Let k_1 be the frequency of E_1 and k_2 be the frequency of E_2. The corresponding relative frequencies are

$$h_1 = k_1/n, \quad h_2 = k_2/n. \tag{13.4.5}$$

Now we ask how many times either E_1 or E_2 occurred. The total frequency is $k_1 + k_2$ since E_1 and E_2 cannot happen simultaneously.

---

[4] An axiom is a well defined basic rule which can be justified, but not proven.

With "E_1 or E_2" we define a new event. As stated in (13.2.3) this event is $E_1 \cup E_2$. We conclude that the relative frequency of $E_1 \cup E_2$ is $h = (k_1 + k_2)/n$. This implies

$$h = h_1 + h_2. \tag{13.4.6}$$

Since probabilities are idealized relative frequencies, we postulate a formula corresponding to (13.4.6). This is expressed by the following axiom.

Axiom 3. Let E_1 and E_2 be two mutually exclusive events belonging to an outcome space Ω. Let $p_1 = P(E_1)$, $p_2 = P(E_2)$, and $p = P(E_1 \cup E_2)$. Then

$$p = p_1 + p_2. \tag{13.4.7}$$

Axiom 3 is also called the *addition rule*. The Venn diagram of Fig. 13.1 depicts the situation of Axiom 3.

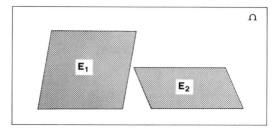

Fig. 13.1. Two mutually exclusive events are represented by two not overlapping point sets. For such events Axiom 3 holds

Example 13.4.1. We call a die balanced if each of the six events $\{1\}, \{2\}, ..., \{6\}$ has probability 1/6. The compound event $\{1, 2\}$ is the union of the two mutually exclusive events $\{1\}$ and $\{2\}$. Hence

$$P(\{1, 2\}) = \frac{1}{6} + \frac{1}{6} = \frac{1}{3}.$$

Example 13.4.2. The assumption that E_1 and E_2 are mutually exclusive is most important as shown by the following counter-example. When a die is thrown, the two events $E_1 = \{1, 2\}$ and $E_2 = \{2, 3\}$ are not mutually exclusive since 2 is a common outcome. The new event $E_1 \cup E_2$ is $\{1, 2, 3\}$. For a balanced die we get $p_1 = P(E_1) = 1/3$, $p_2 = P(E_2) = 1/3$, and $p = P(E_1 \cup E_2) = 1/2$. But $p \neq p_1 + p_2$.

Example 13.4.3. In a problem of genetics we assume that there are only two different alleles A and a at a certain locus. We mate the genotype Aa and Aa according to the following rule:

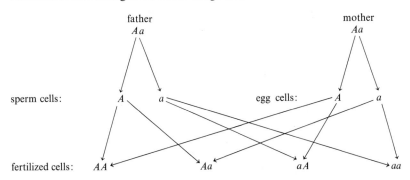

One of the genetic rules states that the four recombinations AA, Aa, aA, aa are equally probable. Therefore, we assign to each of them the probability $1/4$. However, the two recombinations Aa and aA cannot be distinguished biologically. Thus we ask: What is the probability of the compound event "Aa or aA" for which we simply write Aa? Since the two outcomes are mutually exclusive, we get

$$P(Aa) = \frac{1}{4} + \frac{1}{4} = \frac{1}{2}.$$

The result of the cross $Aa \times Aa$ may then be summarized in the form

$$P(AA) = \frac{1}{4} = 0.25,$$

$$P(Aa) = \frac{1}{2} = 0.50, \qquad (13.4.8)$$

$$P(aa) = \frac{1}{4} = 0.25.$$

Returning to the theory, let E and $\bar{E}$ be two complementary events. They satisfy the conditions (13.2.5). Applying Axiom 3 we get

$$P(E \cup \bar{E}) = P(E) + P(\bar{E}).$$

But the event $E \cup \bar{E}$ is identical with Ω and $P(\Omega) = 1$ according to Axiom 2. Hence,

$$P(E) + P(\bar{E}) = 1.$$

Customarily $P(E)$ is denoted by p, and $P(\bar{E})$ by q. Thus

$$p + q = 1. \qquad (13.4.9)$$

The property can easily be generalized to a number, say m, of mutually exclusive and exhaustive events $E_1, E_2, ..., E_m$. The result is stated in the following proposition:

Let $E_1, E_2, ..., E_m$ be m mutually exclusive and exhaustive events of an outcome space Ω with probabilities $p_1, p_2, ..., p_m$, respectively. Then

$$p_1 + p_2 + \cdots + p_m = 1 . \tag{13.4.10}$$

Sometimes the events $E_1, E_2, ..., E_m$ are said to form a *partition* of the outcome space. Figure 13.2 gives a Venn diagram of such a partition.

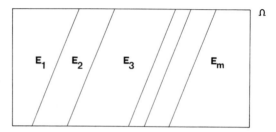

Fig. 13.2. Partition of an outcome space Ω into m mutually exclusive events $E_1, E_2, ..., E_m$

Example 13.4.4. We measure the height H of adult male persons. In Section 13.2 we introduced the sample space $\{H \mid H > 0\}$. For most practical purposes we subdivide the sample space into so-called *groups*. Such a subdivision may be

$$E_1 = \{H \mid H < 150 \text{ cm}\} ,$$
$$E_2 = \{H \mid 150 \text{ cm} \leq H < 160 \text{ cm}\} ,$$
$$E_3 = \{H \mid 160 \text{ cm} \leq H < 170 \text{ cm}\} ,$$
$$\cdots\cdots\cdots\cdots\cdots\cdots\cdots\cdots\cdots\cdots$$
$$E_6 = \{H \mid 190 \text{ cm} \leq H < 200 \text{ cm}\} ,$$
$$E_7 = \{H \mid H \geq 200 \text{ cm}\} .$$

The seven events or groups are mutually exclusive and exhaustive. Hence

$$P(E_1) + P(E_2) + \cdots + P(E_7) = 1 .$$

13.5. Conditional Probabilities

We will introduce the concept of conditional probability by two illustrative examples.

Example 13.5.1. Consider the distribution of genotypes AA, Aa, aa in a plant or an animal population. Assume that individuals are selected at random. This means that each individual has the same chance of being selected. The outcome space is $\{AA, Aa, aa\}$. Each of the three events has its own probability. For instance, in a population of 500 individuals, 180 are of genotype AA, 240 of genotype Aa, and 80 of genotype aa. Then the probability of selecting at random an AA individual is $180/500 = 0.36$, an Aa individual $240/500 = 0.48$, and an aa individual $80/500 = 0.16$. Briefly

$$P(AA) = 0.36, \quad P(Aa) = 0.48, \quad P(aa) = 0.16. \quad (13.5.1)$$

The three probabilities add up to 1 according to formula (13.4.10).

Let us make a further assumption: The distribution (13.5.1) is true only for young individuals. From a certain age on, the gene a causes death. If A is dominant over a, the AA and Aa individuals remain healthy, whereas the aa individuals die. The question now arises: What are the probabilities of AA and Aa individuals after the removal of all aa individuals? We can no longer say that the frequency of AA individuals is 36% and the frequency of Aa individuals 48% since the sum is less than 100%. We are forced to make an adjustment. Consider the *relative occurrence* expressed by the ratio

$$(0.36) : (0.48).$$

It can be reduced to $3 : 4$. We have to find two probabilities satisfying this ratio which add up to one. The adjustment is simply made by dividing 3 and 4 by their sum 7. Thus $P(AA) = 3/7$ and $P(Aa) = 4/7$. The same operation could be performed with the original decimal fractions. Thus the adjusted probabilities for the events AA and Aa turn out to be

$$\frac{0.36}{0.36 + 0.48} \approx 0.43, \quad \frac{0.48}{0.36 + 0.48} \approx 0.57, \quad (13.5.2)$$

respectively. The adjusted values are called *conditional probabilities*. Together with the original probabilities they are shown in Fig. 13.3.

The word "conditional" requires an explanation. In our example the condition is that only AA and Aa individuals survive. This condition is characterized by the compound event

$$E = \{AA, Aa\}.$$

The set E may be considered as a *new outcome space*. Within this space we define new probabilities which are denoted by $P(AA|E)$ and $P(Aa|E)$. They are called *conditional probabilities*. The vertical bar is used in a similar way as in set theory (see Section 2.4). We may read the bar "Under the condition that" or briefly "given that". With the new notation in mind,

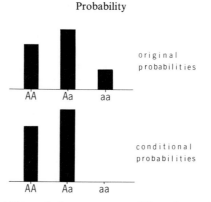

Fig. 13.3. The probabilities of three genotypes. When the *aa* individuals die, the probabilities have to be adjusted

the adjustment which led to (13.5.2) can be written in the form

$$P(AA|E) = \frac{P(AA)}{P(E)}, \quad P(Aa|E) = \frac{P(Aa)}{P(E)}. \tag{13.5.3}$$

The probabilities on the right side, $P(AA)$, $P(Aa)$, $P(E)$, belong to the original outcome space $\{AA, Aa, aa\}$, whereas $P(AA|E)$ and $P(Aa|E)$ refer to the reduced outcome space $\{AA, Aa\}$.

Example 13.5.2. Conditional probabilities occur also in a slightly more general situation. Consider a loaded die with six faces numbered $1, 2, \ldots, 6$. The outcome space is $\{1, 2, 3, 4, 5, 6\}$, but the probabilities of the six outcomes are not equal to $1/6$. We denote them by $p_1, p_2, \ldots, p_6$, respectively. We introduce the following two compound events:

$$A = \{2, 3, 4, 5\}, \quad B = \{4, 5, 6\}. \tag{13.5.4}$$

Event A means: Upper face shows 2 or 3 or 4 or 5. Event B means: Upper face shows 4 or 5 or 6. The two events are partly "overlapping", that is, the intersection is $A \cap B = \{4, 5\}$. The probabilities of the events are

$$P(A) = p_2 + p_3 + p_4 + p_5, \quad P(B) = p_4 + p_5 + p_6,$$
$$P(A \cap B) = p_4 + p_5. \tag{13.5.5}$$

Now we think of a gambler who is interested in a high score, say in the event B. However, when the die is thrown he is given only partial knowledge: He is told that event A has occurred. His question is: What is the probability of event B given that A has occurred? The problem is to find $P(B|A)$.

Since A has occurred, the adjusted or conditional probabilities of the special outcomes 1 and 6 are zero. Hence, in order to obtain the probabilities for the outcomes 2, 3, 4, 5 we have to divide p_2, p_3, p_4, p_5 by their sum $p_2 + p_3 + p_4 + p_5 = P(A)$. The individual probabilities are

$$\frac{p_2}{P(A)}, \quad \frac{p_3}{P(A)}, \quad \frac{p_4}{P(A)}, \quad \frac{p_5}{P(A)}.$$

Their sum is $(p_2 + p_3 + p_4 + p_5)/P(A) = 1$.

The gambler is only interested in the event $\{4, 5\} = A \cap B$. Therefore

$$P(B|A) = \frac{p_4 + p_5}{P(A)} = \frac{P(A \cap B)}{P(A)}.$$

The result of Example 13.5.2. is applicable in a large variety of problems:

Given an outcome space, let A and B be any two of its events. Then the probability of event B, given that event A has occurred, is the probability of simultaneous occurrence of A and B, divided by the probability of A (provided that the denominator does not vanish).

$$\boxed{P(B|A) = \frac{P(A \cap B)}{P(A)}.}$$
(13.5.6)

It is essential not to confuse $P(B|A)$ with $P(A \cap B)$. The probability $P(A \cap B)$ refers to the original outcome space Ω, whereas $P(B|A)$ is a probability defined in the reduced outcome space A. We depict this relationship by a Venn diagram (Fig. 13.4.a).

Assume that B_1, B_2, B_3, B_4 form a *partition* of Ω as shown in Fig. 13.4b and that A is any event of Ω with $P(A) \neq 0$. By Eq. (13.5.6) we may calculate the conditional probabilities $P(B_i|A)$, $i = 1, 2, 3, 4$. Since they are all probabilities in the reduced outcome space A, and since B_1, B_2, B_3, B_4 are exhaustive, the four conditional probabilities should add up to 1. Indeed,

$$P(B_1|A) + P(B_2|A) + P(B_3|A) + P(B_4|A)$$
$$= \frac{P(A \cap B_1)}{P(A)} + \frac{P(A \cap B_2)}{P(A)} + \frac{P(A \cap B_3)}{P(A)} + \frac{P(A \cap B_4)}{P(A)}$$
$$= \frac{P(A \cap \Omega)}{P(A)} = \frac{P(A)}{P(A)} = 1.$$

Example 13.5.3. Consider the *probability of death* in our society. Table 13.1 contains some of the pertinent information. The outcome

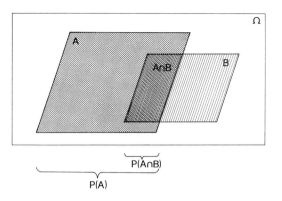

Fig. 13.4 a. The conditional probability $P(B|A)$ is related to the new outcome space A whereas $P(A \cap B)$ belongs to the original outcome space Ω

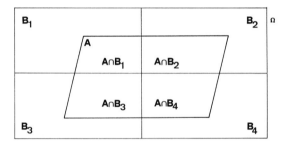

Fig. 13.4 b. Venn diagram illustrating the intersections of a set A (event A) with four not overlapping sets B_1, B_2, B_3, B_4 (mutually exclusive events)

space consists of the mutually exclusive and exhaustive events "death in the first decade" (between birth and the tenth birthday), "death in the second decade", etc. The last event is "death after the 80th birthday". The probabilities are estimated values. Here we are not concerned with the problem of estimating the probabilities from observed data. Rather we take the probabilities for granted.

What is the probability that a person who is now 20 years old will die before he reaches his 30th birthday? To answer the question we cannot simply take the death rate for the third decade (1.21%) from the table. Instead we have to find a conditional probability. We know that the person has already survived the first two decades. Therefore, we have to adjust the probabilities of our outcome space.

Table 13.1. *Probability of death at different ages for the United States*[5]

Age (years)	Probability of death (%)
0–10	3.23
10–20	0.65
20–30	1.21
30–40	1.84
40–50	4.31
50–60	9.69
60–70	18.21
70–80	27.28
80 over	33.58
Total	100.00

The reduced outcome space consists of the event "death after the second decade" which we may interpret as event A in Eq. (13.5.6). We get

$$P(A) = 1.21 + 1.84 + \cdots + 33.58 = 96.12 .$$

If B denotes the event "death before the fourth decade", then $B|A$ is the event of our question. Thus

$$P(B|A) = \frac{P(A \cap B)}{P(A)} = \frac{1.21}{96.12} = 0.0126$$

or 1.26%.

Example 13.5.4. It is well known that *color blindness* is inheritable. Due to the fact that the responsible gene is sex-linked, color blindness occurs more frequently in males than in females.

In a large human population the incidence of red-green color blindness was counted. The relative frequencies are listed in the following table.

Table 13.2. *The incidence of red-green color blindness in a human population*

	Male	Female	Total
color-blind	4.23%	0.65%	4.88%
normal	48.48%	46.64%	95.12%
Total	52.71%	47.29%	100.00%

---

[5] Adapted from "United States Life Tables by Causes of Death: 1959-1961". Vol. 1, 1967, No. 6, p. 15. National Center for Health Statistics, Public Health Service, Washington, D. C.

Disregarding statistical problems, we assume that the relative frequencies in Table 13.2 are so accurate that they can be used as probabilities.

The outcome space consists of the four simple events

> color-blind male,
> normal male,
> color-blind female,
> normal female.

We may obtain these events by cross multiplication (Section 3.2). Let M and F be the events "male" and "female", respectively, and let C and N be the events "color-blind" and "normal", respectively. Then we form the *product set*

$$\{C, N\} \times \{M, F\} = \{CM, CF, NM, NF\}.$$

This is our outcome space. The corresponding probabilities satisfy the axioms of Section 13.4. Indeed, they are numbers between 0% and 100% and add up to 100%.

We may also consider *compound events*. The event "colorblind" is such a compound event:

$$C = \{CM, CF\}.$$

Its probability is 4.88%. Similarly, the event "male" irrespective of any other trait,

$$M = \{CM, NM\},$$

is a compound event. Its probability is 52.71%.

Now we may ask: What is the incidence rate of color blindness for the subpopulation of males? This question leads to conditional probabilities. Formula (13.5.6) states that

$$P(C|M) = \frac{P(C \cap M)}{P(M)}.$$

Using the numerical values of Table 13.2, we obtain $P(C \cap M) = 4.23\%$ and $P(M) = 52.71\%$. Hence $P(C|M) = 4.23/52.71 = 0.0803$ or 8.03%. Thus in males color blindness occurs at a rate of 8.03%.

Similarly we get for the incidence rate of the subpopulation of females

$$P(C|F) = \frac{P(C \cap F)}{P(F)} = \frac{0.65}{47.29} = 0.0137$$

or 1.37%.

13.6. The Multiplication Rule

Assume that A and B are two events that belong to the same outcome space. Then it follows from formula (13.5.6) that

$$P(A \cap B) = P(A) \cdot P(B \mid A). \qquad (13.6.1)$$

In words: *The probability of the simultaneous occurrence of two events A and B is the product of the probability of event A and the conditional probability of event B given A.*

By symmetry, A and B may be interchanged. Therefore, we also have

$$P(A \cap B) = P(B) \cdot P(A \mid B). \qquad (13.6.2)$$

Example 13.6.1. Given that a man celebrates his 70th birthday. What is the probability that he will reach an age of 72? In a life table we find the following conditional probabilities of survival for men:

age x	p_x
70	0.9492
71	0.9444
72	0.9391

Here p_{70} is the probability that a 70 year old man lives until his 71st birthday, p_{71} is the probability that a 71 year old man lives until his 72nd birthday, etc. Applying formula (13.6.1) we conclude that the probability of reaching the 71st as well as the 72nd birthday is

$$0.9492 \times 0.9444 = 0.8964$$

or 89.64%.

A special case of formula (13.6.1) is most important. It may happen that the occurrence of an event A does not influence the outcome of a second event B. This would mean that

$$P(B \mid A) = P(B). \qquad (13.6.3)$$

Whenever (13.6.3) holds, we call the event B independent of the event A. The relationship is symmetric as a comparison of formulas (13.6.1) and (13.6.2) reveals; that is, if B is independent of A, then A is independent of B. They are *independent of each other*. If (13.6.3) is not valid, the event B is said to depend on the event A.

Example 13.6.2. Let us return to color blindness as treated in Example 13.5.4. We obtained

$$P(C \mid M) = 8.03\%, \qquad P(C) = 4.88\%.$$

Hence, $P(C \mid M) \neq P(C)$, that is, color blindness is dependent on sex.

Example 13.6.3. To illustrate independence, we imagine a scientist who wants to know whether there is any dependence between color blindness and deafness in human males. Assume that he is given the following probabilities:

	Deaf	Not deaf	Total
color-blind	0.0004	0.0796	0.0800
not color-blind	0.0046	0.9154	0.9200
Total	0.0050	0.9950	1.0000

Let C be the compound event "color-blind" and D be the compound event "deaf". With $\bar{C}$ and with $\bar{D}$ we denote the complementary events "not color-blind" and "not deaf", respectively. Then from (13.5.6) we get the following conditional probabilities:

$$P(D|C) = \frac{0.0004}{0.0800} = 0.0050 , \qquad P(\bar{D}|C) = \frac{0.0796}{0.0800} = 0.9950 .$$

From the last row of the table we read

$$P(D) = 0.0050 , \qquad P(\bar{D}) = 0.9950 .$$

Hence

$$P(D|C) = P(D) , \qquad P(\bar{D}|C) = P(\bar{D}) ,$$

that is, deafness is independent of color blindness.

In general, whenever (13.6.3) holds, it follows from (13.6.2) that

$$P(A \cap B) = P(A) \cdot P(B) , \qquad (13.6.4)$$

and conversely. This is the *multiplication rule for independent events*. Independence is often defined in terms of the multiplication rule:

Definition. Two events A and B of the same outcome space are said to be independent if the multiplication rule (13.6.4) *holds.*

In Section 13.4 we introduced the addition rule (Axiom 3). There is some formal similarity between the addition and the multiplication rules. However, they should be carefully distinguished from each other. If

$$P(A \cup B) = P(A) + P(B)$$

is to be valid, the two events A and B must be *mutually exclusive*, that is, $A \cap B = \emptyset$. Hence, $P(A \cap B) = 0$. However, if $P(A) \neq 0$, $P(B) \neq 0$, the events A and B cannot be independent since formula (13.6.4) would not be satisfied. The reader should carefully distinguish between the two properties "independent" and "mutually exclusive". These properties occur in different contexts.

The multiplication rule is frequently applied when the same experiment is performed more than once.

Example 13.6.4. An experiment may consist in coin tossing. If H denotes the event "head" and T the event "tail", the outcome space for a single trial is $\{H, T\}$ and for two trials of the experiment

$$\{H, T\} \times \{H, T\} = \{HH, HT, TH, TT\}.$$

The pairs HT and TH are ordered which means that we carefully distinguish between the outcomes of the first and the second trial. Now we introduce the compound event "head at the first trial" and denote it by H_1. Since the coin is assumed to be balanced, we get

$$P(H_1) = \frac{1}{2}.$$

Just for clarity we notice that $H_1 = \{HH, HT\}$. We consider a second compound event "tail at the second trial" and denote it by T_2. We perform the two trials in such a way that the outcome of the second trial is not affected by the first trial. Therefore, we assume that the two events H_1 and T_2 are independent and that

$$P(T_2) = \frac{1}{2}.$$

Applying the multiplication rule we get

$$P(H_1 \cap T_2) = P(H_1) \cdot P(T_2) = \frac{1}{2} \cdot \frac{1}{2} = \frac{1}{4}.$$

Now the event $H_1 \cap T_2$ is identical with HT. Hence,

$$P(HT) = \frac{1}{4}, \tag{13.6.5}$$

a result that we expect intuitively long before we get used to concepts such as compound event, intersection, and independence. In the same way one proves that $P(HH) = P(TH) = P(TT) = 1/4$.

The experiment just described could be performed in a slightly different way. We could toss two distinct coins at the same time and read heads and tails. The outcome space would again be $\{HH, HT, TH, TT\}$. When tossing the two coins we have to make sure that they do not "stick" together or do not attract each other, say by magnetic forces. Otherwise, independence cannot be expected. The multiplication rule leads to the same results as before when we tossed the same coin twice.

Example 13.6.5. Assume that two songbirds (not necessarily of the same species) are sitting on the same tree. During a full hour their song was recorded and duration of songs and breaks measured. The result is presented in the following table:

	Duration		total
	of song	of break	
bird B_1	715 sec	2885 sec	3600 sec
bird B_2	609 sec	2991 sec	3600 sec

When interested in a possible interaction of the two birds we may ask: What are the probabilities that at a given, randomly selected time instant

a) bird B_1 is singing,
b) bird B_2 is singing,
c) birds B_1 and B_2 are singing simultaneously?

From the table it follows that $P(B_1 \text{ singing}) = \dfrac{715}{3600} = 0.199$, $P(B_2 \text{ singing}) = \dfrac{609}{3600} = 0.169$. Question c) can only be answered under some assumption. Let us tentatively assume that the birds sing independently of each other. Eq. (13.6.4) would then yield

$$P(B_1 \text{ and } B_2 \text{ sing.}) = P(B_1 \text{ sing.}) \cdot P(B_2 \text{ sing.})$$
$$= 0.199 \times 0.169 = 0.0336 .$$

In 3.36% of the time the birds would be singing simultaneously. Experience shows, however, that songs overlap only sporadically. We conclude that there must be some dependence between the two events "B_1 singing" and "B_2 singing". The birds try to avoid acoustic interference (cf. Ficken and Hailman, 1974).

The definition of independence can be generalized for more than two events. For instance, three events A, B, C belonging to the same outcome space are said to be independent if

$$P(A \cap B \cap C) = P(A) \cdot P(B) \cdot P(C) . \tag{13.6.6}$$

For an immediate application we watch a gambler. Assume he is tossing a coin over and over again and that he observed the outcome "head" in each of ten consecutive trials. He is very much puzzled since such an event has probability $1/2^{10} = 1/1024$ only. Will the eleventh outcome be head or tail? Which outcome has greater chances? There are surprisingly many people who believe that after a long run of heads the opposite outcome must have a better chance. They expect some kind of "justice" in games of chance. They are likely to bet a large sum and to risk their money. Are these people right? Experience does not support such a belief. If the coin is normal, that is, neither double-headed nor unbalanced, and if no secret mechanism interferes during the motion, the chances for head and tail remain fifty percent. Formula (13.6.6) is then in agreement with our experience, and independence is established.

Scientific applications of Eq. (13.6.6) will be studied in Section 13.8.

13.7. Counting

Many problems in probability require some methods of counting.

Example 13.7.1. How many five-letter words, meaningful and meaningless, can be written with the 26 characters of the alphabet? For the first letter of the word we have a choice among 26 different characters. For the second letter we have the same choice. Combining the choices for the first two letters we get $26 \times 26 = 26^2$ possibilities. For the third letter we have again a choice among 26 characters. Hence, the total number of three-letter words is $26^2 \times 26 = 26^3$. The same argument finally leads to 26^5 different five-letter words.

In general, assume that we are given n sorts of objects, and that an unlimited number of objects of each sort is available. In how many different ways can we fill k distinct spaces, each space with one object? This problem can be solved in much the same way as the preceding example. The result is

$$\underbrace{n \times n \times \cdots \times n}_{k \text{ times}} = n^k . \tag{13.7.1}$$

Example 13.7.2. Each letter of the Morse alphabet consists of two kinds of symbol, dots and dashes. How many different letters can be composed by four such symbols? In this example we have four spaces to fill and two sorts of objects with unlimited supply. It follows from (13.7.1) that we can compose $2^4 = 16$ letters.

Example 13.7.3. There are different kinds of inherited anemia such as spherocytosis, thalassemia, sickle-cell anemia, ovalocytosis, and Fanconi's syndrome. It is believed that abnormal alleles at five different gene loci are responsible (Neel and Schull, 1958, p. 13). Denote the normal alleles at these five loci by A, B, C, D, E and the corresponding abnormal alleles by a, b, c, d, e. At each locus we can distinguish between three genotypes (at the first locus AA, Aa, aa, at the second locus $BB, Bb,$ bb, etc.). The total number of all genetic arrangements is therefore $3^5 = 243$. To relate this example to the general theory, we remark that the loci are the spaces and the three different gene combinations "both normal", "one normal-one abnormal", "both abnormal" are the objects.

In the problems just discussed, each object is available in unlimited number. We speak of *unlimited repetition*.

We draw now our attention to problems of counting arrangements *without repetition*.

Example 13.7.4. Consider five cars which are competing for only three parking spaces. In how many ways can the cars be parked? Let us number the three spaces by $1, 2, 3$ in an arbitrary order. The first space can be taken by anyone of the five cars. For the second space we have only four possibilities since one car is already parked in the first space. We have to combine the 5 possibilities of occupying the first space with the 4 possibilities of occupying the second space. Thus we obtain 5×4 possibilities for the first two spaces together. To park a car on the third space, we have only a choice among the three remaining cars. Combining the three possibilities with the 5×4 previous possibilities we get

$$5 \times 4 \times 3 = 60$$

different ways of parking. Two cars remain without parking space.

In general, assume that we are given k distinct spaces and n different objects without repetition. In order that each space can be filled with exactly one object we have to assume that

$$n \geq k. \tag{13.7.2}$$

Now we number the k spaces in an arbitrary manner. The first space can be filled in n different ways. For the second space only $n-1$ objects are available. Hence, the first two spaces can be filled in

$$n(n-1)$$

ways. For the third space we have a choice among the $n-2$ remaining objects. For each consecutive space the number of available objects decreases by 1. For the kth space, the choice is among

$$n-k+1 \tag{13.7.3}$$

objects. The total number of possibilities to fill the k spaces with n objects is therefore $n(n-1)(n-2)\cdots(n-k+1)$. We call each arrangement a *permutation of n objects taken k at a time*. A convenient notation for the total number of permutations is $_nP_k$. Our result is

$$_nP_k = n(n-1)(n-2)\cdots(n-k+1). \qquad (13.7.4)$$

Example 13.7.5. We imagine a geographic map containing four countries. Each country is to be painted with a different color. There are seven colors available. Then the painting can be done in $_7P_4 = 7 \times 6 \times 5 \times 4 = 840$ different ways.

If the number n of objects is equal to the number of spaces, we obtain an important special case. Each arrangement is simply called a *permutation of the n objects*. From (13.7.4) it follows that there are

$$_nP_n = n(n-1)(n-2)\cdots 3\cdot 2\cdot 1 = n! \qquad (13.7.5)$$

permutations of n objects. The symbol $n!$ is read "n factorial". We introduced it earlier in the book in quite a different context (see Section 10.10). Notice that $1!=1$, $2!=2$, $3!=6$, $4!=24$, $5!=120$, etc.

Example 13.7.6. We consider the four letters e, n, o, t. In how many ways can we arrange them in a line? The answer is $4! = 4 \times 3 \times 2 \times 1$. We list all 24 permutations:

enot	neot	oent	teno
ento	neto	oetn	teon
eont	noet	onet	tneo
eotn	note	onte	tnoe
eton	nteo	oten	toen
etno	ntoe	otne	tone

Notice that very few of these permutations form a word of our language.

With the notation introduced for factorials we can rewrite formula (13.7.4) for the number of permutations on n objects taken k at a time. We multiply by $(n-k)(n-k-1)\cdots 3\cdot 2\cdot 1$ which is the same as $(n-k)!$. Then in order to compensate for the change, we divide by the same product. Thus

$$_nP_k = n(n-1)\cdots(n-k+1) = \frac{n(n-1)\cdots(n-k+1)(n-k)\cdots 2\cdot 1}{(n-k)!}.$$

The numerator is nothing else than $n!$. Hence, the number of permutations of n objects taken k at a time is

$$_nP_k = \frac{n!}{(n-k)!}. \qquad (13.7.6)$$

This formula comprises the special case where $k = n$ if we define $(n-n)!$ or $0!$ to be

$$0! = 1. \tag{13.7.7}$$

A somewhat different problem of counting deals with *selections*. For an experiment, three animals are selected out of five. In how many ways can this be done? We may think of an analogy: When three parking spaces are available for five cars, we found $5 \times 4 \times 3 = 60$ parking arrangements. However, the analogy is not complete. In the parking problem the three parking spaces are distinct, whereas in our selection problem the *order* in which the three animals are arranged is *irrelevant*. Denote the three parked cars by a, b, c. Then we carefully distinguished between the six arrangements $abc, acb, bac, bca, cab, cba$. Whatever the selection of the three cars is, we count each selection six times. Conversely, if we are not interested in the order of the cars (or animals), we have to divide 60 by 6. Thus we find 10 selections of three cars (or animals) taken out of five.

The same argument can be used to derive a general result: In how many ways can we select k out of n different objects? If the order in which the k objects appear is relevant, the result is stated in formula (13.7.4). If not, we have to eliminate the $k!$ permutations of the k selected objects. We call each selection disregarding the order a *combination* and denote the total number of combinations by $_nC_k$. In formula (13.7.4) each combination is counted $k!$ times. Hence,

$$_nC_k \times k! = {_nP_k} = n(n-1)(n-2)\cdots(n-k+1).$$

This implies

$$_nC_k = \frac{n(n-1)\cdots(n-k+1)}{k!}. \tag{13.7.8}$$

This lengthy formula is usually abbreviated by the symbol

$$\binom{n}{k}.$$

There is no standard way of reading this symbol. A suitable way would be "n above k". The symbol should be carefully distinguished from "n over k" which means the fraction n/k.

Example 13.7.7. In how many ways can three out of five animals be selected? Equation (13.7.8) gives the answer:

$$_5C_3 = \binom{5}{3} = \frac{5 \times 4 \times 3}{1 \times 2 \times 3} = 10.$$

If we denote the five animals arbitrarily by a, b, c, d, e, the ten combinations are

$$abc, \quad abd, \quad abe, \quad acd, \quad ace,$$
$$ade, \quad bcd, \quad bce, \quad bde, \quad cde.$$

Notice that dce or edc do not occur in this list since they are only permutations of cde.

Similarly, we obtain

$$_5C_1 = \binom{5}{1} = \frac{5}{1} = 5, \quad _5C_2 = \binom{5}{2} = \frac{5 \cdot 4}{1 \cdot 2} = 10,$$

$$_5C_4 = \binom{5}{4} = \frac{5 \cdot 4 \cdot 3 \cdot 2}{1 \cdot 2 \cdot 3 \cdot 4} = 5.$$

Notice that $\binom{5}{5} = 1$ since we can select 5 objects only in one way. We may also observe that

$$\binom{5}{1} = \binom{5}{4}, \quad \binom{5}{2} = \binom{5}{3}.$$

This property has a particular meaning. If we select one object out of five, the remaining four objects also form a selection. To each selection of one object there corresponds exactly one selection of four objects and conversely. Therefore, their number must be equal. The same is true for selections of two and three objects out of five. In general, it follows that

$$\binom{n}{k} = \binom{n}{n-k} \quad \text{for} \quad k = 1, 2, \ldots, n-1. \tag{13.7.9}$$

Notice that in (13.7.9) the symbol $\binom{n}{n}$ has no counterpart. Formally it would be $\binom{n}{0}$. However, there exists nothing like a combination of zero objects out of n. It is only for the sake of symmetry that we introduce $\binom{n}{0}$. To satisfy (13.7.9) we define

$$\binom{n}{0} = \binom{n}{n} = 1 \tag{13.7.10}$$

Here, n is any positive integer. But we go a step further and allow n to be zero. Thus we define $\binom{0}{0} = 1$.

The numbers $\binom{n}{k}$ are often arranged in such a way that they form *Pascal's triangle* [6]:

$$\binom{0}{0}$$ 1

$$\binom{1}{0} \quad \binom{1}{1}$$ 1   1

$$\binom{2}{0} \quad \binom{2}{1} \quad \binom{2}{2}$$ 1   2   1

$$\binom{3}{0} \quad \binom{3}{1} \quad \binom{3}{2} \quad \binom{3}{3}$$ 1   3   3   1

$$\binom{4}{0} \quad \binom{4}{1} \quad \binom{4}{2} \quad \binom{4}{3} \quad \binom{4}{4}$$ 1   4   6   4   1

$$\binom{5}{0} \quad \binom{5}{1} \quad \binom{5}{2} \quad \binom{5}{3} \quad \binom{5}{4} \quad \binom{5}{5}$$ 1   5   10   10   5   1

There is an easy way to calculate the numbers in Pascal's triangle. We observe that each number is the sum of the two nearest numbers in the row immediately above. Thus $5 = 1 + 4$, $10 = 4 + 6$, etc. The next row, not printed here, will therefore contain the numbers $1 + 5 = 6$, $5 + 10 = 15$, $10 + 10 = 20$, etc. We leave the proof of this property to the reader. The property may be illustrated by the so-called *Roman fountain* (Fig. 13.5).

Formula (13.7.8) for $\binom{n}{k}$ may be rewritten in terms of factorials. For this purpose we replace the numerator of the formula by the expression (13.7.6). Thus we get

$$_nC_k = \binom{n}{k} = \frac{n!}{k!(n-k)!}. \tag{13.7.11}$$

For instance, with $n = 5$ and $k = 2$ we obtain

$$_5C_2 = \binom{5}{2} = \frac{5!}{2! \times 3!} = \frac{120}{2 \times 6} = 10$$

in agreement with an earlier statement.

Formula (13.7.9) can be quickly verified by applying (13.7.11).

---

[6] Blaise Pascal (1623-1662), French mathematician, physicist and philosopher.

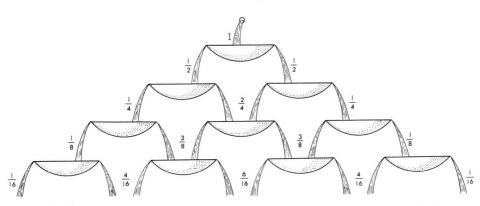

Fig. 13.5. The Roman fountain. Into a basin water runs at rate 1 (meaning one unit of weight per unit of time). On two sides of the basin the water overflows symmetrically at a rate of $\frac{1}{2}$ each and runs into two similar basins below. The water from these two basins overflows also and runs into three symmetrically arranged basins below. The center basin receives water at a rate of $\frac{1}{4} + \frac{1}{4} = \frac{2}{4}$, whereas the outer basins receive water at a rate of only $\frac{1}{4}$ each. The numerators are 1, 2, 1, that is, the second row in Pascal's triangle. The process is then repeated over and over again. In each row the water flows at rates proportional to the corresponding row in Pascal's triangle

Example 13.7.8. We add an example from genetics. Assume that there are six different alleles possible at the same gene locus. We denote them by $A_1, A_2, \ldots, A_6$. How many gene combinations or genotypes are possible? We first count all combinations of two different alleles selected out of the six alleles, such as $A_1 A_2, A_1 A_3$, etc. There are $_6C_2 = \binom{6}{2} = 15$ of them.

Second, we also count the combinations with repetition, that is, the six pairs $A_1 A_1, A_2 A_2$, etc. The total of all combinations is $15 + 6 = 21$.

Example 13.7.9. In a cage there are 20 mice labeled $1, 2, \ldots, 20$. According to the rules of Example 13.3.5, five mice are randomly selected. What is the probability

a) of a particular selection,
b) of a selection of mice with labels less than or equal to 10?

To answer these questions we determine the number of combinations of 5 out of 20 mice. This number is

$$\binom{20}{5} = \frac{20 \cdot 19 \cdot 18 \cdot 17 \cdot 16}{1 \cdot 2 \cdot 3 \cdot 4 \cdot 5} = 15\,504\,.$$

Since all combinations are equally probable, each particular combination has probability $1/15\,504$. This answers the first question. For the second question, we notice that

$$\binom{10}{5} = \frac{10 \cdot 9 \cdot 8 \cdot 7 \cdot 6}{1 \cdot 2 \cdot 3 \cdot 4 \cdot 5} = 252$$

combinations of five mice have labels less than or equal to 10. Hence, the probability of such a sample is

$$252/15\,504 = 0.0163 ,$$

that is, surprisingly small (less than 2%).

An important application of the numbers $\binom{n}{k}$ is made in algebra. We consider powers of a binomial with positive integers as exponents:

$$(a+b)^2 = a^2 + 2ab + b^2 ,$$
$$(a+b)^3 = a^3 + 3a^2b + 3ab^2 + b^3 ,$$
$$(a+b)^4 = a^4 + 4a^3b + 6a^2b^2 + 4ab^3 + b^4 , \quad \text{etc.}$$

The coefficients form the rows in Pascal's triangle. In general, we may write

$$(a+b)^n = \binom{n}{0}a^n + \binom{n}{1}a^{n-1}b + \binom{n}{2}a^{n-2}b^2 + \cdots + \binom{n}{n}b^n . \qquad (13.7.12)$$

One way of proving this formula applies combinations. To explain the idea we confine ourselves to the special case $n=4$. We consider the product

$$(a_1 + b_1)(a_2 + b_2)(a_3 + b_3)(a_4 + b_4) .$$

When working out the multiplications, we get a sum of many terms. Each of the terms contains exactly four factors, such as $a_1b_2b_3b_4$ or $a_3a_4b_1b_2$. All combinations of the a's and b's appear. There are $_4C_3 = \binom{4}{3}$ of those with three b's, $_4C_2 = \binom{4}{2}$ of those with two b's, etc. When we later equate $a_1 = a_2 = a_3 = a_4 = a$ and $b_1 = b_2 = b_3 = b_4 = b$, the term ab^3 appears $\binom{4}{3}$ times, the term a^2b^2 appears $\binom{4}{2}$ times, etc. This explains the formula for $(a+b)^4$. The same argument may be used for any other integral exponent n $(n > 0)$.

Formula (13.7.12) is called the *binomial theorem*. The coefficients $\binom{n}{k}$ are known as *binomial coefficients*. When working out $(a+b)^n$ in a sum of terms $\binom{n}{k} a^{n-k} b^k$, we also say that we *expand* $(a+b)^n$.

Example 13.7.10. Using the binomial theorem we expand

$$(p-2)^5 .$$

For this purpose we identify $p=a, (-2)=b, \; 5=n$ in Eq. (13.7.12). Thus we obtain

$$(p-2)^5 = \binom{5}{0} p^5 + \binom{5}{1} p^4 (-2) + \binom{5}{2} p^3 (-2)^2 + \binom{5}{3} p^2 (-2)^3$$

$$+ \binom{5}{4} p(-2)^4 + \binom{5}{5} (-2)^5$$

$$= p^5 - 10p^4 + 40p^3 - 80p^2 + 80p - 32 .$$

As a *check of calculation* we may let $p=1$ and verify that

$$(-1)^5 = 1 - 10 + 40 - 80 + 80 - 32 .$$

13.8. Binomial Distribution

In this section we will further develop probability theory by using methods of counting.

Example 13.8.1. As an introductory example we study the re-combination of genes. Assume that a fish population pools its reproductive cells. Consider a special locus with alleles A and a. Each reproductive cell (sperm or egg cell) contains exactly one of the two alleles, either A or a. Let p be the probability that a sperm cell contains A and $q=1-p$ be the probability of a. Assume further that the egg cells have the same distribution, that is,

$$P(A)=p, \qquad P(a)=q, \qquad (p+q=1). \tag{13.8.1}$$

The outcome space is $\{A, a\}$.

When sperm cells have fertilized egg cells, we have to consider the new outcome space $\{AA, Aa, aa\}$. Assume that the reproductive cells meet each other at random, that is, that the process is *independent* of the genetic content of each cell. Independence is formulated in mathematical terms by the multiplication rule (13.6.4). Hence, we obtain

$$P(AA)=P(A)\cdot P(A)=p^2 ,$$

$$P(aa)=P(a)\cdot P(a)=q^2 .$$

The heterozygous genotype Aa is formed in two ways. The allele A is either from the sperm or from the egg cell. If the order did count, we would have $P(Aa) = pq$ and $P(aA) = qp$. However, the two cases cannot be distinguished biologically. Hence, disregarding the order we have to add the probabilities: $pq + qp = 2pq$. To sum up, recombination of genes leads to [7]

$$P(AA) = p^2, \quad P(Aa) = 2pq, \quad P(aa) = q^2. \qquad (13.8.2)$$

Since $\{AA, Aa, aa\}$ is our outcome space, the three probabilities should add up to 1. Indeed,

$$p^2 + 2pq + q^2 = (p + q)^2 = 1^2 = 1.$$

Example 13.8.2. We consider the number of boys and girls in a family. We denote a male birth by M and a female birth by F. The probabilities for the occurrence of M and F are not exactly $\frac{1}{2}$. The sex ratio varies slightly from country to country. We assume that

$$P(M) = p = 0.52, \quad P(F) = q = 0.48.$$

We omit the possibility of twin or multiple births. Then experience shows that the outcome of each birth is independent of the outcome of previous births in the same family. Therefore, we apply the multiplication rule. For two children with the outcome space $\{MM, MF, FM, FF\}$ we obtain

$$P(MM) = p^2 = (0.52)^2 = 0.2704,$$
$$P(MF) = pq = (0.52) \times (0.48) = 0.2496,$$
$$P(FM) = qp = (0.48) \times (0.52) = 0.2496,$$
$$P(FF) = q^2 = (0.48)^2 = 0.2304.$$

The total probability is 1. If we disregard the birth order, we get

$$P(MM) = p^2, \quad P(MF) = 2pq, \quad P(FF) = q^2. \qquad (13.8.3)$$

Formally, the result is the same as in (13.8.2).

The example can be easily extended to families with three children. As long as we observe the birth order, the outcome space is

$$\{MMM, MMF, MFM, FMM, MFF, FMF, FFM, FFF\}.$$

---

[7] Some geneticists write the result in the following symbolic way: $p^2 AA + 2pq Aa + q^2 aa$. There is no similar symbolism in other areas of mathematics.

The multiplication rule yields the corresponding probabilities:

$$P(MMM) = ppp = p^3, \qquad P(MFF) = pqq = pq^2,$$
$$P(MMF) = ppq = p^2q, \qquad P(FMF) = qpq = pq^2,$$
$$P(MFM) = pqp = p^2q, \qquad P(FFM) = qqp = pq^2,$$
$$P(FMM) = qpp = p^2q, \qquad P(FFF) = qqq = q^3.$$

However, disregarding the birth order, the outcome space is reduced to

$$\{MMM, MMF, MFF, FFF\},$$

and the corresponding probabilities turn out to be

$$P(MMM) = p^3, \qquad P(MFF) = 3pq^2,$$
$$P(MMF) = 3p^2q, \qquad P(FFF) = q^3. \tag{13.8.4}$$

The four probabilities are the terms in the expansion of $(p+q)^3$. The coefficients are the binomial coefficients $\binom{3}{0}, \binom{3}{1}, \binom{3}{2}, \binom{3}{3}$.

The last remark gives us a hint for further generalization: Consider families with n children. Then we ask: What is the probability of selecting at random a family with k boys and $n-k$ girls? If we respect the birth order, the probability is

$$p^k q^{n-k}$$

since the factor $p = P(M)$ must appear k times and the factor $q = P(F)$ exactly $n-k$ times. However, when disregarding the order, we have the case of selecting "k boys out of n children", and this can be done in as many ways as we can select k objects out of n distinct objects. Since the order does not count, we are dealing with the number of *combinations of n objects, k at a time*. The number is

$$_nC_k = \binom{n}{k}.$$

Hence we have to take the term $p^k q^{n-k}$ as many times as indicated by $_nC_k$. The result is

$$P(k \text{ boys}, n-k \text{ girls}) = \binom{n}{k} p^k q^{n-k}. \tag{13.8.5}$$

This probability is a term in the expansion of $(q+p)^n$.

As a special case we consider $n = 4$. It follows from (13.8.5) that

$$P(MMMM) = \binom{4}{4} p^4 q^0 = p^4 = (0.52)^4 = 0.0731 \,,$$

$$P(MMMF) = \binom{4}{3} p^3 q^1 = 4p^3 q = 4(0.52)^3 (0.48) = 0.2700 \,,$$

$$P(MMFF) = \binom{4}{2} p^2 q^2 = 6p^2 q^2 = 6(0.52)^2 (0.48)^2 = 0.3738 \,,$$

$$P(MFFF) = \binom{4}{1} p^1 q^3 = 4pq^3 = 4(0.52)(0.48)^3 = 0.2300 \,,$$

$$P(FFFF) = \binom{4}{0} p^0 q^4 = q^4 = (0.48)^4 = 0.0531 \,.$$

In our population, families with four boys occur at a rate of 7.31 % but families with two boys and two girls at the much higher rate of 37.38 %.

In general, let E be an event of an outcome space and $\bar{E}$ the complementary event. Thus we may write

$$P(E) = p \,, \qquad P(\bar{E}) = q \quad (p + q = 1) \,. \tag{13.8.6}$$

We perform the experiment n times in such a way that each consecutive outcome is independent of all previous outcomes. Then the probability that E occurs exactly k times is

$$P(k \text{ times}) = \binom{n}{k} p^k q^{n-k} \quad (0 \leqq k \leqq n) \,. \tag{13.8.7}$$

The set of probabilities (13.8.7) for all $k = 0, 1, 2, \ldots, n$ is called a *binomial distribution*. This distribution was investigated by Jacob Bernoulli[8]. Experiments which result in a simple alternative such as "yes-no", "success-failure", "positive-negative reaction" are often called *Bernoulli trials*.

The numerical calculation of the terms (13.8.7) is laborious. Work is greatly facilitated by tables (see Table G of the Appendix and quotations at the end of this chapter).

Before studying further applications, we investigate the special case

$$p = \frac{1}{2} \,, \qquad q = \frac{1}{2} \,.$$

---

[8] Jacob Bernoulli (1654-1705), Swiss mathematician.

As a model experiment we may choose coin tossing with outcomes head (H) and tail (T). Formula (13.8.7) reduces to

$$P(k \text{ times}) = \binom{n}{k} \frac{1}{2^n}. \qquad (13.8.8)$$

For n given, $P(k \text{ times})$ for $k = 0, 1, 2, \ldots, n$ is proportional to $\binom{n}{k}$, that is, to a row in Pascal's triangle (Section 13.7). For coin tossing, the following probabilities are valid:

$$P(HH) = \tfrac{1}{4}, \quad P(HT) = \tfrac{2}{4}, \quad P(TT) = \tfrac{1}{4},$$

$$P(HHH) = \tfrac{1}{8}, \quad P(HHT) = \tfrac{3}{8}, \quad P(HTT) = \tfrac{3}{8}, \quad P(TTT) = \tfrac{1}{8},$$

$$P(HHHH) = \tfrac{1}{16}, \quad P(HHHT) = \tfrac{4}{16}, \quad P(HHTT) = \tfrac{6}{16}, \quad P(HTTT) = \tfrac{4}{16}, \quad P(TTTT) = \tfrac{1}{16}.$$

The histograms of Fig. 13.6 illustrate these special distributions.

A chance mechanism especially suitable for demonstrating the particular binomial distributions with $p = \tfrac{1}{2}$, $q = \tfrac{1}{2}$ is the so-called *binomiator*. Details are explained in Fig. 13.7.

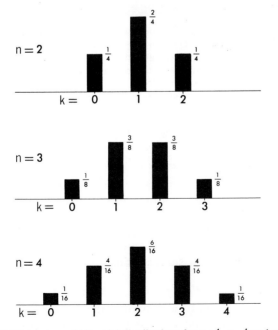

Fig. 13.6. Histograms of binomial distributions for $p = \tfrac{1}{2}$, $q = \tfrac{1}{2}$ and $n = 2, 3, 4$

Fig. 13.7. A model experiment for demonstrating binomial distributions with $p=\frac{1}{2}$, $q=\frac{1}{2}$. Balls are rolling down a slope. On their way they hit obstacles arranged as shown by the black polygons. At each obstacle a ball is thrown either to the left or to the right with probability $\frac{1}{2}$. In our drawing, the probabilities of reaching the compartments below are in this order: 1/64, 6/64, 15/64, 20/64, 15/64, 6/64, 1/64. Thus the numerators form the sixth row in Pascal's triangle. This model experiment was invented by Galton (cf. Fig. 13.20). It is related to the Roman fountain shown in Fig. 13.5

Applications

Example 13.8.3. *Risk of Fatal Effect.* Assume that the probability that a person will die within a month after a certain cancer operation is 18%. What are the probabilities that in three such operations one, two, or all three persons will survive?

Survival means in this connection the opposite of death within a month· after the operation. The outcome space is $\{D, S\}$ where D stands for death and S for survival. The probabilities are

$$P(D) = 0.18, \qquad P(S) = 0.82.$$

If we have good reason to assume that the outcome of one operation is independent of the outcome of the other two operations, we can apply formula (13.8.7). Let k be the number of surviving patients. Then

$$P(k=0) = \binom{3}{0}(0.18)^3 = 0.006 \,,$$

$$P(k=1) = \binom{3}{1}(0.82)(0.18)^2 = 0.080 \,,$$

$$P(k=2) = \binom{3}{2}(0.82)^2(0.18) = 0.363 \,,$$

$$P(k=3) = \binom{3}{3}(0.82)^3 = 0.551 \,.$$

Thus the probability that only one person will survive is 8%, that two survive 36%, and that all three survive 55%, approximately.

Example 13.8.4. *Bioassay.* When an animal is given a treatment, we may ask whether the reaction is positive or not, that is, whether a certain result can be observed or not. Such an outcome is qualitative rather than quantitative. When n like animals are given the same treatment, we may ask how many of the animals react positively. Each single performance of the experiment is called a Bernoulli trial.

Let E be the event "positive reaction" and $\bar{E}$ the complementary event. Let

$$P(E)=p\,, \qquad P(\bar{E})=q\,, \qquad p+q=1\,.$$

Under the assumption that the n trials are independent, the probability that exactly k animals react positively is

$$P(k \text{ pos. reactions}) = \binom{n}{k}p^k q^{n-k}\,.$$

For instance, consider five mice from the same litter all suffering from a vitamin-A deficiency. They are fed a certain dose of carrots. The positive reaction means here recovery from the disease. Assume that the probability of recovery is $p=0.73$. Then we ask: What is the probability that exactly three of the five mice recover? The answer is

$$\binom{5}{3}(0.73)^3(0.27)^2 = 0.284$$

or 28.4%.

Example 13.8.5. *Mutations.* The probability of a mutation per gene and per R (roentgen, unit of intensity of radiation) in mice is approximately 2.5×10^{-7} (see Neel and Schull, 1958, p. 154). What is the probability that in 10 000 genes at least one mutation occurs?

Let k be the number of mutations. The phrase "at least one" means $k=1$ or 2 or 3, etc. It would be extremely laborious to calculate all 10 000 probabilities by formula (13.8.7) and to sum them up. It is much simpler to proceed as follows: The event "at least one" is the complementary event to "no mutation". Thus we calculate first

$$P(k=0)=(1-2.5\times 10^{-7})^{10,000}=0.99999975^{10,000}.$$

The calculation can be performed by logarithms. Few tables, however, will list log 0.99999975. To cope with this difficulty, we recall formula (10.10.12) concerning the expansion of $(1+x)^s$. Let $x=(-2.5)\,10^{-7}$ and $s=10000$. Restricting ourselves to a linear approximation $1+sx$ we obtain

$$P(k=0)=(1-2.5\times 10^{-7})^{10\,000}\approx 1-2.5\times 10^{-7}\times 10^4$$

or

$$P(k=0)\approx 1-2.5\times 10^{-3}.$$

We are interested in the complementary event which is equivalent to $k>0$. Hence,

$$P(k>0)\approx 2.5\times 10^{-3}=0.0025.$$

Our chance to observe a mutation in 10 000 genes which underwent radiation of dose 1R is only one quarter percent.

Example 13.8.6. *Harmful Side Effects.* Assume that a drug causes a serious side effect at a rate of three patients out of one hundred. A pharmacological laboratory wants to test the drug. What is the probability that the side effect occurs in a random sample of ten patients taking the drug?

Before an answer can be given, the question has to be formulated more precisely. Let k be the number of persons who might suffer from the side effect. Then occurrence of the side effect means that $k=1$ or 2 or 3, etc. As in the preceding application, it is easier to treat the case $k=0$ first. The probability that a patient does not suffer from the side effect is $1-0.03=0.97$. Hence[9],

$$P(k=0)=(0.97)^{10},$$

$$\log P(k=0)=10\times\log 0.97=10\times(0.98677-1)$$
$$=9.8677-10=0.8677-1,$$

$$P(k=0)=\text{antilog}(0.8677-1)=0.737,$$

$$P(k>0)=1-P(k=0)=0.263.$$

---

[9] To work out this example we use common logarithms. We could also have applied expansion as in Example 13.8.5., but the accuracy of a linear approximation would not suffice.

Therefore, in a sample of ten randomly selected patients the side effect occurs only with probability 26.3%. When a new drug is being screened, a rare side effect may very well remain undetected.

Example 13.8.7. *A Counterexample.* The binomial distribution is not always applicable when the same experiment is repeated. The reason is that the outcome of a trial could depend on the outcomes of preceding trials. Consider for instance the weather. For simplicity, let us distinguish between rainy and dry days. Assume that at a certain location the rate of rainy days is $1:12$. If we are interested in sequences of rainy days, we may ask: What is the probability that it will rain on July 1, 2, and 3? One is tempted to apply the multiplication rule. This would lead to

$$P(3 \text{ rainy days}) = (1/12)^3 \approx 0.0006,$$

that is, a number which is very small. However, the result is wrong. The weather on one day depends strongly on the weather of the preceding days. The day after a rainy day is likely to be another rainy day. According to our experience the probability of three consecutive rainy days is considerably greater than the result given by the multiplication rule.

*13.9. Random Variables

Events in a random experiment are sometimes *qualitative* in nature. For instance, in genetic experiments with peas, the petals may be white, red, or pink. In science, however, quantification of properties is usually advantageous. In this section we will study several examples of successful quantification. We will also introduce such concepts as random variable, probability distribution, mean and standard deviation of random variables.

Example 13.9.1. We consider the outcome space

$$\{\text{white, pink, red}\}$$

for the petals of experimental peas. According to genetics these colors are due to two alleles, say W and R, at a certain gene locus. The genotype WW produces white flowers, WR pink flowers, and RR red flowers. Thus we may map the outcome space into

$$\{WW, WR, RR\}. \tag{13.9.1}$$

Now we can quantify the three outcomes in a simple and natural way: We count the number of R-alleles in each outcome, thus mapping the outcome space into

$$\{0, 1, 2\}. \tag{13.9.2}$$

Number zero is associated with WW, number one with WR, and number two with RR. In Section 3.4 we called such an association a *function*. A frequently used notation is X. In our example

$$X(WW)=0, \qquad X(WR)=1, \qquad X(RR)=2.$$

The domain of this function is the set (13.9.1), and the range is the set (13.9.2).

Example 13.9.2. We consider an outcome space consisting of only two simple events, say

$$\{\text{success, failure}\}. \tag{13.9.3}$$

At a glance, there seems to be no natural way of quantification. We may map the outcome space into $\{+1, -1\}$ or into $\{1, 0\}$. But why not choose $\{5, \frac{1}{4}\}$? Obviously, we have to find a criterion for the proper selection of mapping. As so often in science, the best choice is the one that serves best. We will show that the mapping

$$\{1, 0\} \tag{13.9.4}$$

is a good choice in quantifying (13.9.3). Here the function X takes on either 1 or 0. Number 1 stands for success, number 0 for failure. Let us perform an experiment with the outcome space (13.9.4) several times. With $X_1, X_2, \ldots, X_n$ we denote the outcomes of the first, second, ..., nth trials, respectively. If, for instance, the sequence is success-failure-failure-success-success-failure, we get $X_1 = 1$, $X_2 = 0$, $X_3 = 0$, $X_4 = 1$, $X_5 = 1$, $X_6 = 0$. Now we form the sum

$$S_n = X_1 + X_2 + \cdots + X_n. \tag{13.9.5}$$

Each term is a zero or a one. We have as many ones as successes. Hence, S_n is equal to the number of successes in n trials (in our example $S_n = 3$). This simple property is used to justify the mapping of (13.9.3) into (13.9.4). The quantity S_n may take any of $n+1$ values $0, 1, 2, \ldots, n$. If we are only interested in the number of successes and not in the order in which the event "success" occurs, the outcome space for n trials is

$$\{0, 1, \ldots, n\}. \tag{13.9.6}$$

Now we introduce probabilities for the outcome space (13.9.3). Instead of writing

$$P(\text{success}) = p, \qquad P(\text{failure}) = q,$$

it is more convenient to use the notation

$$P(X = 1) = p, \quad P(X = 0) = q. \tag{13.9.7}$$

When repeating the experiment we may ask for the probability of k successes in n trials, or briefly, for the probability of the event $S_n = k$. Under the assumption of independent trials, the result was stated in formula (13.8.7). We may rewrite the result in the form

$$P(S_n = k) = \binom{n}{k} p^k q^{n-k}. \tag{13.9.8}$$

Example 13.9.3. Another example where the outcome space can be easily quantified is the life table of Section 13.5 (Table 13.1). The outcome space consists of "death in the first decade", "death in the second decade", etc. We may associate the number k with the event "death in the kth decade". Thus we map the outcome space into $\{1, 2, 3, \ldots\}$. Instead of the phrase "death occurred in the kth decade" we could simply write $X = k$.

When we are dealing with height, weight, pulse rate, blood pressure, etc., the simple events of the outcome space are measurements, and thus already numbers. A mapping into other numbers is usually not required. It is customary to denote variable measurements and countings by *capital letters* such as X, Y, Z, U, V, etc.

In general, when an outcome is mapped into a set of numbers, we introduce a variable quantity, say X, which is a function of the outcomes. Such a quantity is called a *random variable*.

In Section 1.2 we studied the different levels of quantification. It is precisely the *transition from the nominal level to the ordinal, interval, or ratio level* which allows us to introduce a random variable. Thus, a random variable can take on the different scores assigned at an ordinal level. For instance, $X = 0$ could mean failure, $X = 1$ success of a trial. More frequently, however, random variables are used for quantities defined at an interval level such as temperatures in degrees Celsius, time, altitude, electric potential, or for quantities defined at a ratio scale such as length, area, volume, weight.

We assume now that a random variable X can only take on a finite number of values, say $x_1, x_2, \ldots, x_m$. Thus the outcome space is

$$\{x_1, x_2, \ldots, x_m\}. \tag{13.9.9}$$

We denote the probability associated with x_i by p_i $(i = 1, 2, \ldots, m)$. This means that

$$P(X = x_1) = p_1, P(X = x_2) = p_2, \ldots, P(X = x_m) = p_m. \tag{13.9.10}$$

The total probability must be one, that is, $\Sigma p_i = 1$. We call a set of probabilities associated with an outcome space a *probability distribution*.

For many reasons it is desirable to define a suitable average or mean of the random variable X. We cannot simply take the arithmetic mean $\Sigma x_i/m$ since we would neglect the rate at which each x_i occurs in the random experiment. To get the proper idea, let us return to the frequency concept of probability. Each p_i is an idealization of a relative frequency. Assume that the event $X = x_1$ has occurred n_1 times, the event $X = x_2$ has occurred n_2 times, etc. in a total of n performances of the experiment. Then we obtain

$$n_1 + n_2 + \ldots + n_m = n,$$

$$\bar{x} = \frac{n_1 x_1 + n_2 x_2 + \cdots + n_m x_m}{n} = \sum_{i=1}^{m} \frac{n_i}{n} \cdot x_i.$$

Here $\bar{x}$ is the *weighted arithmetic mean* of $x_1, x_2, \ldots, x_m$. The "weights" are $n_1, n_2, \ldots, n_m$. Replacing the relative frequencies n_i/n by probabilities p_i we get the following definition of the mean value of X:

$$\text{mean of } X = \sum_{i=1}^{m} p_i x_i. \tag{13.9.11}$$

If we had to guess or to estimate the outcome of a single performance of the experiment, the mean of X would be a good choice. Our guess

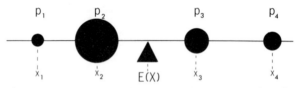

Fig. 13.8. The expectation of X interpreted as a center of gravity

could be too high or too low, but not too far away from the actual outcome. For this reason, we also call the mean the *expected value* or the *expectation* of X. The notation is $E(X)$. Hence (13.9.11) turns into

$$\boxed{E(X) = \sum_{i=1}^{m} p_i x_i.} \tag{13.9.12}$$

In probability theory the words "average", "mean", and "expectation" are used in the same sense.

Formula (13.9.12) has an interesting counterpart in mechanics: Consider each x_i as the abscissa of a point on a straight line and interpret p_i as a mass or a weight concentrated in the point x_i (see Fig. 13.8). Then the point with abscissa $E(X)$ is identical with the *center of gravity*.

Example 13.9.4. Assume for simplicity that the sex ratio is $1:1$. Let the random variable X stand for the number of boys in families with four children. The outcome space is $\{0, 1, 2, 3, 4\}$. The corresponding probabilities are given by formula (13.8.8) for $n = 4$. Thus the probabilities are

$$P(X = 0) = \frac{1}{16}, \qquad P(X = 1) = \frac{4}{16}, \qquad P(X = 2) = \frac{6}{16},$$

$$P(X = 3) = \frac{4}{16}, \qquad P(X = 4) = \frac{1}{16}.$$

Now we ask for the average or the expected number of boys in such families. By symmetry, it is easy to guess that this average is two. We verify the result by applying formula (13.9.12):

$$E(X) = \left(\frac{1}{16} \cdot 0\right) + \left(\frac{4}{16} \cdot 1\right) + \left(\frac{6}{16} \cdot 2\right) + \left(\frac{4}{16} \cdot 3\right) + \left(\frac{1}{16} \cdot 4\right) = 2.$$

Similarly, the average or expected number of boys in families with five children is 2.5.

Example 13.9.5. We study the occurrence of a lethal allele a in a population with genotypes AA, Aa, aa. Let X be the number of a's in each genotype, that is $X = 0$ in AA, $X = 1$ in Aa, $X = 2$ in aa. For the probabilities we introduce the following notations:

$$P(X = 0) = p_1, \qquad P(X = 1) = p_2, \qquad P(X = 2) = p_3 \qquad (p_1 + p_2 + p_3 = 1).$$

According to formula (13.9.12) the expected number of a's is

$$E(X) = p_1 \cdot 0 + p_2 \cdot 1 + p_3 \cdot 2 = p_2 + 2p_3. \tag{13.9.13}$$

If, for instance, $p_1 = 0.7$, $p_2 = 0.2$, $p_3 = 0.1$, we get $E(X) = 0.4$, that is, on the average of the three genotypes the lethal gene occurs 0.4 times.

With the mere knowledge of $E(X)$ we have no idea how much the outcomes $x_1, x_2, \ldots, x_m$ in (13.9.9) differ from the expected value. Some probability distributions are concentrated around $E(X)$, others not. Therefore, it is desirable to have a *measure of dispersion*. As in Section 1.9 we consider the *square of deviation*

$$(x_i - E(X))^2 \qquad (i = 1, 2, \ldots, m) \tag{13.9.14}$$

since we are not interested in the sign of $x_i - E(X)$. To get an average deviation we cannot simply take the ordinary arithmetic mean of all m terms in (13.9.14). We have to observe the rate at which each x_i occurs.

For the same reason as in formula (13.9.11) we form the weighted arithmetic mean with "weights" p_i. The result is called the *variance* of X and is defined by

$$\text{Var}(X) = \sum_{i=1}^{m} p_i(x_i - E(X))^2 . \tag{13.9.15}$$

For most people the formula looks less frightening if we introduce the following standard notations:

$$E(X) = \mu , \qquad \text{Var}(X) = \sigma^2 . \tag{13.9.16}$$

(Greek letters mu and sigma). With this notation, formula (13.9.15) turns into

$$\boxed{\sigma^2 = \sum_{i=1}^{m} p_i(x_i - \mu)^2 .} \tag{13.9.17}$$

As is frequently the case, the x_i are not pure numbers but expressed in a certain unit, such as cm, kg, sec. Then formula (13.9.12) reveals that $E(X)$ is expressed in the same unit. However, σ^2 is measured in the square of such a unit. For instance, if the mean is 85 cm, the variance may be 16 cm². This property of the variance is disadvantageous. Therefore, we take the square root of the variance and get

$$\sigma = \left[\sum_{i=1}^{m} p_i(x_i - \mu)^2 \right]^{\frac{1}{2}} . \tag{13.9.18}$$

This measure of dispersion is widely used and is known as the *standard deviation* of X from the mean.

Example 13.9.6. We consider a random variable X with the outcome space {83 cm, 84 cm, 85 cm, 86 cm, 87 cm} and with probabilities 1/11, 2/11, 5/11, 2/11, 1/11, respectively. With respect to 85 cm this particular distribution is symmetric. If we interpret the probabilities as masses, the center of gravity falls into the point with abscissa 85 cm. The reader may verify that $\mu = E(X)$ is indeed 85 cm. Temporarily omitting "cm" we get for the variance

$$\sigma^2 = \frac{1}{11}(83 - 85)^2 + \frac{2}{11}(84 - 85)^2 + \frac{5}{11}(85 - 85)^2$$

$$+ \frac{2}{11}(86 - 85)^2 + \frac{1}{11}(87 - 85)^2$$

or

$$\sigma^2 = 12/11 \text{ cm}^2 .$$

The standard deviation is

$$\sigma = (12/11 \text{ cm}^2)^{\frac{1}{2}} = 1.044 \text{ cm} .$$

Now we apply the concepts of expectation, variance, and standard deviation to the *binomial distribution*. In Eq. (13.9.12) the x_i are the outcomes $0, 1, 2, ..., n$ and the p_i are the binomial probabilities as given by Eq. (13.9.8) for $i = k$. Therefore, combining Eqs. (13.9.8) and (13.9.12) we obtain

$$\mu = E(S_n) = \binom{n}{0} p^0 q^{n-0} \cdot 0 + \binom{n}{1} p^1 q^{n-1} \cdot 1 + \cdots + \binom{n}{n} p^n q^0 \cdot n$$

$$= np \left(q^{n-1} + \frac{n-1}{1} pq^{n-2} + \frac{(n-1)(n-2)}{1 \cdot 2} p^2 q^{n-3} + \cdots + p^{n-1} \right)$$

$$= np \left(q^{n-1} + \binom{n-1}{1} pq^{n-2} + \binom{n-1}{2} p^2 q^{n-3} + \cdots + \binom{n-1}{n-1} p^{n-1} \right)$$

$$= np(q + p)^{n-1} .$$

Since $q + p = 1$, we get the surprisingly simple result

$$\boxed{\mu = E(S_n) = np .} \tag{13.9.19}$$

A similar, but lengthier calculation which we will skip here leads to the formula for the *variance*

$$\boxed{\sigma^2 = \text{Var}(S_n) = npq .} \tag{13.9.20}$$

Hence, the *standard deviation* of the binomial distribution is

$$\sigma = \sqrt{npq} . \tag{13.9.21}$$

Example 13.9.7. A symmetric die is thrown 12 times. If the upper face is a six, we consider it a success and denote this event by E. Any other event is a failure and denoted by $\bar{E}$. Here

$$p = P(E) = \tfrac{1}{6}, \qquad q = P(\bar{E}) = \tfrac{5}{6} .$$

By S_{12} we denote the number of successes in $n = 12$ trials. This random variable ranges from 0 to 12. According to Eq. (13.9.19) the expectation is

$$\mu = E(S_{12}) = 12 \cdot \tfrac{1}{6} = 2 .$$

The result corresponds to our intuition which suggests that in 12 trials we should expect 2 upper faces "six".

The variance follows from Eq. (13.9.20):

$$\sigma^2 = \text{Var}(S_{12}) = 12 \cdot \tfrac{1}{6} \cdot \tfrac{5}{6} = \tfrac{5}{3} .$$

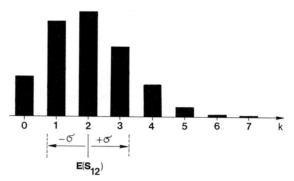

Fig. 13.9. Binomial distribution for $p = \frac{1}{6}$, $n = 12$. Expectation and standard deviation of the random variable S_{12} are shown

From this we get the standard deviation

$$\sigma = \sqrt{\tfrac{5}{3}} = 1.29 \, .$$

The binomial distribution of this example is depicted in Fig. 13.9.

13.10. The Poisson Distribution

In ecology the distribution pattern of plants or animals of the same species over a region (field, forest) has been studied many times. For this purpose the region is subdivided into a large number of so-called *quadrats* (squares or rectangles of equal area). Fig. 13.10 shows an example. In each single quadrat the number of individuals is counted. Among the 20 quadrats shown, there are some empty quadrats and some quadrats which contain 1, 2, 3, or more individuals. A complete breakdown of the distribution is given below:

Number of individuals per quadrat	Number of quadrats	Total number of individuals in each sort of quadrat
0	3	0
1	6	6
2	5	10
3	4	12
4	1	4
5	0	0
6	1	6
Total:	20	38

In some parts of the region we observe an aggregation of plants. Other parts show some kind of emptiness. However, as a whole we feel that the individuals are randomly dispersed.

Can the pattern of random dispersion be described mathematically? This is indeed possible. For simplicity, we replace the plants or animals by balls. Now we drop a ball over the region in such a way that each quadrat has the same probability p of being hit by the ball. In Fig. 13.10, p equals 1/20. Then we repeat the same experiment n times and assume that each trial is independent of all previous trials. Let X be the number of balls which hit a particular quadrat. The random variable X could take on any of the integers $0, 1, 2, \ldots, n$. We have to find the probability that X takes on a specific value k. This probability was established

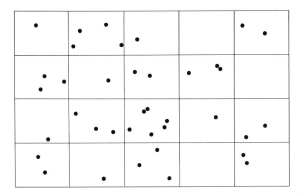

Fig. 13.10. Distribution of individual plants of a species over a region

in Section 13.8. Writing X instead of S_n we get

$$P(X = k) = \binom{n}{k} p^k q^{n-k} \tag{13.10.1}$$

where $q = 1 - p$. In Fig. 13.10, we have $p = 1/20$ and $n = 38$ (total number of individuals). A brief table of values rounded-off to three decimals follows:

k	$P(X = k)$	k	$P(X = k)$	
0	0.142	4	0.081	
1	0.285	5	0.029	
2	0.277	6	0.008	
3	0.175	7	0.002	etc.

Therefore, we expect

$$20 \times 0.142 = 2.84 \quad \text{empty quadrats,}$$
$$20 \times 0.285 = 5.70 \quad \text{quadrats with one ball,}$$
$$20 \times 0.277 = 5.54 \quad \text{quadrats with two balls,}$$
$$20 \times 0.175 = 3.50 \quad \text{quadrats with three balls,}$$
$$20 \times 0.081 = 1.62 \quad \text{quadrats with four balls, etc.}$$

The theoretical result is quite compatible with the countings from Fig. 13.10.

Despite the good agreement between theory and experience, formula (13.10.1) is not the final answer to our problem. The total number n of balls is usually very large, and its exact value is irrelevant. On the other hand, the probability p that a particular quadrat out of a large number of quadrats is hit by a ball is very small. Again it is irrelevant to know the exact value of p. Finally, the calculation of the binomial terms in (13.10.1) is rather laborious.

For all these reasons, it is appropriate to ask for the limiting distribution which we get as n tends to infinity and p to zero. To initiate the limiting process, we consider the mean value or the expectation of X. It was stated in formula (13.9.19). For convenience, we denote this mean here by m rather than by $E(X)$ or μ. Thus,

$$m = np . \tag{13.10.2}$$

In our example $n = 38$ and $p = 1/20$, hence $m = 1.9$. In most applications m ranges from 0 to about 10. Whereas n tends to infinity and p to zero, *we keep the mean m constant*. Rearranging formula (13.10.1) and replacing p by m/n according to (13.10.2) we obtain

$$\binom{n}{k} p^k q^{n-k} = \binom{n}{k} \frac{p^k q^n}{q^k} = \binom{n}{k} \frac{m^k q^n}{n^k q^k}$$
$$= \frac{n(n-1)(n-2) \cdots (n-k+1)}{k!} \cdot \frac{1}{n^k q^k} \cdot m^k \cdot q^n . \tag{13.10.3}$$

As $n \to \infty$, the factors $k!$ and m^k remain constant. The last factor is

$$q^n = (1-p)^n = \left(1 - \frac{m}{n}\right)^n .$$

Replacing a by $-m$ in formula (10.8.4) we get

$$\lim_{n \to \infty} q^n = e^{-m} . \tag{13.10.4}$$

The remaining expression in (13.10.3) is

$$\frac{n(n-1)(n-2)\cdots(n-k+1)}{n^k(1-m/n)^k} = \frac{n(n-1)\cdots(n-k+1)}{(n-m)^k}.$$

Both numerator and denominator of this fraction can be written as polynomials in n of degree k. The term of highest degree is n^k in both cases. Applying the method in Section 8.1 we find that the fraction tends to 1 as n tends to infinity. Notice that k remains constant in the limiting process.

Combining the results we get

$$\lim_{n\to\infty} \binom{n}{k} p^k q^{n-k} = \frac{m^k e^{-m}}{k!} \qquad (k=0, 1, 2, \ldots).$$

With these probabilities we define a new random variable X:

$$P(X=k) = \frac{m^k e^{-m}}{k!}. \tag{13.10.5}$$

Here the outcome space consists of all integers $k = 0, 1, 2, 3, \ldots$. The outcome space is therefore *infinite*.

Formula (13.10.5) establishes the famous *Poisson distribution*[10]. To say that objects are *randomly dispersed* or *randomly distributed* over a region is to say that they follow a Poisson distribution.

It is not laborious to calculate the terms in formula (13.10.5) since k hardly ever exceeds 10 or 20. For greater convenience tables of the Poisson distribution are available (see Table H of the Appendix and citation at the end of this chapter).

In our numerical example we have $m = 1.9$. For this value the probabilities (13.10.5) are

k	$P(X=k)$	k	$P(X=k)$	
0	0.150	4	0.081	
1	0.284	5	0.031	
2	0.270	6	0.010	
3	0.171	7	0.002	etc.

These probabilities differ little from the corresponding values of the binomial distribution listed after formula (13.10.1). A graphical presentation of our special Poisson distribution is given in Fig. 13.11.

---

[10] Siméon Denis Poisson (1781-1840), French mathematician and physicist.

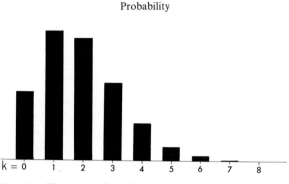

Fig. 13.11. Histogram of the Poisson distribution with mean 1.9

As we already know, the mean of the Poisson distribution (13.10.5) is

$$E(X) = m . \tag{13.10.6}$$

It is easy to find the variance. Formula (13.9.20) states the variance for the binomial distribution. In this formula we replace p by m/n and then let n tend to infinity. Thus

$$npq = n \cdot \frac{m}{n} \left(1 - \frac{m}{n}\right) = m \left(1 - \frac{m}{n}\right) \to m .$$

Hence,

$$\sigma^2 = \mathrm{Var}(X) = m . \tag{13.10.7}$$

Notice that the *mean* and the *variance of a Poisson distribution* have the *same numerical value*. The standard deviation is $\sigma = m^{\frac{1}{2}}$. In Fig. 13.11 the standard deviation is $1.9^{\frac{1}{2}} = 1.38$.

Applications

Example 13.10.1. *Ecology.* The distribution of plants and animals over a region is seldom a Poisson distribution. For some reason, individuals may be *aggregated into clumps.* Then counting the frequency within quadrats we get many empty quadrats and many quadrats with high frequencies. This increases the variance. The equality (13.10.7) is no longer valid. Instead we find

$$\sigma^2 > m . \tag{13.10.8}$$

The contrary may also occur: Consider the case where there is approximately the same distance from one individual to its neighbor. The distribution is *nearly uniform.* Examples are the trees in an orchard or hairs on the skin. Per quadrat we count almost the same number of individuals. The variance is relatively small. Hence,

$$\sigma^2 < m.$$ (13.10.9)

For more details see MacArthur and Connell (1966, p. 44ff.), Pielou (1969, Part 2). For a mathematical theory of clumping we refer the reader also to Roach (1968).

Example 13.10.2. *Bacteria and Blood Counts.* On a small glass plate with a square grid a liquid containing single cells is spread homogeneously. Great care is taken that the liquid has a constant thickness. Under the microscope we count the number of squares with no cells, with one cell, with two cells, etc. The observed distribution should be close to a Poisson distribution. This is very often the case. Large deviations may occur due to random fluctuations, or more likely, due to a factor causing clusters of cells.

There are also instruments for counting cells automatically. The liquid passes through a narrow glass tube. Each passing cell causes a change in transparency. An electronic eye is able to register the cells with high speed, but with restricted accuracy. When countings for equal amounts of liquid are compared, the distribution should again resemble a Poisson distribution.

For tests of the hypothesis of a Poisson distribution we refer the reader to books on statistics.

Example 13.10.3. *Mutations.* A large number of agar plates are treated with an antibiotic. Bacteria which are spread over the plates cannot multiply except for those rare mutants that are antibiotic-resistant. They form colonies. If the experimental conditions can be kept fairly constant for all agar plates, the counting of colonies should approximately result in a Poisson distribution.

Example 13.10.4. *Radioactive Decay.* In a radioactive substance, disintegration of nuclei occurs spontaneously, that is, disintegration is not caused by factors outside of the nucleus. The number of pulses in a Geiger counter during time intervals of a fixed length are recorded. There are intervals with no pulse, with one pulse, with two pulses, etc. The hypothesis of a Poisson distribution is well supported by experimental facts.

In this application the time intervals play the same role as the quadrats in ecology or the agar plates in bacteriology. The Poisson distribution is applicable to time intervals as well as to sections of space.

Example 13.10.5. *Daily Life.* When a rain storm starts, the first drops on a paved road are distributed according to a Poisson law. The same is true for the misprints in a book, the number of letters or of telephone calls that we receive on weekdays. In a hospital the number of births or deaths per day, at a street intersection the monthly number of traffic accidents, in industry the number of defective items produced per hour, these all follow more or less the Poisson law.

13.11. Continuous Distributions

Let us consider an introductory example.

Example 13.11.1. The α-globulin content x in the blood plasma of a large number of healthy, human adults is measured. Each measurement is expressed in grams per 100 milliliters. We are interested in the relative frequency of occurrence. For a rough survey we may subdivide the measurements into three groups of width 0.18 g/100 ml with midpoints at 0.60, 0.78, 0.96 g/100 ml. The histogram (a) of Fig. 13.12 depicts the result. The area of each bar is numerically equal to the corresponding relative frequency. The total area of the histogram is 1 or 100%.

For more accurate information we may use a larger number of groups. The histogram (b) of Fig. 13.12 shows groups of width 0.06 g/100 ml with midpoints at 0.60, 0.66, ..., 1.02 g/100 ml. Again the area of each bar is numerically equal to the corresponding relative frequency.

The actual measurements can be made so precisely that an even finer distinction is possible as demonstrated by histogram (c) of Fig. 13.12.

Due to the imperfection of our instruments, however, precision cannot be increased indefinitely. For the presentation of empirical data we will always be forced to use a finite number of groups. The distribution of relative frequency is necessarily *discrete*.

On the other hand, for a theoretical distribution we hesitate to use a rather arbitrary subdivision into groups. We feel that our quantity x varies continuously. In other words: We consider x as a *continuous random variable* and denote it by X. At the same time we replace relative frequencies by probabilities. We no longer have a histogram with single bars. Instead we draw a smooth curve whose ordinates indicate the *density of probability* at each point of the x axis (Fig. 13.12d). (For the notion of density see end of Section 9.1.) Consider the curve as the graph of a certain function $y = f(x)$. Then $f(x)$ is called the *probability density function*. A distribution which has a density function is said to be *continuous*[11].

---

[11] In strict mathematical terminology such a distribution is called *absolutely continuous*.

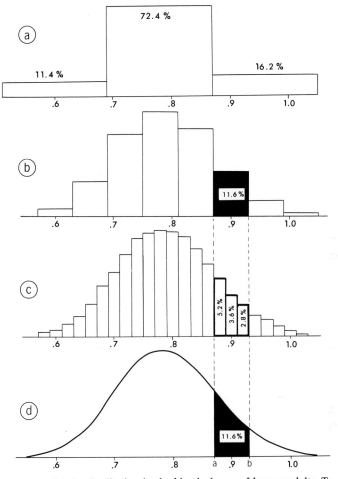

Fig. 13.12. The α-globulin distribution in the blood plasma of human adults. To obtain higher precision the number of groups is increased. In the limit the distribution is continuous

What is the relationship between probability density and probability itself? The transition from a discrete to a continuous distribution gives us a hint. We fix an interval $[a, b]$ on the x axis and ask for the relative frequency of all measurements that fall into this interval. As we see from Fig. 13.12c we have to add the relative frequency of all groups that fall into the interval, that is, $5.2\% + 3.6\% + 2.8\% = 11.6\%$. In other words: We must find the total *area* of all bars of the histogram above $[a, b]$. Quite correspondingly we obtain the probability that the random variable X falls into the interval $[a, b]$ by calculating the area between $[a, b]$ and the curve (Fig. 13.8d). Hence,

$$P(a \leq X \leq b) = \int_a^b f(x)\,dx, \tag{13.11.1}$$

that is, the probability of the compound event $X \in [a, b]$ is the integral of the density function over the interval $[a, b]$.

Why did we not determine the probability of a *simple event* first? Let a be a fixed number. Then $X = a$ would be a simple event. From (13.11.1) it follows that

$$P(X = a) = \int_a^a f(x)\,dx = 0 \tag{13.11.2}$$

since the area over a point of the x axis is zero. The result may look surprising. The event $X = a$ is by no means impossible. Yet its probability vanishes. However, there is no contradiction. Let $a = 2/3 = 0.666...$ be a precise and not a rounded-off number. It is extremely unlikely that a quantity takes on this particular value. In addition, there are infinitely many other numbers in the neighborhood of a so that the probability of a simple event from a continuous distribution must vanish.

In formula (13.11.1) we have expressed a probability in terms of the probability density function. Is the converse also possible, that is, can we derive the probability density function from the knowledge of probabilities? The answer is yes. As a preliminary step in this direction we introduce a new function. Assume for simplicity that the total probability 1 is contained in a finite interval, say from A to B on the x axis (Fig. 13.13). This means that

$$\int_A^B f(x)\,dx = 1. \tag{13.11.3}$$

We consider the probability that the random variable X falls into an interval $[A, x]$ where the upper bound x is variable. According to (13.11.1) this probability is

$$P(X \leq x) = \int_A^x f(x)\,dx.$$

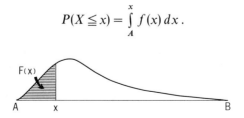

Fig. 13.13. The distribution function of a continuous probability distribution

It is somewhat confusing that x appears as the variable of integration and, at the same time, as the variable upper limit of the integral. Therefore, we change the notation of the variable of integration and choose t instead of x (cf. formula (9.5.11)). The probability $P(X \leq x)$ is a function of x and will be denoted by $F(x)$. Our formula may then be rewritten in the form

$$F(x) = P(X \leq x) = \int_A^x f(t)\, dt . \qquad (13.11.4)$$

The new function $F(x)$ is called the *distribution function*. It has the following properties:

a) $F(A) = 0$ by virtue of (13.11.2),
b) $F(B) = 1$ by virtue of (13.11.3),
c) $F(x)$ is monotone increasing from A to B.

From formula (9.5.5) we know that integration and differentiation are inverse operations. Hence, formula (13.11.4) implies that

$$F'(x) = f(x), \qquad (13.11.5)$$

that is, *the probability density function is the derivative of the distribution function*. Thus the probability density function is the derivative of the probability associated with the interval $[A, x]$.

The knowledge of the distribution function $F(x)$ is quite helpful in all applications of probability theory and statistics. For instance, if $[a, b]$ denotes an interval of the x axis which belongs to $[A, B]$, then we get

$$P(X \leq a) = F(a),$$
$$P(X \leq b) = F(b),$$
$$P(a < X \leq b) = P(X \leq b) - P(X \leq a), \qquad (13.11.6)$$
$$P(a < X \leq b) = F(b) - F(a)^{12} .$$

The result is depicted in Fig. 13.14.

Example 13.11.2. *The Uniform Distribution.* In a study of animal behavior, birds were released one at a time under circumstances that made orientation very difficult. It was expected that the birds would choose *random directions*. What do we mean by "random" in this connection? Directions may be defined by the azimuth α, that is, by the angle between north and the direction measured clockwise. The direction is said to be random if each azimuth from $0°$ to $360°$ has the same chance of

---

[12] Since for a continuous distribution $P(X = a) = 0$ and $P(X = b) = 0$, we may also write $P(a < X < b) = F(b) - F(a)$ or $P(a \leq X \leq b) = F(b) - F(a)$.

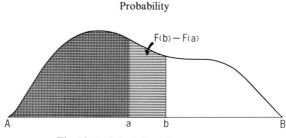

Fig. 13.14. Calculation of $P(a < X \leq b)$

being chosen, or more precisely, if each azimuth has the same probability density. The probability density function $f(\alpha)$ is therefore a constant over the interval $[0°, 360°]$. Since the area between the interval
$[0°, 360°]$ and the graph of $f(\alpha)$ must be 1, the constant value of $f(\alpha)$
is 1/360 (Fig. 13.15). According to (13.11.4) the distribution function is

$$F(\alpha) = \int_0^\alpha \frac{1}{360} \, dt = \alpha/360 \quad (0 \leq \alpha \leq 360°).$$

The example can be generalized for any finite interval $[A, B]$ of the
x axis. A distribution with constant probability density is called a
uniform distribution.

Example 13.11.3. We already know that in radioactivity the
scintillations follow a Poisson distribution (see Example 13.10.4).
The time instants when nuclei decay are denoted by $t_1, t_2, \ldots$ and plotted
as dots on a time axis in Fig. 13.16. In addition, time intervals between
consecutive t-values are denoted by $\Delta t_1, \Delta t_2, \ldots$. We may treat Δt as a
random variable and introduce a new and simpler notation

$$X = \Delta t .$$

This random variable is typically *continuous* as X can take any value
from zero up to infinity.

We are interested in the distribution of X, especially in the probability
density function $f(x)$. A theoretical argument which is not presented

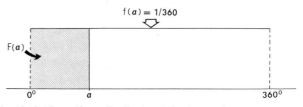

Fig. 13.15. The uniform distribution defined over the interval $[0°, 360°]$

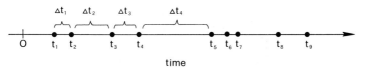

time

Fig. 13.16. Time intervals between subsequent nucleic decays. The probability distribution of these intervals is exponential

here reveals that

$$f(x) = \frac{1}{\mu} e^{-x/\mu} \qquad (x \geq 0)$$

where $\mu = E(X)$ is the *mean duration* between consecutive time instants of decaying nuclei. A distribution with density $f(x)$ is called an *exponential distribution*. For the particular value $\mu = 2.5$ sec it is depicted in Fig. 13.17.

By Eq. (13.11.4) we get the following formula for the distribution function:

$$F(x) = \int_0^x f(t)\, dt = \int_0^x \frac{1}{\mu} e^{-t/\mu}\, dt = -e^{-t/\mu}\Big|_0^x$$

or

$$F(x) = 1 - e^{-x/\mu} \qquad (x \geq 0)$$

We may easily check that $F(0) = 0$ and $\lim_{x \to \infty} F(x) = 1$.

Example 13.11.4. *A Special Normal Distribution.* Many distributions are bell-shaped and fairly symmetric with respect to the mean. A frequently used theoretical model for this kind of distribution is defined by the density function

$$\phi(x) = (2\pi)^{-\frac{1}{2}} \exp(-x^2/2). \qquad (13.11.7)$$

We have introduced the exponential function in Chapter 10. Notice that the exponent $-x^2/2$ is negative and that x is squared. The

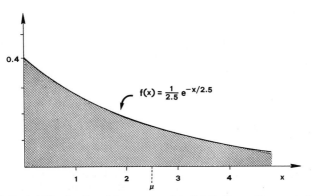

$$f(x) = \frac{1}{2.5} e^{-x/2.5}$$

Fig. 13.17. Probability density of the exponential distribution with expectation $\mu = 2.5$

coefficient of the exponential function is $(2\pi)^{-\frac{1}{2}} = 1/\sqrt{2\pi} = 0.39894$. The function (13.11.7) is defined for the whole real axis.

We add a few statements which we cannot prove in this book:

1) The total probability equals 1 as required, that is,

$$\int\limits_{-\infty}^{\infty} \phi(t)\, dt = (2\pi)^{-\frac{1}{2}} \int\limits_{-\infty}^{\infty} e^{-t^2/2}\, dt = 1 .$$

2) The mean or the expectation of the distribution equals 0.
3) The standard deviation equals 1.

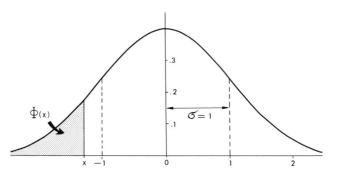

Fig. 13.18. The normal distribution with mean zero and standard deviation one

This distribution is called the *normal or Gaussian*[13] *distribution* with mean zero and standard deviation one. A graph of the density function (13.11.7) is shown in Fig. 13.18. The density function reaches only one maximum. We say that the distribution has only *one mode* or that it is *unimodal*.

Since the normal distribution ranges from $-\infty$ to $+\infty$, it does not appear to be a distribution suitable for applications in the natural sciences. No real quantity can reach an infinite value. However, we should judge the normal distribution from a different point of view. Due to the exponential function in (13.11.7) the density function $f(x)$ tapers off very fast as $|x|$ increases. The probability of a value outside of the interval $[-3, +3]$ is 0.0027, and of a value outside of $[-4, +4]$ only 0.00004. Such an event is practically impossible.

---

[13] The normal distribution was introduced long before C. F. Gauss (1777-1855) brought the distribution into general use in applied mathematics. Cf. last footnote in this section.

The *normal distribution function* is denoted by $\Phi(x)$ and according to (13.11.4) defined by

$$\Phi(x) = \int_{-\infty}^{x} (2\pi)^{-\frac{1}{2}} \exp(-t^2/2)\, dt. \qquad (13.11.8)$$

This integral cannot be expressed by a finite number of functions that are known to us. Fortunately, there are tables available containing numerical values of $\Phi(x)$ and of $\Phi'(x)$ (see Table I of the Appendix and references at the end of this chapter).

Example 13.11.5. *The General Normal Distribution.* The graph in Fig. 13.18 may be "squeezed" or "stretched" to get a standard deviation that differs from one. It may also be shifted along the x axis to provide for a mean value μ which differs from zero. The new density function is

$$(2\pi)^{-\frac{1}{2}}\sigma^{-1} \exp(-(x-\mu)^2/2\sigma^2). \qquad (13.11.9)$$

Again we omit the proof since it would require tools which go beyond the scope of this book. The density function (13.11.9) is plotted in Fig. 13.19 for two different values of the mean and two values of the standard deviation. The two graphs have different heights since the area between the x axis and each graph must be one.

Numerous quantities in the natural sciences seem to be normally distributed, and many statistical procedures are based on the assumption of underlying normal distribution. The question arises: Is the frequent

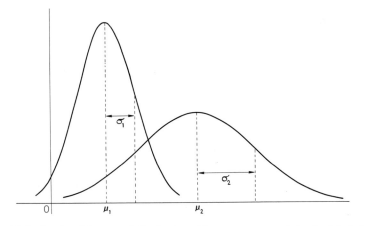

Fig. 13.19. Graphs of the normal distribution with means μ_1, μ_2 and standard deviations σ_1, σ_2, respectively. Geometrically, the standard deviation is the horizontal distance from the mean to one of the points of inflection (cf. Problem 13.11.11)

occurrence of the normal distribution a purely empirical fact, or has it a theoretical basis?

To answer this question let us first consider an example: the height of adult persons. It is well-known that the body height is essentially determined by genetic factors. There is good reason to assume that several genes located at different loci contribute to the body height. Some of them are sex-related. Each gene depends on chance. Before and after birth many other factors also contribute to the body height. Such factors are nutrition, environment, health status, work, and exercise. Some of the many genetic and nongenetic factors tend to increase, some to decrease the height. Most of the factors act independently. Now we form a mathematical model for the height of adult persons. Denote the height by H and the contributions of various factors to H by $X_1, X_2, X_3, \ldots$. Then

$$H = X_1 + X_2 + X_3 + \cdots . \tag{13.11.10}$$

The height H as well as the components X_i are random variables. We do not know the distributions of $X_1, X_2, \ldots$ in detail. Some of the components may take on only positive or only negative values, others may be capable of both signs. We merely know from experience that all these components are reasonably limited in size. We may think that in the absence of any precise knowledge very little can be said about the distribution of H. Surprisingly, this is not the case. Mathematicians are able to prove that *the distribution of H is approximately normal.*

In this connection the so-called *central limit theorem* states[14]:

Let $X_1, X_2, X_3, \ldots$ be an infinite sequence of random variables. Assume that

a) $X_1, X_2, X_3, \ldots$ *are mutually independent,*

b) *each X_i takes on only values from a finite interval $[-A, A]$ where A denotes a constant,*

c) the sum of variances

$$\sum_{i=1}^{n} \text{Var}(X_i) = \sigma_1^2 + \sigma_2^2 + \cdots + \sigma_n^2$$

tends to infinity as $n \to \infty$.

Then the distribution of the partial sum

$$S_n = X_1 + X_2 + X_3 + \cdots + X_n \tag{13.11.11}$$

tends to the normal distribution as $n \to \infty$.

---

[14] The central limit theorem has been proven under a variety of assumptions. We have chosen a form which seems to be especially useful for applications in the natural sciences.

Let us consider this basic theorem and its consequences in detail. The assumption (a) is not always satisfied in applications. In the example of body height, there might be some interaction between genetic factors or between environmental factors. With regard to assumption (b), there is no difficulty whatsoever since there is no restriction imposed on the size of A, and since all quantities in the natural sciences are necessarily finite. Assumption (c) implies that there are infinitely many components whose variances are not too small. This assumption is hardly ever violated in the natural sciences. A restriction is imposed by formula (13.11.11): The components X_i are assumed to be additive. This assumption need not be satisfied. We may for instance think of components which must be multiplied.

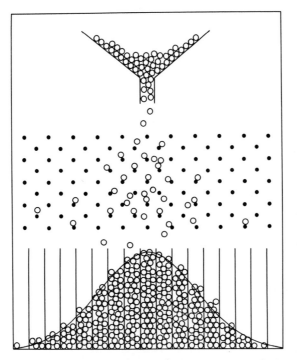

Fig. 13.20. The Galton board. Through a funnel small balls (e.g. shot) enter a board which is inclined to the horizontal. On their way down, the balls strike nails which are placed on the board in many rows. Each ball is deviated either to the right or to the left whenever it collides with a nail or another ball. At the foot of the board are many equally spaced compartments which collect the balls. The compartments near the center receive the most balls. To the sides the frequency tapers off. The distribution resembles closely a normal distribution. The reason is the joint effect of a large number of independent random deviations X_i imposed on the balls. The distribution of a single random variable X_i is not known

Notice that the components may have discrete or continuous distributions. Some of the components may have symmetric, others skew distributions, some either unimodal or multimodal distributions. Whatever the distributions of the components may be, the result is always the same: With increasing n, the distribution of S_n approaches a normal distribution. The central limit theorem was first proved by Ljapunov[15].

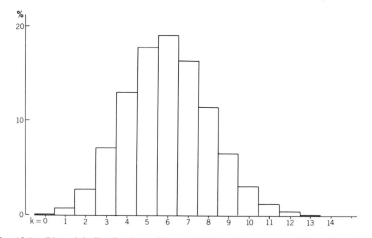

Fig. 13.21. Binomial distribution with probability of success $p = 0.3$ and number of trials $n = 20$. The abscissa k denotes the number of successes. As n increases, the binomial distribution approaches a normal distribution (de Moivre)

The frequent occurrence of distributions that resemble the normal distribution is attributed to the central limit theorem. A model experiment for this theorem is the Galton board[16] as shown in Fig. 13.20.

Finally, let us apply the central limit theorem to formula (13.9.5) for the number of successes in n Bernoulli trials. The random variables $X_1, X_2, ..., X_n$ take on the values 0 and 1 only and are mutually independent. All assumptions of the central limit theorem are satisfied. Hence, the distribution of S_n, known as *binomial distribution*, tends to the *normal distribution* as n tends to infinity. This particular result was found by de Moivre back in 1733[17]. Fig. 13.21 depicts a binomial distribution with probability of success $p = 0.3$ and $n = 20$ trials. We see

[15] Alexander Michailowicz Ljapunov (1857-1918), Russian mathematician.

[16] Sir Francis Galton (1822-1911), English explorer, meteorologist, anthropologist, geneticist, and statistician.

[17] Abraham de Moivre (1667-1754), English mathematician of French origin. He was among the first mathematicians who knew the normal distribution.

that the distribution is slightly skew, but with increasing number of trials the skewness will disappear.

Recommended tables:

a) *Binomial Distribution.* National Bureau of Standards (1950): seven decimal places, *n* ranging from 2 to 50, *p* in steps of 0.01. Individual and cumulative terms.

Romig (1953): six decimal places, *n* ranging from 50 to 100, *p* in steps of 0.01. Individual and cumulative terms.

b) *Poisson Distribution.* Beyer (1966): four decimal places, *m* ranging from 0 to 10. Individual and cumulative terms.

Pearson and Hartley (1966): six decimal places, *m* ranging from 0 to 15 in steps of 0.1.

General Electric Company (1962): eight decimal places, *m* ranging 0 to 205. Individual and cumulative terms.

c) *Normal Density and Distribution Functions.* Beyer (1966): four decimal places.

Diem *et al.* (1970): five decimal places.

Pearson and Hartley (1966): seven decimal places.

National Bureau of Standards (1942): fifteen decimal places.

Recommended for further reading:

Feller (1968); Gelbaum and March (1969); Goldberg (1960); Gnedenko and Khinchin (1961); Grossman and Turner (1974); Hammond and Householder (1963); Hodges and Lehmann (1964); Lefort (1967); Mosimann (1968); Mosteller, Rourke, and Thomas (1961); C. A. B. Smith (1969, Vol. 2).

Problems for Solution

13.2.1. For the systolic blood pressure of a person we may distinguish between the following three cases, called simple events:

A: blood pressure less than 120 mm,

B: blood pressure between 120 mm and 150 mm,

C: blood pressure above 150 mm.

The outcome space is $\{A, B, C\}$. List all possible compound events.

13.2.2. Children are classified according to frame (small $= S$, medium $= M$, large $= L$) and according to height (tall $= T$, short $= $ Sh). Establish an outcome space containing all combinations.

13.2.3. For the sex distribution in families with two children, what are the intersection and the union of the two events "first child a boy" and "second child a boy"? (Hint: Consider the outcome space $\{BB, BG, GB, GG\}$.)

13.2.4. Within a ten-minute-interval a cuckoo can be heard $n = 0, 1, 2, \ldots,$ or 50 times. Let A be the event $\{n \leq 15\}$. What is $\bar{A}$?

13.2.5. Let x denote the variable weight of glucose in 100 ml of blood plasma and let x_0 and x_1 be two fixed values $(x_0 < x_1)$. Define the following four events:

$$A : x < x_0, \quad B : x > x_0. \quad C : x < x_1, \quad D : x > x_1 .$$

Which of the events are mutually exclusive?

13.3.1. In an experiment 14 out of 20 animals react positively. What are the relative frequencies of this event and of the complementary event?

13.3.2. In a behavioral study animals are classified active-hungry (88 individuals), active-fed (41 individuals), resting-hungry (17 individuals), resting-fed (108 individuals). Calculate the relative frequency of each group.

13.3.3. A roulette wheel has 37 positions numbered $0, 1, \ldots, 36$. Assume that the ball comes to rest at each position with equal probabilities. What is the probability
a) of an even number,
b) of a number greater than 30,
c) of a number which is at most 10?

13.3.4. Fifteen similar mice were weighed. In grams the weights are

$$
\begin{array}{ccccc}
x_i = 28 & 31 & 26 & 26 & 29 \\
31 & 30 & 27 & 25 & 30 \\
28 & 28 & 23 & 32 & 30 .
\end{array}
$$

What are the relative frequencies of the following events:

$$E : x < 26 , \quad F : x \leq 26 , \quad G : 26 < x < 31 ?$$

13.3.5. In a cage are twenty rabbits. Six of them have a lethal blood mutation and three others a bone mutation. If a rabbit is selected at random, what is the probability that it has no mutation?

13.3.6. In a human population 35 are of blood group A, 47 of blood group B, 21 of blood group AB, and 4 of blood group 0. What is the probability that a randomly selected individual is of blood group AB?

13.3.7. A pigeon is trained to be rewarded by food only if it pecks three keys A, B, C in the particular order $C\ B\ A$. Describe the sample space of all possible orders in three consecutive peckings assuming that the pigeon hits each key only once. If all these orders are equally probable, what is the probability to get food in the first trial?

13.4.1. At a certain gene locus two alleles C and D can occur. Assume that the possible genotypes have the following probabilities:

$$P(CC) = 0.46, \quad P(CD) = 0.31, \quad P(DD) = 0.23.$$

What is the probability that a genotype contains

(a) the allele C,   (b) the allele D?

13.4.2. Assume that in a population the blood groups A (antigen A present), B (antigen B present), AB (both antigens present), 0 (neither antigen present) occur with the following probabilities:

$$P(A) = 0.35, \quad P(B) = 0.42, \quad P(AB) = 0.18, \quad P(0) = 0.05.$$

What is the probability that a randomly selected individual has a) antigen A, b) antigen B, c) neither antigen?

13.4.3. Let A and B be two events belonging to the same outcome space. Show that

$$P(A) + P(B) = P(A \cup B) + P(A \cap B).$$

13.4.4. In a human population the probability of being deaf is estimated to be 0.0050 and the probability of being blind 0.0085. Both infirmities combined occur with probability 0.0006. What is the probability of being blind and/or deaf? (Hint: Use the result of the previous problem.)

13.5.1. Assume that the sex ratio is $1:1$. It is known to us that a certain family has two children and that one of the children is a girl. What is the probability that the other child is also a girl? (Hint: assign conditional probabilities to the outcome space {BB, BG, GB, GG}).

13.5.2. In the preceding problem, assume that the family has three children and that one of them is a girl. What is the probability that the other two children are a) both boys, b) both girls?

13.5.3. Using Table 13.1 in Section 13.5 find the probability that a) a ten year old child will die during his second decade, b) a fifty year old person will live for at least another decade.

13.5.4. Dealing with MN blood groups, an individual has either an M antigen or an N antigen or both of them (the case "neither M nor N" does not exist). In a population the relative frequencies are:

> antigen M alone 42%,
>
> antigen N alone 33%,
>
> antigen M and N combined 25%.

A randomly selected individual is found to have antigen M. What is the probability that this individual also has antigen N?

13.5.5. In a population the relative frequencies in AB0 blood groups are:

> antigen A present 39%,
>
> antigen B present 48%.

In 15% of all individuals both antigens are present. a) What is the relative frequency of individuals lacking both antigens (blood group 0)? b) If a randomly selected individual has antigen B, what is the probability that he lacks antigen A?

13.5.6. Assume that the occurrence of deafness is independent of sex. Calculate the four missing probabilities in the following table:

	Deaf	Not deaf	Total
male			0.527
female			0.473
total	0.005	0.995	1.000

13.5.7. In a bag there are, well mixed, three red balls and five white balls. What is the probability of drawing red balls in both a first and a second trial? (Notice that the first ball drawn is not replaced.)

13.5.8. A beam of neutrons irradiates two layers of tissue. The probability that a neutron is absorbed by the first layer is 8% and the probability of absorption by the second layer (after passage through the first layer) is 15%. What is the probability that a neutron passes through both layers?

13.5.9. At a locus of a certain pair of chromosomes the alleles A and a may occur. The genotypes AA, Aa, aa have the probabilities

$$p_{AA} = 0.11, \qquad p_{Aa} = 0.37, \qquad \cdot\, p_{aa} = 0.52.$$

At a locus of another pair of chromosomes the alleles B and b may occur. The genotypes BB, Bb, bb have the probabilities

$$p_{BB} = 0.35, \qquad p_{Bb} = 0.25, \qquad p_{bb} = 0.40.$$

Find the probabilities of the gene combinations a) AA together with bb, b) Aa together with Bb.

13.6.1. A symmetric die is thrown twice. What is the probability that a) the two faces are at least 5, b) the sum of the two faces is exactly 11, c) the sum of the two faces is at least 11?

13.6.2. A coin is tossed four times. What is the probability for the event head-head-tail-tail (in this order)?

13.6.3. The AB0-blood groups are determined by three alleles a, b, o at the same gene locus. Assume that the father is of genotype ao and the mother of genotype bo. What is the probability that a child is of genotype a) ab, b) ao, c) bo, d) oo, e) aa? (Cf. Section 3.2.)

13.6.4. In a human population, let p, q, r be the probabilities of the alleles a, b, o, respectively $(p+q+r=1)$. Assume random mating. What is the probability of each of the genotypes oo, ab, ao, aa, bo, bb? Calculate the results for the numerical example $p = 29\%$, $q = 7\%$, $r = 64\%$.

13.6.5. In the preceding problem there are only four phenotypes:
  1. blood group 0   (genotype oo),
  2. blood group AB (genotype ab),
  3. blood group A   (genotypes aa and ao),
  4. blood group B   (genotypes bb and bo).

Let p_1, p_2, p_3, p_4 be the probabilities of these four blood groups, respectively. Verify the formulas

$$p_1 + p_3 = (r+p)^2 = (1-q)^2, \qquad p_1 + p_4 = (r+q)^2 = (1-p)^2.$$

13.7.1. Can we unambiguously code a) five hundred, b) one thousand patients by a two-letter word when the letters are chosen from a 25-letter alphabet?

13.7.2. Nucleic acids (DNA, RNA) are sequences of only four bases. Some RNA-molecules consist only of A = adenine, C = cytosine, G = guanine, U = uracil. How many sequences of ten such constituents can be formed, if unlimited repetition is allowed?

13.7.3. In electronic counters and calculators digits are often represented by illuminating some of the seven bars assembled in the pattern ⊟. Thus

$$1 = |, \quad 2 = ⊐, \quad 3 = ⊒, \quad 4 = ⊔, \quad 5 = ⊏, \quad 6 = ⊔, \text{ etc.}$$

How many different symbols can be expressed by the seven bars?

13.7.4. Find

a) $7!$ b) $7!/6!$,   c) $8!/6!$,   d) $9!/11!$,
e) $7!/(3!\,4!)$,   f) $98!\,2!/100!$,   g) $0!/1!$,   h) $4!/(2!\,1!\,1!\,0!)$.

13.7.5. Calculate $\binom{5}{2}, \binom{6}{3}, \binom{8}{3}, \binom{9}{8}, \binom{9}{9}, \binom{11}{4}$.

13.7.6. Using the symmetry property, Eq. (13.7.9), calculate

$$\binom{10}{8}, \quad \binom{100}{98}, \quad \binom{9}{6}, \quad \binom{20}{17}.$$

13.7.7. Using Pascal's triangle calculate $\binom{10}{k}$, $k = 1, 2, 3, \ldots$ from known values $\binom{9}{0} = 1, \binom{9}{1} = 9, \binom{9}{2} = 36, \binom{9}{3} = 84, \binom{9}{4} = 126, \ldots$

13.7.8. Five animals are assigned five different treatments. In how many ways can this be done?

13.7.9. Inbred mice are used for an experiment. Out of a total of 38 mice three mice are selected. How many such selections are possible?

13.7.10. In a study of behavior four animals are assigned one of six tasks. In how many ways can this be done if the same task can occur a) repeatedly, b) only once?

13.7.11. Due to a shortage of vaccine only four out of 10 persons can be vaccinated. In how many ways could the four persons be chosen?

13.7.12. How many distinct permutations are there of the digits of the following numbers:

a) 1975,   b) 3844,   c) 173112?

13.7.13. An experimenter has 5 cages with rats and 3 cages with mice. In how many ways can he place the cages in a row if he does not distinguish among the cages with rats nor among the cages with mice?

13.7.14. Out of 10 male and 7 female experimental cats, a sample consisting of two males and two females was chosen. In how many ways could this be done?

13.7.15. Consider an experimental colony consisting of ten worker bees and a queen. In how many ways can a group of four bees be chosen if a) the queen must belong to the group, b) the queen must not belong to the group?

13.7.16. Generalize the preceding problem for a colony consisting of n worker bees and one queen. The group to be selected should contain k bees. With the result conclude that

$$\binom{n}{k-1} + \binom{n}{k} = \binom{n+1}{k}.$$

13.7.17. Using the binomial theorem (13.7.12) expand the following expressions:

a) $(1+x)^4$ b) $(1-x)^4$ (Hint: let $a=1$ and $b=-x$)
c) $(2+p)^5$ d) $(z+\frac{1}{2})^3$ e) $(a+1)^6 - (a-1)^6$.

13.7.18. Show that

a) $\binom{n}{0} + \binom{n}{1} + \binom{n}{2} + \cdots + \binom{n}{n} = 2^n$,

b) $\binom{n}{0} - \binom{n}{1} + \binom{n}{2} - \cdots \pm \binom{n}{n} = 0$.

(Hint: expand $(1+1)^n$ and $(1-1)^n$.)

13.7.19. Simplify

a) $\binom{n+1}{k} / \binom{n}{k}$,   b) $(k-1)! \binom{m}{k}$,   c) $\binom{N}{4} + \binom{N}{5}$.

13.7.20. Using the binomial theorem (13.7.12) expand the following expressions:

a) $(1+x)^5$,   b) $(1+x)^m$ (m = natural number),
c) $(3u+7)^3$,   d) $(7r-4)^3$,   e) $(m-n)^5$.

13.7.21. Expand

a) $(a+b)^{n-1}$,   b) $(a+b)^{2n}$.

13.8.1. Assume that the sex ratio is $1:1$. What is the probability that a family with five children has a) 2 boys, 3 girls, b) no girls, c) at least one boy, d) at most two boys?

13.8.2. From a table of random numbers ten two-digit numbers are read (without previous inspection). What is the probability that exactly five of them are even and five odd? Using Table K perform a repeated experiment of this sort and compare the outcome with your theoretical result.

13.8.3. Let a game consist of 24 throws of a pair of symmetric dice. Is it more advantageous to bet that no "double six" occurs or to bet that at least one "double six" occurs? (Problem of the French nobleman and passionate gambler Chevalier de Méré around 1650.)

13.8.4. From many years of observation it is estimated that in a particular region snowfall (of at least 1 cm) per winter occurs with probability $p=0.57$, and that the event is independent from winter to winter. Find the probability that snowfall occurs at least once in three consecutive winters.

13.8.5. A device to catch wasps was tested. Only 128 out of 720 wasps attracted by the bait could be caught. Therefore, the probability of catching a wasp is estimated to be $p=128/720=0.178$. What is the probability that out of three randomly chosen wasps a) none is caught, b) at least one is caught?

13.8.6. In a certain population, left-handedness occurs with probability $p=0.18$. What is the probability that in a class of 10 children exactly 2 are left-handed?

13.8.7. If both parents are of genotype Aa (heterozygous), their children are of genotypes AA, Aa, aa with probabilities

$$P(AA)=\tfrac{1}{4}, \qquad P(Aa)=\tfrac{1}{2}, \qquad P(aa)=\tfrac{1}{4}$$

(see Example 13.4.3). What is the probability that out of four children a) exactly one is of genotype aa, b) at least one is of genotype aa?

13.8.8. Assume that the sex ratio is $1:1$. It is known that a certain family with four children has at least a girl. What is the probability that all four children are girls?

13.8.9. Solve the same problem when it is known that at least two of the four children are girls.

13.8.10. A farmer's family wants to have at least one male child who could inherit the farm. Do three children give a chance of 90% (or more) that one child is a boy?

13.8.11. In many developing countries aging parents are economically dependent on male descendants. Assuming that one out of five children dies before reaching adulthood, what is the probability that at least one male child survives when five children were born? (Assume for simplicity that boys and girls are equally likely and that they have the same mortality.)

13.9.1. Consider an outcome space $\{E_1, E_2, E_3\}$ with three elementary events and probabilities

$$P(E_1) = 0.5, \qquad P(E_2) = 0.3, \qquad P(E_3) = 0.2.$$

Define a random variable X by

$$X(E_1) = 1, \qquad X(E_2) = 5, \qquad X(E_3) = 10.$$

a) Plot a histogram of this distribution,
b) Plot the distribution function $F(x)$,
c) Calculate the expectation and the standard deviation of X.

13.9.2. A certain game of chance results in a winning $ 100 with probability 0.1 and a winning $ 50 with probability 0.25. With the remaining probability there is nothing to win. a) What is the probability of winning at least $ 50? b) Calculate the average winning (expectation).

13.9.3. A certain treatment is able to immunize 78% of rabbits against a disease. A new sample of 50 rabbits is tested. Let X denote the number of animals that will become immune. What are the expectation and the standard deviation of X?

13.9.4. Using Table G of the Appendix plot a histogram of the binomial distribution with $n = 7$ and $p = 0.35$. Also plot the distribution function.

13.9.5. In Table G of the Appendix find the terms of the binomial distribution with $p = 0.05$ and $n = 6$. From these terms calculate the expectation of S_6 and check whether the result is in agreement with Eq. (13.9.19).

13.9.6. A symmetric die is thrown 24 times. Consider the desired event $E = \{\text{face six}\}$. Let X be the frequency at which E occurs. Find $E(X), \text{Var}(X)$, and σ. If someone obtains seven "sixes" in such a game, does this indicate any unusual ability? (Hint: Identify X with S_n in Eqs. (13.9.19) and (13.9.20).)

13.9.7. Assume that the sex ratio is $1:1$. Determine the probabilities that in a randomly selected family with five children the number of boys is $X = 0$ or 1 or 2 or ...5. Plot a graph of the distribution function $F(x)$ of the random variable X. Also find $E(X)$ and Var X. (Hint: Identify X with S_n in Eqs. (13.9.19) and (13.9.20).)

13.9.8. Two hundred families with three children are sampled at random. How many families do we expect with a) no girl, b) one girl, c) two girls, provided that the sex ratio is $1:1$?

13.9.9. Consider children of parents of genotype Aa. The distribution is

$$P(AA) = 0.25, \qquad P(Aa) = 0.50, \qquad P(aa) = 0.25$$

(see Example 13.4.3). 240 such children are sampled at random. Let N_1, N_2, N_3 be the number of children of genotypes AA, Aa, aa, respectively. a) Find $E(N_1), E(N_2), E(N_3)$. b) Find $\text{Var}(N_1)$, $\text{Var}(N_2), \text{Var}(N_3)$ and the standard deviations.

13.9.10. Find an alternative proof of Eq. (13.9.19) using differential calculus. Consider the function

$$f(x) = (x + q)^n = \binom{n}{0} q^n + \binom{n}{1} xq^{n-1} + \binom{n}{2} x^2 q^{n-2} + \cdots .$$

Calculate $f'(x)$ and show that $E(S_n) = pf'(p)$.

13.10.1. For $m = 0.5$ calculate the terms of the Poisson distribution. For e^{-m} use Table E of the Appendix. Compare the result with Table H.

13.10.2. For $m = 5$ show that the terms $P(X = 4)$ and $P(X = 5)$ of the Poisson distribution are exactly equal.

13.10.3. Let $m = 2$. Calculate $E(X)$ and Var X for the Poisson distribution using the individual terms of Table H of the Appendix. Show that the results are in agreement with Eqs. (13.10.6) and (13.10.7).

13.10.4. Let $m = 1.8$. Assuming Poisson distribution find a) $P(X \geq 7)$, b) $P(X < 2)$, c) $P(X > 0)$ by means of Table H.

13.10.5. In an experiment with radioactive tracer, 4.4 scintillations per second were observed on the average. a) What is the standard deviation of the actual number of scintillations per second? b) Find the probability that at least one scintillation occurs in a given time interval of length 1 sec.

13.10.6. An electronic bacteria counter registers on the average 375 bacteria per cm^3 of a liquid. Assuming Poisson distribution, what is the standard deviation of the number of bacteria per cm^3?

13.10.7. When counting erythrocytes (red blood cells) a square grid may be used, over which a drop of blood is evenly distributed. Under the microscope an average of 8 erythrocytes is observed per single square. Is it reasonable that countings per single square frequently deviate by 1 or 2 from the average 8, and occasionally by 3 or more?

13.10.8. Eight experimental rats move "randomly" on a floor which is subdivided into twenty quadrats of equal size numbered 1, 2, ..., 20. A picture is taken. Assuming Poisson distribution a) what is the probability that exactly one rat is found in quadrat No. 1, b) how many quadrats are expected to contain 0, 1, 2, etc. rats? (Hint: Calculate the average number of rats per quadrat and use Table H.)

13.10.9. A hundred birds are given a chance, one at a time, to select between twenty like cages which are arranged in a circle. We assume that the birds have no directional preference. What is the probability that a specific cage is chosen 0, 1, 2, ... times? (Hint: use the Poisson distribution.)

13.10.10. In a 48 hour period a total of 28 snails could be caught in 20 traps. Assuming Poisson distribution, how many traps are expected to contain 0, 1, 2, ... snails? (Hint: Use Table H.)

13.11.1. When studying biological rhythms, an experimenter registers the time instant each morning when a hamster begins its activity. He finds that the time instant varies between $5:35$ and $6:10$. Assuming that the distribution is uniform, plot a graph a) of the probability density function, b) of the distribution function.

13.11.2. The life span of a sort of light bulbs is exponentially distributed. If the average life span (expectation) is 3.6 months, establish the probability density function. (Hint: cf. Example 13.11.3.)

13.11.3. Falling leaves from a tree are distributed by winds in all directions. The probability of finding leaves at a distance r from the trunk decreases with increasing r. Assume that the probability distribution is exponential that is,

$$f(r) = \frac{1}{\mu} e^{-r/\mu} \qquad (r \geq 0)$$

(cf. Example 13.11.3). With a mean distance of $\mu = 8$ m plot this density function.

13.11.4. In the preceding problem find the probability that a leaf falls inside a circle of radius $r = 8$ m.

13.11.5. Using Table I plot the density function of the normal distribution with $\mu = 0$ and $\sigma = 1$.

13.11.6. Using Table I plot the density function of the normal distribution with $\mu = 0$ and $\sigma = 0.5$.

13.11.7. Using Table I plot the density function of the normal distribution with $\mu = 8$ and $\sigma = 2$.

13.11.8. Using Table I plot a graph of the normal distribution function
with $\mu = 0$ and $\sigma = 1$.

13.11.9. Using Table I plot a graph of the normal distribution function
with $\mu = 0$ and $\sigma = 0.5$.

13.11.10. Assume normal distribution with $\mu = 0$ and $\sigma = 1$. Using Table I
find the following probabilities:

a) $P(X \leqq 1)$ b) $P(X \leqq 2.7)$ c) $P(X > 2)$ d) $P(X \leqq -2)$
e) $P(0 < X \leqq 1)$ f) $P(1 < X \leqq 2)$ g) $P(-1 < X \leqq 1)$

13.11.11. By means of the second derivative determine the points of
inflection of the probability density function (13.11.7).

Chapter 14

Matrices and Vectors

14.1. Notations

In the last decades matrices have become indispensable in various applications of mathematics to biology, especially in statistics. We will deal with such applications in Sections 14.3 and 14.9.

An important special case of matrices is vectors. We will concentrate on vectors and their many applications to life sciences in Sections 14.4 and 14.5.

Prerequisites for understanding this chapter are Chapters 1, 2, 3 and 5, for some applications also Chapters 11 and 13.

Matrices are *rectangular arrays of numbers*. Thus for simultaneous equations such as

$$a_1 x + b_1 y + c_1 z = d_1 ,$$

$$a_2 x + b_2 y + c_2 z = d_2 ,$$

(14.1.1)

we may form three matrices

$$\begin{pmatrix} a_1 & b_1 & c_1 \\ a_2 & b_2 & c_2 \end{pmatrix}, \quad \begin{pmatrix} x \\ y \\ z \end{pmatrix}, \quad \begin{pmatrix} d_1 \\ d_2 \end{pmatrix}.$$

(14.1.2)

The first matrix contains the known coefficients of x, y, z in their proper order. The second matrix contains the unknown quantities, and the last matrix the known quantities on the right side of the equations.

In a similar way we may form matrices for other types of equations such as

$$a_0 x^3 + a_1 x^2 + a_2 x + a_3 = 0$$

or

$$\frac{dx}{dt} = ax + by, \quad \frac{dy}{dt} = cx + dy .$$

It does not matter whether matrices are written with parentheses, brackets, or braces. Thus by

$$\begin{pmatrix} a & b \\ c & d \end{pmatrix}, \quad \begin{bmatrix} a & b \\ c & d \end{bmatrix}, \quad \begin{Bmatrix} a & b \\ c & d \end{Bmatrix}$$

we mean the same thing. However, matrices should not be confused with symbols of a similar kind. For instance, $\begin{pmatrix} n \\ k \end{pmatrix}$ may be a binomial coefficient as defined by (13.7.8) or (13.7.11) rather than a matrix. The symbol

$$\begin{vmatrix} a & b \\ c & d \end{vmatrix}$$

with vertical bars does not stand for a matrix, but for the quantity $ad - bc$. This symbol is called a *determinant* (see Section 14.6).

A useful notation is *double subscripts*. Suppose a research worker performs an experiment repeatedly under various conditions using different treatments. We denote the number of treatments by m and the number of measurements at each treatment by n. If n and m are not particularly small, the alphabet would not contain enough letters to denote the single measurements, nor would single subscripts such as in $x_1, x_2, \ldots$ be helpful. It is convenient to introduce a second subscript. Thus by

$$x_{23} \ (\text{read}: x \text{ two, three})$$

we mean the third measurement of the second treatment. The measurements form the matrix

Treatment No.	Measurement No.			
	1	2	3	... n
1	x_{11}	x_{12}	$x_{13} \ldots x_{1n}$	
2	x_{21}	x_{22}	$x_{23} \ldots x_{2n}$	
...				
m	x_{m1}	x_{m2}	$x_{m3} \ldots x_{mn}$	

Each x is called an *element* or a *component* of the matrix. The general element in the ith row and the jth columns is x_{ij}. The first subscript designates the row, the second the column in which the element is located. In our notation, i ranges from 1 to m, and j ranges from 1 to n.

For brevity, matrices are denoted by single letters. This may cause some misunderstanding since letters frequently stand for single numbers and not for arrays of numbers. Therefore, some distinction is desirable. In print, we use *boldface letters* as notations for matrices. For instance, we may write

$$A = \begin{pmatrix} 3 & 5 & -1 \\ 0 & 7 & 2 \end{pmatrix}, \quad y = \begin{pmatrix} y_1 \\ y_2 \\ y_3 \end{pmatrix}.$$

For clarity, the number of rows and columns is often added to the notation. Thus, A is a "2 by 3" matrix and y a "3 by 1" matrix, symbolically

$$A = A^{2 \times 3}, \quad y = y^{3 \times 1}.$$

A matrix consisting of a single row or a single column is frequently called a *vector* for reasons that will be explained in Section 14.4. Thus

$$u = u^{1 \times 4} = (u_1, u_2, u_3, u_4)$$

is a *row vector* and

$$s = s^{3 \times 1} = \begin{pmatrix} s_1 \\ s_2 \\ s_3 \end{pmatrix}$$

a *column vector*. Notice that the elements of row vectors are often separated by commas.

A matrix with as many rows as columns is said to be a *square matrix*. Examples are

$$Q^{2 \times 2} = \begin{pmatrix} 10 & -2 \\ -4 & 10 \end{pmatrix}, \quad X^{n \times n} = \begin{pmatrix} x_{11} & x_{12} & \cdots & x_{1n} \\ x_{21} & x_{22} & \cdots & x_{2n} \\ \multicolumn{4}{c}{\dotfill} \\ x_{n1} & x_{n2} & \cdots & x_{nn} \end{pmatrix}.$$

The elements $x_{11}, x_{22}, \ldots, x_{nn}$ $(i = j)$ form the so-called *main diagonal* or simply the *diagonal* of $X^{n \times n}$. The other elements x_{ij} with $i \neq j$ are either above or below the diagonal. They are the *off-diagonal* elements. Only square matrices have a diagonal. An $n \times n$ matrix is sometimes called a matrix of *order n*.

14.2. Matrix Algebra

The concept of a matrix would be trivial if it consisted merely in a notation for special arrays of numbers. The importance of matrices stems from the fact that very useful operations on matrices can be defined.

Two matrices are said to be equal if corresponding elements coincide. Equal matrices must have the same number of rows and columns. Thus

$$\begin{pmatrix} a_{11} & a_{12} \\ a_{21} & a_{22} \end{pmatrix} = \begin{pmatrix} 1 & 0 \\ -2 & 0 \end{pmatrix}$$

implies that $a_{11} = 1$, $a_{12} = 0$, $a_{21} = -2$, $a_{22} = 0$. Notice that

$$\begin{pmatrix} u \\ v \\ w \end{pmatrix} \neq \begin{pmatrix} v \\ u \\ w \end{pmatrix} \text{ if } u \neq v, \quad \begin{pmatrix} a_{11} & a_{12} & 0 \\ a_{21} & a_{22} & 0 \end{pmatrix} \neq \begin{pmatrix} a_{11} & a_{12} \\ a_{21} & a_{22} \end{pmatrix}.$$

To introduce the *addition of matrices* we consider an example: It is well-known that the ability to taste phenylthiocarbamide is inheritable. The following sample was collected:

Parental type	Number of children	
	taster	non-taster
taster × taster	88	13
taster × non-taster	52	25
non-taster × non-taster	0	19

The data forms a 3×2 matrix which we denote by A. Later a larger sample was collected. With the same arrangement, the new matrix became

$$B = \begin{pmatrix} 122 & 18 \\ 102 & 73 \\ 0 & 91 \end{pmatrix}.$$

It is quite natural to pool the data, that is, to add corresponding frequencies. This operation is known as addition of the two matrices. Thus

$$A + B = \begin{pmatrix} 88 & 13 \\ 52 & 25 \\ 0 & 19 \end{pmatrix} + \begin{pmatrix} 122 & 18 \\ 102 & 73 \\ 0 & 91 \end{pmatrix} = \begin{pmatrix} 210 & 31 \\ 154 & 98 \\ 0 & 110 \end{pmatrix}.$$

In general, we can only add matrices that have the same number of rows and the same number of columns. Let

$$A = \begin{pmatrix} a_{11} & a_{12} & \cdots & a_{1n} \\ a_{21} & a_{22} & \cdots & a_{2n} \\ \cdots\cdots\cdots\cdots\cdots\cdots \\ a_{m1} & a_{m2} & \cdots & a_{mn} \end{pmatrix}, \quad B = \begin{pmatrix} b_{11} & b_{12} & \cdots & b_{1n} \\ b_{21} & b_{22} & \cdots & b_{2n} \\ \cdots\cdots\cdots\cdots\cdots\cdots \\ b_{m1} & b_{m2} & \cdots & b_{mn} \end{pmatrix}.$$

Then the sum of A and B is defined by

$$A + B = \begin{pmatrix} a_{11} + b_{11} & a_{12} + b_{12} & \cdots & a_{1n} + b_{1n} \\ a_{21} + b_{21} & a_{22} + b_{22} & \cdots & a_{2n} + b_{2n} \\ \hdotsfor{4} \\ a_{m1} + b_{m1} & a_{m2} + b_{m2} & \cdots & a_{mn} + b_{mn} \end{pmatrix}. \tag{14.2.1}$$

From formula (1.14.1) we conclude that matrix addition obeys the following laws:

$$A + B = B + A \text{ (commutative law)}, \tag{14.2.2}$$

$$A + (B + C) = (A + B) + C \text{ (associative law)}. \tag{14.2.3}$$

In particular, $A + A$ leads to a matrix in which each element is doubled. It is natural to write $2A$ for this matrix. Similarly, let k be any number. Then we define

$$kA = k \begin{pmatrix} a_{11} & \cdots & a_{1n} \\ \hdotsfor{3} \\ a_{m1} & \cdots & a_{mn} \end{pmatrix} = \begin{pmatrix} ka_{11} & \cdots & ka_{1n} \\ \hdotsfor{3} \\ ka_{m1} & \cdots & ka_{mn} \end{pmatrix}. \tag{14.2.4}$$

For $k = -1$ we get $(-1)A$ which will be denoted by $-A$. With this matrix, we may define *matrix subtraction* by

$$B - A = B + (-A). \tag{14.2.5}$$

Less easy to learn is the *multiplication of one matrix by another matrix*. We consider first a special case, the multiplication of a row vector by a column vector with equal number of components. Let

$$A = (a_1, a_2, \ldots, a_n), \qquad B = \begin{pmatrix} b_1 \\ b_2 \\ \vdots \\ b_n \end{pmatrix}.$$

Then the product of A and B is defined by

$$AB = (a_1 b_1 + a_2 b_2 + \cdots + a_n b_n) = \left(\sum_{i=1}^{n} a_i b_i \right). \tag{14.2.6}$$

The result is a matrix consisting of a single element, the number $a_1 b_1 + \cdots + a_n b_n$. It is customary but strictly speaking not correct to interpret and to treat such a matrix as an ordinary number. As a consequence, the parentheses in (14.2.6) are usually omitted. For instance, assume that in a survey 88 families have no children, 217 families have one child, and 370 families have two children. Then the total number of children follows by the matrix multiplication

$$(88, 217, 370) \cdot \begin{pmatrix} 0 \\ 1 \\ 2 \end{pmatrix} = 88 \times 0 + 217 \times 1 + 370 \times 2 = 957 \,.$$

It is worth stating formula (14.2.6) in words: *A row vector with components numbered* $1, \dots, n$ *is multiplied by a column vector with components numbered* $1, \dots, n$ *by multiplying equally numbered components and then adding the products. The result is a single number.* Such a sum of products of corresponding components is often called an *inner product* of two vectors. A reader who is not familiar with this concept should work out a good number of self-made numerical examples before he proceeds with the text.

A word of caution is appropriate. It is quite correct if the elements of A are measured in one type of units and the elements of B in another type of units. For instance,

$$AB = (3 \text{ kp}, 4 \text{ kp}, 1 \text{ kp}) \begin{pmatrix} 2\,\text{m} \\ 0\,\text{m} \\ 5\,\text{m} \end{pmatrix} = 6 \text{ kpm} + 5 \text{ kpm} = 11 \text{ kpm}$$

is all right since the addition can be performed, but

$$AB = (3 \text{ kp}, 7 \text{ m}, 4) \begin{pmatrix} 3 \\ 1 \\ 2 \end{pmatrix} = 9 \text{ kp} + 7 \text{ m} + 8$$

would be nonsense.

Now we slightly generalize the rule by multiplying two matrices which we have already listed in formula (14.1.2):

$$C = \begin{pmatrix} a_1 & b_1 & c_1 \\ a_2 & b_2 & c_2 \end{pmatrix} \qquad u = \begin{pmatrix} x \\ y \\ z \end{pmatrix}$$

We calculate the inner products of each row of C with the column vector u and form a new matrix:

$$Cu = \begin{pmatrix} a_1 & b_1 & c_1 \\ a_2 & b_2 & c_2 \end{pmatrix} \begin{pmatrix} x \\ y \\ z \end{pmatrix} = \begin{pmatrix} a_1 x + b_1 y + c_1 z \\ a_2 x + b_2 y + c_2 z \end{pmatrix} \,.$$

Let us compare the result with formula (14.1.1). Using the notation

$$d = \begin{pmatrix} d_1 \\ d_2 \end{pmatrix}$$

we get

$$Cu = d. \tag{14.2.7}$$

This brief matrix equation is equivalent to the system (14.1.1) of simultaneous equations.

In general, let

$$A = A^{m \times r} = \begin{pmatrix} a_{11} & a_{12} & \cdots & a_{1r} \\ \cdots\cdots\cdots\cdots\cdots\cdots \\ a_{m1} & a_{m2} & \cdots & a_{mr} \end{pmatrix}, \quad B = B^{r \times n} = \begin{pmatrix} b_{11} & b_{12} & \cdots & b_{1n} \\ \cdots\cdots\cdots\cdots\cdots\cdots \\ b_{r1} & b_{r2} & \cdots & b_{rn} \end{pmatrix}.$$

Then we form the inner products of each row of A with each column of B. The rule of thumb

$$\boxed{\text{row by column}}$$

has to be carefully observed. The first inner product is

$$a_{11}b_{11} + a_{12}b_{21} + \cdots + a_{1r}b_{r1}.$$

As we have m rows of A to multiply by n columns of B, we get mn entries for the product matrix. The product of A and B is therefore an $m \times n$ matrix which we denote by $C = C^{m \times n}$. Hence

$$\boxed{A^{m \times r} \, B^{r \times n} = C^{m \times n}.} \tag{14.2.8}$$

The inner product of the ith row of A and the jth column of B stands at the intersection of the ith row and the jth column of C. Thus this element is

$$c_{ij} = a_{i1}b_{1j} + a_{i2}b_{2j} + \cdots + a_{ir}b_{rj}.$$

In order that two matrices can be multiplied by each other we require that *the number of columns of the first matrix correspond to the number of rows of the second matrix.* Thus, in formula (14.2.8), A has r columns and B has r rows. When this requirement is not satisfied, we cannot form the inner products, and therefore no matrix product exists.

A numerical example may be useful at this point:

$$\begin{pmatrix} 3 & 0 & 2 \\ -1 & 3 & 5 \end{pmatrix} \begin{pmatrix} 8 & 0 & 5 \\ 1 & 7 & -3 \\ 1 & 2 & 0 \end{pmatrix} = \begin{pmatrix} 26 & 4 & 15 \\ 0 & 31 & -14 \end{pmatrix}.$$

The elements of the product matrix are

$$3 \times 8 + 0 \times 1 + 2 \times 1 = 26,$$
$$3 \times 0 + 0 \times 7 + 2 \times 2 = 4,$$
$$3 \times 5 + 0 \times (-3) + 2 \times 0 = 15,$$
$$(-1) \times 8 + 3 \times 1 + 5 \times 1 = 0,$$
$$(-1) \times 0 + 3 \times 7 + 5 \times 2 = 31,$$
$$(-1) \times 5 + 3 \times (-3) + 5 \times 0 = -14.$$

In searching for laws of multiplication we easily detect that the *commutative law cannot be valid*. For instance, if

$$A = (a_1, a_2, a_3), \qquad B = \begin{pmatrix} b_1 \\ b_2 \\ b_3 \end{pmatrix},$$

we get as in (14.2.6)

$$A\,B = a_1 b_1 + a_2 b_2 + a_3 b_3,$$

whereas

$$B\,A = \begin{pmatrix} b_1 \\ b_2 \\ b_3 \end{pmatrix} (a_1, a_2, a_3) = \begin{pmatrix} b_1 a_1 & b_1 a_2 & b_1 a_3 \\ b_2 a_1 & b_2 a_2 & b_2 a_3 \\ b_3 a_1 & b_3 a_2 & b_3 a_3 \end{pmatrix}.$$

Hence,

$$\boxed{A\,B \neq B\,A}.$$
(14.2.9)

The associative law, however, is valid:

$$A\,(B\,C) = (A\,B)\,C.$$
(14.2.10)

We omit the proof since it is rather cumbersome. We confine ourselves to working out an example. Let

$$u = \begin{pmatrix} x \\ y \end{pmatrix}, \qquad u' = (x, y), \qquad A = \begin{pmatrix} a & b \\ c & d \end{pmatrix}.$$
(14.2.11)

The matrix u' is a row vector containing the same components in the same order as the column vector u. The matrix A is square, that is, it has the same number of rows and columns. We now form the product

$$u'A = (x, y)\begin{pmatrix} a & b \\ c & d \end{pmatrix} = (ax + cy, bx + dy)$$

and multiply the result by u from the right:

$$(u'A)u = (ax + cy, bx + dy)\begin{pmatrix} x \\ y \end{pmatrix} = ax^2 + cxy + bxy + dy^2. \quad (14.2.12)$$

According to the associative law we could have obtained the same result by first forming $A u$ and then multiplying from the left by u'. Indeed,

$$A u = \begin{pmatrix} a & b \\ c & d \end{pmatrix}\begin{pmatrix} x \\ y \end{pmatrix} = \begin{pmatrix} ax + by \\ cx + dy \end{pmatrix},$$

$$u'(A u) = (x, y)\begin{pmatrix} ax + by \\ cx + dy \end{pmatrix} = ax^2 + bxy + cxy + dy^2$$

which coincides with (14.2.12). The example is also of interest from a different point of view. It shows that quadratic functions can be written as matrix products.

Finally, we will search for a law connecting addition and multiplication. For ordinary numbers we know the *distributive law* (see formula (1.14.3)). Fortunately, the same law is also valid for matrices:

$$A(B + C) = AB + AC. \quad (14.2.13)$$

Again we omit the proof since it would be lengthy. An example is given in Problem 14.2.17.

In (14.2.11) we defined two vectors u and u' with the same elements in the same order. The row vector u' is called the *transpose* of the column vector u. In matrix algebra it frequently occurs that we have to change columns into rows or rows into columns. Thus

$$C' = \begin{pmatrix} a_1 & a_2 \\ b_1 & b_2 \\ c_1 & c_2 \end{pmatrix}$$

is defined to be the transpose[1] of the matrix C used in (14.2.7). The forming of a transpose is an operation on matrices which does not occur in ordinary numbers. Notice that a second application of this operation

---

[1] Some authors prefer the notation C^T, C', or $'C$.

generates the original matrix:

$$(C')' = C. \tag{14.2.14}$$

It is easy to prove that if A and B are $m \times n$ matrices, then

$$(A + B)' = A' + B'.$$

The transpose of a square matrix is again square. Thus

$$\begin{pmatrix} 3 & -1 & 2 \\ 0 & 4 & 0 \\ 5 & 8 & -1 \end{pmatrix}' = \begin{pmatrix} 3 & 0 & 5 \\ -1 & 4 & 8 \\ 2 & 0 & -1 \end{pmatrix}.$$

The diagonal elements 3, 4, -1 do not change their position.

A square matrix is called *symmetric* if it is equal to its transpose. Thus

$$M = \begin{pmatrix} 7 & 0 & 5 \\ 0 & 3 & -2 \\ 5 & -2 & 0 \end{pmatrix}$$

is symmetric since $M' = M$.

A very special case of a symmetric matrix is the so-called *unit matrix* or *identity matrix*

$$I = I^{n \times n} = \begin{pmatrix} 1 & 0 & 0 & \cdots & 0 \\ 0 & 1 & 0 & \cdots & 0 \\ 0 & 0 & 1 & \cdots & 0 \\ & & \cdots\cdots\cdots & \\ 0 & 0 & 0 & \cdots & 1 \end{pmatrix} \tag{14.2.15}$$

whose diagonal elements are 1 and whose off-diagonal elements are 0. The name stems from the fact that any matrix with an admissible number of rows or columns is reproduced when multiplied by I. For instance, using C from (14.2.7) we get

$$IC' = \begin{pmatrix} 1 & 0 & 0 \\ 0 & 1 & 0 \\ 0 & 0 & 1 \end{pmatrix} \begin{pmatrix} a_1 & a_2 \\ b_1 & b_2 \\ c_1 & c_2 \end{pmatrix} = \begin{pmatrix} a_1 & a_2 \\ b_1 & b_2 \\ c_1 & c_2 \end{pmatrix} = C'$$

and

$$CI = \begin{pmatrix} a_1 & b_1 & c_1 \\ a_2 & b_2 & c_2 \end{pmatrix} \begin{pmatrix} 1 & 0 & 0 \\ 0 & 1 & 0 \\ 0 & 0 & 1 \end{pmatrix} = \begin{pmatrix} a_1 & b_1 & c_1 \\ a_2 & b_2 & c_2 \end{pmatrix} = C.$$

Notice further that if I is the identity matrix of order n and A is any square matrix of order n, then $IA = AI = A$. In other words, the identity matrix commutes with any other square matrix of the same order.

14.3. Applications

Example 14.3.1. *Graph theory.* It frequently occurs that objects are related to certain characteristics and that we have to study what we called in Section 3.3 a *relation*. A useful tool in the study of relations is *graph theory*. A well-known example are chemical graphs or structural formulas. The vertices are atoms and the edges chemical bonds. We form a matrix in the following way: Atoms are numbered in an arbitrary order. If there is a bond between two atoms, we assign the number one, otherwise the number zero. The numbers are collected in a square matrix as shown in Fig. 14.1 (adapted from Rouvray, 1973). Such matrices are called incidence matrices.

Graph theory is today used in various areas such as ecology, taxonomy, physiology, and epidemiology (see e.g. Goel *et al.*, 1971; Jardine and Sibson, 1971; Laue, 1970).

Example 14.3.2. *Population Dynamics.* In Section 11.3 we have studied a deterministic birth and death process. In this model we did not account for the fact that the birth and death rates are dependent on the age of individuals.

Lewis (1942) and Leslie (1945) introduced a deterministic model which takes the age structure of the population into consideration. To simplify the mathematical treatment, we consider the birth and death process in steps of constant duration. Let Δt be a suitably chosen time interval. For a human population we may choose $\Delta t = 5$ years or, for a more precise analysis, $\Delta t = 1$ year. With the chosen value of Δt we consider the age structure of the population at times $t = 0$, Δt, $2\Delta t$, etc. We

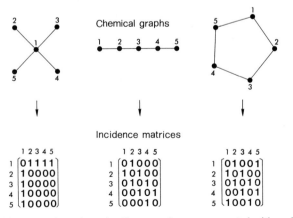

Fig. 14.1. Incidence matrices of graphs. If two vertices are connected with each other, the number one is assigned, otherwise the number zero

also introduce *age groups* $x = 0, 1, 2, \ldots$: Group $x = 0$ contains the ages from 0 to Δt, group $x = 1$ the ages from Δt to $2\Delta t$, group $x = 2$ the ages from $2\Delta t$ to $3\Delta t$, etc. The last possible age group is denoted by $x = m$. Notice that the group interval has the same length as the interval between consecutive time instants of the population.

In a bisexual population we need only consider the *females*. The age structure of the males is of minor importance.

At times $t = 0$, $t = \Delta t$, $t = 2\Delta t$, $\ldots$ the size of the female population is represented by the vectors

$$
\boldsymbol{n}_0 = \begin{pmatrix} n_{00} \\ n_{10} \\ n_{20} \\ \vdots \\ n_{m0} \end{pmatrix}, \quad
\boldsymbol{n}_1 = \begin{pmatrix} n_{01} \\ n_{11} \\ n_{21} \\ \vdots \\ n_{m1} \end{pmatrix}, \quad
\boldsymbol{n}_2 = \begin{pmatrix} n_{02} \\ n_{12} \\ n_{22} \\ \vdots \\ n_{m2} \end{pmatrix}, \ldots \qquad (14.3.1)
$$

respectively. Here, n_{xk} designates the number of females of age group x at time $t = k \cdot \Delta t$.

Let $F_0, F_1, F_2, \ldots, F_m$ be the average number of daughters born in time interval Δt to a female of age group $0, 1, 2, \ldots, m$, respectively. The females may not be reproductive at all ages. There may exist a prereproductive and a postreproductive phase. Thus, some of the F_x, say F_0, F_1 and F_m, may be zero. The total number of daughters born during the first time interval from $t = 0$ to $t = \Delta t$ is then

$$
F_0 n_{00} + F_1 n_{10} + \cdots + F_m n_{m0}. \qquad (14.3.2)
$$

Now we assume that by $F_0, F_1, F_2, \ldots$ we have only counted those daughters that survive until the time interval has passed in which they were born. At the end of this interval they will all be considered as of age 0. Hence, the number n_{01} of females of group 0 at time $t = \Delta t$ is equal to the result in formula (14.3.2).

By P_x we denote the probability that a female of age group x survives and will enter the age group $x + 1$. Hence, for $x = 0, 1, \ldots, m-1$ we obtain

$$
P_x n_{xk} = n_{x+1, k+1}. \qquad (14.3.3)
$$

Notice that $P_m = 0$. In the special case $k = 0$ we get $P_x n_{x0} = n_{x+1,1}$.

The transition of the population from $t = 0$ to $t = \Delta t$ may be summarized by matrix multiplication as follows

$$
\begin{pmatrix}
F_0 & F_1 & \cdots & F_{m-1} & F_m \\
P_0 & 0 & \cdots & 0 & 0 \\
0 & P_1 & \cdots & 0 & 0 \\
\multicolumn{5}{c}{\dotfill} \\
0 & 0 & \cdots & P_{m-1} & 0
\end{pmatrix}
\begin{pmatrix} n_{00} \\ n_{10} \\ n_{20} \\ \vdots \\ n_{m0} \end{pmatrix}
=
\begin{pmatrix} n_{01} \\ n_{11} \\ n_{21} \\ \vdots \\ n_{m1} \end{pmatrix}, \qquad (14.3.4)
$$

or in matrix notation

$$M n_0 = n_1 \qquad (14.3.5)$$

where M denotes the first factor in (14.3.4).

Assuming that F_x and P_x remain constant as time advances, we may repeat the same procedure. Thus

$$M n_1 = n_2, \qquad M n_2 = n_3, \qquad \text{etc.}$$

It follows that

$$n_2 = M (M n_0) = M^2 n_0$$

and by induction that

$$n_k = M^k n_0 . \qquad (14.3.6)$$

Given the initial population size and age structure and given the so-called *projection matrix* M, we can calculate size and age structure of the future population by (14.3.6). Given n_0 only, we may ask: What properties must M have if the population is to be *stable* in size, that is, if $n_0 = n_1 = n_2 = ...$? We leave it to the reader to answer this question.

We may also be interested in a *stable age distribution*. By this notion we mean constant proportions between different age groups. This question leads to the equation

$$n_{k+1} = \lambda n_k \qquad \text{or} \qquad M n_k = \lambda n_k \qquad (14.3.7)$$

where λ denotes a positive number which accounts for possible population growth. If we are given M, the solution of (14.3.7) for n_k and λ requires advanced tools of matrix theory such as the concept of characteristic equation of a matrix (see Section 14.9).

For more details and recent advances see Crow and Kimura (1970), Goodman (1969), Keyfitz (1968), Leslie (1945, 1948, 1959), Lopez (1961), Pielou (1969), J. H. Pollard (1966 and 1973), Searle (1966), Skellam (1967), Sykes (1969), Williamson (1972).

Example 14.3.3. *Ecology.* The following example is adapted from Thrall, Mortimer, Rebman, and Baum (1967, PL 5).

It is well known that poisonous pollutants (such as DDT or mercury) accumulate in *food chains*. We choose to study a particular ecological system with the following three links of a chain:

1. Vegetation providing food for herbivores. The different species of plants are denoted by $p_1, p_2, ..., p_r$.

2. Herbivorous animals feeding on the plants described in 1. The different species of herbivores are denoted by $a_1, a_2, ... a_s$.

3. Carnivorous animals living on herbivores as described in 2. The different species of carnivores are denoted by $c_1, c_2, ..., c_t$.

We may ask: What is the amount of plant p_i that is eaten indirectly by carnivore c_j during a particular season? To answer this question we introduce the following matrix X for the transition from link 1 to link 2:

$$
\begin{array}{c}
\\
\\
p_1 \\
p_2 \\
\vdots \\
p_r
\end{array}
\begin{array}{cccc}
a_1 & a_2 & \cdots & a_s \\
\\
\left(\begin{array}{cccc}
x_{11} & x_{12} & \cdots & x_{1s} \\
x_{21} & x_{22} & \cdots & x_{2s} \\
\multicolumn{4}{c}{\dotfill} \\
x_{r1} & x_{r2} & \cdots & x_{rs}
\end{array}\right) = X
\end{array}
\tag{14.3.8}
$$

Here x_{11} denotes the average amount (in g or kg) of plant p_1 eaten by each individual of species a_1 during the season. Generally, x_{ik} is the average amount of plant p_i eaten by each individual of species a_k.

We also define a matrix Y for the transition from link 2 to link 3:

$$
\begin{array}{c}
\\
\\
a_1 \\
a_2 \\
\vdots \\
a_s
\end{array}
\begin{array}{cccc}
c_1 & c_2 & \cdots & c_t \\
\\
\left(\begin{array}{cccc}
y_{11} & y_{12} & \cdots & y_{1t} \\
y_{21} & y_{22} & \cdots & y_{2t} \\
\multicolumn{4}{c}{\dotfill} \\
y_{s1} & y_{s2} & \cdots & y_{st}
\end{array}\right) = Y
\end{array}
\tag{14.3.9}
$$

Here y_{11} denotes the number of animals of species a_1 devoured by all individuals of species c_1 together. Generally, y_{kj} is the number of animals of species a_k devoured by carnivores of species c_j during the season. Note that y_{kj} is a number, whereas x_{ik} is a quantity.

Consider now the animals of species c_1. By feeding on species a_1 they consume indirectly the amount $x_{11}y_{11}$ of plant p_1. By feeding on species a_2 they consume $x_{12}y_{21}$ of plant p_1, etc. The average total amount of plant p_1 indirectly consumed by all carnivores of species c_1 is therefore

$$x_{11}y_{11} + x_{12}y_{21} + x_{13}y_{31} + \cdots + x_{1s}y_{s1}.$$

The result is an inner product, more precisely, the first row of X multiplied by the first column of Y. The result can be quickly generalized. The amount of plant p_i consumed indirectly by carnivore c_j is the product of the ith row of X and the jth column of Y. We get all particular results from the matrix product

$$XY. \tag{14.3.10}$$

This answers our question.

Example 14.3.4. *Genetics.* Consider a particular gene locus with possible alleles A and a. Assume that in a population the allele A is present with probability p and the allele a with probability $q = 1 - p$. Let R_1 and R_2 be two relatives for whom we know for sure that they have one allele in common. The relatives could be parent (R_1) and child (R_2) or, in the reverse order, child (R_1) and parent (R_2). We are interested in the probability of a certain genotype for R_2 given the genotype of R_1. The results are anticipated in the following matrix which we denote by T:

Genotype of R_1	Genotype of R_2		
	AA	Aa	aa
AA	p	q	0
Aa	$\frac{1}{2}p$	$\frac{1}{2}$	$\frac{1}{2}q$
aa	0	p	q

$$\begin{pmatrix} p & q & 0 \\ \tfrac{1}{2}p & \tfrac{1}{2} & \tfrac{1}{2}q \\ 0 & p & q \end{pmatrix} = T \qquad (14.3.11)$$

If R_1 is of genotype AA, then R_2 must have the allele A in common with R_1. Since the second allele is independent of the known allele, this second allele is either A with probability p or a with probability q. The relative R_2 cannot be of genotype aa. These arguments prove the first row of T.

If R_1 is of genotype Aa, then R_2 must have either the allele A or the allele a in common with R_1, either one with probability $\frac{1}{2}$. Since the second allele is independent of the first allele, this second allele is either A with probability p or a with probability q. Hence R_2 is of genotype AA with probability $\frac{1}{2}p$, of genotype Aa with probability $\frac{1}{2}p + \frac{1}{2}q = \frac{1}{2}$, or of genotype aa with probability $\frac{1}{2}q$. These arguments prove the second row of T. The proof of the third row is quite analogous.

Now we go a step further and investigate the relationship between grandparents and grandchildren. Assume for instance that a grandparent is of genotype AA. Then we easily derive the probabilities for the different genotypes for the grandchildren by the same method:

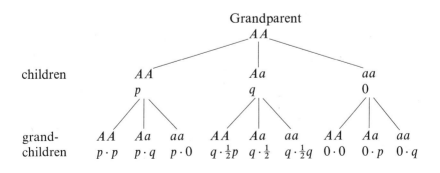

Hence, for grandchildren

$$
\begin{aligned}
P(AA) &= p \cdot p + q \cdot \tfrac{1}{2} p + 0 \cdot 0, \\
P(Aa) &= p \cdot q + q \cdot \tfrac{1}{2} + 0 \cdot p, \\
P(aa) &= p \cdot 0 + q \cdot \tfrac{1}{2} q + 0 \cdot q.
\end{aligned}
$$

These expressions can be interpreted as "first row of T multiplied by the first, second, third column of T, respectively". In general, we get the probabilities of the genotypes AA, Aa, aa for grandchildren simply by

$$
TT = T^2. \tag{14.3.12}
$$

The result can be written in the form

$$
T^2 = \begin{pmatrix}
p^2 + \tfrac{1}{2} pq & pq + \tfrac{1}{2} q & \tfrac{1}{2} q^2 \\
\tfrac{1}{2} p^2 + \tfrac{1}{4} p & pq + \tfrac{1}{4} & \tfrac{1}{4} q + \tfrac{1}{2} q^2 \\
\tfrac{1}{2} p^2 & \tfrac{1}{2} p + pq & \tfrac{1}{2} pq + q^2
\end{pmatrix}. \tag{14.3.13}
$$

Proceeding to great-grandchildren we would have to elaborate the matrix multiplication by considering

$$
TT^2 = T^3. \tag{14.3.14}
$$

Li (1958, Chap. 3) shows that by introducing two other matrices the burden of matrix multiplication can be substantially reduced. In addition, with his so-called ITO-method it is easy to get corresponding results for relationships such as sib-sib, sib-half sib, uncle-nephew, cousin-cousin (cf. Problem 14.3.3).

Example 14.3.5. *Markov Chain.* One of the RNA molecules has become famous for being able to replicate in a test tube without the help of living cells (Mills *et al.*, 1973). Like other RNA molecules it is a long chain of only four constituents, the so-called *bases:*

$$
\begin{aligned}
A &= \text{adenine}, \\
C &= \text{cytosine}, \\
G &= \text{guanine}, \\
U &= \text{uracil}.
\end{aligned}
$$

The molecule is depicted in Fig. 14.2. At a glance, we recognize regularities in the sequence of bases at the beginning (left) and at the end (right). Otherwise, the bases seem to follow each other in a "random" order. To investigate the type of randomness we may use a technique which was invented by A. A. Markov[2]. We consider the chain as result of a *stochastic process* where each link of the chain is in one of the four

---

[2] Andrei Andreevich Markov (1856—1922), Russian mathematician.

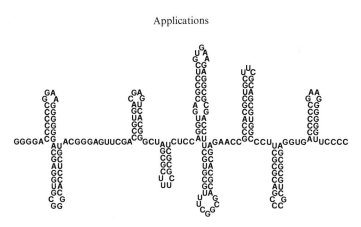

Fig. 14.2. An RNA molecule, 218 nucleotides in length. The four bases are A(adenine), C(cytosine), G(guanine), U(uracil). As shown in the text, the transition of consecutive nucleotides cannot be explained by a Markov process

states A, C, G, U. The change from one link to the next, that is from one state to another, is called *transition*. With four states we have 16 different transitions $A \to A$, $A \to C$, $A \to G$, $A \to U$, $C \to A$, $C \to C$, etc. The 16 possible transitions occur with the following *frequencies*:

	A	C	G	U	total
$A \to$	4	14	13	0	31
$C \to$	7	23	32	11	73
$G \to$	18	25	26	12	81
$U \to$	2	12	9	9	32

Most frequently occurring transitions are $C \to G$, $G \to G$, $G \to C$, and $C \to C$, whereas $A \to A$ and $U \to A$ are rare. The transition $A \to U$ never occurs in our special RNA sequence. From the (absolute) frequencies we derive the *relative frequencies* in each row:

	A	C	G	U	total
$A \to$	0.129	0.452	0.419	0.000	1.000
$C \to$	0.096	0.315	0.438	0.151	1.000
$G \to$	0.222	0.309	0.321	0.148	1.000
$U \to$	0.063	0.375	0.281	0.281	1.000

These relative frequencies form a matrix called the *transition matrix*:

$$F = \begin{pmatrix} 0.129 & 0.452 & 0.419 & 0.000 \\ 0.096 & 0.315 & 0.438 & 0.151 \\ 0.222 & 0.309 & 0.321 & 0.148 \\ 0.063 & 0.375 & 0.281 & 0.281 \end{pmatrix}. \qquad (14.3.15)$$

Note that the *sum of elements in each row equals* 1.

We may theorize, and think of a chemical process which generates such transitions with certain probabilities, the *transition probabilities*. They form a theoretical transition matrix

$$P = \begin{pmatrix} p_{11} & p_{12} & p_{13} & p_{14} \\ p_{21} & p_{22} & p_{23} & p_{24} \\ p_{31} & p_{32} & p_{33} & p_{34} \\ p_{41} & p_{42} & p_{43} & p_{44} \end{pmatrix}. \qquad (14.3.16)$$

p_{11} is the probability of transition $A \to A$ and is estimated by the empirical value 0.129. Likewise, p_{12} is the probability of transition $A \to C$ and is estimated by the empirical value 0.452, etc. For each row we postulate that

$$\sum_{k=1}^{4} p_{ik} = 1 \qquad (i = 1, 2, 3, 4). \qquad (14.3.17)$$

Now we extend our investigation to two-step transitions such as $A \to C \to A$, $G \to C \to U$, etc. and may ask: *What are the relative frequencies and/or probabilities of the 16 two-step transitions* $A \to \cdot \to A$, $A \to \cdot \to C$, etc? The dot means that we are not interested in the particular state in between. The transition $A \to \cdot \to A$ actually contains the following four cases

$$A \to A \to A$$
$$A \to C \to A$$
$$A \to G \to A$$
$$A \to U \to A.$$

In Markov chains we assume that consecutive transitions are independent of each other in the sense of probability theory (Section 13.6). This is the *Markov property*. On a time scale, the Markov property means that future events depend only on the present time, not on the past.

Under the assumption of Markov chains the *multiplication rule* holds, and we obtain

$$P(A \to A \to A) = P(A \to A) \cdot P(A \to A) = p_{11} p_{11},$$
$$P(A \to C \to A) = P(A \to C) \cdot P(C \to A) = p_{12} p_{21},$$
$$P(A \to G \to A) = P(A \to G) \cdot P(G \to A) = p_{13} p_{31},$$
$$P(A \to U \to A) = P(A \to U) \cdot P(U \to A) = p_{14} p_{41}.$$

We get the probability of $A \to \cdots \to A$ by using the *addition rule* of Section 13.4:

$$P(A \to \cdots \to A) = p_{11} p_{11} + p_{12} p_{21} + p_{13} p_{31} + p_{14} p_{41}.$$

This is nothing else than the inner product of the first row by the first column of the transition matrix P. By the same argument we can calculate the probability of any other two-step transition. The result is always comprized in the formula "row $\times$ column". All 16 probabilities are therefore generated by the *matrix multiplication*

$$P^2 = \begin{pmatrix} p_{11} \cdots p_{14} \\ \vdots \\ p_{41} \cdots p_{44} \end{pmatrix} \begin{pmatrix} p_{11} \cdots p_{14} \\ \vdots \\ p_{41} \cdots p_{44} \end{pmatrix}. \tag{14.3.18}$$

We may check whether our particular RNA molecule has the Markov property. For this purpose we replace the unknown transition probabilities of Eq. (14.3.16) by the empirical values of Eq. (14.3.15) and P^2 by F^2. Thus we get

$$F^2 = \begin{pmatrix} 0.153 & 0.330 & 0.387 & 0.130 \\ 0.149 & 0.335 & 0.361 & 0.155 \\ 0.139 & 0.352 & 0.373 & 0.136 \\ 0.124 & 0.339 & 0.360 & 0.177 \end{pmatrix}. \tag{14.3.19}$$

Finally we confront the result of this calculation with the relative frequencies of two-step transitions which we get by direct counting from Fig. 14.2:

$$\begin{pmatrix} 0.065 & 0.323 & 0.548 & 0.064 \\ 0.153 & 0.444 & 0.208 & 0.195 \\ 0.161 & 0.222 & 0.457 & 0.160 \\ 0.156 & 0.437 & 0.313 & 0.094 \end{pmatrix} \tag{14.3.20}$$

The agreement is poor. A statistical treatment, not reproduced here, reveals that there is a significant difference between the matrices (14.3.19) and (14.3.20). We conclude that our RNA molecule does not have the Markov property which means in this case that subsequent transitions do depend on each other.

Markov chain theory plays an increasingly important role in such areas as population dynamics, genetics, evolution, ecology, physiology, epidemiology, animal behavior. See Bailey (1964), Chiang (1968), Gani (1973), Iosifescu and Tăutu (1973), Jacquard (1974), Jacquez (1972), Karlin (1973), Khazanie (1968), Kingman (1969), Lillestøl (1968), Pielou (1969), Pollard (1973).

Example 14.3.6. *Statistics.* Matrix algebra is indispensable in such areas as multivariate statistical analysis, design of experiments, and the analysis of variance and covariance.

To give an idea of such a statistical application we consider a regression model. Suppose that we are given a scatter diagram consisting of points (x_i, y_i), $i = 1, 2, \ldots, n$ where x_i and y_i are measurements. Suppose further that the dots are close to a certain curve and that we want to fit a quadratic parabola. The equation is

$$y = a + bx + cx^2 \tag{14.3.21}$$

where the constants a, b, c are unknown. The coordinates x_i, y_i of the dots do not exactly satisfy an equation of type (14.3.21). Therefore, we write

$$y_i = a + bx_i + cx_i^2 + e_i \quad (i = 1, 2, \ldots, n) \tag{14.3.22}$$

where e_i is an error term. The system of Eqs. (14.3.22) may be written in matrix notation in the form

$$y = Xb + e \tag{14.3.23}$$

by introducing the matrices

$$y = \begin{pmatrix} y_1 \\ y_2 \\ \vdots \\ y_n \end{pmatrix}, \quad X = \begin{pmatrix} 1 & x_1 & x_1^2 \\ 1 & x_2 & x_2^2 \\ \cdots\cdots\cdots \\ 1 & x_n & x_n^2 \end{pmatrix}, \quad b = \begin{pmatrix} a \\ b \\ c \end{pmatrix}, \quad e = \begin{pmatrix} e_1 \\ e_2 \\ \vdots \\ e_n \end{pmatrix}. \tag{14.3.24}$$

The Eqs. (14.3.22) are linear in the unknown coefficients a, b, c. Hence we call (14.3.22) or (14.3.23) a *linear model*.

To find suitable coefficients we apply the method of "*least squares*". We minimize the sum of squares of errors

$$e_1^2 + e_2^2 + \cdots + e_n^2 = e'e. \tag{14.3.25}$$

In view of formula (14.3.23) this means that the unknown coefficients a, b, c are determined in such a way that

$$e'e = (y - Xb)'(y - Xb) \tag{14.3.26}$$

takes the smallest possible value. The solution can be found by methods of matrix algebra. Or we may apply differential calculus (cf. example at the end of Section 12.3.) For a full treatment of this problem we refer the reader to books on statistics.

14.4. Vectors in Space

We have defined vectors as row and column matrices. Vectors with two or three elements can be interpreted geometrically. On the one hand, this interpretation provides for a better understanding of matrix operations. On the other hand, geometrically interpreted vectors can be applied to various problems of life sciences.

We first consider a rectangular xy-coordinate system in the *plane*. Let (x, y) be an arbitrary point in the plane. With the point (x, y) we associate a directed line segment, also called an *arrow*. Its tail is at the origin O of the coordinate system, and the tip coincides with the point (x, y). Let a be the column vector

$$a = \begin{pmatrix} x \\ y \end{pmatrix}. \tag{14.4.1}$$

Then there is a one-to-one correspondence between the vector a and the arrow described above. It is customary to use the word "vector" for both the matrix a and the arrow[3] (see Fig. 14.3). We call x and y the *coordinates* or the *components* of the vector a. Let

$$a_1 = \begin{pmatrix} x_1 \\ y_1 \end{pmatrix}, \quad a_2 = \begin{pmatrix} x_2 \\ y_2 \end{pmatrix} \tag{14.4.2}$$

be two vectors.

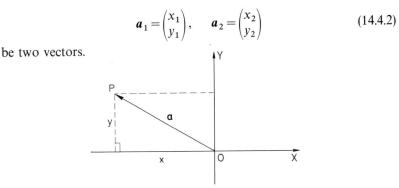

Fig. 14.3. A vector a is represented by an arrow with coordinates x, y

---

[3] The original meaning of "vector" was a *directed quantity* such as a force, a velocity, or an acceleration. Undirected quantities are called *scalars*. For instance, mass, density, and temperature are scalars. Later, the word "vector" was also used in the sense of a directed line segment or an arrow. In recent decades, the meaning has shifted again, namely to row and column matrices. Quite similarly we may call (x, y) either a matrix, a vector, or a point. In matrix algebra *column vectors are preferred* to row vectors.

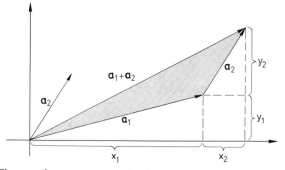

Fig. 14.4. The sum of two vectors a_1 and a_2 is a vector which can be constructed by means of a triangle. The components of $a_1 + a_2$ are $x_1 + x_2$ and $y_1 + y_2$

We now try to find a geometric meaning for the vector sum

$$a_1 + a_2 = \begin{pmatrix} x_1 + x_2 \\ y_1 + y_2 \end{pmatrix}. \tag{14.4.3}$$

In Fig. 14.4 the three vectors a_1, a_2, and $a_1 + a_2$ are depicted. It is also shown how $a_1 + a_2$ can be constructed. For this purpose we move the arrow a_2 temporarily, parallel to itself, so that its tail coincides with the tip of a_1. Now the two arrows form a broken line. The sum $a_1 + a_2$ is represented by the arrow which joins the origin O with the tip of the shifted arrow a_2. The three vectors are said to form a *vector triangle*. Due to the commutative law of matrix addition we could also move the arrow a_1 so that it forms a broken line with arrow a_2.

It is easy to generalize vector addition to three or more vectors. If we have to add a_1, a_2, ..., a_n, we form a broken line beginning with a_1, say, then proceeding with the shifted arrows a_2, a_3, ..., a_n. The arrow which joins the origin O with the tip of a_n represents the vector

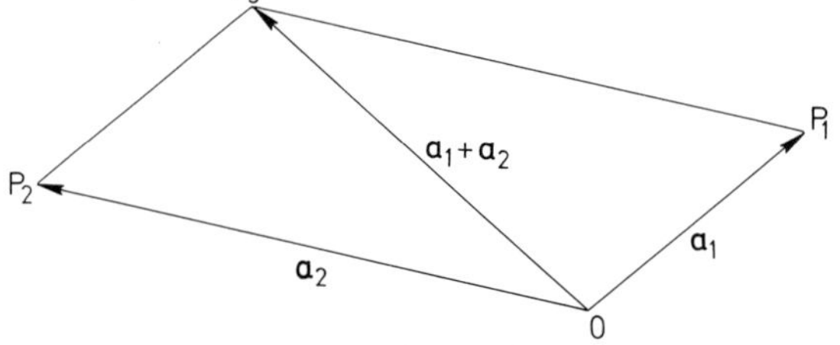

Fig. 14.5. The construction of a vector sum by means of a parallelogram. Line $P_1 P_3$ is parallel to line OP_2, and line $P_2 P_3$ is parallel to line OP_1

$a_1 + a_2 + \cdots + a_n$. In applications the sum is frequently called the *resultant vector* and the single vectors a_1, a_2, ..., a_n its *components*. The figure consisting of a_1, a_2, ..., a_n and the resultant vector is called a *vector polygon*. It is a generalization of the vector triangle.

Sometimes it is preferable to construct the sum of two vectors by means of a *parallelogram*. Fig. 14.5 explains the procedure. However, this construction is not practical for more than two vectors.

Now we consider a multiple of a vector a. The geometric meaning of

$$a + a = 2a \, ,$$

$$a + a + a = 3a \, , \quad \text{etc.}$$

is easy to understand (Fig. 14.6). We may also multiply a vector by a fractional number, say by $2/3$, or by a negative number, say by -2. According to the definition given in formula (14.2.4) we obtain

$$ka = k\binom{x}{y} = \binom{kx}{ky}. \tag{14.4.4}$$

Both coordinates, x and y, have to be multiplied by k (Fig. 14.6).

Subtraction is defined by

$$a - b = a + (-b) \, , \tag{14.4.5}$$

that is, subtraction of b is equivalent to adding the *opposite vector* $-b$. Fig. 14.7 shows the construction and the geometric meaning of $a - b$.

Notice that

$$a - a = \binom{x}{y} - \binom{x}{y} = \binom{0}{0}. \tag{14.4.6}$$

The result is called the *zero vector*. Both tip and tail of this vector fall into the origin O. This particular vector has no direction.

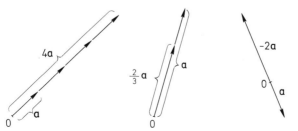

Fig. 14.6. Multiplication of a vector a by a number k. If we multiply a vector by a positive number, its direction remains unchanged. However, multiplication by a negative number yields a vector pointing in the opposite direction

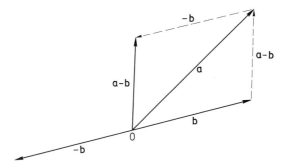

Fig. 14.7. The three vectors a, b, and $a-b$ form a vector triangle such that $b+(a-b)=a$

Notice that we did not need to make an assumption about the units in which x and y are measured. All the preceding results remain valid if, for instance, x is measured in grams and y in seconds. This will no longer be true when we introduce the absolute value of a vector. From now on we assume that *both coordinates x and y are measured in the same unit.* We define the *absolute value* of a vector $a = \begin{pmatrix} x \\ y \end{pmatrix}$ by

$$|a| = \sqrt{x^2 + y^2} \ . \tag{14.4.7}$$

If a is the zero vector, its absolute value is zero; otherwise the absolute value is a positive number. In the applications, $|a|$ is also called the *magnitude* of the vector a.

In the special case where x and y are measured in a unit of length, then $|a|$ is the distance from O to the point (x, y) by virtue of the Pythagorean theorem (Fig. 14.3). In such a case, $|a|$ is also called the *length* of the vector a.

The absolute value of a vector could be calculated by matrix multiplication. Let $a' = (x, y)$ be the transpose of a. Then we form the inner product

$$a'a = (x, y)\begin{pmatrix} x \\ y \end{pmatrix} = x^2 + y^2 \ .$$

Hence,

$$|a|^2 = a'a \ . \tag{14.4.8}$$

More generally, let

$$a_1 = \begin{pmatrix} x_1 \\ y_1 \end{pmatrix}, \quad a_2 = \begin{pmatrix} x_2 \\ y_2 \end{pmatrix}$$

be any two vectors. Then we may ask whether the inner product

$$a_1' a_2 = x_1 x_2 + y_1 y_2$$

has any geometric meaning. Notice that

$$a_2' a_1 = a_1' a_2 . \tag{14.4.9}$$

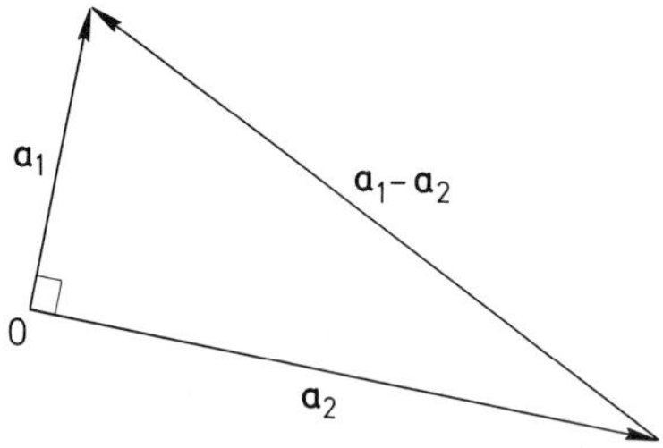

Fig. 14.8. Proof of formula (14.4.10a)

This is the *commutative law for the inner product.* As we already know the commutative law is in general not valid for matrix multiplication. To approach geometry we consider first a special case. We assume that the arrows a_1 and a_2 are *perpendicular* (Fig. 14.8). The vectors a_1, a_2, and $a_1 - a_2$ form a right triangle and it follows from the Pythagorean theorem that

$$|a_1 - a_2|^2 = |a_1|^2 + |a_2|^2 . \tag{14.4.10a}$$

Notice that (14.4.10a) holds if, and only if, a_1 and a_2 are perpendicular. Applying formulas (14.4.8) and (14.4.9) we also get

$$\begin{aligned}
|a_1 - a_2|^2 &= (a_1 - a_2)' (a_1 - a_2) = (a_1' - a_2')(a_1 - a_2) \\
&= a_1' a_1 - a_1' a_2 - a_2' a_1 + a_2' a_2 \\
&= |a_1|^2 + |a_2|^2 - 2a_1' a_2 .
\end{aligned} \tag{14.4.10b}$$

Notice that (14.4.10b) holds whether or not a_1 and a_2 are perpendicular. Comparison of (14.4.10a) and (14.4.10b) leads to

$$a_1' a_2 = 0 . \tag{14.4.11a}$$

Using coordinates we may also write

$$\boxed{x_1 x_2 + y_1 y_2 = 0 .} \tag{14.4.11b}$$

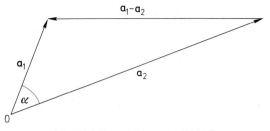

Fig. 14.9. Proof of formula (14.4.12)

This equation holds if, and only if, the two nonzero vectors a_1 and a_2 are perpendicular or orthogonal. The condition (14.4.11a) or the equivalent condition (14.4.11b) is therefore called the *orthogonality condition*.

The result can be extended to the case where a_1 and a_2 form an arbitrary angle α (Fig. 14.9). We have to replace the Pythagorean theorem by the *law of cosines* which states in our notation

$$|a_1 - a_2|^2 = |a_1|^2 + |a_2|^2 - 2|a_1||a_2|\cos\alpha. \qquad (14.4.12)$$

Comparison with (14.4.10b) yields

$$\boxed{a_1' a_2 = |a_1||a_2|\cos\alpha.} \qquad (14.4.13)$$

In words: *The inner product of two vectors a_1 and a_2 is equal to the product of three factors: the absolute values of a_1 and a_2 and the cosine of the angle between the two vectors.*

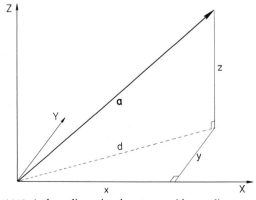

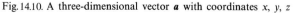

Fig. 14.10. A three-dimensional vector a with coordinates x, y, z

If one of the vectors is the zero vector, the product is zero, and α is undetermined. If the *angle* α is *acute*, then $\cos \alpha > 0$, and the product (14.4.13) is *positive*. If, however, the *angle* α is *obtuse*, then $\cos \alpha < 0$, and the product (14.4.13) is negative. We have already discussed the case $\alpha = 90°$ which implies $\cos \alpha = 0$.

Formula (14.4.13) is often used to calculate the angle α between two nonzero vectors. In view of $a'_1 a_2 = x_1 x_2 + y_1 y_2$ we obtain

$$\cos \alpha = \frac{x_1 x_2 + y_1 y_2}{|a_1| |a_2|}, \qquad |a_1| \neq 0, \qquad |a_2| \neq 0. \qquad (14.4.14)$$

So far we have considered vectors in the two-dimensional space. Fortunately, there is no difficulty in extending the results to the *three-dimensional space*. Let (x, y, z) be a point in a rectangular xyz-coordinate system. Let a be the column vector

$$a = \begin{pmatrix} x \\ y \\ z \end{pmatrix}. \qquad (14.4.15)$$

Then there is a one-to-one correspondence between the vector a and the arrow whose tail is at the origin O and whose tip coincides with the point (x, y, z) (see Fig. 14.10). Therefore it is convenient to use the word "vector" also for an arrow.

Addition of two vectors is explained in the same way as in formula (14.4.3) and in Fig. 14.4. Algebraically we add two vectors by adding corresponding coordinates. Geometrically, the vector sum, also called the resultant vector, is represented by the "third" side in the vector triangle.

Multiplication of a vector by an ordinary number is defined in the same way as in (14.4.4). Fig. 14.6 may serve without any change as illustration. Likewise Fig. 14.7 can immediately be used in the three-dimensional space.

The *absolute value*[4] or the *magnitude* of the vector (14.4.15) is defined by

$$|a| = \sqrt{x^2 + y^2 + z^2} \qquad (14.4.16)$$

provided that x, y, z are measured in the same unit. If this unit is a length, then $|a|$ has a simple geometric meaning. In Fig. 14.10 we first calculate the length d of the hypotenuse of the right triangle with legs x, y:

$$d^2 = x^2 + y^2 .$$

---

[4] In modern mathematics the absolute value of a vector is a special case of what is called *norm* of a vector. In this connection the symbol $|a|$ is replaced by $\|a\|$.

Second, we apply the Pythagorean theorem to the right triangle with legs d and z and get

$$d^2 + z^2 = x^2 + y^2 + z^2$$

for the square of the hypotenuse. In view of formula (14.4.16) the last expression is $|a|^2$. Hence, $|a|$ is the *length* of vector a.

Furthermore, formulas (14.4.8), (14.4.9), (14.4.10a), (14.4.10b), (14.4.11a) remain valid in the three-dimensional space since the same proofs can be applied as above. The *orthogonality condition* (14.4.11b) changes into

$$a_1' a_2 = x_1 x_2 + y_1 y_2 + z_1 z_2 = 0 . \tag{14.4.17}$$

Moreover, formulas (14.4.12) and (14.4.13) remain unaltered. Only formula (14.4.14) has to be adjusted to become

$$\cos\alpha = \frac{x_1 x_2 + y_1 y_2 + z_1 z_2}{|a_1| \, |a_2|}, \quad |a_1| \neq 0, \quad |a_2| \neq 0. \tag{14.4.18}$$

In fact, the extension of the results from the two- to the three-dimensional space is so simple that mathematicians could not refrain from inventing the *four-dimensional space*. The lack of geometric intuition was no barrier. The step from the three- to the four-dimensional space was so tempting that it had to be undertaken. We invite the reader to perform this step without receiving any further help. (See also Example 14.5.7).

Not only was it possible to work with vectors in the four-dimensional space with ease, applications in the natural sciences became also quite successful. Further generalization of the geometry of vectors to n-dimensional spaces with $n = 5, 6, \ldots$ was a matter of routine. Around 1900, even infinitely dimensional spaces were invented.

Today vector algebra in n-dimensional spaces is a fruitful tool in multivariate statistical analysis and in the analysis of variance and covariance.

14.5. Applications

Example 14.5.1. *Inclined Plane.* In Fig. 14.11 a human body is resting on an inclined plane. What is the force trying to pull the body down along the plane, and what is the force pressing the body toward the plane?

Let F_1 be the force which pulls the body downward along the plane and F_2 the force perpendicular to the plane which presses the body toward the plane. In mechanics we learn that these two forces are caused by gravitation and that their vector sum must be the total gravitational force denoted by F. Hence,

$$F_1 + F_2 = F . \tag{14.5.1}$$

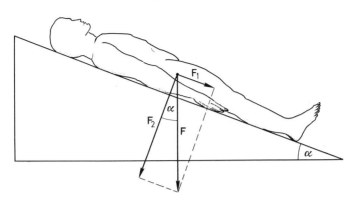

Fig. 14.11. A body lying on an inclined plane. Vector addition relates the forces F_1 and F_2 to the force F of gravitation

The forces F_1 and F_2 are called the *components* of F. Let α be the angle of inclination of the plane. Since F_2 is perpendicular to the plane, α is also the angle between F_2 and F. From Fig. 14.11 we obtain the following formulas for the magnitudes of the components of F:

$$|F_1| = |F| \sin\alpha, \qquad |F_2| = |F| \cos\alpha. \qquad (14.5.2)$$

For instance, if $\alpha = 30°$, we obtain $\sin\alpha = 0.500$, $\cos\alpha = 0.866$. Thus $|F_1|$ is only 50% and $|F_2|$ only 86.6% of the body weight. On the one hand, to prevent gliding of the body, a force of magnitude $\frac{1}{2}|F|$ has to act in the direction opposite to F_1. On the other hand, the pressure against the plane is relieved by 13.4% as compared with the body lying on a horizontal plane.

In a similar way, the forces acting on a broken bone can be studied (Fig. 14.12). For a detailed account see Pauwels (1965, p. 10ff.).

Example 14.5.2. *Levers.* As an example of a lever we take the forearm (Fig. 14.13). Its fulcrum is the elbow. When the angle between upper arm and forearm is not 90°, the force F generated by the forearm flexor has to be decomposed into a component F_1 perpendicular to the forearm and a component F_2 parallel to the forearm. The force F_2 does not generate any rotation of the forearm. Only F_1 acts in the proper direction. If the angle between upper arm and forearm approaches 180°, the magnitude of F_1 is considerably smaller than that of F. Thus part of the muscle force is "lost". By vector algebra we may write

$$F_1 = F - F_2. \qquad (14.5.3)$$

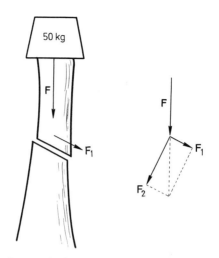

Fig. 14.12. Forces acting on a broken bone. The force F may be caused by gravitation or by a muscle. We call F_1 a shearing force

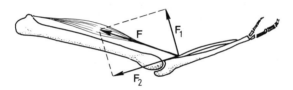

Fig. 14.13. The force F_1 which lifts the forearm may be considerably smaller in magnitude than the force F generated by the forearm flexor

The theory of levers is applied to the skeleton in books such as Gray (1968), Ricci (1967), Williams and Lissner (1962).

Example 14.5.3. *Leg Traction.* Fig. 14.14 shows how a leg may be stretched by a pulley line for therapeutic purposes. We denote by F_1 the vertical force of the weight. The string of the pulley line has everywhere the same tension. Hence, the forces F_2 and F_3 in Fig. 14.14a or b have the same magnitude as F_1, that is,

$$|F_2| = |F_3| = |F_1|. \tag{14.5.4}$$

The stretching force F is the resultant of F_2 and F_3. Hence,

$$F = F_2 + F_3. \tag{14.5.5}$$

When the angle between F_2 and F_3 tends to zero, then $|F|$ increases and tends to $|F_2| + |F_3| = 2|F_1|$ as is easily seen from Fig. 14.14a. When,

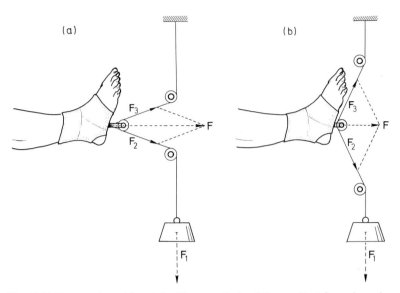

Fig. 14.14. Leg traction with weight. The magnitude of the resultant force depends on the angle between the forces F_2 and F_3. The figure is redrawn from Fig. 3.5 in Williams and Lissner (1962, p. 23)

however, the angle between F_2 and F_3 tends to $180°$, $|F|$ decreases and tends to zero (Fig. 14.14b). Hence,

$$0 \leqq |F| \leqq 2|F_1| . \tag{14.5.6}$$

Example 14.5.4. *Center of Gravity (or Mass).* A body consists of a very large number of particles which are all subject to gravitation. To find the center of gravity we begin with the simple case of two points P_1 and P_2 of equal mass (Fig. 14.15). Then, by symmetry, the center of gravity is the midpoint of P_1 and P_2. We denote it by C. To find the midpoint C by vector algebra, we denote the arrows with tail at the origin O and tips at P_1 and P_2 by a_1 and a_2, respectively. Using the parallelogram of Fig. 14.5 we determine the vector sum of a_1 and a_2. Since the diagonals of a parallelogram bisect each other, the vector c which points to C is

$$c = \frac{a_1 + a_2}{2} . \tag{14.5.7}$$

This is simply the *arithmetic mean* of the two vectors a_1 and a_2. Let $a_1 = \begin{pmatrix} x_1 \\ y_1 \end{pmatrix}$ and $a_2 = \begin{pmatrix} x_2 \\ y_2 \end{pmatrix}$. Then c has coordinates $(x_1 + x_2)/2$ and $(y_1 + y_2)/2$.

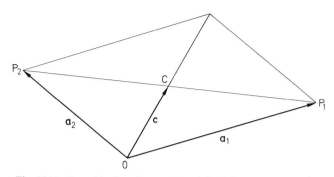

Fig. 14.15. The midpoint of two points defined by vectors a_1 and a_2

If we are given n points of equal mass by vectors $a_1, a_2, \ldots, a_n$, mechanics shows that the center of gravity is given by the vector

$$c = \frac{a_1 + a_2 + \cdots + a_n}{n} = \frac{1}{n} \sum_{i=1}^{n} a_i. \tag{14.5.8}$$

The result is simply an extension of formula (14.5.7). The same formula is of basic importance in the *statistical analysis of directions* as applied for instance to homing and migrating birds. For details see Batschelet (1965).

A further generalization occurs when different masses M_i are located at the points given by vectors a_i $(i = 1, 2, \ldots, n)$. In this case, the ordinary arithmetic mean turns into the *weighted arithmetic mean*

$$c = \frac{M_1 a_1 + M_2 a_2 + \cdots + M_n a_n}{M_1 + M_2 + \cdots + M_n}. \tag{14.5.9}$$

Fig. 14.16 illustrates the case of three masses. The formula is valid in the two-and the three-dimensional space. It is also useful in spaces of higher dimensions although it loses its physical meaning. An important application of (14.5.9) is made in *multivariate statistical analysis*.

Example 14.5.5. *Kinetics.* There are numerous biological applications of vectors in the study of *animal locomotion*. We will confine ourselves to discussing the jumping of grasshoppers (Fig. 14.15).

We call the time interval from the initiation of muscle action until the feet leave the ground the *take-off* or the *acceleration period*. During this period the center of gravity moves from a point C to a certain point C'. The two points C and C' form a directed line segment or a vector which we denote by s. Let α be the angle between s and the horizontal line. Then $|s| \sin \alpha$ is the vertical component of s.

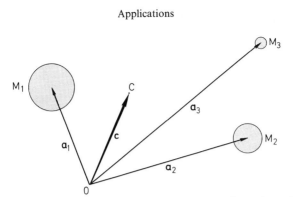

Fig. 14.16. The center of gravity C of three masses M_1, M_2, M_3 located at points given by vectors a_1, a_2, a_3, respectively

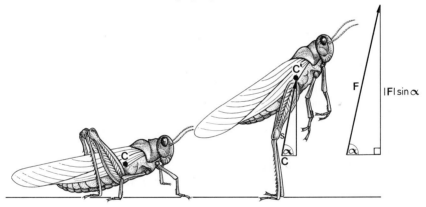

Fig. 14.17. Jumping grasshopper

During the take-off period the force that accelerates the grasshopper's body does not remain constant in magnitude, but we may operate with an *average force* which we denote by F. This force F is called the *thrust*.

Now we want to calculate the maximum height h over the ground reached by the jumping grasshopper. On the one hand, the part of kinetic energy that is transformed into potential energy is

vertical component of F times vertical component of s

$$= |F| \sin \alpha \cdot |s| \sin \alpha = |F| \, |s| \sin^2 \alpha \, .$$

On the other hand, the potential energy reached at the highest point is

$$m \cdot g \cdot h$$

where m is the mass of the animal (in grams) and g the acceleration due to gravity ($= 981$ cm sec^{-2}).

If we neglect the slight loss in energy due to air resistance, we equate the two energies. Then dividing by $m \cdot g$ we obtain

$$h = |F| \, |s| \sin^2 \alpha / m \cdot g \,. \tag{14.5.10}$$

This formula may also be used for calculating $|F|$ when the other quantities are known from measurements.

The reader will find more mathematics concerning jumping animals in Alexander (1968), Gray (1968), Hoyle (1955), and Hughes (1965).

Example 14.5.6. *Navigation.* A velocity is a quantity with a direction and can therefore be represented by a vector. The velocity of an air or a water current is measured from a fixed point on the ground. We may also think of a velocity which a body maintains relative to the flowing medium (air or water).

Thus when a bird flies at a speed of 10 m/sec relative to the air and exactly opposite to the direction of the wind, and when the wind has a velocity of 3 m/sec, then the resultant traveling speed of the bird is 10 m/sec + (− 3 m/sec) = 7 m/sec.

This is a special case of a more general situation. We consider a bird headed for a certain destination while flying against high wind under a certain angle, or the analogous situation for a fish (Fig. 14.18).

Let v be the vector of velocity of the animal relative to the air or to the water, and let w denote the vector of velocity of the wind or the current relative to ground. Then the vector of traveling velocity u (relative to ground) is

$$u = v + w \,. \tag{14.5.11}$$

The vector u points toward the destination of the animal. However, the animal's body is not oriented straight toward the destination, but in the direction in which vector v points.

A numerical example may illustrate the use of formula (14.5.11). As in Fig. 14.18 we fix an x axis opposite to the direction of the wind and a y axis perpendicular to the x axis. Then the x component of w is negative and the y component vanishes. Temporarily dropping the unit m/sec of velocity, let

$$v = \begin{pmatrix} 8 \\ 3 \end{pmatrix}, \quad w = \begin{pmatrix} -6 \\ 0 \end{pmatrix}.$$

From (14.5.11) it follows that

$$u = \begin{pmatrix} 8 \\ 3 \end{pmatrix} + \begin{pmatrix} -6 \\ 0 \end{pmatrix} = \begin{pmatrix} 2 \\ 3 \end{pmatrix}.$$

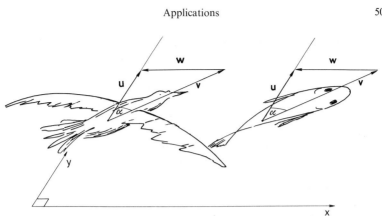

Fig. 14.18. Animals headed for a certain destination under the influence of high wind or of a strong current

Hence, the speed of the bird relative to the air has magnitude $|v| = (8^2 + 3^2)^{\frac{1}{2}} = 73^{\frac{1}{2}} = 8.54$ m/sec, but the speed toward the destination amounts only to $|u| = (2^2 + 3^2)^{\frac{1}{2}} = 13^{\frac{1}{2}} = 3.61$ m/sec.

If we are interested in the angle α between the vectors u and v, we apply formula (14.4.14). Again dropping m/sec temporarily we obtain

$$v = \binom{x_1}{y_1} = \binom{8}{3}, \quad u = \binom{x_2}{y_2} = \binom{2}{3},$$
$$x_1 x_2 + y_1 y_2 = 16 + 9 = 25, \quad |v| = 8.54, \quad |u| = 3.61.$$

Hence,

$$\cos\alpha = \frac{25}{8.54 \times 3.61} = 0.812, \quad \alpha = 35.8°.$$

Example 14.5.7. *Genetics.* Gene frequencies in the ABO blood group have been investigated in various populations. If we distinguish among four alleles A_1, A_2, B, O, the following relative frequencies, f_{ki}, have been reported (Cavalli-Sforza and Edwards, 1967, p. 250):

Allele	Eskimo f_{1i}	Bantu f_{2i}	English f_{3i}	Korean f_{4i}
A_1	0.2914	0.1034	0.2090	0.2208
A_2	0.0000	0.0866	0.0696	0.0000
B	0.0316	0.1200	0.0612	0.2069
0	0.6770	0.6900	0.6602	0.5723
Total	1.0000	1.0000	1.0000	1.0000

The question arises: How near is one population to another? In other words: We should find a suitable measure for a *genetic distance*.

A method using vector algebra was proposed by Cavalli-Sforza and Edwards (1967). In a first step we represent each population by a *unit vector*, that is, a vector of absolute value 1. For this purpose we take the square root of each frequency, say

$$x_{ki} = \sqrt{f_{ki}}.$$ (14.5.12)

Since

$$\sum_{i=1}^{4} f_{ki} = 1$$ (14.5.13)

for each of the four populations, we get

$$\sum_{i=1}^{4} x_{ki}^2 = 1$$ (14.5.14)

which means that each of the four vectors

| (Eskimo) | (Bantu) | (English) | (Korean) |

$$a_1 = \begin{pmatrix} x_{11} \\ x_{12} \\ x_{13} \\ x_{14} \end{pmatrix} \qquad a_2 = \begin{pmatrix} x_{21} \\ x_{22} \\ x_{23} \\ x_{24} \end{pmatrix} \qquad a_3 = \begin{pmatrix} x_{31} \\ x_{32} \\ x_{33} \\ x_{34} \end{pmatrix} \qquad a_4 = \begin{pmatrix} x_{41} \\ x_{42} \\ x_{43} \\ x_{44} \end{pmatrix}$$

is a unit vector, that is, $|a_k| = 1$.

In the four-dimensional space the tips of these vectors are located on a sphere of radius 1.

Now it is plausible to use the angle between two vectors as measure of the *distance* between the corresponding populations. If we denote the angle between a_1 (Eskimo) and a_2 (Bantu) by θ, then Eq. (14.4.18) yields

$$\cos\theta = a_1' a_2$$ (14.5.15)

since $|a_1| = 1$, $|a_2| = 1$.

Here are the numerical details:

$$a_1 = \begin{pmatrix} 0.5398 \\ 0.0000 \\ 0.1778 \\ 0.8228 \end{pmatrix}, \qquad a_2 = \begin{pmatrix} 0.3216 \\ 0.2943 \\ 0.3464 \\ 0.8307 \end{pmatrix},$$

$$\cos\theta = a_1' a_2 = (0.5398)(0.3216) + \cdots = 0.9187,$$

$$\theta = 23.2°.$$

In the same way we find the following *genetic distances*[5]:

---

[5] In the original paper all angles are divided by 90°. For alternative measurements of distance see Jacquard (1974, Chapter 14).

	Eskimo	Bantu	English	Korean
Eskimo	0°	23.2°	16.4°	16.8°
Bantu	23.2°	0°	9.8°	20.4°
English	16.4°	9.8°	0°	19.6°
Korean	16.8°	20.4°	19.6°	0°

The smallest genetic distance between different populations occur at Bantu-English, the largest at Eskimo-Bantu.

14.6. Determinants

Consider the system of linear equations

$$ax_1 + bx_2 = u \qquad \text{or} \qquad \begin{pmatrix} a & b \\ c & d \end{pmatrix} \begin{pmatrix} x_1 \\ x_2 \end{pmatrix} = \begin{pmatrix} u \\ v \end{pmatrix}. \tag{14.6.1}$$
$$cx_1 + dx_2 = v$$

Its solution is

$$x_1 = \frac{ud - vb}{ad - bc}, \qquad x_2 = \frac{av - cu}{ad - bc} \tag{14.6.2}$$

provided that the common denominator $ad - bc$ differs from zero.

The formation of terms in the numerators and denominators follows the same rule. For this reason we introduce *determinants* by the following definition:

$$\det \begin{pmatrix} a & b \\ c & d \end{pmatrix} = \begin{vmatrix} a & b \\ c & d \end{vmatrix} = ad - bc. \tag{14.6.3}$$

Note that a *matrix* is a *rectangular array of numbers* whereas a *determinant* is a *single number*, the result of a calculation.

With Eq. (14.6.3) in mind, the solution (14.6.2) can be written in the form

$$x_1 = \frac{\begin{vmatrix} u & b \\ v & d \end{vmatrix}}{\begin{vmatrix} a & b \\ c & d \end{vmatrix}}, \qquad x_2 = \frac{\begin{vmatrix} a & u \\ c & v \end{vmatrix}}{\begin{vmatrix} a & b \\ c & d \end{vmatrix}}. \tag{14.6.4}$$

To find an easy rule, we introduce the following notations

$$A = \begin{pmatrix} a & b \\ c & d \end{pmatrix}, \qquad u = \begin{pmatrix} u \\ v \end{pmatrix}.$$

The denominators in Eqs. (14.6.4) are $\det A$. To find the numerators we apply *Cramer's rule*[6]:

---

[6] Gabriel Cramer (1704—1752), Swiss mathematician.

1. To get the numerator of x_1 replace the first column of A by the column vector u.

2. To get the numerator of x_2 replace the second column of A by the column vector u.

Example 14.6.1. Solve the system

$$8x_1 - 15x_2 = 7,$$
$$3x_1 + 22x_2 = 5.$$

Here

$$A = \begin{pmatrix} 8 & -15 \\ 3 & 22 \end{pmatrix}, \qquad u = \begin{pmatrix} 7 \\ 5 \end{pmatrix}.$$

In Eqs. (14.6.4) the denominator is

$$\det A = \begin{vmatrix} 8 & -15 \\ 3 & 22 \end{vmatrix} = 8 \times 22 - 3 \times (-15) = 221.$$

To find the numerators we apply Cramer's rule: Replacing $\begin{pmatrix} 8 \\ 3 \end{pmatrix}$ by $\begin{pmatrix} 7 \\ 5 \end{pmatrix}$ we obtain the numerator of x_1:

$$\begin{vmatrix} 7 & -15 \\ 5 & 22 \end{vmatrix} = 7 \times 22 - 5 \times (-15) = 229,$$

hence

$$x_1 = \frac{229}{221}.$$

Similarly, when $\begin{pmatrix} -15 \\ 22 \end{pmatrix}$ is replaced by $\begin{pmatrix} 7 \\ 5 \end{pmatrix}$, the numerator of x_2 turns out to be

$$\begin{vmatrix} 8 & 7 \\ 3 & 5 \end{vmatrix} = 8 \times 5 - 3 \times 7 = 19,$$

hence

$$x_2 = \frac{19}{221}.$$

Linear systems of more than two equations can also be solved by determinants. We will show the details for three equations with three unknowns:

$$a_{11}x_1 + a_{12}x_2 + a_{13}x_3 = u_1$$
$$a_{21}x_1 + a_{22}x_2 + a_{23}x_3 = u_2 \qquad (14.6.5)$$
$$a_{31}x_1 + a_{32}x_2 + a_{33}x_3 = u_3$$

or

$$A x = u \qquad (14.6.6)$$

when using the following matrix notations:

$$A = \begin{pmatrix} a_{11} & a_{12} & a_{13} \\ a_{21} & a_{22} & a_{23} \\ a_{31} & a_{32} & a_{33} \end{pmatrix}, \quad x = \begin{pmatrix} x_1 \\ x_2 \\ x_3 \end{pmatrix}, \quad u = \begin{pmatrix} u_1 \\ u_2 \\ u_3 \end{pmatrix}.$$

A lengthy calculation leads to

$$x_1 = \frac{u_1 a_{22} a_{33} - u_1 a_{32} a_{23} - u_2 a_{12} a_{33}}{a_{11} a_{22} a_{33} - a_{11} a_{32} a_{23} - a_{21} a_{12} a_{33}} \tag{14.6.7}$$
$$\frac{+ u_2 a_{32} a_{13} + u_3 a_{12} a_{23} - u_3 a_{22} a_{13}}{+ a_{21} a_{32} a_{13} + a_{31} a_{12} a_{23} - a_{31} a_{22} a_{13}}$$

provided that the denominator differs from zero. By means of determinants we can rewrite the formula:

$$x_1 = \frac{u_1 \begin{vmatrix} a_{22} & a_{23} \\ a_{32} & a_{33} \end{vmatrix} - u_2 \begin{vmatrix} a_{12} & a_{13} \\ a_{32} & a_{33} \end{vmatrix} + u_3 \begin{vmatrix} a_{12} & a_{13} \\ a_{22} & a_{23} \end{vmatrix}}{a_{11} \begin{vmatrix} a_{22} & a_{23} \\ a_{32} & a_{33} \end{vmatrix} - a_{21} \begin{vmatrix} a_{12} & a_{13} \\ a_{32} & a_{33} \end{vmatrix} + a_{31} \begin{vmatrix} a_{12} & a_{13} \\ a_{22} & a_{23} \end{vmatrix}} \tag{14.6.8}$$

Introducing a 3×3 determinant by the definition

$$\det A = \begin{vmatrix} a_{11} & a_{12} & a_{13} \\ a_{21} & a_{22} & a_{23} \\ a_{31} & a_{32} & a_{33} \end{vmatrix}$$
$$= a_{11} \begin{vmatrix} a_{22} & a_{23} \\ a_{32} & a_{33} \end{vmatrix} - a_{21} \begin{vmatrix} a_{12} & a_{13} \\ a_{32} & a_{33} \end{vmatrix} + a_{31} \begin{vmatrix} a_{12} & a_{13} \\ a_{22} & a_{23} \end{vmatrix} \tag{14.6.9}$$

we simply obtain

$$x_1 = \frac{\begin{vmatrix} u_1 & a_{12} & a_{13} \\ u_2 & a_{22} & a_{23} \\ u_3 & a_{32} & a_{33} \end{vmatrix}}{\det A}. \tag{14.6.10}$$

Cramer's rule is again applicable: To get the numerator we replace the first column of A by the column matrix u. Similarly we obtain

$$x_2 = \frac{\begin{vmatrix} a_{11} & u_1 & a_{13} \\ a_{21} & u_2 & a_{23} \\ a_{31} & u_3 & a_{33} \end{vmatrix}}{\det A}, \quad x_3 = \frac{\begin{vmatrix} a_{11} & a_{12} & u_1 \\ a_{21} & a_{22} & u_2 \\ a_{31} & a_{32} & u_3 \end{vmatrix}}{\det A}. \tag{14.6.11}$$

Equation (14.6.9) shows how a 3×3 determinant can be expanded. The first factors are the elements of the first column. The second factors are 2×2 determinants, so-called *minors* of A. The minor corresponding to a_{11} is generated by omitting the row and column where a_{11} occurs. Similarly, the minor belonging to a_{21} is generated by omitting row and column of a_{21}, etc. Note that the signs in Eq. (14.6.9) alternate: $+ - +$.

Expansion of a 3×3 determinant can be done in many different ways. Instead of expanding along the first column, any other column is appropriate as can be verified. In the same way a 3×3 determinant can also be expanded along rows. We confine ourselves to write down two of the six possible expansions:

(along
second
column)
$$- a_{12} \begin{vmatrix} a_{21} & a_{23} \\ a_{31} & a_{33} \end{vmatrix} + a_{22} \begin{vmatrix} a_{11} & a_{13} \\ a_{31} & a_{33} \end{vmatrix} - a_{32} \begin{vmatrix} a_{11} & a_{13} \\ a_{21} & a_{23} \end{vmatrix},$$

(along
first
row)
$$a_{11} \begin{vmatrix} a_{22} & a_{23} \\ a_{32} & a_{33} \end{vmatrix} - a_{12} \begin{vmatrix} a_{21} & a_{23} \\ a_{31} & a_{33} \end{vmatrix} + a_{13} \begin{vmatrix} a_{21} & a_{22} \\ a_{31} & a_{32} \end{vmatrix}.$$

The plus and minus signs are generated by factors which alternate according to the rule:

$$\begin{matrix} +1 & -1 & +1 \\ -1 & +1 & -1 \\ +1 & -1 & +1 \end{matrix} \qquad (14.6.12)$$

Correspondingly we distinguish between *even* and *odd elements*. Thus

$$a_{11}, a_{13}, a_{22}, a_{31}, a_{33} \quad \text{are even},$$
$$a_{12}, a_{21}, a_{23}, a_{32} \qquad \text{are odd}.$$

We assign the factor $+1$ *to even elements and the factor* -1 *to odd elements.*

Example 14.6.2.
$$3x_1 + 2x_2 - x_3 = 18$$
$$2x_1 - 5x_2 - 7x_3 = -5$$
$$x_1 + 4x_2 + x_3 = 2$$

Cramer's rule yields immediately

$$x_1 = \frac{\begin{vmatrix} 18 & 2 & -1 \\ -5 & -5 & -7 \\ 2 & 4 & 1 \end{vmatrix}}{\begin{vmatrix} 3 & 2 & -1 \\ 2 & -5 & -7 \\ 1 & 4 & 1 \end{vmatrix}}.$$

Expansion along the first column leads to

$$x_1 = \frac{18\begin{vmatrix} -5 & -7 \\ 4 & 1 \end{vmatrix} - (-5)\begin{vmatrix} 2 & -1 \\ 4 & 1 \end{vmatrix} + 2\begin{vmatrix} 2 & -1 \\ -5 & -7 \end{vmatrix}}{3\begin{vmatrix} -5 & -7 \\ 4 & 1 \end{vmatrix} - 2\begin{vmatrix} 2 & -1 \\ 4 & 1 \end{vmatrix} + 1\begin{vmatrix} 2 & -1 \\ -5 & -7 \end{vmatrix}} = \frac{406}{38}.$$

Similarly, we get

$$x_2 = \frac{\begin{vmatrix} 3 & 18 & -1 \\ 2 & -5 & -7 \\ 1 & 2 & 1 \end{vmatrix}}{\begin{vmatrix} 3 & 2 & -1 \\ 2 & -5 & -7 \\ 1 & 4 & 1 \end{vmatrix}} = \frac{-144}{38}, \quad x_3 = \frac{\begin{vmatrix} 3 & 2 & 18 \\ 2 & -5 & -5 \\ 1 & 4 & 2 \end{vmatrix}}{\begin{vmatrix} 3 & 2 & -1 \\ 2 & -5 & -7 \\ 1 & 4 & 1 \end{vmatrix}} = \frac{246}{38}.$$

Example 14.6.3. Calculate

$$\begin{vmatrix} 13 & -713 & 25 \\ 0 & 9 & 0 \\ -2 & 106 & 8 \end{vmatrix}.$$

We may expand this determinant along any row or column. However, in view of the two zeros it is most economic to expand along the second row. The element 9 is even, hence the factor is $+1$. Thus we get

$$9\begin{vmatrix} 13 & 25 \\ -2 & 8 \end{vmatrix} = 9(104 + 50) = 1386.$$

Note that the result is independent of the two elements -713 and 106.

There are various *rules* which are quite helpful when dealing with determinants:

1. *A determinant does not change its value if rows and columns are interchanged:*

$$\det A' = \det A. \tag{14.6.13}$$

Here A' denotes the *transpose* of A. To prove the rule we may expand A' along an arbitrary row and A along the corresponding column.

As a consequence, all properties of rows can also be applied to columns, and vice versa.

2. *If two rows (or two columns) are interchanged, then the value of a determinant changes by the factor* -1.

For instance,

$$
\begin{vmatrix} a_1 & a_2 & a_3 \\ b_1 & b_2 & b_3 \\ c_1 & c_2 & c_3 \end{vmatrix} = - \begin{vmatrix} b_1 & b_2 & b_3 \\ a_1 & a_2 & a_3 \\ c_1 & c_2 & c_3 \end{vmatrix} \qquad (14.6.14)
$$

as easily checked by expansion along the third row.

3. *If two rows (or two columns) are identical, the value of a determinant is zero.*

This follows from the previous rule by interchanging the identical rows (or columns).

4. *If each element of a row (or column) is multiplied by a factor λ, then the value of the determinant is also multiplied by λ.*

For instance,

$$
\begin{vmatrix} \lambda a_1 & \lambda a_2 & \lambda a_3 \\ b_1 & b_2 & b_3 \\ c_1 & c_2 & c_3 \end{vmatrix} = \lambda \begin{vmatrix} a_1 & a_2 & a_3 \\ b_1 & b_2 & b_3 \\ c_1 & c_2 & c_3 \end{vmatrix}. \qquad (14.6.15)
$$

This follows immediately by expanding the determinant along the first row.

Notice that this rule *differs considerably from the corresponding rule for matrices* as stated in Eq. (14.2.4).

5. *Let*

$$
A = \begin{pmatrix} a_1 & a_2 & a_3 \\ u_1 & u_2 & u_3 \\ v_1 & v_2 & v_3 \end{pmatrix}, \qquad B = \begin{pmatrix} b_1 & b_2 & b_3 \\ u_1 & u_2 & u_3 \\ v_1 & v_2 & v_3 \end{pmatrix}
$$

be two square matrices which differ only in one row (or column). Then

$$
\det A + \det B = \begin{vmatrix} a_1 + b_1 & a_2 + b_2 & a_3 + b_3 \\ u_1 & u_2 & u_3 \\ v_1 & v_2 & v_3 \end{vmatrix}. \qquad (14.6.16)
$$

The proof of Eq. (14.6.16) follows immediately upon expansion along the first row.

Notice that only the elements of a single row (or column) are added whereas all other elements remain unaltered. This rule *differs strikingly from the rule for matrix addition* as stated in Eq. (14.2.1).

6. *If to each row (or column) a multiple of another row (or column) is added, the value of a determinant remains unaltered.*

For instance,

$$\begin{vmatrix} a_1 & a_2 & a_3 \\ b_1 & b_2 & b_3 \\ c_1 & c_2 & c_3 \end{vmatrix} = \begin{vmatrix} a_1 + \lambda b_1 & a_2 + \lambda b_2 & a_3 + \lambda b_3 \\ b_1 & b_2 & b_3 \\ c_1 & c_2 & c_3 \end{vmatrix}.$$

To prove this statement we split the right determinant into two determinants using Eq. 14.6.16. Upon factoring λ the second determinant will vanish since two rows are identical.

7. The determinant of a product of square matrices is equal to the product of their determinants, that is,

$$\det (AB) = \det A \cdot \det B \tag{14.6.17}$$

The proof is laborious and shall be omitted here.

Determinants of higher order are defined iteratively in the same way as shown by Eq. (14.6.9). For instance, expanding along the first row we get

$$\begin{vmatrix} a_1 & a_2 & a_3 & a_4 \\ b_1 & b_2 & b_3 & b_4 \\ c_1 & c_2 & c_3 & c_4 \\ d_1 & d_2 & d_3 & d_4 \end{vmatrix} = a_1 \begin{vmatrix} b_2 & b_3 & b_4 \\ c_2 & c_3 & c_4 \\ d_2 & d_3 & d_4 \end{vmatrix} - a_2 \begin{vmatrix} b_1 & b_3 & b_4 \\ c_1 & c_3 & c_4 \\ d_1 & d_3 & d_4 \end{vmatrix}$$

$$+ a_3 \begin{vmatrix} b_1 & b_2 & b_4 \\ c_1 & c_2 & c_4 \\ d_1 & d_2 & d_4 \end{vmatrix} - a_4 \begin{vmatrix} b_1 & b_2 & b_3 \\ c_1 & c_2 & c_3 \\ d_1 & d_2 & d_3 \end{vmatrix}.$$

All the rules stated above are also correct for higher order determinants.

Example 14.6.4. Calculate

$$\begin{vmatrix} 8 & 0 & -4 & 1 \\ 2 & 3 & 2 & 2 \\ 1 & -3 & 5 & 7 \\ -4 & 0 & 2 & 5 \end{vmatrix}.$$

The second column contains already two zeros. A third zero can be generated by adding the third row to the second row. Thus we get

$$\begin{vmatrix} 8 & 0 & -4 & 1 \\ 3 & 0 & 7 & 9 \\ 1 & -3 & 5 & 7 \\ -4 & 0 & 2 & 5 \end{vmatrix}.$$

The element -3 of the second column is odd which calls for the factor -1. Hence, expanding along the second column we obtain

$$(-1)(-3)\begin{vmatrix} 8 & -4 & 1 \\ 3 & 7 & 9 \\ -4 & 2 & 5 \end{vmatrix}.$$

Multiplying the third row by 2 and adding it to the first row yields

$$(-1)(-3)\begin{vmatrix} 0 & 0 & 11 \\ 3 & 7 & 9 \\ -4 & 2 & 5 \end{vmatrix} = 3 \cdot 11 \begin{vmatrix} 3 & 7 \\ -4 & 2 \end{vmatrix} = 1122 .$$

14.7. Inverse of a Matrix

In the preceding section we have learned how to solve linear equations by explicit formulas. In this section we will improve the presentation by using matrix algebra.

We return to Eqs. (14.6.5) which we may also write in matrix form:

$$Ax = u . \tag{14.7.1}$$

It is tempting to apply ordinary algebra, that is, to "divide" by A and to write

$$x = A^{-1}u . \tag{14.7.2}$$

Whether such a simple algebra is possible, has to be investigated. A glance at Eq. (14.6.8) shows that x_1 (and similarly x_2 and x_3) are linear functions of u_1, u_2, u_3. Thus it is possible to establish a matrix A^{-1} which satisfies (14.7.2). From Eq. (14.6.8) we deduce

$$A^{-1} = \frac{1}{\det A} \begin{pmatrix} +\begin{vmatrix} a_{22} & a_{23} \\ a_{32} & a_{33} \end{vmatrix} & -\begin{vmatrix} a_{12} & a_{13} \\ a_{32} & a_{33} \end{vmatrix} & +\begin{vmatrix} a_{12} & a_{13} \\ a_{22} & a_{23} \end{vmatrix} \\ -\begin{vmatrix} a_{21} & a_{23} \\ a_{31} & a_{33} \end{vmatrix} & +\begin{vmatrix} a_{11} & a_{13} \\ a_{31} & a_{33} \end{vmatrix} & -\begin{vmatrix} a_{11} & a_{13} \\ a_{21} & a_{23} \end{vmatrix} \\ +\begin{vmatrix} a_{21} & a_{22} \\ a_{31} & a_{32} \end{vmatrix} & -\begin{vmatrix} a_{11} & a_{12} \\ a_{31} & a_{32} \end{vmatrix} & +\begin{vmatrix} a_{11} & a_{12} \\ a_{21} & a_{22} \end{vmatrix} \end{pmatrix} \tag{14.7.3}$$

The numerator is a 3×3 matrix. Its elements are minors of A properly arranged and provided with plus and minus signs. We may establish A^{-1} by proceeding in four steps:

1. Form a matrix by replacing each element a_{ik} with the corresponding minor.

2. Multiply each minor by $+1$ or -1 depending on whether a_{ik} is even or odd.

3. Transpose the matrix.

4. Divide the matrix by det A.

This rule is not restricted to 3×3 matrices; it can be used for any square matrix $A^{n \times n}$ provided that det $A \neq 0$.

Once A^{-1} is ready, the solution of (14.7.1) proceeds as follows: We multiply by A^{-1} from the left and get

$$A^{-1}Ax = A^{-1}u.$$

According to (14.7.2) we conclude that the left side is identical to x and, therefore, that $A^{-1}A$ is the identity matrix I which we introduced in Eq. (14.2.15). Thus

$$\boxed{A^{-1}A = I.}\tag{14.7.4}$$

If we had to retrace our steps, that is, to solve Eq. (14.7.2) with respect to u, we would multiply by A from the left,

$$Ax = AA^{-1}u,$$

and by comparison with (14.7.1) conclude that

$$\boxed{AA^{-1} = I}.\tag{14.7.5}$$

Not only is A^{-1} the inverse of A, but A is also the inverse of A^{-1}, a result that can be written in the form

$$(A^{-1})^{-1} = A.\tag{14.7.6}$$

Whereas the commutative law for matrix multiplication does not hold in general, the two factors A and A^{-1} can be interchanged.

Example 14.7.1. Let

$$A = \begin{pmatrix} 4 & 1 & 2 \\ 5 & -1 & 3 \\ 2 & 2 & 0 \end{pmatrix}.$$

To find the inverse matrix we first calculate all minors in the original order.

first row: $\quad \begin{vmatrix} -1 & 3 \\ 2 & 0 \end{vmatrix} = -6, \quad \begin{vmatrix} 5 & 3 \\ 2 & 0 \end{vmatrix} = -6, \quad \begin{vmatrix} 5 & -1 \\ 2 & 2 \end{vmatrix} = 12,$

second row: $\quad \begin{vmatrix} 1 & 2 \\ 2 & 0 \end{vmatrix} = -4, \quad \begin{vmatrix} 4 & 2 \\ 2 & 0 \end{vmatrix} = -4, \quad \begin{vmatrix} 4 & 1 \\ 2 & 2 \end{vmatrix} = 6,$

third row: $\quad \begin{vmatrix} 1 & 2 \\ -1 & 3 \end{vmatrix} = 5, \quad \begin{vmatrix} 4 & 2 \\ 5 & 3 \end{vmatrix} = 2, \quad \begin{vmatrix} 4 & 1 \\ 5 & -1 \end{vmatrix} = -9.$

So far we have the matrix

$$\begin{pmatrix} -6 & -6 & 12 \\ -4 & -4 & 6 \\ 5 & 2 & -9 \end{pmatrix}.$$

Second, we multiply even elements by $+1$ and odd elements by -1:

$$\begin{pmatrix} -6 & 6 & 12 \\ 4 & -4 & -6 \\ 5 & -2 & -9 \end{pmatrix}.$$

Third, we transpose the matrix:

$$\begin{pmatrix} -6 & 4 & 5 \\ 6 & -4 & -2 \\ 12 & -6 & -9 \end{pmatrix}.$$

Fourth, we divide by $\det A = 6$ and obtain the final result

$$A^{-1} = \tfrac{1}{6} \begin{pmatrix} -6 & 4 & 5 \\ 6 & -4 & -2 \\ 12 & -6 & -9 \end{pmatrix}.$$

The reader is invited to check the result by using Eq. (14.7.4) and/or Eq. (14.7.5).

Example 14.7.2. Find the inverse of

$$A = \begin{pmatrix} 4 & 3 \\ 5 & 7 \end{pmatrix}.$$

Here, the minors need no special calculation. We confine ourselves to list the results of the four steps:

1st step: $\begin{pmatrix} 7 & 5 \\ 3 & 4 \end{pmatrix},$

2nd step: $\begin{pmatrix} 7 & -5 \\ -3 & 4 \end{pmatrix},$

3rd step: $\begin{pmatrix} 7 & -3 \\ -5 & 4 \end{pmatrix},$

4th step: $\tfrac{1}{13} \begin{pmatrix} 7 & -3 \\ -5 & 4 \end{pmatrix} = A^{-1}.$

14.8. Linear dependence

Consider the two linear equations

$$3x - 2y = 7,$$
$$-6x + 4y = -14.$$

We quickly verify that the second equation is a multiple of the first equation. It contains no new information. Anyone of the infinitely many solutions of the first equation is also a solution of the second equation, and vice versa. We say that the two equations are *linearly dependent*.

The notion of linear dependence is most important not only for equations, but also for a variety of other applications.

To begin with the theory, we consider a vector and some multiples of it:

$$\boldsymbol{a} = \begin{pmatrix} 6 \\ 2 \\ -4 \end{pmatrix}, \quad \tfrac{3}{2}\boldsymbol{a} = \begin{pmatrix} 9 \\ 3 \\ -6 \end{pmatrix}, \quad (-3)\boldsymbol{a} = \begin{pmatrix} -18 \\ -6 \\ 12 \end{pmatrix}.$$

Such vectors are called linearly dependent on each other. If we interpret vectors as directed line segments, then two dependent vectors have the same or opposite directions as shown in Fig. 14.19.

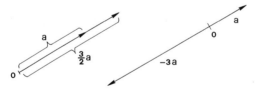

Fig. 14.19. Pairs of linearly dependent vectors

More generally, *two vectors $\boldsymbol{a}_1$ and $\boldsymbol{a}_2$ are said to be linearly dependent, if two numbers λ_1 and λ_2 exist which are not both zero so that*

$$\lambda_1 \boldsymbol{a}_1 + \lambda_2 \boldsymbol{a}_2 = \boldsymbol{0} \tag{14.8.1}$$

holds. Here, $\boldsymbol{0}$ denotes the *zero vector*. Otherwise, two vectors are said to be *linearly independent*.

Example 14.8.1. Are the two vectors

$$\boldsymbol{a}_1 = \begin{pmatrix} 9 \\ -3 \\ 12 \\ 6 \end{pmatrix}, \quad \boldsymbol{a}_2 = \begin{pmatrix} -15 \\ 5 \\ -20 \\ -10 \end{pmatrix}$$

linearly dependent? The answer is yes. By inspection we find the numbers $\lambda_1 = 5$ and $\lambda_2 = 3$ so that

$$\lambda_1 \boldsymbol{a}_1 + \lambda_2 \boldsymbol{a}_2 = 5 \begin{pmatrix} 9 \\ -3 \\ 12 \\ 6 \end{pmatrix} + 3 \begin{pmatrix} -15 \\ 5 \\ -20 \\ -10 \end{pmatrix} = \begin{pmatrix} 0 \\ 0 \\ 0 \\ 0 \end{pmatrix}.$$

Clearly $\boldsymbol{a}_2$ is a multiple of $\boldsymbol{a}_1$.

Now we are prepared to understand the following definition:

Three vectors $\boldsymbol{a}_1, \boldsymbol{a}_2, \boldsymbol{a}_3$ are said to be linearly dependent, if numbers $\lambda_1, \lambda_2, \lambda_3$ exist which are not all zero so that

$$\lambda_1 \boldsymbol{a}_1 + \lambda_2 \boldsymbol{a}_2 + \lambda_3 \boldsymbol{a}_3 = \boldsymbol{0} \qquad (14.8.2)$$

holds.

There is an easy geometric interpretation of Eq. (14.8.2). Assume that $\lambda_3 \neq 0$. Then we can solve the equation for $\boldsymbol{a}_3$:

$$\boldsymbol{a}_3 = (-\lambda_1/\lambda_3) \boldsymbol{a}_1 + (-\lambda_2/\lambda_3) \boldsymbol{a}_2 . \qquad (14.8.3)$$

Here, $\boldsymbol{a}_3$ is a *linear combination* of $\boldsymbol{a}_1$ and $\boldsymbol{a}_2$. This means geometrically that $\boldsymbol{a}_3$ must fall into the plane spanned by $\boldsymbol{a}_1$ and $\boldsymbol{a}_2$. This is shown in Fig. 14.20.

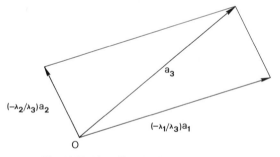

Fig. 14.20. Three linearly dependent vectors $\boldsymbol{a}_1, \boldsymbol{a}_2, \boldsymbol{a}_3$

Three *linearly independent* vectors cannot fall into the same plane.

Linear dependence can be detected by determinants as we will illustrate by numerical examples.

Example 14.8.2. Given the two row vectors

$$\boldsymbol{a}_1 = (4, -2, 10), \qquad \boldsymbol{a}_2 = (6, -3, 15).$$

We combine the two rows to give a rectangular matrix

$$\begin{pmatrix} 4 & -2 & 10 \\ 6 & -3 & 15 \end{pmatrix}$$

and recognize that the two rows are *proportional.* Therefore, any minor that we can form must be zero:

$$\begin{vmatrix} 4 & -2 \\ 6 & -3 \end{vmatrix} = 0, \qquad \begin{vmatrix} -2 & 10 \\ -3 & 15 \end{vmatrix} = 0, \qquad \begin{vmatrix} 4 & 10 \\ 6 & 15 \end{vmatrix} = 0.$$

Conversely, the vanishing determinants indicate that the two rows are proportional or linearly dependent.

Example 14.8.3. Are the three row vectors

$$a_1 = (2, 1, 5, -1), \qquad a_2 = (5, 0, 10, 0), \qquad a_3 = (1, -2, 0, 3)$$

linearly dependent? This would be the case only if all 3×3 minors of the matrix

$$\begin{pmatrix} 2 & 1 & 5 & -1 \\ 5 & 0 & 10 & 0 \\ 1 & -2 & 0 & 3 \end{pmatrix}$$

were zero. A first minor is

$$\begin{vmatrix} 2 & 1 & 5 \\ 5 & 0 & 10 \\ 1 & -2 & 0 \end{vmatrix} = \begin{vmatrix} 2 & 1 & 5 \\ 1 & -2 & 0 \\ 1 & -2 & 0 \end{vmatrix}$$

and equal to zero, since two rows are identical. A second minor is

$$\begin{vmatrix} 2 & 1 & -1 \\ 5 & 0 & 0 \\ 1 & -2 & 3 \end{vmatrix} = 5(-1) \begin{vmatrix} 1 & -1 \\ -2 & 3 \end{vmatrix} = -5.$$

Obviously, the condition is not satisfied. Therefore, the three vectors are linearly independent.

Now, we return to *linear equations.* We will have to distinguish among several cases. This can best be explained by examples.

Example 14.8.4. We have already seen that the two equations

$$3x - 2y = 7$$
$$-6x + 4y = -14$$

are linearly dependent so that one of the equations can be dropped. Now we replace one of the numbers on the right side by another number, say -14 by 11:

$$3x - 2y = 7$$
$$-6x + 4y = 11.$$

The left sides are still linearly dependent since

$$(-6x + 4y) = (-2)(3x - 2y)$$

no matter what particular values are assigned to x and y. But the right sides do no longer satisfy the same proportion:

$$11 \neq (-2)\,7\,.$$

Therefore, the two equations contradict each other. No values of x and y can ever satisfy both equations simultaneously. There exists no solution.

Notice that, with matrix notation

$$\begin{pmatrix} 3 & -2 \\ -6 & 4 \end{pmatrix}\begin{pmatrix} x \\ y \end{pmatrix} = \begin{pmatrix} 7 \\ 11 \end{pmatrix},$$

the determinant of $\begin{pmatrix} 3 & -2 \\ -6 & 4 \end{pmatrix}$ is zero and, therefore, that Cramer's rule cannot be applied.

A geometric interpretation of the two cases is given in Fig. 14.21. In a Cartesian coordinate system, a linear equation in x and y is represented by a straight line. Equations that are merely multiples of each other are represented by two coinciding lines. They have infinitely many points in common (Fig. 14.21a). However, if we change one of the numbers on the right side of the equations, the two lines become parallel and have therefore no point in common (Fig. 14.21b).

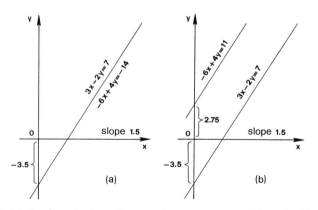

Fig. 14.21. (a) Two linearly dependent equations are represented by coinciding graphs. (b) If only the left sides of the equations are linearly dependent, the two graphs are parallel straight lines.

Example 14.8.5. Consider the equations

$$3x - 2y = 0$$

$$x + 5y = 0$$

where all numbers on the right side are zero. Such a system is called *homogeneous*, whereas any system with at least one number $\neq 0$ on the right is called *nonhomogeneous*. Obviously in a homogeneous system, there exist always the particular solution

$$x = 0, \qquad y = 0.$$

It is called the *trivial solution*.

Whether or not a *nontrivial* solution exists, depends on the determinant of the coefficients. In our case

$$\begin{vmatrix} 3 & -2 \\ 1 & 5 \end{vmatrix} \neq 0$$

so that Cramer's rule can be applied and yields a trivial solution only.

However, if homogeneous equations are given such as

$$3x - 2y = 0,$$

$$15x - 10y = 0,$$

the determinant vanishes. The two equations are linearly dependent so that one of them can be dropped. There exists a nontrivial solution, e.g. $x = 4$, $y = 6$.

The *general theory* of n linear equations with n unknowns contains precisely the cases considered in the previous examples. We write the system in matrix form

$$A x = u .\tag{14.8.4}$$

Case 1: $\det A \neq 0$. Cramer's rule can be applied to solve Eq. (14.8.4). There exists a unique solution.

Subcase 1a: $u = 0$ (homogeneous equations). There exists only the trivial solution $x = 0$.

Subcase 1b: $u \neq 0$ (nonhomogeneous equations). There exists a unique solution which is not trivial, that is, at least one x_i is different from zero.

Case 2: $\det A = 0$. Cramer's rule cannot be applied. The left sides of Eq. (14.8.4) are linearly dependent.

Subcase 2a: The equations (left and right sides combined) are linearly dependent. At least one equation can be dropped. If a solution exists at all, then there exist infinitely many solutions.

Subcase 2b: The equations (left and right sides combined) are not linearly dependent. They contradict each other. There exists no solution whatsoever.

Example 14.8.6. Given

$$x_1 - 2x_2 - x_3 = 0,$$
$$2x_1 + x_2 = 0,$$
$$x_1 - 3x_2 + x_3 = 0.$$

This system of equations is homogeneous. The determinant is

$$\begin{vmatrix} 1 & -2 & -1 \\ 2 & 1 & 0 \\ 1 & -3 & 1 \end{vmatrix} = 12 \neq 0.$$

This is Subcase 1a. There exists only the trivial solution $x_1 = 0$, $x_2 = 0$, $x_3 = 0$.

However, if the third equation were

$$x_1 + 3x_2 + x_3 = 0,$$

the determinant would be

$$\begin{vmatrix} 1 & -2 & -1 \\ 2 & 1 & 0 \\ 1 & 3 & 1 \end{vmatrix} = 0.$$

In addition, the three equations (left and right sides combined) are linearly dependent. One of the equations, in this case the third, can be dropped. The remaining equations

$$x_1 - 2x_2 - x_3 = 0,$$
$$2x_1 + x_2 = 0$$

have an infinite number of nontrivial solutions. We obtain them by choosing an arbitrary value for x_1, say

$$x_1 = p.$$

Then

$$x_2 = -2p,$$
$$x_3 = 5p.$$

14.9. Eigenvalues and Eigenvectors

We now turn our attention to a most peculiar application of matrices and determinants. Let

$$A = \begin{pmatrix} a & b \\ c & d \end{pmatrix}$$

be a given square matrix. Upon multiplication, A transforms a vector

$$x = \begin{pmatrix} x \\ y \end{pmatrix}$$

into a new vector

$$u = \begin{pmatrix} u \\ v \end{pmatrix}.$$

Thus the *transformation* is

$$A x = u. \tag{14.9.1}$$

In geometric as well as nongeometric applications the question often arises whether a vector x exists which by Eq. (14.9.1) is transformed into itself or at least into a multiple of itself. A multiple of x would be written as

$$\lambda x$$

where λ is any real number. The question leads to the equation

$$A x = \lambda x. \tag{14.9.2}$$

The problem of solving this equation is by no means routine. We have to find a suitable λ as well as a suitable vector x. Any suitable λ is called an *eigenvalue* and a corresponding vector x (different from $\mathbf{0}$) an *eigenvector*[7]. Alternative, but less frequently used words are *characteristic value* for λ and *characteristic vector* for x.

Example 14.9.1. Given the matrix

$$A = \begin{pmatrix} 3 & -2 \\ 2 & -2 \end{pmatrix}.$$

Solve Eq. (14.9.2), that is,

$$3x - 2y = \lambda x,$$
$$2x - 2y = \lambda y.$$

---

[7] The strange word "eigenvalue" is an incomplete translation of the German word "Eigenwert" meaning a value inherent in the problem. The adjective "eigen" (pronounced īgən) has the same origin and meaning as the English word "own". This explanation holds also for "eigenvector".

For this purpose we collect all terms with x and y on the left side:

$$(3 - \lambda) x - 2y = 0,$$
$$2x - (2 + \lambda) y = 0. \tag{14.9.3}$$

These two equations in x and y are *homogeneous*. There exists a trivial solution $x = 0$, $y = 0$. However, a nontrivial solution can only exist, if the determinant of coefficients vanishes as stated in the previous section (Subcase 2a). Therefore, we have to satisfy

$$\begin{vmatrix} 3 - \lambda & -2 \\ 2 & -(2 + \lambda) \end{vmatrix} = 0 \quad \text{or} \quad \lambda^2 - \lambda - 2 = 0$$

which is called the *characteristic equation* of the problem. Solutions are

$$\lambda_1 = -1, \qquad \lambda_2 = 2.$$

Thus we have found two different eigenvalues.

With either eigenvalue we can return to Eqs. (14.9.3). With $\lambda_1 = -1$ we get

$$4x - 2y = 0,$$
$$2x - y = 0.$$

One of the equation can be dropped since they are linearly dependent. A solution corresponding to λ_1 would be

$$x_1 = 1, \qquad y_1 = 2 \quad \text{or} \quad \boldsymbol{x}_1 = \begin{pmatrix} 1 \\ 2 \end{pmatrix}.$$

Thus $\boldsymbol{x}_1$ is an eigenvector. Of course, any multiple of $\boldsymbol{x}_1$ would also serve as an eigenvector corresponding to λ_1.

Similarly, for $\lambda_2 = 2$ we have to satisfy the equation

$$x - 2y = 0$$

which is possible by the eigenvector

$$\boldsymbol{x}_2 = \begin{pmatrix} 2 \\ 1 \end{pmatrix}$$

or any multiple of it.

To get a more general idea of the eigenvalue problem, we let

$$A = \begin{pmatrix} a_{11} & a_{12} & a_{13} \\ a_{21} & a_{22} & a_{23} \\ a_{31} & a_{32} & a_{33} \end{pmatrix}, \qquad \boldsymbol{x} = \begin{pmatrix} x_1 \\ x_2 \\ x_3 \end{pmatrix}.$$

Eq. (14.9.2) leads to

$$
\begin{aligned}
(a_{11} - \lambda) x_1 + \quad a_{12} x_2 \quad + \quad a_{13} x_3 \quad &= 0, \\
a_{21} x_1 \quad + (a_{22} - \lambda) x_2 + \quad a_{23} x_3 \quad &= 0, \\
a_{31} x_1 \quad + \quad a_{32} x_2 \quad + (a_{33} - \lambda) x_3 &= 0.
\end{aligned}
\tag{14.9.4}
$$

This is a system of homogeneous linear equations in x_1, x_2, x_3. A nontrivial solution exists only if the determinant of coefficients vanishes, that is, if

$$
\begin{vmatrix}
a_{11} - \lambda & a_{12} & a_{13} \\
a_{21} & a_{22} - \lambda & a_{23} \\
a_{31} & a_{32} & a_{33} - \lambda
\end{vmatrix} = 0.
\tag{14.9.5}
$$

Expansion of the determinant leads to a polynomial of degree three. Hence Eq. (14.9.5) is a cubic equation which may be written in the form

$$
\lambda^3 + p\lambda^2 + q\lambda + r = 0.
\tag{14.9.6}
$$

This equation has at most three different roots $\lambda_1, \lambda_2, \lambda_3$, the eigenvalues of the problem. In a numerical problem it might be quite laborious to calculate eigenvalues and, if required, also eigenvectors. Fortunately, there are computer programs available to complete the task.

Example 14.9.2. In Section 11.7 we dealt with a system of simultaneous differential equations

$$
\frac{dx}{dt} = a \cdot x + b \cdot y,
$$
$$
\frac{dy}{dt} = c \cdot x + d \cdot y.
\tag{14.9.7}
$$

Applying the method of trial functions we let

$$
x = A e^{\lambda t}, \quad y = B e^{\lambda t}
\tag{14.9.8}
$$

and had to satisfy the equations

$$
(a - \lambda) A + bB = 0,
$$
$$
cA + (d - \lambda) B = 0.
\tag{14.9.9}
$$

Typically, this is an eigenvalue problem. Since Eqs. (14.9.9) in A and B are homogeneous, and since we seek a nontrivial solution, the determinant of the coefficients has to vanish, that is,

$$
\begin{vmatrix}
a - \lambda & b \\
c & d - \lambda
\end{vmatrix} = 0
\tag{14.9.10}
$$

or

$$
\lambda^2 - (a + d)\,\lambda + (ad - bc) = 0.
\tag{14.9.11}
$$

Eq. (14.9.10), or (14.9.11), is called characteristic equation, its roots λ_1 and λ_2 characteristic values or eigenvalues. Any suitable pair (A, B) is then an eigenvector. Details are worked out in Section 11.7.

Example 14.9.3. Assume that a three-compartment system can well be described by linear differential equations:

$$dx/dt = a_{11}x + a_{12}y + a_{13}z,$$
$$dy/dt = a_{21}x + a_{22}y + a_{23}z,$$
$$dz/dt = a_{31}x + a_{32}y + a_{33}z.$$

By means of trial functions

$$x = Ae^{\lambda t}, \quad y = Be^{\lambda t}, \quad z = Ce^{\lambda t}$$

we are led to a homogeneous system of linear equations in the unknowns A, B, C:

$$(a_{11} - \lambda)A + \quad a_{12}B + \quad a_{13}C = 0,$$
$$a_{21}A + (a_{22} - \lambda)B + \quad a_{23}C = 0,$$
$$a_{31}A + \quad a_{32}B + (a_{33} - \lambda)C = 0.$$

A nontrivial solution is possible only if the characteristic equation

$$\begin{vmatrix} a_{11} - \lambda & a_{12} & a_{13} \\ a_{21} & a_{22} - \lambda & a_{23} \\ a_{31} & a_{32} & a_{33} - \lambda \end{vmatrix} = 0 \qquad (14.9.12)$$

holds. This is a cubic equation in λ. The roots denoted by $\lambda_1, \lambda_2, \lambda_3$ are the eigenvalues of the systems. The final solution follows along the lines presented in Section 11.7.

Example 14.9.4. Calculate the eigenvalues of the matrix

$$A = \begin{pmatrix} 1 & -2 & 8 \\ 0 & 10 & -3 \\ 0 & 7 & 0 \end{pmatrix}.$$

First, from each element of the diagonal we have to subtract λ (as in the preceding examples). This operation may be performed by using the identity matrix I:

$$A - \lambda I = \begin{pmatrix} 1 & -2 & 8 \\ 0 & 10 & -3 \\ 0 & 7 & 0 \end{pmatrix} - \lambda \begin{pmatrix} 1 & 0 & 0 \\ 0 & 1 & 0 \\ 0 & 0 & 1 \end{pmatrix} = \begin{pmatrix} 1 - \lambda & -2 & 8 \\ 0 & 10 - \lambda & -3 \\ 0 & 7 & -\lambda \end{pmatrix}.$$

Second, we have to equate the determinant of the new matrix to zero:

$$\det(A - \lambda I) = \begin{vmatrix} 1-\lambda & -2 & 8 \\ 0 & 10-\lambda & -3 \\ 0 & 7 & -\lambda \end{vmatrix} = 0.$$

Expanding the determinant along the first column leads to

$$(1 - \lambda) \begin{vmatrix} 10-\lambda & -3 \\ 7 & -\lambda \end{vmatrix} = 0$$

and further to

$$(1 - \lambda)(\lambda^2 - 10\lambda + 21) = 0.$$

The three roots or eigenvalues are

$$\lambda_1 = 1, \quad \lambda_2 = 3, \quad \lambda_3 = 7.$$

Notice that the elements -2 and 8 of A do not contribute to the eigenvalues.

Example 14.9.5. In Example 14.3.2 on *population dynamics* we have studied the concept of *stable age distribution*. We arrived at Eq. (14.3.7). Now we know that λ is an eigenvalue and n_k an eigenvector. λ can be found by solving the characteristic equation

$$\det(M - \lambda I) = 0. \tag{14.9.13}$$

Recommended for further reading: Grossman and Turner (1974), Hadeler (1974), Magar (1972), Nahikian (1964), Seal (1964), Searle (1966), C. A. B. Smith (1969, Vol. 2).

Problems for Solution

14.1.1. In a genetic study the number of individuals are counted which are of genotypes AA, Aa, aa at one locus and of genotype BB, Bb, bb at another locus. Assign $i = 0, 1, 2$ to AA, Aa, aa, respectively, and $k = 0, 1, 2$ to BB, Bb, bb, respectively. Form a matrix of the number n_{ik} of individuals having genotype i, k.

14.1.2. During five days, at three time instants a day, the temperature of ambient air is measured. Let t_{ik} be the temperature at the i-th time instant and on the k-th day. Form a 3×5 matrix of the measurements.

14.2.1. Let

$$a = (1, 3, -1, 5), \quad b = (4, 4, 2, 5), \quad c = (3, 1, 3, 0).$$

Find

$$\text{a) } \boldsymbol{a} + \boldsymbol{b} + \boldsymbol{c}, \quad \text{b) } \boldsymbol{a} + \boldsymbol{b} - \boldsymbol{c},$$
$$\text{c) } \boldsymbol{a} - \boldsymbol{b} + \boldsymbol{c}, \quad \text{d) } \boldsymbol{a} - \boldsymbol{b} - \boldsymbol{c}.$$

14.2.2. If

$$\boldsymbol{u} = \begin{pmatrix} x_1 \\ x_2 \end{pmatrix}, \quad \boldsymbol{v} = \begin{pmatrix} y_1 \\ y_2 \end{pmatrix}, \quad \boldsymbol{w} = \begin{pmatrix} z_1 \\ z_2 \end{pmatrix},$$

find

$$3\boldsymbol{u} - \boldsymbol{v} + \boldsymbol{w}.$$

14.2.3. Let

$$A = \begin{pmatrix} 3 & -2 & 1 \\ 0 & 5 & 4 \end{pmatrix}, \quad B = \begin{pmatrix} -2 & 2 & 0 \\ 5 & 1 & 1 \end{pmatrix}.$$

Evaluate

$$\text{a) } A + B, \quad \text{b) } A - B,$$
$$\text{c) } A + 2B, \quad \text{d) } 3A - B.$$

14.2.4. Let A and B be two matrices which have the same number of rows and the same number of columns, and let λ and μ be arbitrary numbers. Show that

$$(\lambda + \mu) A = \lambda A + \mu A,$$
$$\lambda (A + B) = \lambda A + \lambda B,$$
$$\lambda (\mu A) = (\lambda \mu) A.$$

14.2.5. Let

$$\boldsymbol{a} = (4, 4, 5), \quad \boldsymbol{b} = (6, -2, 7), \quad \boldsymbol{c} = (2, 0, 1).$$

Evaluate the inner products

$$\text{a) } \boldsymbol{a} \boldsymbol{b}', \quad \text{b) } \boldsymbol{a} \boldsymbol{c}', \quad \text{c) } \boldsymbol{b} \boldsymbol{c}', \quad \text{d) } \boldsymbol{b} \boldsymbol{a}',$$
$$\text{e) } \boldsymbol{a} \boldsymbol{a}', \quad \text{f) } \boldsymbol{b} \boldsymbol{b}', \quad \text{g) } \boldsymbol{c} \boldsymbol{c}'.$$

Why is $\boldsymbol{a} \boldsymbol{b}' = \boldsymbol{b} \boldsymbol{a}'$?

14.2.6. If

$$\boldsymbol{u} = \begin{pmatrix} u_1 \\ u_2 \\ u_3 \end{pmatrix}, \quad \boldsymbol{v} = \begin{pmatrix} 3 \\ -2 \\ 5 \end{pmatrix}, \quad \boldsymbol{w} = \begin{pmatrix} w_1 \\ w_2 \\ 0 \end{pmatrix},$$

find

$$\text{a) } \boldsymbol{u}' \boldsymbol{v}, \quad \text{b) } \boldsymbol{u}' \boldsymbol{w}, \quad \text{c) } \boldsymbol{v}' \boldsymbol{w},$$
$$\text{d) } \boldsymbol{u}' \boldsymbol{u}, \quad \text{e) } \boldsymbol{v}' \boldsymbol{v}, \quad \text{f) } \boldsymbol{w}' \boldsymbol{w}.$$

14.2.7. A forest was subdivided into small squares of 1 m^2 area each called *quadrat*. The number of ferns per quadrat was counted. The result was given in vector form:

$$
\begin{array}{cc}
\text{number of ferns} & \text{number of such} \\
\text{per quadrat} & \text{quadrats}
\end{array}
$$

$$
f = \begin{pmatrix} 0 \\ 1 \\ 2 \\ 3 \\ 4 \\ 5 \end{pmatrix}, \qquad
q = \begin{pmatrix} 23 \\ 17 \\ 8 \\ 6 \\ 8 \\ 2 \end{pmatrix}.
$$

(The meaning is: 23 quadrats contain 0 ferns, 17 quadrats contain 1 fern, etc.) Find the total number of ferns counted by a matrix operation.

14.2.8. Write Eq. (9.4.4) by means of an inner product.

14.2.9. Write Eq. (13.9.12) by means of an inner product.

14.2.10. Write the following expressions as inner products of two vectors:

a) $a_1 x_1 + a_2 x_2 + a_3 x_3 + a_4 x_4$,

b) $1 + 2p + 3p^2 + \cdots + np^{n-1}$,

c) $a \cos \alpha + b \sin \alpha$.

14.2.11. Evaluate

a) $\begin{pmatrix} 4 & 0 & -1 \\ 2 & 2 & 3 \end{pmatrix} \begin{pmatrix} 4 \\ 2 \\ 1 \end{pmatrix}$,

b) $(2, 3, -1) \begin{pmatrix} 5 & 0 \\ 0 & 1 \\ 5 & 2 \end{pmatrix}$,

c) $\begin{pmatrix} 3 & 3 & 0 \\ 2 & 1 & 4 \end{pmatrix} \begin{pmatrix} 3 & 0 \\ 2 & -1 \\ 1 & 5 \end{pmatrix}$,

d) $\begin{pmatrix} 4 & 7 \\ 0 & -1 \end{pmatrix} \begin{pmatrix} 5 & 2 \\ 3 & 4 \end{pmatrix}$.

14.2.12. Let

$$
M = \begin{pmatrix} 2 & 1 \\ 0 & -1 \end{pmatrix}, \qquad
N = \begin{pmatrix} 0 & 3 \\ 5 & 2 \end{pmatrix}.
$$

Show that $MN \neq NM$.

14.2.13. Let

$$
R = \begin{pmatrix} 1 & 1 & 1 & 1 \\ -1 & 1 & -1 & 1 \end{pmatrix}, \qquad
s = \begin{pmatrix} a \\ b \\ c \\ d \end{pmatrix}.
$$

Calculate

    a) Rs,   b) $s'R'$,   and show that $(Rs)' = s'R'$.

14.2.14. Write the column matrix

$$\begin{pmatrix} 3x+2y+z \\ 5x+3y+z \end{pmatrix}$$

as a product of two matrices $A = A^{2 \times 3}$ and $v = v^{3 \times 1}$.

14.2.15. Let

$$u = \begin{pmatrix} x \\ y \end{pmatrix}, \quad S = \begin{pmatrix} 3 & -2 \\ 1 & 1 \end{pmatrix}.$$

Find $u'Su$.

14.2.16. Work out the matrix product

$$(x, y, 1)\begin{pmatrix} 3 & 1 & 0 \\ 1 & 2 & -1 \\ 0 & -1 & 4 \end{pmatrix}\begin{pmatrix} x \\ y \\ 1 \end{pmatrix}$$

in two ways using the associative law (14.2.10).

14.2.17. Let

$$A = \begin{pmatrix} a & b \\ c & d \end{pmatrix}, \quad B = \begin{pmatrix} 2 & 5 \\ 0 & 4 \end{pmatrix}, \quad C = \begin{pmatrix} -1 & -2 \\ 3 & 0 \end{pmatrix}.$$

Work out $A(B+C)$ in two ways according to the distributive law (14.2.13).

14.2.18. Let

$$A = \begin{pmatrix} 2 & 0 \\ 1 & 2 \end{pmatrix}, \quad B = \begin{pmatrix} -1 & 2 \\ 3 & 2 \end{pmatrix}, \quad C = \begin{pmatrix} 5 & 1 \\ 1 & 7 \end{pmatrix}.$$

Calculate

    a) $A(B+C)$,   b) $A(B-C)$,

    c) $(B+C)A$,   d) ABC.

14.2.19. Let $A = \begin{pmatrix} 1 & 2 \\ 0 & 1 \end{pmatrix}$. Calculate

$AA = A^2$, $AAA = A^3$ and, by induction, A^n for any natural n.

14.2.20. Let

$$S = \begin{pmatrix} 1 & -7 & 0 \\ 0 & 1 & -7 \\ 0 & 0 & 1 \end{pmatrix}.$$

Find $SS = S^2$, $SSS = S^3$.

14.2.21. Let

$$U = \begin{pmatrix} 1 & 0 & 0 \\ 0 & 2 & 0 \\ 0 & 0 & 3 \end{pmatrix}.$$

Find a formula for U^n where n is a natural number.

14.2.22. Let

$$M = \begin{pmatrix} -1 & 5 & 5 \\ 3 & -1 & 5 \\ 3 & 3 & -1 \end{pmatrix}.$$

Evaluate $M'M$.

14.2.23. Let

$$A = \tfrac{1}{2}\begin{pmatrix} 8 & 5 \\ 6 & 5 \end{pmatrix}, \quad B = \tfrac{1}{5}\begin{pmatrix} 5 & -5 \\ -6 & 8 \end{pmatrix}.$$

Show that $AB = I$ where I denotes the identity matrix.

14.2.24. Let

$$A = \begin{pmatrix} 4 & -2 \\ -5 & 3 \end{pmatrix}, \quad B = \tfrac{1}{2}\begin{pmatrix} 3 & 2 \\ 5 & 4 \end{pmatrix}.$$

Show that $AB = I$ and $BA = I$ where I denotes the identity matrix as defined in (14.2.15). A is called the *inverse matrix* of B and B the inverse matrix of A. We write $B = A^{-1}$, $A = B^{-1}$.

14.2.25. Let

$$A = \tfrac{1}{5}\begin{pmatrix} 3 & 4 \\ -4 & 3 \end{pmatrix}.$$

Show that $A'A = I$ and $AA' = I$. According to the preceding problem, A' is the inverse matrix of A. A matrix A with the property $AA' = A'A = I$ is called an *orthogonal matrix*.

14.2.26. Show that

$$A = \begin{pmatrix} \sqrt{2}/2 & -1/2 & -1/2 \\ \sqrt{2}/2 & 1/2 & 1/2 \\ 0 & \sqrt{2}/2 & -\sqrt{2}/2 \end{pmatrix}$$

satisfies $A'A = I$ and that, therefore, A is an orthogonal matrix.

14.2.27. Solve the matrix equation

$$\begin{pmatrix} x & 3 \\ y & 4 \end{pmatrix}\begin{pmatrix} 2 & -1 \\ 3 & -2 \end{pmatrix} = \begin{pmatrix} 43 & -23 \\ -2 & -1 \end{pmatrix}$$

with respect to x and y.

14.2.28. Solve the matrix equation

$$\begin{pmatrix} 1 & 2 \\ -2 & 3 \end{pmatrix} \begin{pmatrix} u & 6 \\ 1 & v \end{pmatrix} = \begin{pmatrix} 10 & -8 \\ -13 & -33 \end{pmatrix}$$

with respect to u and v.

14.2.29. Let

$$a = (3, 1, 1), \qquad b = (4, 2, 1).$$

Find a) ab', b) $a'b$.

14.2.30. Given

$$u = \begin{pmatrix} u_1 \\ u_2 \\ u_3 \end{pmatrix}, \qquad v = \begin{pmatrix} v_1 \\ v_2 \\ v_3 \end{pmatrix}.$$

Calculate a) uv', b) $u'v$.

14.2.31. Using

$$A = \begin{pmatrix} a_1 & a_2 & a_3 \\ b_1 & b_2 & b_3 \end{pmatrix}, \qquad B = \begin{pmatrix} x_1 & y_1 \\ x_2 & y_2 \\ x_3 & y_3 \end{pmatrix}$$

verify that $(AB)' = B'A'$.

14.3.1. Apply the transition of Eq. (14.3.4) to an insect population of three age groups of sizes $n_{00} = 500$, $n_{10} = 300$, $n_{20} = 200$ (females only). Assume that the average number of "daughters" born to each female is $F_0 = 0$ (age group 0), $F_1 = 80$ (age group 1), $F_2 = 50$ (age group 2) during a particular season of length Δt. The probabilities of survival are assumed to be $P_0 = 0.6$ and $P_1 = 0.8$, respectively. Find the size of the next female population.

14.3.2. Solve the same problem in case of four age groups with $n_{00} = 800$, $n_{10} = 500$, $n_{20} = 300$, $n_{30} = 150$, $F_0 = 0$, $F_1 = 80$, $F_2 = 50$, $F_3 = 0$, $P_0 = 0.6$, $P_1 = 0.4$, $P_2 = 0.3$.

14.3.3. Let T be the matrix defined in (14.3.11) and let

$$O = \begin{pmatrix} p^2 & 2pq & q^2 \\ p^2 & 2pq & q^2 \\ p^2 & 2pq & q^2 \end{pmatrix} \qquad (p + q = 1).$$

(Letter O used after Li, 1958). Verify that

a) $TO = O$, b) $T^2 = \frac{1}{2}T + \frac{1}{2}O$, c) $T^3 = \frac{1}{4}T + \frac{3}{4}O$.

14.3.4. Use the result of the preceding problem to find a formula expressing $T^n (n = 2, 3, 4, ...)$ linearly by T and 0. What is the limiting matrix as n tends to infinity?

14.3.5. A signal operated by a laboratory mouse has only two faces: R = red, G = green. At each trial the mouse may or may not change the signal. Suppose that the following transition probabilities are given:

$$R \to R: p_{11} = 0.4,$$
$$R \to G: p_{12} = 0.6,$$
$$G \to R: p_{21} = 0.2,$$
$$G \to G: p_{22} = 0.8.$$

Assume further that each trial is independent of past experience. Then the outcomes of each trial form a *Markov chain* with two states (R and G). Establish the transition matrix and calculate the two-step probabilities using the equivalent of Eq. (14.3.18).

14.3.6. Let

$$P = \begin{pmatrix} 0.7 & 0.2 & 0.1 \\ 0.3 & 0.4 & 0.3 \\ 0.1 & 0.3 & 0.6 \end{pmatrix}$$

be the transition matrix of a *Markov chain*. Find the transition matrix of two-step transitions.

14.4.1. Plot the three vectors

$$a_1 = \begin{pmatrix} 3 \\ 1 \end{pmatrix}, \quad a_2 = \begin{pmatrix} 2 \\ 2 \end{pmatrix}, \quad a_3 = \begin{pmatrix} -1 \\ 3 \end{pmatrix}$$

in a rectangular coordinate system. Find their sum a) algebraically, b) geometrically.

14.4.2. Find the sum of

$$a_1 = \begin{pmatrix} 7 \\ 2 \end{pmatrix}, \quad a_2 = \begin{pmatrix} 0 \\ 3 \end{pmatrix}, \quad a_3 = \begin{pmatrix} -8 \\ 1 \end{pmatrix}, \quad a_4 = \begin{pmatrix} -3 \\ 2 \end{pmatrix}$$

geometrically by using a vector polygon. Check the result by algebraic addition.

14.4.3. Given the vectors

$$a = \begin{pmatrix} 7 \\ -2 \end{pmatrix}, \quad b = \begin{pmatrix} 5 \\ 5 \end{pmatrix}.$$

Plot the sum and the difference of a and b. Also plot the vectors $-a, -b, \frac{1}{2}a, -\frac{3}{5}b$.

14.4.4. Given

$$v = \begin{pmatrix} -1 \\ 5 \end{pmatrix}, \quad w = \begin{pmatrix} 8 \\ 8 \end{pmatrix}.$$

Plot a) $v + w$, b) $v - w$, c) $-v + w$, d) $-v - w$.

14.4.5. Given

$$a = \begin{pmatrix} 3 \\ -4 \end{pmatrix}, \quad b = \begin{pmatrix} 7 \\ 2 \end{pmatrix}.$$

Find the midpoint between the two points which are represented by a and b. Show that the corresponding vector is $\frac{1}{2}(a + b)$.

14.4.6. Three forces

$$F_1 = \begin{pmatrix} 5 \\ -8 \end{pmatrix}, \quad F_2 = \begin{pmatrix} 3 \\ 10 \end{pmatrix}, \quad F_3 = \begin{pmatrix} -8 \\ -2 \end{pmatrix}$$

act upon the point O. Show that they are in equilibrium.

14.4.7. Given

$$a = \begin{pmatrix} 4 \\ -3 \end{pmatrix}, \quad b = \begin{pmatrix} 2 \\ 7 \end{pmatrix}.$$

Calculate a) the absolute values of a and b, b) the angle between a and b.

14.4.8. Find the angle between the vectors

$$p = \begin{pmatrix} 12 \\ -5 \end{pmatrix} \quad \text{and} \quad q = \begin{pmatrix} 6 \\ 8 \end{pmatrix}.$$

14.4.9. Given the three vectors u, v, w by

$$u' = (5, 3), \quad v' = (-1, 5), \quad w' = (6, -10).$$

Are any two of the given vectors orthogonal?

14.4.10. Let

$$a_1 = \begin{pmatrix} 5 \\ -8 \\ 6 \end{pmatrix}, \quad a_2 = \begin{pmatrix} -3 \\ 3 \\ 0 \end{pmatrix}, \quad a_3 = \begin{pmatrix} 6 \\ 1 \\ 6 \end{pmatrix}.$$

Calculate

a) $c = a_1 + a_2 + a_3$,

b) $d = a_1 - a_2 + a_3$,

c) $e = -a_1 + a_2 - a_3$.

14.4.11 Let

$$F_1 = \begin{pmatrix} 8 \\ -4 \\ 5 \end{pmatrix}, \quad F_2 = \begin{pmatrix} -2 \\ 0 \\ 7 \end{pmatrix}, \quad F_3 = \begin{pmatrix} -6 \\ 2 \\ -2 \end{pmatrix}$$

be three forces acting upon the point O. Find an additional force which restores the equilibrium.

14.4.12. Calculate the absolute values of the vectors

$$a = \begin{pmatrix} 3 \\ 4 \\ 0 \end{pmatrix}, \quad b = \begin{pmatrix} 24 \\ 0 \\ 10 \end{pmatrix}, \quad c = \begin{pmatrix} 1 \\ -3 \\ 1 \end{pmatrix}.$$

14.4.13. Given the three-dimensional vectors u, v, w by

$$u' = (3, -1, 0), \quad v' = (-2, 3, 1), \quad w' = (5, 2, 4).$$

Are any two of the three given vectors orthogonal?

14.4.14. Let

$$u = \begin{pmatrix} 3 \\ 8 \\ -2 \end{pmatrix}, \quad v = \begin{pmatrix} -x \\ -1 \\ 5 \end{pmatrix}.$$

Determine x so that u and v are orthogonal.

14.4.15. Given the three-dimensional vectors a and b by $a' = (3, 2, -1)$, $b' = (4, 0, 5)$. Calculate the angle between the two vectors.

14.4.16. Find the angle between the vectors

$$p = \begin{pmatrix} 8 \\ -1 \\ 3 \end{pmatrix}, \quad q = \begin{pmatrix} -3 \\ 1 \\ -2 \end{pmatrix}.$$

14.4.17. Assume that the vectors a and b point in opposite directions. Show that

$$a'b = -|a|\,|b|.$$

14.4.18. In the four-dimensional space let

$$a = \begin{pmatrix} 2 \\ 1 \\ 0 \\ 3 \end{pmatrix}, \quad b = \begin{pmatrix} 3 \\ 0 \\ 4 \\ 4 \end{pmatrix}.$$

Find   a) $a + b$,   b) $a - b$,

c) $2a - 3b$,   d) $\frac{1}{2}(a + b)$,

e) $a'b$,   f) $|a|, |b|$,

g) the angle between a and b.

14.5.1. Assume that three equal masses are concentrated in points given by the vectors

$$a_1 = \begin{pmatrix} 3 \\ 2 \\ 5 \end{pmatrix}, \quad a_2 = \begin{pmatrix} 6 \\ 0 \\ 3 \end{pmatrix}, \quad a_3 = \begin{pmatrix} 8 \\ 7 \\ 1 \end{pmatrix}.$$

Find the center of mass.

14.5.2. In the preceding problem suppose that the three masses are not equal but

$$M_1 = 8 \text{ kg}, \quad M_2 = 5 \text{ kg}, \quad M_3 = 2 \text{ kg}.$$

Calculate the center of mass.

14.5.3. In Fig. 14.14 calculate $|F|$ if $|F_1|$ and the angle α between the vectors F_2 and F_3 are given.

14.5.4. Assume that a grasshopper weighs 2 g and that it reaches a maximum height of 30 cm above the horizontal ground when taking off at an angle of $60°$ to the ground. Assume further that the grasshopper's center of gravity moves 3 cm during the acceleration period. Using formula (14.5.10) find the magnitude of the thrust.

14.6.1. Calculate

$$\text{a)} \begin{vmatrix} 3 & 2 \\ 4 & 5 \end{vmatrix}, \quad \text{b)} \begin{vmatrix} 3 & 1 \\ -2 & 4 \end{vmatrix}, \quad \text{c)} \begin{vmatrix} 18 & 11 \\ 6 & -2 \end{vmatrix}.$$

14.6.2. Calculate

$$\text{a)} \begin{vmatrix} u & 7 \\ v & 2 \end{vmatrix}, \quad \text{b)} \begin{vmatrix} 3u & -4 \\ -u & +3 \end{vmatrix}, \quad \text{c)} \begin{vmatrix} 4a & -2b \\ 3a & -5b \end{vmatrix}.$$

14.6.3. Using Cramer's rule solve:

a) $6x_1 + 5x_2 = 8$ b) $-2x + 3y = 2r$

$5x_1 + 7x_2 = 13$, $8x - 13y = 5s$.

14.6.4. Using Cramer's rule solve:

a) $5A - 7B = 3$ b) $3\lambda + 10\mu - 5 = 0$

$2A + 5B = -8$, $4\lambda - \mu + 3 = 0$.

14.6.5. Evaluate

$$\text{a)} \begin{vmatrix} 4 & 7 & 5 \\ 3 & -2 & 8 \\ 4 & 0 & 0 \end{vmatrix}, \qquad \text{b)} \begin{vmatrix} 4 & 0 & 5 \\ -2 & 1 & 1 \\ 3 & 7 & 0 \end{vmatrix}.$$

14.6.6. Evaluate

$$\text{a)} \begin{vmatrix} 13 & 2 & 0 \\ 4 & -7 & 3 \\ 8 & 8 & 0 \end{vmatrix}, \qquad \text{b)} \begin{vmatrix} 0 & 2 & -4 \\ -2 & 0 & -1 \\ -4 & 1 & 0 \end{vmatrix}.$$

14.6.7. Using Cramer's rule solve:

$$\begin{aligned} x_1 - 2x_2 + x_3 &= 8 \\ 3x_1 + x_2 - x_3 &= 1 \\ 5x_1 - 3x_2 + 2x_3 &= -3. \end{aligned}$$

14.6.8. Using Cramer's rule solve:

$$\begin{aligned} A - B \quad\;\; &= 0 \\ B - 5C &= 1 \\ 2A \quad\;\; + 3C &= 4. \end{aligned}$$

14.6.9. Evaluate most economically

$$\text{a)} \begin{vmatrix} 3 & -1 & -1 \\ 5 & 18 & 16 \\ 6 & -2 & 2 \end{vmatrix}, \qquad \text{b)} \begin{vmatrix} 1 & 12 & 3 \\ 5 & -2 & 5 \\ 8 & 7 & 8 \end{vmatrix},$$

$$\text{c)} \begin{vmatrix} 5 & 3 & -4 \\ 9 & -6 & 8 \\ 23 & 1 & 7 \end{vmatrix}, \qquad \text{d)} \begin{vmatrix} 26 & 39 & -13 \\ 4 & -2 & 1 \\ 7 & 21 & -49 \end{vmatrix}.$$

14.6.10. Evaluate most economically

$$\begin{vmatrix} 3 & -1 & 5 & 8 \\ -3 & 1 & 0 & 7 \\ 6 & -2 & -5 & 3 \\ 1 & 0 & 38 & 54 \end{vmatrix}.$$

14.6.11. Show that

$$\begin{vmatrix} 1 & a & a^2 \\ 1 & b & b^2 \\ 1 & c & c^2 \end{vmatrix} = (a-b)(b-c)(c-a).$$

(Hint: Subtract lines from each other)

14.6.12. Add the determinants

$$\begin{vmatrix} a & b & c \\ 3 & 3 & 4 \\ 1 & 2 & 2 \end{vmatrix}, \quad \begin{vmatrix} p & q & r \\ 3 & 3 & 4 \\ 1 & 2 & 2 \end{vmatrix}$$

to one single determinant.

14.6.13. Let

$$A = \begin{pmatrix} 4 & 2 & 1 \\ 3 & -4 & 5 \\ 2 & 2 & 2 \end{pmatrix}.$$

Find $\det(\lambda A)$.

14.6.14. Using

$$A = \begin{pmatrix} 3 & 2 \\ 4 & 1 \end{pmatrix}, \quad B = \begin{pmatrix} -1 & 3 \\ 1 & 5 \end{pmatrix}$$

show that $\det(AB) = \det A \cdot \det B$.

14.7.1. By solving

$$\begin{pmatrix} 3 & 7 \\ 2 & 5 \end{pmatrix} \begin{pmatrix} x & y \\ u & v \end{pmatrix} = \begin{pmatrix} 1 & 0 \\ 0 & 1 \end{pmatrix}$$

find the inverse matrix of the first factor.

14.7.2. Find the inverse matrix of

$$A = \begin{pmatrix} 3 & 7 \\ 2 & 5 \end{pmatrix}$$

by using the rules stated below Eq. (14.7.3).

14.7.3. Find the inverse matrix of

$$A = \begin{pmatrix} 3 & -1 & 0 \\ 0 & 4 & 2 \\ 3 & 0 & 1 \end{pmatrix}.$$

14.7.4. Find the inverse matrix of

$$B = \begin{pmatrix} 1 & 1 & 1 \\ 0 & 0 & 3 \\ -1 & 1 & 0 \end{pmatrix}.$$

14.7.5. The following matrices are called *triangular matrices* since all elements on one side of the main diagonal are zero:

$$\text{a) } A = \begin{pmatrix} 1 & 4 & 6 \\ 0 & 1 & 3 \\ 0 & 0 & 1 \end{pmatrix}, \quad \text{b) } B = \begin{pmatrix} 3 & 0 & 0 \\ 1 & 1 & 0 \\ 0 & 5 & -2 \end{pmatrix}.$$

Find A^{-1} and B^{-1}.

14.7.6. Let

$$A = \begin{pmatrix} a & b \\ c & d \end{pmatrix}.$$

Show that $(A')^{-1} = (A^{-1})'$.

14.7.7. Let

$$A = \begin{pmatrix} 4 & 5 \\ 5 & 7 \end{pmatrix}.$$

Show that

$$(A^2)^{-1} = (A^{-1})^2.$$

14.7.8. Let

$$x = \begin{pmatrix} x_1 \\ x_2 \end{pmatrix}, \quad A = \begin{pmatrix} 4 & 7 \\ 3 & 5 \end{pmatrix}, \quad u = \begin{pmatrix} p \\ q \end{pmatrix}.$$

Solve the matrix equation

$$Ax = u$$

for x by using the inverse of A.

14.7.9. Let

$$x = \begin{pmatrix} x_1 \\ x_2 \\ x_3 \end{pmatrix}, \quad A = \begin{pmatrix} 3 & 0 & 1 \\ 3 & -1 & 0 \\ 0 & 4 & 2 \end{pmatrix}, \quad y = \begin{pmatrix} y_1 \\ y_2 \\ y_3 \end{pmatrix}.$$

Solve the matrix equation

$$Ax = y$$

for x by means of A^{-1}.

14.7.10. Let

$$S = \begin{pmatrix} a & b \\ c & d \end{pmatrix}, \quad U = \begin{pmatrix} 2 & 1 \\ 5 & 3 \end{pmatrix}.$$

Find

$$L = USU^{-1}.$$

14.8.1. Are the three vectors

$$a = (3, -1, 5, 2),$$
$$b = (0, 0, 5, 2),$$
$$c = (3, -1, 0, 0)$$

linearly dependent?

14.8.2. Are the three vectors

$$u = (1, 1, 0, 2, 5),$$
$$v = (0, -1, 0, 4, 7),$$
$$w = (0, 0, 1, -5, 3)$$

linearly dependent? (Hint: use determinants beginning on the left).

14.8.3. Discuss the character of the solutions for the following systems of equations:

a) $3x - y = 7$ b) $3x - y = 5$
      $-6x + 2y = -14,$ $-6x + 2y = 4,$

c) $3x - y = 0$ d) $3x - y = 5$
      $-6x + 2y = 0,$ $-6x + y = 4,$

e) $3x - y = 0$ f) $ax + by = 0$ $(a \neq 0,$
      $-6x + y = 0,$ $ax - by = 0$ $b \neq 0).$

14.8.4. Which of the following systems of equations have 1) a unique solution, 2) an infinite number of solutions, 3) only a trivial solution, 4) no solution?

a) $A + 3B = 0$ b) $5\lambda - 3\mu = 7$
     $2A - B = 0,$ $\lambda + \mu = -1,$

c) $8m - 6n = 4$ d) $12r + 10s = 7$
     $4m - 3n = 2,$ $18r + 15s = 6,$

e) $28p + 16q = 0$ f) $x + y - 4 = 0$
    $21p + 12q = 0,$ $-x - y + 5 = 0.$

14.8.5. Determine λ in such a way that the system

$$\lambda x_1 + 12x_2 = 0$$
$$3x_1 + \lambda x_2 = 0$$

has a nontrivial solution.

14.8.6. Find κ (Greek kappa) such that the system

$$\kappa x + \quad 3y - \quad 2z = 0$$
$$(\kappa - 1)y + \quad 7z = 0$$
$$(\kappa + 2)z = 0$$

has a nontrivial solution.

14.9.1. Find eigenvalues and corresponding eigenvectors for the following matrices:

a) $\begin{pmatrix} 8 & 1 \\ -5 & 2 \end{pmatrix}$, b) $\begin{pmatrix} 7 & 3 \\ 4 & 3 \end{pmatrix}$, c) $\begin{pmatrix} 1 & -4 \\ 5 & 13 \end{pmatrix}$.

14.9.2. Given the matrices

$$A = \begin{pmatrix} 8 & 9 \\ 3 & 2 \end{pmatrix}, \quad B = \begin{pmatrix} -9 & 5 \\ -3 & -1 \end{pmatrix}, \quad C = \begin{pmatrix} 7 & 3 \\ 1 & 5 \end{pmatrix}.$$

Find eigenvalues and eigenvectors for A, B, C.

14.9.3. Find eigenvalues and corresponding eigenvectors for

$$A = \begin{pmatrix} 2 & 0 & 0 \\ 1 & 7 & 3 \\ 5 & 4 & 3 \end{pmatrix}.$$

14.9.4. Find eigenvalues and eigenvectors for

$$S = \begin{pmatrix} 1 & 5 & 0 \\ 7 & 3 & 0 \\ 4 & 3 & -1 \end{pmatrix}.$$

14.9.5. Let

$$A = \begin{pmatrix} p_1 & q_1 \\ p_2 & q_2 \end{pmatrix}$$

be the transition matrix of a *Markov chain* $(p_1 + q_1 = 1, p_2 + q_2 = 1)$. Show that $\lambda = 1$ is an eigenvalue of A.

14.9.6. Generalize the preceding problem for a $n \times n$ transition matrix A. (Hint: Add all columns of $A - I$ together and show that the columns are linearly dependent).

14.9.7. In the study of recurrent sib mating, Crow and Kimura (1970, p. 88) are confronted with the characteristic equation

$$\begin{vmatrix} 0 - \lambda & 1 \\ \frac{1}{4} & \frac{1}{2} - \lambda \end{vmatrix} = 0.$$

Find the characteristic values λ_1 and λ_2.

14.9.8. Crow and Kimura (1970, p. 91) study a genetic problem of inbreeding which leads to the characteristic equation

$$\begin{vmatrix} -\lambda & 1 & 0 \\ 0 & \frac{1}{2}-\lambda & \frac{1}{2} \\ \frac{1}{4} & \frac{1}{2} & -\lambda \end{vmatrix} = 0.$$

Verify that the polynomial equation is

$$\lambda^3 - \tfrac{1}{2}\lambda^2 - \tfrac{1}{4}\lambda - \tfrac{1}{8} = 0$$

and that one characteristic value is $\lambda \approx 0.9196$.

14.9.9. In a problem of population dynamics (Example 14.3.2) suppose that the matrix M is given by

$$M = \begin{pmatrix} 0 & 80 & 50 & 0 \\ 0.6 & 0 & 0 & 0 \\ 0 & 0.4 & 0 & 0 \\ 0 & 0 & 0.3 & 0 \end{pmatrix}.$$

Work out the equation $\det(M - \lambda I) = 0$ which is pertinent for the existence of a stable age distribution.

Chapter 15

Complex Numbers

15.1. Introduction

If x is any positive or negative number, the square of x is always positive. Therefore, no real number satisfies the quadratic equation

$$x^2 = -1. \qquad (15.1.1)$$

However, nobody likes a result that states "it is impossible". Mathematicians began early to search for a new sort of numbers. One could formally write $x = \sqrt{-1}$, but it is not possible to state whether $\sqrt{-1}$ is greater or smaller than a given real number. For a long time people thought that it is a necessary attribute of numbers to have a "size" with a specific order. Consequently, $\sqrt{-1}$ could not be called a number. On the other hand, algebraic operations with $\sqrt{-1}$ could be performed easily. The situation finally led to a compromise: $\sqrt{-1}$ was called an *imaginary number*. The first letter of "imaginary" was proposed as a notation:

$$i = \sqrt{-1}. \qquad (15.1.2)$$

Today the idea that numbers can necessarily be ordered according to their size is abandoned. There is nothing mysterious about imaginary numbers. They can be added, subtracted, multiplied, divided. Together with the real numbers they form the set of *complex numbers*. Each number is of the form

$$a + bi$$

where a and b are real numbers.

Complex numbers are not only useful for mathematical investigations. They also serve immediate practical purposes in a variety of problems in physics and engineering[1]. More and more often they enter the biological literature. For instance they can be found in biological books such as K. S. Cole (1968), Grodins (1963), Jacquard (1974), Lotka (1956), Milhorn (1966), Rosen (1967), Sollberger (1965).

---

[1] In engineering, the notation i is replaced by j in order to avoid confusion with the notation i or I for the electric current.

15.2. The Complex Plane

To provide complex numbers with some kind of "reality", we map these numbers into points of a plane.

First, we plot the *real number line* as a horizontal axis. From now on we call it the *real axis* and denote it by X (Fig. 15.1). This axis corresponds to the x axis in an ordinary xy-coordinate system. Second, we draw a vertical axis through the point representing the number 0 and choose the same unit of length as on the horizontal axis. We associate

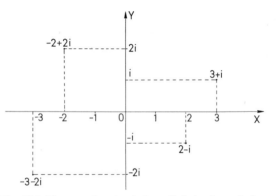

Fig. 15.1. The complex plane, also called the Argand plane

the imaginary number i with the point that corresponds to $x = 0$, $y = 1$ in an ordinary xy-coordinate system. Therefore, the vertical axis is called the *imaginary axis* and denoted by Y. It also becomes clear why i is known as the *imaginary unit*.

We may form positive and negative multiples of i and associate them with the points on the imaginary axis. A complex number such as $3 + i$ is represented by a point having coordinates $x = 3$, $y = 1$ in an xy-coordinate system (Fig. 15.1). In general, given two real numbers x and y, the complex number

$$z = x + iy \qquad (15.2.1)$$

is associated with the point (x, y) in an xy-coordinate system.

A complex number consists of a *real part* and an *imaginary part*. Thus in $z = x + iy$ we call x the *real part* and y the *imaginary part*. We write

$$\text{Re } z = x, \qquad \text{Im } z = y. \qquad (15.2.2)$$

A real number turns out to be a special case of a complex number, namely $z = x + i \cdot 0$. Similarly, an imaginary number is another special case, namely $z = 0 + iy$. To emphasize that the real part vanishes, we sometimes call $z = iy$ a *purely imaginary* number.

There is a one-to-one correspondence between the pairs (x, y) of real numbers and the complex numbers $z = x + iy$, symbolically

$$(x, y) \leftrightarrow z = x + iy. \tag{15.2.3}$$

When we interpret (x, y) as a vector (Section 14.4), say

$$\boldsymbol{a} = \begin{pmatrix} x \\ y \end{pmatrix},$$

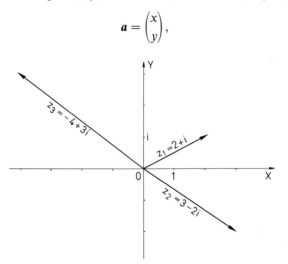

Fig. 15.2. Complex numbers may be interpreted as vectors

then we also obtain a one-to-one correspondence:

$$\boldsymbol{a} = \begin{pmatrix} x \\ y \end{pmatrix} \leftrightarrow z = x + iy. \tag{15.2.4}$$

Fig. 15.2 illustrates the vector interpretation of complex numbers.

The *absolute value or modulus* of $z = x + iy$ is defined by

$$|z| = (x^2 + y^2)^{\frac{1}{2}}. \tag{15.2.5}$$

This is a positive number except for the particular case $z = 0$ where $|z| = 0$. Formula (15.2.5) corresponds to (14.4.7) for vectors. Geometrically, $|z|$ is the length of the vector $\boldsymbol{a} = \begin{pmatrix} x \\ y \end{pmatrix}$ or the distance of the point $x + iy$ from the point $z = 0$. If, for instance, $z = -12 + 5i$, then $|z| = (144 + 25)^{\frac{1}{2}} = 13$.

It is also convenient to introduce *polar coordinates* for complex numbers. In Section 5.5 we considered the polar coordinates r, α of a point P in the xy-plane. By r we mean the distance of P from the origin O and by α the angle between the positive x axis and the line OP measured in the counter-clockwise direction. If P coincides with O, then $r = 0$, and α is not determined. Let x, y be the rectangular coordinates of P and assume that x and y are measured and plotted in the same unit. Then we obtain

$$x = r \cos \alpha, \qquad y = r \sin \alpha. \tag{15.2.6}$$

In the same way we may introduce polar coordinates of a point in the complex plane (Fig. 15.3). The distance r is identical with the absolute value of z, that is,

$$r = |z|. \tag{15.2.7}$$

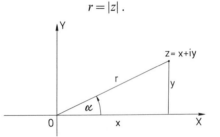

Fig. 15.3. The polar coordinates r, α of a complex number $z = x + iy$

The *polar angle* α, also called *argument* or *phase angle*, ranges from 0 to 2π radians (or from $0°$ to $360°$). For $z = 0$, the polar angle α is not determined. For a real number $z = x \neq 0$, the polar angle α is 0 if $x > 0$ and $\pi (= 180°)$ if $x < 0$. For a purely imaginary number iy we get $\alpha = \pi/2 (= 90°)$ if $y > 0$ and $\alpha = 3\pi/2 (= 270°)$ if $y < 0$. Sometimes it is convenient to use negative polar angles. Thus a polar angle of $-\pi/2$ means that the point is located on the negative side of the imaginary axis.

Using formulas (15.2.6) we may rewrite $z = x + iy$ in the form

$$z = r(\cos \alpha + i \sin \alpha). \tag{15.2.8}$$

Example 15.2.1. Consider the complex number $z = -7 + 5i$ with components $x = -7$ and $y = 5$. The corresponding point is located in the second quadrant. The absolute value is

$$r = |z| = (49 + 25)^{\frac{1}{2}} = \sqrt{74} = 8.602\ldots.$$

From formula (15.2.6) it follows that

$$\cos \alpha = \frac{x}{r} = -0.8137\ldots, \qquad \sin \alpha = \frac{y}{r} = 0.5812\ldots$$

and from a table of trigonometric functions that

$$\alpha = 144.4°.$$

15.3. Algebraic Operations

If

$$z_1 = x_1 + iy_1, \qquad z_2 = x_2 + iy_2 \qquad (15.3.1)$$

are two complex numbers, then we mean by the *sum* and the *difference* of z_1 and z_2

$$z_1 + z_2 = (x_1 + x_2) + i(y_1 + y_2), \qquad (15.3.2)$$

$$z_1 - z_2 = (x_1 - x_2) + i(y_1 - y_2), \qquad (15.3.3)$$

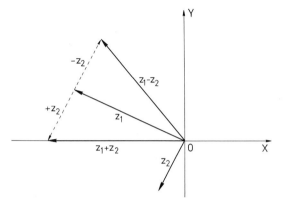

Fig. 15.4. Graphical addition and subtraction of complex numbers

respectively. These operations are quite *analogous to vector addition and subtraction* as considered in Section 14.4. For this reason, it is convenient to map complex numbers into vectors. Fig. 15.4 illustrates how addition and subtraction of complex numbers are performed graphically.

The commutative law

$$z_1 + z_2 = z_2 + z_1 \qquad (15.3.4)$$

and the associative law

$$z_1 + (z_2 + z_3) = (z_1 + z_2) + z_3 \qquad (15.3.5)$$

for the operation of addition are obviously valid.

To define *multiplication* we proceed in two steps. First, we treat the imaginary number i in the same way as if it were a real number. For instance, we multiply the complex numbers $3 + 2i$ and $4 - 5i$ using the commutative, associative, and distributive laws:

$$(3 + 2i)(4 - 5i) = 12 - 15i + 8i - 10i^2 .$$

Second, we remember that i was introduced to satisfy equation (15.1.1). Hence we equate

$$i^2 = -1. \qquad (15.3.6)$$

In our example we obtain

$$(3 + 2i)(4 - 5i) = 12 - 15i + 8i - 10(-1) = 22 - 7i.$$

The result is a new complex number, the product of $3 + 2i$ and $4 - 5i$.

In general, if $z_1 = x_1 + iy_1$ and $z_2 = x_2 + iy_2$, we get

$$\begin{aligned}
z_1 z_2 &= (x_1 + iy_1)(x_2 + iy_2) \\
&= x_1 x_2 + ix_1 y_2 + ix_2 y_1 + i^2 y_1 y_2 \qquad (15.3.7) \\
&= (x_1 x_2 - y_1 y_2) + i(x_1 y_2 + x_2 y_1).
\end{aligned}$$

It is easy but laborious to show that the commutative law

$$z_1 z_2 = z_2 z_1, \qquad (15.3.8)$$

the associative law

$$z_1(z_2 z_3) = (z_1 z_2) z_3, \qquad (15.3.9)$$

and the distributive law

$$z_1(z_2 + z_3) = z_1 z_2 + z_1 z_3 \qquad (15.3.10)$$

are all valid.

We postpone the geometric meaning of complex multiplication to Section 15.4.

Before we can deal with division, we have to introduce the notion of *complex conjugate numbers*. Let $z = x + iy$. Then we change the sign of the imaginary part. The new number is called the complex conjugate of z and denoted by $\bar{z}$. Thus,

$$\bar{z} = x - iy. \qquad (15.3.11)$$

Notice that if we repeat the operation on $\bar{z}$, the result is z. Hence, the complex conjugate of $\bar{z}$ is z itself. Fig. 15.5 depicts two complex conjugate numbers.

The *product of two complex conjugate numbers is always a real number.* Indeed,

$$z \cdot \bar{z} = (x + iy)(x - iy) = x^2 - i^2 y^2 = x^2 + y^2. \qquad (15.3.12)$$

In view of formula (15.2.5) it follows that

$$z \cdot \bar{z} = |z|^2. \qquad (15.3.13)$$

Now we find the *reciprocal* $1/z$ of a nonzero complex number $z = x + iy$ by multiplying numerator and denominator by $\bar{z}$. Thus, using formula (15.3.12), we obtain

$$\frac{1}{z} = \frac{1}{z} \cdot \frac{\bar{z}}{\bar{z}} = \frac{\bar{z}}{z\bar{z}} = \frac{x - iy}{x^2 + y^2}. \qquad (15.3.14)$$

For instance, if $z = 7 + 3i$, we get

$$\frac{1}{7 + 3i} = \frac{7 - 3i}{(7 + 3i)(7 - 3i)} = \frac{7 - 3i}{49 + 9} = \frac{7}{58} - \frac{3}{58} i.$$

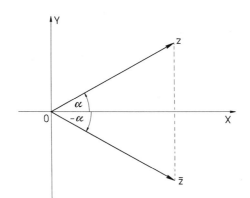

Fig. 15.5. Two complex numbers z and $\bar{z}$ are complex conjugate if the corresponding points in the complex plane are symmetric about the real axis. They have the same absolute value but the opposite polar angle

As a special case we notice that

$$1/i = -i. \qquad (15.3.15)$$

Division of two complex numbers z_1 and z_2 can be reduced to a multiplication of z_1 by the reciprocal of z_2. Hence,

$$\frac{z_1}{z_2} = z_1 \cdot \frac{1}{z_2}.$$

The second factor is treated as in formula (15.3.14), and the multiplication follows the rules applied in formula (15.3.7). For example,

$$\frac{1 + 2i}{1 + i} = (1 + 2i) \frac{1}{1 + i} \frac{1 - i}{1 - i} = (1 + 2i) \frac{1 - i}{2} = \frac{3 + i}{2}.$$

To sum up, *the four operations of addition, subtraction, multiplication and division with complex numbers are performed by applying the ordinary rules of algebra. Whenever the square i^2 appears, it is replaced by -1.*

In Section 14.4 we studied the multiplication of two-dimensional vectors

$$a_1 = \begin{pmatrix} x_1 \\ y_1 \end{pmatrix}, \quad a_2 = \begin{pmatrix} x_2 \\ y_2 \end{pmatrix}$$

and defined the inner product by

$$a_1' a_2 = (x_1, y_1) \begin{pmatrix} x_2 \\ y_2 \end{pmatrix} = x_1 x_2 + y_1 y_2 .$$

A comparison with formula (15.3.7) shows that the multiplication of two complex numbers is quite a different operation.

Only when multiplying a vector $\begin{pmatrix} x \\ y \end{pmatrix}$ or a complex number $x + iy$ by a real number is the analogy of the two operations guaranteed. Indeed, if k denotes a real number, we get

$$k \begin{pmatrix} x \\ y \end{pmatrix} = \begin{pmatrix} kx \\ ky \end{pmatrix}, \quad k(x + iy) = (kx) + i(ky) .$$

In other words, *addition or subtraction of complex numbers, and the multiplication of a complex number by a real factor follow the same rules as the corresponding operations on two-dimensional vectors, but otherwise the analogy breaks down.*

15.4 Exponential Functions of Complex Variables

In Section 10.10 we learned that the series expansion

$$e^x = 1 + x + \frac{x^2}{2!} + \frac{x^3}{3!} + \dots \tag{15.4.1}$$

converges for all real numbers x. We may also use this expansion as a definition of e^x. Now we will define e^z where $z = x + iy$ is an arbitrary complex number. First, we define e^{iy} where iy is a purely imaginary number. Formally replacing x by iy in Eq. (15.4.1) we obtain

$$e^{iy} = 1 + (iy) + \frac{(iy)^2}{2!} + \frac{(iy)^3}{3!} + \frac{(iy)^4}{4!} + \frac{(iy)^5}{5!} + \dots$$

$$= 1 + iy - \frac{y^2}{2!} - i\frac{y^3}{3!} + \frac{y^4}{4!} + i\frac{y^5}{5!} \pm \dots$$

$$= \left(1 - \frac{y^2}{2!} + \frac{y^4}{4!} \pm \dots \right) + i\left(y - \frac{y^3}{3!} + \frac{y^5}{5!} \pm \dots \right).$$

The real and the imaginary parts are series expansions of $\cos y$ and $\sin y$, respectively (cf. Problem 10.10.3). These expansions converge for all real values of y. Hence, e^{iy} is defined by

$$e^{iy} = \cos y + i \sin y. \qquad (15.4.2)$$

The variable y plays the role of an angle. For this reason, y is often replaced by a Greek letter. We may rewrite Eq. (15.4.2) in the form

$$\boxed{e^{i\alpha} = \cos \alpha + i \sin \alpha} \qquad (15.4.3)$$

The famous formula is due to Euler (cf. footnote in Section 8.2).

With Eqs. (15.4.1) and (15.4.2) in mind, we can now define e^z:

$$e^z = e^{x+iy} = e^x \cdot e^{iy}. \qquad (15.4.4)$$

This result leads us to the series expansion

$$e^z = 1 + z + \frac{z^2}{2!} + \frac{z^3}{3!} + \dots \qquad (15.4.5)$$

which is formally the same as (15.4.1). With tools that go beyond the scope of this book, it can be shown that the series expansion (15.4.5) converges for all complex numbers z.

We return to Euler's formula (15.4.3) and consider some special cases:

$$\alpha = 0, \qquad e^{i0} = 1,$$

$$\alpha = \frac{\pi}{2}, \qquad e^{i\frac{\pi}{2}} = i,$$

$$\alpha = \pi, \qquad e^{i\pi} = -1, \qquad (15.4.6)$$

$$\alpha = \frac{3\pi}{2}, \qquad e^{i\frac{3\pi}{2}} = -i.$$

Let z be any complex number with absolute value r and polar angle α. Then it follows from formulas (15.2.8) and (15.4.3) that

$$z = re^{i\alpha}. \tag{15.4.7}$$

With this result in mind we are now able to find a *geometric interpretation of complex multiplication*. Let

$$z_1 = r_1 e^{i\alpha_1}, \qquad z_2 = r_2 e^{i\alpha_2}$$

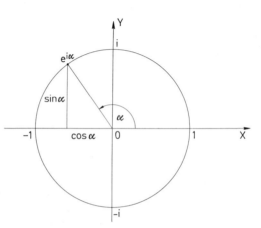

Fig. 15.6. The unit circle in the complex plane. Every number located on the circumference of the unit circle can be written in the form $e^{i\alpha}$

be two complex numbers. Then

$$z_1 z_2 = (r_1 e^{i\alpha_1})(r_2 e^{i\alpha_2}) = r_1 r_2 e^{i(\alpha_1 + \alpha_2)}. \tag{15.4.8}$$

The absolute value of the product $z_1 z_2$ is therefore

$$|z_1 z_2| = r_1 r_2 = |z_1|\,|z_2|, \tag{15.4.9}$$

that is, the *absolute value of the product of complex numbers is the product of the absolute values of these numbers*.

We can also read the *polar angle of the product* from formula (15.4.8). This angle is

$$\alpha_1 + \alpha_2$$

if contained in the interval from 0 to 2π (or from 0° to 360°). Otherwise, we simply reduce the sum by 2π (or 360°) as explained at the end of Section 5.3.

Example 15.4.1. Let

$$z_1 = 2e^{i\pi/6}, \quad z_2 = 3e^{i\pi/3}.$$

The absolute values are 2 and 3 and the polar angles $\pi/6$ and $\pi/3$, respectively. Hence,

$$z_1 z_2 = 2 \cdot 3 e^{i(\pi/6 + \pi/3)} = 6 e^{i\pi/2}.$$

The product has absolute value 6 and polar angle $\pi/2$ (or 90°). The example is depicted in Fig. 15.7.

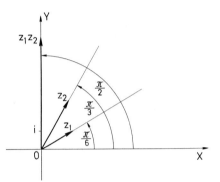

Fig. 15.7. When multiplying two complex numbers, the polar angle of the product is the sum of the polar angles of the two factors

Of special importance is the multiplication of a complex number $z = re^{i\alpha}$ by a complex number of absolute value 1, say by $e^{i\beta}$:

$$z e^{i\beta} = r e^{i\alpha} e^{i\beta} = r e^{i(\alpha + \beta)}. \tag{15.4.10}$$

If we interpret z as a vector, the result simply means that we have to *rotate the vector z by the angle β in the counter-clockwise direction.*

Using Euler's formula (15.4.3), we obtain

$$e^{i\alpha} = \cos\alpha + i\sin\alpha,$$
$$e^{i\beta} = \cos\beta + i\sin\beta,$$
$$e^{i(\alpha+\beta)} = \cos(\alpha+\beta) + i\sin(\alpha+\beta).$$

It follows from formula (15.4.10) that

$$\cos(\alpha+\beta) + i\sin(\alpha+\beta) = (\cos\alpha + i\sin\alpha)(\cos\beta + i\sin\beta)$$
$$= \cos\alpha\cos\beta + i\cos\alpha\sin\beta + i\sin\alpha\cos\beta + i^2\sin\alpha\sin\beta$$
$$= (\cos\alpha\cos\beta - \sin\alpha\sin\beta) + i(\sin\alpha\cos\beta + \cos\alpha\sin\beta).$$

Separating real and imaginary parts we obtain

$$\cos(\alpha + \beta) = \cos\alpha\cos\beta - \sin\alpha\sin\beta ,$$
$$\sin(\alpha + \beta) = \sin\alpha\cos\beta + \cos\alpha\sin\beta .$$

Thus we have verified formulas (5.7.2).

Finally, Euler's formula may be used to express trigonometric functions in terms of exponential functions. Since $\sin(-\alpha) = -\sin\alpha$

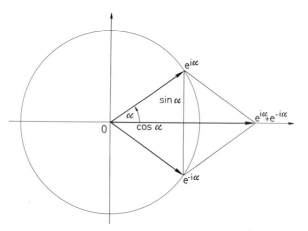

Fig. 15.8. Verification of formulas (15.4.12). The quantities $e^{i\alpha}$ and $e^{-i\alpha}$ are represented by *unit vectors*, that is, by vectors of length 1. Their sum and difference can be easily determined by vector addition and subtraction, respectively

and $\cos(-\alpha) = \cos\alpha$, we obtain the two complex conjugate numbers

$$e^{i\alpha} = \cos\alpha + i\sin\alpha ,$$
$$e^{-i\alpha} = \cos\alpha - i\sin\alpha .$$

(15.4.11)

We may consider these two equations as simultaneous linear equations in the unknown quantities $\cos\alpha$ and $\sin\alpha$. The solution is

$$\cos\alpha = \frac{1}{2}(e^{i\alpha} + e^{-i\alpha}),$$

$$\sin\alpha = \frac{1}{2i}(e^{i\alpha} - e^{-i\alpha}).$$

(15.4.12)

The result may be verified geometrically using Fig. 15.8. Notice the striking similarity between formulas (15.4.12) and formulas (10.11.1), (10.11.2) for the hyperbolic cosine and sine.

15.5. Quadratic Equations

Complex numbers occur frequently in connection with quadratic equations. The standard form of the quadratic equation is

$$ax^2 + bx + c = 0. \tag{15.5.1}$$

We assume that the coefficients are real numbers and that $a \neq 0$.

In Section 4.6 we stated the solution as

$$x = \frac{-b \pm (b^2 - 4ac)^{\frac{1}{2}}}{2a}. \tag{15.5.2}$$

The nature of the roots depends very much on the *discriminant*

$$D = b^2 - 4ac. \tag{15.5.3}$$

If $D > 0$, Eq. (15.5.1) has two different real roots x_1 and x_2. If $D = 0$, the roots are still real, but they fall together into the single value $x_1 = x_2 = -b/2a$. It remains to study the case

$$D < 0. \tag{15.5.4}$$

Here $D = -|D|$ and

$$(b^2 - 4ac)^{\frac{1}{2}} = \sqrt{-|D|} = \pm i\sqrt{|D|}.$$

Hence we obtain the two complex roots

$$x = \frac{-b \pm i\sqrt{|D|}}{2a}. \tag{15.5.5}$$

The real part, that is, $-b/2a$ is the same for both roots. The two roots differ, however, in the sign of the imaginary part. Therefore, the two roots are *complex conjugate*.

Example 15.5.1. The discriminant of the equation

$$4x^2 - 12x + 25 = 0$$

is $D = 12^2 - (4 \times 4 \times 25) = -256$. Hence, the roots are the two complex conjugate numbers

$$\frac{12 \pm \sqrt{-256}}{8} = \frac{12 \pm 16i}{8} = \frac{3 \pm 4i}{2}.$$

Example 15.5.2. Let

$$A = \begin{pmatrix} 6 & 5 \\ -1 & 4 \end{pmatrix}$$

be a 2×2 matrix. Frequently we have to find the *eigenvalues* of such a matrix. According to Section 14.9 the problem of eigenvalues leads to the equation

$$\begin{vmatrix} 6-\lambda & 5 \\ -1 & 4-\lambda \end{vmatrix} = \lambda^2 - 10\lambda + 29 = 0 \,.$$

Hence the eigenvalues are

$$\lambda_1 = \frac{10+\sqrt{-16}}{2} = 5+2i \,, \qquad \lambda_2 = \frac{10-\sqrt{-16}}{2} = 5-2i \,.$$

15.6. Oscillations

Most, perhaps even all, plants, animals, microorganisms, single organs or pieces of tissue participate in *cyclical processes* known as *biological rhythms*. Among the best known biological rhythms are brain waves, heart beats, breathing, daily variations in the activity of the kidneys and the liver, daily movements of leaves, the monthly cycle of menstruation, and yearly breeding patterns. For a survey of biological rhythms and for references see Sollberger (1965).

Cyclical processes may be initiated by external signals and forces such as light, changes in temperature and humidity, chemical stimuli, and corpuscular radiation. Very often biological rhythms continue when the signals or the forces in consideration are removed. Thus leaves continue their up-and-down movement when kept in constant darkness. Or hamsters do not interrupt their daily rhythm of activity and rest when held under constant light. Hence, organisms must have an *internal timing system* which is known as *biological clock*.

Little is known about the nature of biological clocks. One theory claims that such a clock would read and count incoming rhythmic signals from the physical environment and would apply the information to control the organism. In principle, the clock would run like an electric clock driven by the sixty cycles per second of an alternate current (cf. F. A. Brown, 1965). Another theory states that a biological clock is more complete. In principle it would resemble an ordinary clock which contains its own oscillating system. Such a clock could run without periodic signals from outside (see e.g. Pittendrigh and Bruce, 1957).

In this section we will concentrate on the mathematics of oscillating systems. There are various kinds of oscillating systems which we will call *oscillators* from now on.

In mechanics we know the pendulum, the balance of a wrist watch, vibrating strings and membranes, the tuning fork and other oscillators. Electrical oscillators are radar and radio transmitters and receivers. Oscillations may require only little space such as in a variety of molecular and atomic oscillators. Less known are thermal oscillators; a familiar example is a geyser. We also know a good number of chemical systems that are able to generate oscillations (for references see Sollberger, 1965, p. 59).

As an instructive example we will study the oscillations of a body which is attached to a spring (Fig. 15.9). We allow the body to move only in the vertical direction. We consider the vertical line as an x axis with the positive side downward and the origin $x = 0$ at the equilibrium point of the body where the weight of the body and the elastic force of the spring cancel each other.

Let $F = F(x)$ be the resulting force as a function of x. If the body is below the equilibrium point, say at x_1, the force $F(x_1) = F_1$ is negative since it accelerates the body in the negative x direction. Conversely, if the body is above the equilibrium point, say at x_2, the force $F(x_2) = F_2$

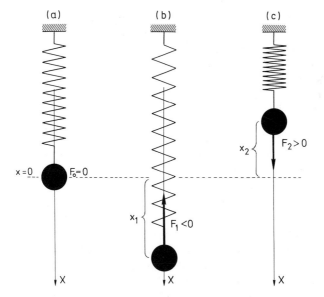

Fig. 15.9. A body hanging on a spiral spring as an example of a mechanical oscillator. The force F acting on the body is the resultant of the weight of the body and the elastic force of the spring

is positive. Hence, x and $F(x)$ have opposite signs. For moderate displacements x from the equilibrium it is an experimental fact (Hooke's law) that $F(x)$ is a *linear function* of x. Thus we obtain

$$F(x) = -kx \qquad (15.6.1)$$

where k denotes a positive constant.

The displacement x is a function of time. Hence, we may write $x = x(t)$. Now we consider the velocity $v = v(t) = dx/dt$ and the acceleration $dv/dt = d^2x/dt^2$ of the body. Newton's second law of dynamics states that

$$M\frac{d^2x}{dt^2} = F(x) \qquad (15.6.2)$$

where M denotes the mass of the body.

From (15.6.2) and (15.6.1) it follows that

$$\boxed{M\frac{d^2x}{dt^2} + kx = 0} \qquad (M>0, k>0). \qquad (15.6.3)$$

We have tacitly assumed that there are no other forces present. In particular, we have excluded friction and assumed that the mass of the spring is negligible.

Formula (15.6.3) is a second-order differential equation for the unknown function $x = x(t)$. The equation is linear, and the coefficients are constants. Such a differential equation is solved by the *method of trial functions* which we already applied in Section 11.7. We try the exponential function

$$x = Ae^{\lambda t}. \qquad (15.6.4)$$

Substituting x in (15.6.3) by this function we get

$$M\lambda^2 Ae^{\lambda t} + kAe^{\lambda t} = (M\lambda^2 + k)\, Ae^{\lambda t} = 0.$$

The new equation holds if, and only if, either $A = 0$ or

$$\boxed{M\lambda^2 + k = 0}. \qquad (15.6.5)$$

We disregard the trivial case $A = 0$. The quadratic equation (15.6.5) is called the *characteristic equation*. It has the two imaginary roots

$$\lambda = \pm\sqrt{-k/M} = \pm i\sqrt{k/M}.$$

Let

$$\omega = \sqrt{k/M}. \qquad (15.6.6)$$

Then $\lambda_1 = i\omega$ and $\lambda_2 = -i\omega$ are the two roots. Both $e^{\lambda_1 t} = e^{i\omega t}$ and $e^{\lambda_2 t} = e^{-i\omega t}$ are solutions of the differential equation (15.6.3). The general solution is

$$A_1 e^{i\omega t} + A_2 e^{-i\omega t} \qquad (15.6.7)$$

with arbitrary real or complex constants A_1 and A_2. In general, this is a complex function of time. For our mechanical problem, however, only a real function is useful. Hence, we break (15.6.7) into a real and an imaginary part and choose the constants A_1 and A_2 in such a way that the imaginary part vanishes. The real part will then be our function $x = x(t)$. By means of formula (15.4.11) we can replace $e^{i\omega t}$ and $e^{-i\omega t}$ by linear aggregates of $\cos \omega t$ and $\sin \omega t$. Hence, the real part of (15.6.7) is of the form

$$x = a \cos \omega t + b \sin \omega t \qquad (15.6.8)$$

where a and b denote arbitrary real constants. It is convenient to rewrite this function in the form

$$\boxed{x = c \cos \omega(t - t_0)} \qquad (c > 0), \qquad (15.6.9)$$

a function which we introduced in Section 5.9. There we called ω the *angular frequency*, c the *amplitude* and t_0 the *acrophase*. The relationships between the constants a, b, c, ω and t_0 are given by

$$a = c \cos \omega t_0, \qquad b = c \sin \omega t_0. \qquad (15.6.10)$$

A graph of $x(t)$ is shown in Fig. 5.21c. The differential equation (15.6.3) leads to a sinusoidal oscillation. The inertia of the body keeps the system going, and in the absence of friction, the amplitude remains constant. For this reason the oscillating system is called an *undamped harmonic oscillator*.

Now we will make our oscillating system more realistic by adding *friction*, for instance air resistance. For moderate velocities it is an experimental fact that the frictional force is proportional to the velocity $v = dx/dt$. Since friction is reducing velocity, the frictional force has a sign opposite to the velocity v. Hence, the frictional force is $-fv = -f \, dx/dt$ where f is a positive constant. To get the total force $F(x)$ acting upon the body, we have to add this amount to the right side of (15.6.1). Thus,

$$F(x) = -kx - f \, dx/dt. \qquad (15.6.11)$$

From (15.6.2) and (15.6.11) we derive the differential equation

$$M \frac{d^2 x}{dt^2} + f \frac{dx}{dt} + kx = 0 \qquad (M > 0, f > 0, k > 0). \quad (15.6.12)$$

This equation is again of the second order and linear with constant coefficients. Using the trial function (15.6.4) we obtain

$$M \lambda^2 A e^{\lambda t} + f \lambda A e^{\lambda t} + k A e^{\lambda t} = 0.$$

This equation is satisfied if, and only if, either $A = 0$ (trivial case) or

$$M \lambda^2 + f \lambda + k = 0 \qquad (15.6.13)$$

holds. The quadratic equation (15.6.13) is called the *characteristic equation*. The two roots λ_1 and λ_2 are

$$\frac{1}{2M} (-f \pm \sqrt{f^2 - 4Mk}). \qquad (15.6.14)$$

The nature of the solution depends on the *discriminant*

$$D = f^2 - 4Mk.$$

If $D > 0$, that is, if the frictional constant f is relatively large, the roots (15.6.14) are real. Since

$$\sqrt{f^2 - 4Mk} < f,$$

λ_1 and λ_2 are both negative. Therefore, the solution of our differential equation is of the form $c_1 e^{-|\lambda_1|t} + c_2 e^{-|\lambda_2|t}$. This function of t is monotone decreasing which means that the body "creeps" toward its equilibrium.

   As we are interested in oscillations, we abandon the case $D \geq 0$ and require therefore that

$$D = f^2 - 4Mk < 0. \qquad (15.6.15)$$

Now we obtain two complex conjugate roots

$$\frac{1}{2M} (-f \pm i \sqrt{|D|})$$

which we denote by $-\alpha+i\omega$ and $-\alpha-i\omega$, respectively. The constant $\alpha=f/2M$ is positive. Moreover we have

$$\omega = \frac{1}{2M}\sqrt{|D|} = \frac{1}{2M}(4Mk - f^2)^{\frac{1}{2}}. \qquad (15.6.16)$$

With the new notation the general solution of our differential equation (15.6.12) is

$$A_1 e^{(-\alpha+i\omega)t} + A_2 e^{(-\alpha-i\omega)t} = e^{-\alpha t}(A_1 e^{i\omega t} + A_2 e^{-i\omega t}).$$

As we have to restrict ourselves to a real solution, we may treat the expression in parentheses in the same way as formula (15.6.7). Hence the displacement $x(t)$ is of the form

$$x(t) = ce^{-\alpha t} \cos\omega(t - t_0) \qquad (\alpha > 0). \qquad (15.6.17)$$

When comparing this solution with (15.6.9) we see that $x(t)$ still oscillates with angular frequency ω, but that the amplitude $ce^{-\alpha t}$ is monotone decreasing. Therefore the motion will taper off as t tends to infinity. Our mechanical system is then called a *damped harmonic oscillator*. Applying formula (5.9.2) we obtain for the period l the constant value

$$l = 2\pi/\omega.$$

A graph of $x(t)$ is shown in Fig. 15.10.

There is an important connection between the differential equation (15.6.12) and the system of first-order differential equations which we considered in Section 11.7. To find this connection we replace dx/dt by $v = v(t)$ and d^2x/dt^2 by dv/dt in Eq. (15.6.12). Thus we obtain

$$\frac{dx}{dt} = v,$$

$$\frac{dv}{dt} = -\frac{k}{M}x - \frac{f}{M}v. \qquad (15.6.18)$$

These two equations are of the form (11.7.1). We have only to identify y with v and to equate

$$a = 0, \qquad b = 1,$$
$$c = -k/M, \qquad d = -f/M.$$

Conversely, when eliminating v from (15.6.18) we get the second-order Eq. (15.6.12). Hence, (15.6.12) and (15.6.18) are equivalent statements. We leave it to the reader to prove that the characteristic Eq. (11.7.5) reduces to (15.6.13).

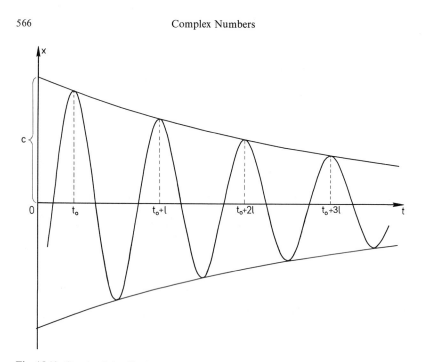

Fig. 15.10. Graph of the displacement $x(t)$ in the case of a damped harmonic oscillator. The amplitude is decreasing. The acrophase is denoted by t_0 and the period by l

Damped harmonic oscillators do not occur only in mechanics. To give the reader an idea of the wide applicability of this concept, we mention a typical electric oscillator (Fig. 15.11). An electric current $I = I(t)$ oscillates in an open circuit which contains a condenser of capacity C, a resistor of resistance R and a coil of self-inductance L. Using the laws of electricity the following differential equation can be

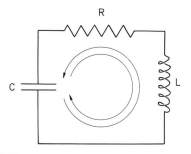

Fig. 15.11. An electric circuit oscillator. It consists of a condenser (C farads) a coil (L henries), and a resistor (R ohms) in series. The electric current $I = I(t)$ is alternating

derived:

$$L\frac{d^2I}{dt^2} + R\frac{dI}{dt} + \frac{1}{C}I = 0.$$ (15.6.19)

This equation is formally the same as Eq. (15.6.12). Therefore, if a condition equivalent to (15.6.15) is satisfied, the solution is of the form (15.6.17). Comparing the electrical oscillator with the mechanical analogue, we see that the resistor plays the role of a frictional force, the coil the role of the inert mass, and the condenser the role of the elastic spring.

A further generalization of the oscillator model is possible by allowing *external forces* to interfere. By external forces we mean forces which do not depend on the present state of the oscillator. Biologically speaking, an external force could originate outside of an organism, but it could also be caused by the organism itself and imposed on the oscillator.

Let $E(t)$ denote the external force. Then the differential equation (15.6.12) turns into

$$\boxed{M\frac{d^2x}{dt^2} + f\frac{dx}{dt} + kx = E(t)}.$$ (15.6.20)

The new differential equation is still linear in x but no longer homogeneous. We call (15.6.12) *homogeneous* since with each solution $x(t)$, a constant multiple $Ax(t)$ is also a solution. The new Eq. (15.6.20) loses this property if $E(t) \neq 0$ for some values of t and is therefore called *inhomogeneous*. In biology we call the system *free running* if $E(t) = 0$ for all values of t and *forced* in the opposite case.

Inhomogeneous differential equations are much more difficult to solve than their homogeneous counterparts. For this reason we will only consider a special case. We return to undamped oscillations by omitting friction. Using formula (15.6.6) and replacing k/M by ω^2 in (15.6.3), we get the homogeneous differential equation

$$\frac{d^2x}{dt^2} + \omega^2 x = 0.$$

Moreover, we require the external force to be harmonic with amplitude A and angular frequency ω_1, say

$$E(t) = A\cos\omega_1 t.$$ (15.6.21)

Then we obtain the inhomogeneous differential equation

$$\frac{d^2 x}{dt^2} + \omega^2 x = r \cos \omega_1 t \tag{15.6.22}$$

where r stands for A/M. We confine ourselves to stating the solution:

$$x(t) = c \cos \omega(t - t_0) + \frac{r}{\omega^2 - \omega_1^2} \cos \omega_1 t \tag{15.6.23a}$$

if $\omega_1 \neq \omega$, and

$$x(t) = c \cos \omega(t - t_0) + \frac{rt}{2\omega} \sin \omega t \tag{15.6.23b}$$

if $\omega_1 = \omega$. The constants c and t_0 are arbitrary.

The reader is invited to verify the solution by substituting $x(t)$ into (15.6.22).

The two cases $\omega_1 \neq \omega$ and $\omega_1 = \omega$ have to be carefully distinguished:

Case (a): $\omega_1 \neq \omega$. Formula (15.6.23a) consists of two cosine terms with different amplitudes and different angular frequencies. If ω_1 deviates only slightly from ω, then $\omega^2 - \omega_1^2$ is a small quantity, and the amplitude $r/(\omega^2 - \omega_1^2)$ is large. This amplitude could become so large that the first term with frequency ω is negligible. In this case, $x(t)$ represents approximately an undamped harmonic oscillation. Its angular frequency is ω_1, the frequency of the external force, and not the frequency ω of the free running oscillator. We say therefore that the oscillator is *synchronized* or *entrained* by the external force. The process of *synchronization* or *entrainment* seems to be important in biological systems where *periods in free running systems can be slightly changed by external harmonic forces*.

Case (b): $\omega_1 = \omega$. Here the frequency of the external force is exactly equal to the frequency of the free running oscillator. Formula (15.6.23b) contains two harmonic terms with the same angular frequency ω. However, the *second amplitude $rt/2\omega$ tends to infinity* as t increases. The system is a harmonic oscillator with increasing amplitude. Hence, the system is *unstable*. This particular phenomenon is known as *resonance*.

For further details see Defares et al. (1973), Klotter (1960), Milhorn (1966), Pavlidis (1973), Rosen (1967). Biological applications are treated in Bailey (1957, p. 136ff.), K. S. Cole (1968, p. 9), J. M. Smith (1968), and Sollberger (1965).

In Section 11.8 we treated a *nonlinear model* which also resulted in oscillations. This model, known as the Lotka-Volterra model, originated in ecology. It is likely that future research in biological rhythms will imply nonlinear models (cf. Goodwin, 1963, p. 34; Pavlidis, 1973; Wever, 1972).

Problems for Solution

15.2.1. Find the absolute values and the polar angles of the following complex numbers:

a) $3 + 4i$ b) $3 - 4i$ c) $-1 - i$ d) $5 - 5i$

15.2.2. Find complex numbers whose absolute values and polar angles are given by

a) $r = 5, \alpha = 30°$,   b) $r = 3, \ \alpha = 120°$,

c) $r = 1, \alpha = 270°$,   d) $r = 10, \alpha = -45°$.

15.3.1. Evaluate

a) $(3 + 7i) - (2 - 5i)$,   b) $\frac{3}{2}(-8 + 5i)$,

c) $5i \cdot 2i$,   d) $(-3i)(9i)$,

e) $(\pm 4i)^2$,   f) $(8 + 3i)(7 + 5i)$,

g) $(4 - 5i)(5 - 6i)$,   h) $(2 + 3i)^2$,

i) $(4 - 5i)^2$,   k) $(8 - 3i)(8 + 3i)$.

15.3.2. Evaluate

a) $\frac{5}{3}(4 - 12i)$,   b) $(-8 + 3i) - (-1 - 4i)$,   c) $(4i)(-3i)$,

d) $i^2 \cdot i$,   e) $i^3 \cdot i^2$,   f) $(-i^3)(-i)^2$,

g) $(1 - 2i)(1 + 2i)$,   h) $(4i + 5)^2$,   i) $(i - 6)^2$,

k) $(3 + 5i)(4 + 2i)$,   l) $(4 - 8i)(1 - 2i)$,   m) $(1 + i)^3$.

15.3.3. Evaluate

a) $\dfrac{1}{2 + 5i}$,   b) $\dfrac{1}{a + bi}$,

c) $\dfrac{1}{3 - i}$,   d) $\dfrac{5 - 3i}{2 + 4i}$.

15.3.4. Find

a) $\dfrac{2}{2 + i}$,   b) $\dfrac{3}{5 - i}$,   c) $\dfrac{1}{(2 + i)(2 - i)}$,

d) $\dfrac{1}{(2+i)^2}$,   e) $\dfrac{i}{2-i}$,   f) $\dfrac{3}{(1-2i)^2}$,

g) $\dfrac{1-8i}{1+8i}$,   h) $\dfrac{3+2i}{2+3i}$,   i) $\dfrac{9+3i}{11-2i}$.

15.4.1. Write the following complex numbers in the form $re^{i\alpha}$:

a) $3+4i$,   b) $3-4i$,   c) $2+2i$,

d) $-4+i$,   e) $-12-i$,   f) $-6+8i$.

15.4.2. Rewrite the following numbers in the form $a+bi$:

a) $e^{2\pi i/3}$,   b) $5e^{\pi i}$,   c) $\frac{1}{3}e^{-\pi i/2}$.

15.4.3. Evaluate

a) $e^{i\pi/2}\cdot e^{i\pi/3}$ b) $e^{-i\pi/3}\cdot e^{i\pi/2}$

c) $e^{2\pi i}\cdot e^{-2\pi i/3}$ d) $e^{i\pi}/e^{i\pi/2}$.

15.4.4. Multiply the following numbers by $i=e^{i\pi/2}$ and check the result graphically:

a) $3+2i$ b) $-1+i$ c) $-2-2i$

d) $-i$ e) -3 f) $e^{-i\pi/3}$.

15.5.1. Solve the following quadratic equations:

a) $x^2+9=0$,   b) $x^2+6x+25=0$,

c) $\lambda^2=6\lambda-18$,   d) $p(p+12)+61=0$,

e) $2u+\dfrac{6}{u}=5$,   f) $2s-50=s^2$.

15.5.2. Solve the following equations with respect to x:

a) $x^2+p^2=0$,   b) $x^2+ax+2a^2=0$,

c) $2x^2=s(x+3s)$,   d) $x+\dfrac{r^2+1}{x}=r$.

15.5.3. Find the eigenvalues of the following matrices:

a) $\begin{pmatrix} 4 & 1 \\ -2 & 2 \end{pmatrix}$,   b) $\begin{pmatrix} 7 & 13 \\ -1 & 3 \end{pmatrix}$.

15.5.4. Let

$$A=\begin{pmatrix} 15 & -4 \\ 5 & 7 \end{pmatrix}, \qquad B=\begin{pmatrix} 2 & 13 \\ -2 & -8 \end{pmatrix}.$$

Find the eigenvalues of A and B.

15.6.1. Solve the following differential equations:

a) $2\dfrac{d^2x}{dt^2} + 3x = 0$, b) $4\dfrac{d^2x}{dt^2} + 4\dfrac{dx}{dt} + 5x = 0$.

15.6.2. Solve the following differential equations:

a) $p^2\dfrac{d^2x}{dt^2} = -x$, b) $\dfrac{d^2x}{dt^2} + 4a\dfrac{dx}{dt} + 5a^2x = 0$.

Appendix

(Tables A–K, see pp. 572–586)

Table A: Square of x

x	0	1	2	3	4	5	6	7	8	9
1.0	1.000	1.020	1.040	1.061	1.082	1.103	1.124	1.145	1.166	1.188
1.1	1.210	1.232	1.254	1.277	1.300	1.323	1.346	1.369	1.392	1.416
1.2	1.440	1.464	1.488	1.513	1.538	1.563	1.588	1.613	1.638	1.664
1.3	1.690	1.716	1.742	1.769	1.796	1.823	1.850	1.877	1.904	1.932
1.4	1.960	1.988	2.016	2.045	2.074	2.103	2.132	2.161	2.190	2.220
1.5	2.250	2.280	2.310	2.341	2.372	2.403	2.434	2.465	2.496	2.528
1.6	2.560	2.592	2.624	2.657	2.690	2.723	2.756	2.789	2.822	2.856
1.7	2.890	2.924	2.958	2.993	3.028	3.063	3.098	3.133	3.168	3.204
1.8	3.240	3.276	3.312	3.349	3.386	3.423	3.460	3.497	3.534	3.572
1.9	3.610	3.648	3.686	3.725	3.764	3.803	3.842	3.881	3.920	3.960
2.0	4.000	4.040	4.080	4.121	4.162	4.203	4.244	4.285	4.326	4.368
2.1	4.410	4.452	4.494	4.537	4.580	4.623	4.666	4.709	4.752	4.796
2.2	4.840	4.884	4.928	4.973	5.018	5.063	5.108	5.153	5.198	5.244
2.3	5.290	5.336	5.382	5.429	5.476	5.523	5.570	5.617	5.664	5.712
2.4	5.760	5.808	5.856	5.905	5.954	6.003	6.052	6.101	6.150	6.200
2.5	6.250	6.300	6.350	6.401	6.452	6.503	6.554	6.605	6.656	6.708
2.6	6.760	6.812	6.864	6.917	6.970	7.023	7.076	7.129	7.182	7.236
2.7	7.290	7.344	7.398	7.453	7.508	7.563	7.618	7.673	7.728	7.784
2.8	7.840	7.896	7.952	8.009	8.066	8.123	8.180	8.237	8.294	8.352
2.9	8.410	8.468	8.526	8.585	8.644	8.703	8.762	8.821	8.880	8.940
3.0	9.000	9.060	9.120	9.181	9.242	9.303	9.364	9.425	9.486	9.548
3.1	9.610	9.672	9.734	9.797	9.860	9.923	9.986	10.05	10.11	10.18
3.2	10.24	10.30	10.37	10.43	10.50	10.56	10.63	10.69	10.76	10.82
3.3	10.89	10.96	11.02	11.09	11.16	11.22	11.29	11.36	11.42	11.49
3.4	11.56	11.63	11.70	11.76	11.83	11.90	11.97	12.04	12.11	12.18
3.5	12.25	12.32	12.39	12.46	12.53	12.60	12.67	12.74	12.82	12.89
3.6	12.96	13.03	13.10	13.18	13.25	13.32	13.40	13.47	13.54	13.62
3.7	13.69	13.76	13.84	13.91	13.99	14.06	14.14	14.21	14.29	14.36
3.8	14.44	14.52	14.59	14.67	14.75	14.82	14.90	14.98	15.05	15.13
3.9	15.21	15.29	15.37	15.44	15.52	15.60	15.68	15.76	15.84	15.92
4.0	16.00	16.08	16.16	16.24	16.32	16.40	16.48	16.56	16.65	16.73
4.1	16.81	16.89	16.97	17.06	17.14	17.22	17.31	17.39	17.47	17.56
4.2	17.64	17.72	17.81	17.89	17.98	18.06	18.15	18.23	18.32	18.40
4.3	18.49	18.58	18.66	18.75	18.84	18.92	19.01	19.10	19.18	19.27
4.4	19.36	19.45	19.54	19.62	19.71	19.80	19.89	19.98	20.07	20.16
4.5	20.25	20.34	20.43	20.52	20.61	20.70	20.79	20.88	20.98	21.07
4.6	21.16	21.25	21.34	21.44	21.53	21.62	21.72	21.81	21.90	22.00
4.7	22.09	22.18	22.28	22.37	22.47	22.56	22.66	22.75	22.85	22.94
4.8	23.04	23.14	23.23	23.33	23.43	23.52	23.62	23.72	23.81	23.91
4.9	24.01	24.11	24.21	24.30	24.40	24.50	24.60	24.70	24.80	24.90
5.0	25.00	25.10	25.20	25.30	25.40	25.50	25.60	25.70	25.81	25.91
5.1	26.01	26.11	26.21	26.32	26.42	26.52	26.63	26.73	26.83	26.94
5.2	27.04	27.14	27.25	27.35	27.46	27.56	27.67	27.77	27.88	27.98
5.3	28.09	28.20	28.30	28.41	28.52	28.62	28.73	28.84	28.94	29.05
5.4	29.16	29.27	29.38	29.48	29.59	29.70	29.81	29.92	30.03	30.14

x	0	1	2	3	4	5	6	7	8	9
5.5	30.25	30.36	30.47	30.58	30.69	30.80	30.91	31.02	31.14	31.25
5.6	31.36	31.47	31.58	31.70	31.81	31.92	32.04	32.15	32.26	32.38
5.7	32.49	32.60	32.72	32.83	32.95	33.06	33.18	33.29	33.41	33.52
5.8	33.64	33.76	33.87	33.99	34.11	34.22	34.34	34.46	34.57	34.69
5.9	34.81	34.93	35.05	35.16	35.28	35.40	35.52	35.64	35.76	35.88
6.0	36.00	36.12	36.24	36.36	36.48	36.60	36.72	36.84	36.97	37.09
6.1	37.21	37.33	37.45	37.58	37.70	37.82	37.95	38.07	38.19	38.32
6.2	38.44	38.56	38.69	38.81	38.94	39.06	39.19	39.31	39.44	39.56
6.3	39.69	39.82	39.94	40.07	40.20	40.32	40.45	40.58	40.70	40.83
6.4	40.96	41.09	41.22	41.34	41.47	41.60	41.73	41.86	41.99	42.12
6.5	42.25	42.38	42.51	42.64	42.77	42.90	43.03	43.16	43.30	43.43
6.6	43.56	43.69	43.82	43.96	44.09	44.22	44.36	44.49	44.62	44.76
6.7	44.89	45.02	45.16	45.29	45.43	45.56	45.70	45.83	45.97	46.10
6.8	46.24	46.38	46.51	46.65	46.79	46.92	47.06	47.20	47.33	47.47
6.9	47.61	47.75	47.89	48.02	48.16	48.30	48.44	48.58	48.72	48.86
7.0	49.00	49.14	49.28	49.42	49.56	49.70	49.84	49.98	50.13	50.27
7.1	50.41	50.55	50.69	50.84	50.98	51.12	51.27	51.41	51.55	51.70
7.2	51.84	51.98	52.13	52.27	52.42	52.56	52.71	52.85	53.00	53.14
7.3	53.29	53.44	53.58	53.73	53.88	54.02	54.17	54.32	54.46	54.61
7.4	54.76	54.91	55.06	55.20	55.35	55.50	55.65	55.80	55.95	56.10
7.5	56.25	56.40	56.55	56.70	56.85	57.00	57.15	57.30	57.46	57.61
7.6	57.76	57.91	58.06	58.22	58.37	58.52	58.68	58.83	58.98	59.14
7.7	59.29	59.44	59.60	59.75	59.91	60.06	60.22	60.37	60.53	60.68
7.8	60.84	61.00	61.15	61.31	61.47	61.62	61.78	61.94	62.09	62.25
7.9	62.41	62.57	62.73	62.88	63.04	63.20	63.36	63.52	63.68	63.84
8.0	64.00	64.16	64.32	64.48	64.64	64.80	64.96	65.12	65.29	65.45
8.1	65.61	65.77	65.93	66.10	66.26	66.42	66.59	66.75	66.91	67.08
8.2	67.24	67.40	67.57	67.73	67.90	68.06	68.23	68.39	68.56	68.72
8.3	68.89	69.06	69.22	69.39	69.56	69.72	69.89	70.06	70.22	70.39
8.4	70.56	70.73	70.90	71.06	71.23	71.40	71.57	71.74	71.91	72.08
8.5	72.25	72.42	72.59	72.76	72.93	73.10	73.27	73.44	73.62	73.79
8.6	73.96	74.13	74.30	74.48	74.65	74.82	75.00	75.17	75.34	75.52
8.7	75.69	75.86	76.04	76.21	76.39	76.56	76.74	76.91	77.09	77.26
8.8	77.44	77.62	77.79	77.97	78.15	78.32	78.50	78.68	78.85	79.03
8.9	79.21	79.39	79.57	79.74	79.92	80.10	80.28	80.46	80.64	80.82
9.0	81.00	81.18	81.36	81.54	81.72	81.90	82.08	82.26	82.45	82.63
9.1	82.81	82.99	83.17	83.36	83.54	83.72	83.91	84.09	84.27	84.46
9.2	84.64	84.82	85.01	85.19	85.38	85.56	85.75	85.93	86.12	86.30
9.3	86.49	86.68	86.86	87.05	87.24	87.42	87.61	87.80	87.98	88.17
9.4	88.36	88.55	88.74	88.92	89.11	89.30	89.49	89.68	89.87	90.06
9.5	90.25	90.44	90.63	90.82	91.01	91.20	91.39	91.58	91.78	91.97
9.6	92.16	92.35	92.54	92.74	92.93	93.12	93.32	93.51	93.70	93.90
9.7	94.09	94.28	94.48	94.67	94.87	95.06	95.26	95.45	95.65	95.84
9.8	96.04	96.24	96.43	96.63	96.83	97.02	97.22	97.42	97.61	97.81
9.9	98.01	98.21	98.41	98.60	98.80	99.00	99.20	99.40	99.60	99.80

Table B: Square root of x

x	0	1	2	3	4	5	6	7	8	9
1.0	1.000	1.005	1.010	1.015	1.020	1.025	1.030	1.034	1.039	1.044
1.1	1.049	1.054	1.058	1.063	1.068	1.072	1.077	1.082	1.086	1.091
1.2	1.095	1.100	1.105	1.109	1.114	1.118	1.122	1.127	1.131	1.136
1.3	1.140	1.145	1.149	1.153	1.158	1.162	1.166	1.170	1.175	1.179
1.4	1.183	1.187	1.192	1.196	1.200	1.204	1.208	1.212	1.217	1.221
1.5	1.225	1.229	1.233	1.237	1.241	1.245	1.249	1.253	1.257	1.261
1.6	1.265	1.269	1.273	1.277	1.281	1.285	1.288	1.292	1.296	1.300
1.7	1.304	1.308	1.311	1.315	1.319	1.323	1.327	1.330	1.334	1.338
1.8	1.342	1.345	1.349	1.353	1.356	1.360	1.364	1.367	1.371	1.375
1.9	1.378	1.382	1.386	1.389	1.393	1.396	1.400	1.404	1.407	1.411
2.0	1.414	1.418	1.421	1.425	1.428	1.432	1.435	1.439	1.442	1.446
2.1	1.449	1.453	1.456	1.459	1.463	1.466	1.470	1.473	1.476	1.480
2.2	1.483	1.487	1.490	1.493	1.497	1.500	1.503	1.507	1.510	1.513
2.3	1.517	1.520	1.523	1.526	1.530	1.533	1.536	1.539	1.543	1.546
2.4	1.549	1.552	1.556	1.559	1.562	1.565	1.568	1.572	1.575	1.578
2.5	1.581	1.584	1.587	1.591	1.594	1.597	1.600	1.603	1.606	1.609
2.6	1.612	1.616	1.619	1.622	1.625	1.628	1.631	1.634	1.637	1.640
2.7	1.643	1.646	1.649	1.652	1.655	1.658	1.661	1.664	1.667	1.670
2.8	1.673	1.676	1.679	1.682	1.685	1.688	1.691	1.694	1.697	1.700
2.9	1.703	1.706	1.709	1.712	1.715	1.718	1.720	1.723	1.726	1.729
3.0	1.732	1.735	1.738	1.741	1.744	1.746	1.749	1.752	1.755	1.758
3.1	1.761	1.764	1.766	1.769	1.772	1.775	1.778	1.780	1.783	1.786
3.2	1.789	1.792	1.794	1.797	1.800	1.803	1.806	1.808	1.811	1.814
3.3	1.817	1.819	1.822	1.825	1.828	1.830	1.833	1.836	1.838	1.841
3.4	1.844	1.847	1.849	1.852	1.855	1.857	1.860	1.863	1.865	1.868
3.5	1.871	1.873	1.876	1.879	1.881	1.884	1.887	1.889	1.892	1.895
3.6	1.897	1.900	1.903	1.905	1.908	1.910	1.913	1.916	1.918	1.921
3.7	1.924	1.926	1.929	1.931	1.934	1.936	1.939	1.942	1.944	1.947
3.8	1.949	1.952	1.954	1.957	1.960	1.962	1.965	1.967	1.970	1.972
3.9	1.975	1.977	1.980	1.982	1.985	1.987	1.990	1.992	1.995	1.997
4.0	2.000	2.002	2.005	2.007	2.010	2.012	2.015	2.017	2.020	2.022
4.1	2.025	2.027	2.030	2.032	2.035	2.037	2.040	2.042	2.045	2.047
4.2	2.049	2.052	2.054	2.057	2.059	2.062	2.064	2.066	2.069	2.071
4.3	2.074	2.076	2.078	2.081	2.083	2.086	2.088	2.090	2.093	2.095
4.4	2.098	2.100	2.102	2.105	2.107	2.110	2.112	2.114	2.117	2.119
4.5	2.121	2.124	2.126	2.128	2.131	2.133	2.135	2.138	2.140	2.142
4.6	2.145	2.147	2.149	2.152	2.154	2.156	2.159	2.161	2.163	2.166
4.7	2.168	2.170	2.173	2.175	2.177	2.179	2.182	2.184	2.186	2.189
4.8	2.191	2.193	2.195	2.198	2.200	2.202	2.205	2.207	2.209	2.211
4.9	2.214	2.216	2.218	2.220	2.223	2.225	2.227	2.229	2.232	2.234
5.0	2.236	2.238	2.241	2.243	2.245	2.247	2.249	2.252	2.254	2.256
5.1	2.258	2.261	2.263	2.265	2.267	2.269	2.272	2.274	2.276	2.278
5.2	2.280	2.283	2.285	2.287	2.289	2.291	2.293	2.296	2.298	2.300
5.3	2.302	2.304	2.307	2.309	2.311	2.313	2.315	2.317	2.319	2.322
5.4	2.324	2.326	2.328	2.330	2.332	2.335	2.337	2.339	2.341	2.343

x	0	1	2	3	4	5	6	7	8	9
5.5	2.345	2.347	2.349	2.352	2.354	2.356	2.358	2.360	2.362	2.364
5.6	2.366	2.369	2.371	2.373	2.375	2.377	2.379	2.381	2.383	2.385
5.7	2.387	2.390	2.392	2.394	2.396	2.398	2.400	2.402	2.404	2.406
5.8	2.408	2.410	2.412	2.415	2.417	2.419	2.421	2.423	2.425	2.427
5.9	2.429	2.431	2.433	2.435	2.437	2.439	2.441	2.443	2.445	2.447
6.0	2.449	2.452	2.454	2.456	2.458	2.460	2.462	2.464	2.466	2.468
6.1	2.470	2.472	2.474	2.476	2.478	2.480	2.482	2.484	2.486	2.488
6.2	2.490	2.492	2.494	2.496	2.498	2.500	2.502	2.504	2.506	2.508
6.3	2.510	2.512	2.514	2.516	2.518	2.520	2.522	2.524	2.526	2.528
6.4	2.530	2.532	2.534	2.536	2.538	2.540	2.542	2.544	2.546	2.548
6.5	2.550	2.551	2.553	2.555	2.557	2.559	2.561	2.563	2.565	2.567
6.6	2.569	2.571	2.573	2.575	2.577	2.579	2.581	2.583	2.585	2.587
6.7	2.588	2.590	2.592	2.594	2.596	2.598	2.600	2.602	2.604	2.606
6.8	2.608	2.610	2.612	2.613	2.615	2.617	2.619	2.621	2.623	2.625
6.9	2.627	2.629	2.631	2.632	2.634	2.636	2.638	2.640	2.642	2.644
7.0	2.646	2.648	2.650	2.651	2.653	2.655	2.657	2.659	2.661	2.663
7.1	2.665	2.666	2.668	2.670	2.672	2.674	2.676	2.678	2.680	2.681
7.2	2.683	2.685	2.687	2.689	2.691	2.693	2.694	2.696	2.698	2.700
7.3	2.702	2.704	2.706	2.707	2.709	2.711	2.713	2.715	2.717	2.718
7.4	2.720	2.722	2.724	2.726	2.728	2.729	2.731	2.733	2.735	2.737
7.5	2.739	2.740	2.742	2.744	2.746	2.748	2.750	2.751	2.753	2.755
7.6	2.757	2.759	2.760	2.762	2.764	2.766	2.768	2.769	2.771	2.773
7.7	2.775	2.777	2.778	2.780	2.782	2.784	2.786	2.787	2.789	2.791
7.8	2.793	2.795	2.796	2.798	2.800	2.802	2.804	2.805	2.807	2.809
7.9	2.811	2.812	2.814	2.816	2.818	2.820	2.821	2.823	2.825	2.827
8.0	2.828	2.830	2.832	2.834	2.835	2.837	2.839	2.841	2.843	2.844
8.1	2.846	2.848	2.850	2.851	2.853	2.855	2.857	2.858	2.860	2.862
8.2	2.864	2.865	2.867	2.869	2.871	2.872	2.874	2.876	2.877	2.879
8.3	2.881	2.883	2.884	2.886	2.888	2.890	2.891	2.893	2.895	2.897
8.4	2.898	2.900	2.902	2.903	2.905	2.907	2.909	2.910	2.912	2.914
8.5	2.915	2.917	2.919	2.921	2.922	2.924	2.926	2.927	2.929	2.931
8.6	2.933	2.934	2.936	2.938	2.939	2.941	2.943	2.944	2.946	2.948
8.7	2.950	2.951	2.953	2.955	2.956	2.958	2.960	2.961	2.963	2.965
8.8	2.966	2.968	2.970	2.972	2.973	2.975	2.977	2.978	2.980	2.982
8.9	2.983	2.985	2.987	2.988	2.990	2.992	2.993	2.995	2.997	2.998
9.0	3.000	3.002	3.003	3.005	3.007	3.008	3.010	3.012	3.013	3.015
9.1	3.017	3.018	3.020	3.022	3.023	3.025	3.027	3.028	3.030	3.032
9.2	3.033	3.035	3.036	3.038	3.040	3.041	3.043	3.045	3.046	3.048
9.3	3.050	3.051	3.053	3.055	3.056	3.058	3.059	3.061	3.063	3.064
9.4	3.066	3.068	3.069	3.071	3.072	3.074	3.076	3.077	3.079	3.081
9.5	3.082	3.084	3.085	3.087	3.089	3.090	3.092	3.094	3.095	3.097
9.6	3.098	3.100	3.102	3.103	3.105	3.106	3.108	3.110	3.111	3.113
9.7	3.114	3.116	3.118	3.119	3.121	3.122	3.124	3.126	3.127	3.129
9.8	3.130	3.132	3.134	3.135	3.137	3.138	3.140	3.142	3.143	3.145
9.9	3.146	3.148	3.150	3.151	3.153	3.154	3.156	3.158	3.159	3.161

Table B (cont.): Square root of x

x	0	1	2	3	4	5	6	7	8	9
10	3.162	3.178	3.194	3.209	3.225	3.240	3.256	3.271	3.286	3.302
11	3.317	3.332	3.347	3.362	3.376	3.391	3.406	3.421	3.435	3.450
12	3.464	3.479	3.493	3.507	3.521	3.536	3.550	3.564	3.578	3.592
13	3.606	3.619	3.633	3.647	3.661	3.674	3.688	3.701	3.715	3.728
14	3.742	3.755	3.768	3.782	3.795	3.808	3.821	3.834	3.847	3.860
15	3.873	3.886	3.899	3.912	3.924	3.937	3.950	3.962	3.975	3.987
16	4.000	4.012	4.025	4.037	4.050	4.062	4.074	4.087	4.099	4.111
17	4.123	4.135	4.147	4.159	4.171	4.183	4.195	4.207	4.219	4.231
18	4.243	4.254	4.266	4.278	4.290	4.301	4.313	4.324	4.336	4.347
19	4.359	4.370	4.382	4.393	4.405	4.416	4.427	4.438	4.450	4.461
20	4.472	4.483	4.494	4.506	4.517	4.528	4.539	4.550	4.561	4.572
21	4.583	4.593	4.604	4.615	4.626	4.637	4.648	4.658	4.669	4.680
22	4.690	4.701	4.712	4.722	4.733	4.743	4.754	4.764	4.775	4.785
23	4.796	4.806	4.817	4.827	4.837	4.848	4.858	4.868	4.879	4.889
24	4.899	4.909	4.919	4.930	4.940	4.950	4.960	4.970	4.980	4.990
25	5.000	5.010	5.020	5.030	5.040	5.050	5.060	5.070	5.079	5.089
26	5.099	5.109	5.119	5.128	5.138	5.148	5.158	5.167	5.177	5.187
27	5.196	5.206	5.215	5.225	5.235	5.244	5.254	5.263	5.273	5.282
28	5.292	5.301	5.310	5.320	5.329	5.339	5.348	5.357	5.367	5.376
29	5.385	5.394	5.404	5.413	5.422	5.431	5.441	5.450	5.459	5.468
30	5.477	5.486	5.495	5.505	5.514	5.523	5.532	5.541	5.550	5.559
31	5.568	5.577	5.586	5.595	5.604	5.612	5.621	5.630	5.639	5.648
32	5.657	5.666	5.675	5.683	5.692	5.701	5.710	5.718	5.727	5.736
33	5.745	5.753	5.762	5.771	5.779	5.788	5.797	5.805	5.814	5.822
34	5.831	5.840	5.848	5.857	5.865	5.874	5.882	5.891	5.899	5.908
35	5.916	5.925	5.933	5.941	5.950	5.958	5.967	5.975	5.983	5.992
36	6.000	6.008	6.017	6.025	6.033	6.042	6.050	6.058	6.066	6.075
37	6.083	6.091	6.099	6.107	6.116	6.124	6.132	6.140	6.148	6.156
38	6.164	6.173	6.181	6.189	6.197	6.205	6.213	6.221	6.229	6.237
39	6.245	6.253	6.261	6.269	6.277	6.285	6.293	6.301	6.309	6.317
40	6.325	6.332	6.340	6.348	6.356	6.364	6.372	6.380	6.387	6.395
41	6.403	6.411	6.419	6.427	6.434	6.442	6.450	6.458	6.465	6.473
42	6.481	6.488	6.496	6.504	6.512	6.519	6.527	6.535	6.542	6.550
43	6.557	6.565	6.573	6.580	6.588	6.595	6.603	6.611	6.618	6.626
44	6.633	6.641	6.648	6.656	6.663	6.671	6.678	6.686	6.693	6.701
45	6.708	6.716	6.723	6.731	6.738	6.745	6.753	6.760	6.768	6.775
46	6.782	6.790	6.797	6.804	6.812	6.819	6.826	6.834	6.841	6.848
47	6.856	6.863	6.870	6.877	6.885	6.892	6.899	6.907	6.914	6.921
48	6.928	6.935	6.943	6.950	6.957	6.964	6.971	6.979	6.986	6.993
49	7.000	7.007	7.014	7.021	7.029	7.036	7.043	7.050	7.057	7.064
50	7.071	7.078	7.085	7.092	7.099	7.106	7.113	7.120	7.127	7.134
51	7.141	7.148	7.155	7.162	7.169	7.176	7.183	7.190	7.197	7.204
52	7.211	7.218	7.225	7.232	7.239	7.246	7.253	7.259	7.266	7.273
53	7.280	7.287	7.294	7.301	7.308	7.314	7.321	7.328	7.335	7.342
54	7.348	7.355	7.362	7.369	7.376	7.382	7.389	7.396	7.403	7.409

x	0	1	2	3	4	5	6	7	8	9
55	7.416	7.423	7.430	7.436	7.443	7.450	7.457	7.463	7.470	7.477
56	7.483	7.490	7.497	7.503	7.510	7.517	7.523	7.530	7.537	7.543
57	7.550	7.556	7.563	7.570	7.576	7.583	7.589	7.596	7.603	7.609
58	7.616	7.622	7.629	7.635	7.642	7.649	7.655	7.662	7.668	7.675
59	7.681	7.688	7.694	7.701	7.707	7.714	7.720	7.727	7.733	7.740
60	7.746	7.752	7.759	7.765	7.772	7.778	7.785	7.791	7.797	7.804
61	7.810	7.817	7.823	7.829	7.836	7.842	7.849	7.855	7.861	7.868
62	7.874	7.880	7.887	7.893	7.899	7.906	7.912	7.918	7.925	7.931
63	7.937	7.944	7.950	7.956	7.962	7.969	7.975	7.981	7.987	7.994
64	8.000	8.006	8.012	8.019	8.025	8.031	8.037	8.044	8.050	8.056
65	8.062	8.068	8.075	8.081	8.087	8.093	8.099	8.106	8.112	8.118
66	8.124	8.130	8.136	8.142	8.149	8.155	8.161	8.167	8.173	8.179
67	8.185	8.191	8.198	8.204	8.210	8.216	8.222	8.228	8.234	8.240
68	8.246	8.252	8.258	8.264	8.270	8.276	8.283	8.289	8.295	8.301
69	8.307	8.313	8.319	8.325	8.331	8.337	8.343	8.349	8.355	8.361
70	8.367	8.373	8.379	8.385	8.390	8.396	8.402	8.408	8.414	8.420
71	8.426	8.432	8.438	8.444	8.450	8.456	8.462	8.468	8.473	8.479
72	8.485	8.491	8.497	8.503	8.509	8.515	8.521	8.526	8.532	8.538
73	8.544	8.550	8.556	8.562	8.567	8.573	8.579	8.585	8.591	8.597
74	8.602	8.608	8.614	8.620	8.626	8.631	8.637	8.643	8.649	8.654
75	8.660	8.666	8.672	8.678	8.683	8.689	8.695	8.701	8.706	8.712
76	8.718	8.724	8.729	8.735	8.741	8.746	8.752	8.758	8.764	8.769
77	8.775	8.781	8.786	8.792	8.798	8.803	8.809	8.815	8.820	8.826
78	8.832	8.837	8.843	8.849	8.854	8.860	8.866	8.871	8.877	8.883
79	8.888	8.894	8.899	8.905	8.911	8.916	8.922	8.927	8.933	8.939
80	8.944	8.950	8.955	8.961	8.967	8.972	8.978	8.983	8.989	8.994
81	9.000	9.006	9.011	9.017	9.022	9.028	9.033	9.039	9.044	9.050
82	9.055	9.061	9.066	9.072	9.077	9.083	9.088	9.094	9.099	9.105
83	9.110	9.116	9.121	9.127	9.132	9.138	9.143	9.149	9.154	9.160
84	9.165	9.171	9.176	9.182	9.187	9.192	9.198	9.203	9.209	9.214
85	9.220	9.225	9.230	9.236	9.241	9.247	9.252	9.257	9.263	9.268
86	9.274	9.279	9.284	9.290	9.295	9.301	9.306	9.311	9.317	9.322
87	9.327	9.333	9.338	9.343	9.349	9.354	9.359	9.365	9.370	9.375
88	9.381	9.386	9.391	9.397	9.402	9.407	9.413	9.418	9.423	9.429
89	9.434	9.439	9.445	9.450	9.455	9.460	9.466	9.471	9.476	9.482
90	9.487	9.492	9.497	9.503	9.508	9.513	9.518	9.524	9.529	9.534
91	9.539	9.545	9.550	9.555	9.560	9.566	9.571	9.576	9.581	9.586
92	9.592	9.597	9.602	9.607	9.612	9.618	9.623	9.628	9.633	9.638
93	9.644	9.649	9.654	9.659	9.664	9.670	9.675	9.680	9.685	9.690
94	9.695	9.701	9.706	9.711	9.716	9.721	9.726	9.731	9.737	9.742
95	9.747	9.752	9.757	9.762	9.767	9.772	9.778	9.783	9.788	9.793
96	9.798	9.803	9.808	9.813	9.818	9.823	9.829	9.834	9.839	9.844
97	9.849	9.854	9.859	9.864	9.869	9.874	9.879	9.884	9.889	9.894
98	9.899	9.905	9.910	9.915	9.920	9.925	9.930	9.935	9.940	9.945
99	9.950	9.955	9.960	9.965	9.970	9.975	9.980	9.985	9.990	9.995

Table C: Trigonometric functions

degrees	radians	sin	cos	tan	cot		
0	0.0000	0.0000	1.0000	0.0000		1.5708	90
1	0.0175	0.0175	0.9998	0.0175	57.290	1.5533	89
2	0.0349	0.0349	0.9994	0.0349	28.636	1.5359	88
3	0.0524	0.0523	0.9986	0.0524	19.081	1.5184	87
4	0.0698	0.0698	0.9976	0.0699	14.301	1.5010	86
5	0.0873	0.0872	0.9962	0.0875	11.430	1.4835	85
6	0.1047	0.1045	0.9945	0.1051	9.5144	1.4661	84
7	0.1222	0.1219	0.9925	0.1228	8.1443	1.4486	83
8	0.1396	0.1392	0.9903	0.1405	7.1154	1.4312	82
9	0.1571	0.1564	0.9877	0.1584	6.3138	1.4137	81
10	0.1745	0.1736	0.9848	0.1763	5.6713	1.3963	80
11	0.1920	0.1908	0.9816	0.1944	5.1446	1.3788	79
12	0.2094	0.2079	0.9781	0.2126	4.7046	1.3614	78
13	0.2269	0.2250	0.9744	0.2309	4.3315	1.3439	77
14	0.2443	0.2419	0.9703	0.2493	4.0108	1.3265	76
15	0.2618	0.2588	0.9659	0.2679	3.7321	1.3090	75
16	0.2793	0.2756	0.9613	0.2867	3.4874	1.2915	74
17	0.2967	0.2924	0.9563	0.3057	3.2709	1.2741	73
18	0.3142	0.3090	0.9511	0.3249	3.0777	1.2566	72
19	0.3316	0.3256	0.9455	0.3443	2.9042	1.2392	71
20	0.3491	0.3420	0.9397	0.3640	2.7475	1.2217	70
21	0.3665	0.3584	0.9336	0.3839	2.6051	1.2043	69
22	0.3840	0.3746	0.9272	0.4040	2.4751	1.1868	68
23	0.4014	0.3907	0.9205	0.4245	2.3559	1.1694	67
24	0.4189	0.4067	0.9135	0.4452	2.2460	1.1519	66
25	0.4363	0.4226	0.9063	0.4663	2.1445	1.1345	65
26	0.4538	0.4384	0.8988	0.4877	2.0503	1.1170	64
27	0.4712	0.4540	0.8910	0.5095	1.9626	1.0996	63
28	0.4887	0.4695	0.8829	0.5317	1.8807	1.0821	62
29	0.5061	0.4848	0.8746	0.5543	1.8040	1.0647	61
30	0.5236	0.5000	0.8660	0.5774	1.7321	1.0472	60
31	0.5411	0.5150	0.8572	0.6009	1.6643	1.0297	59
32	0.5585	0.5299	0.8480	0.6249	1.6003	1.0123	58
33	0.5760	0.5446	0.8387	0.6494	1.5399	0.9948	57
34	0.5934	0.5592	0.8290	0.6745	1.4826	0.9774	56
35	0.6109	0.5736	0.8192	0.7002	1.4281	0.9599	55
36	0.6283	0.5878	0.8090	0.7265	1.3764	0.9425	54
37	0.6458	0.6018	0.7986	0.7536	1.3270	0.9250	53
38	0.6632	0.6157	0.7880	0.7813	1.2799	0.9076	52
39	0.6807	0.6293	0.7771	0.8098	1.2349	0.8901	51
40	0.6981	0.6428	0.7660	0.8391	1.1918	0.8727	50
41	0.7156	0.6561	0.7547	0.8693	1.1504	0.8552	49
42	0.7330	0.6691	0.7431	0.9004	1.1106	0.8378	48
43	0.7505	0.6820	0.7314	0.9325	1.0724	0.8203	47
44	0.7679	0.6947	0.7193	0.9657	1.0355	0.8029	46
45	0.7854	0.7071	0.7071	1.0000	1.0000	0.7854	45
		cos	sin	cot		tan	radians	degrees

x .	0	1	2	3	4	5	6	7	8	9
10	0000	0043	0086	0128	0170	0212	0253	0294	0334	0374
11	0414	0453	0492	0531	0569	0607	0645	0682	0719	0755
12	0792	0828	0864	0899	0934	0969	1004	1038	1072	1106
13	1139	1173	1206	1239	1271	1303	1335	1367	1399	1430
14	1461	1492	1523	1553	1584	1614	1644	1673	1703	1732
15	1761	1790	1818	1847	1875	1903	1931	1959	1987	2014
16	2041	2068	2095	2122	2148	2175	2201	2227	2253	2279
17	2304	2330	2355	2380	2405	2430	2455	2480	2504	2529
18	2553	2577	2601	2625	2648	2672	2695	2718	2742	2765
19	2788	2810	2833	2856	2878	2900	2923	2945	2967	2989
20	3010	3032	3054	3075	3096	3118	3139	3160	3181	3201
21	3222	3243	3263	3284	3304	3324	3345	3365	3385	3404
22	3424	3444	3464	3483	3502	3522	3541	3560	3579	3598
23	3617	3636	3655	3674	3692	3711	3729	3747	3766	3784
24	3802	3820	3838	3856	3874	3892	3909	3927	3945	3962
25	3979	3997	4014	4031	4048	4065	4082	4099	4116	4133
26	4150	4166	4183	4200	4216	4232	4249	4265	4281	4298
27	4314	4330	4346	4362	4378	4393	4409	4425	4440	4456
28	4472	4487	4502	4518	4533	4548	4564	4579	4594	4609
29	4624	4639	4654	4669	4683	4698	4713	4728	4742	4757
30	4771	4786	4800	4814	4829	4843	4857	4871	4886	4900
31	4914	4928	4942	4955	4969	4983	4997	5011	5024	5038
32	5051	5065	5079	5092	5105	5119	5132	5145	5159	5172
33	5185	5198	5211	5224	5237	5250	5263	5276	5289	5302
34	5315	5328	5340	5353	5366	5378	5391	5403	5416	5428
35	5441	5453	5465	5478	5490	5502	5514	5527	5539	5551
36	5563	5575	5587	5599	5611	5623	5635	5647	5658	5670
37	5682	5694	5705	5717	5729	5740	5752	5763	5775	5786
38	5798	5809	5821	5832	5843	5855	5866	5877	5888	5899
39	5911	5922	5933	5944	5955	5966	5977	5988	5999	6010
40	6021	6031	6042	6053	6064	6075	6085	6096	6107	6117
41	6128	6138	6149	6160	6170	6180	6191	6201	6212	6222
42	6232	6243	6253	6263	6274	6284	6294	6304	6314	6325
43	6335	6345	6355	6365	6375	6385	6395	6405	6415	6425
44	6435	6444	6454	6464	6474	6484	6493	6503	6513	6522
45	6532	6542	6551	6561	6571	6580	6590	6599	6609	6618
46	6628	6637	6646	6656	6665	6675	6684	6693	6702	6712
47	6721	6730	6739	6749	6758	6767	6776	6785	6794	6803
48	6812	6821	6830	6839	6848	6857	6866	6875	6884	6893
49	6902	6911	6920	6928	6937	6946	6955	6964	6972	6981
50	6990	6998	7007	7016	7024	7033	7042	7050	7059	7067
51	7076	7084	7093	7101	7110	7118	7126	7135	7143	7152
52	7160	7168	7177	7185	7193	7202	7210	7218	7226	7235
53	7243	7251	7259	7267	7275	7284	7292	7300	7308	7316
54	7324	7332	7340	7348	7356	7364	7372	7380	7388	7396

Table D (cont.): Common logarithm of x

x	0	1	2	3	4	5	6	7	8	9
55	7404	7412	7419	7427	7435	7443	7451	7459	7466	7474
56	7482	7490	7497	7505	7513	7520	7528	7536	7543	7551
57	7559	7566	7574	7582	7589	7597	7604	7612	7619	7627
58	7634	7642	7649	7657	7664	7672	7679	7686	7694	7701
59	7709	7716	7723	7731	7738	7745	7752	7760	7767	7774
60	7782	7789	7796	7803	7810	7818	7825	7832	7839	7846
61	7853	7860	7868	7875	7882	7889	7896	7903	7910	7917
62	7924	7931	7938	7945	7952	7959	7966	7973	7980	7987
63	7993	8000	8007	8014	8021	8028	8035	8041	8048	8055
64	8062	8069	8075	8082	8089	8096	8102	8109	8116	8122
65	8129	8136	8142	8149	8156	8162	8169	8176	8182	8189
66	8195	8202	8209	8215	8222	8228	8235	8241	8248	8254
67	8261	8267	8274	8280	8287	8293	8299	8306	8312	8319
68	8325	8331	8338	8344	8351	8357	8363	8370	8376	8382
69	8388	8395	8401	8407	8414	8420	8426	8432	8439	8445
70	8451	8457	8463	8470	8476	8482	8488	8494	8500	8506
71	8513	8519	8525	8531	8537	8543	8549	8555	8561	8567
72	8573	8579	8585	8591	8597	8603	8609	8615	8621	8627
73	8633	8639	8645	8651	8657	8663	8669	8675	8681	8686
74	8692	8698	8704	8710	8716	8722	8727	8733	8739	8745
75	8751	8756	8762	8768	8774	8779	8785	8791	8797	8802
76	8808	8814	8820	8825	8831	8837	8842	8848	8854	8859
77	8865	8871	8876	8882	8887	8893	8899	8904	8910	8915
78	8921	8927	8932	8938	8943	8949	8954	8960	8965	8971
79	8976	8982	8987	8993	8998	9004	9009	9015	9020	9025
80	9031	9036	9042	9047	9053	9058	9063	9069	9074	9079
81	9085	9090	9096	9101	9106	9112	9117	9122	9128	9133
82	9138	9143	9149	9154	9159	9165	9170	9175	9180	9186
83	9191	9196	9201	9206	9212	9217	9222	9227	9232	9238
84	9243	9248	9253	9258	9263	9269	9274	9279	9284	9289
85	9294	9299	9304	9309	9315	9320	9325	9330	9335	9340
86	9345	9350	9355	9360	9365	9370	9375	9380	9385	9390
87	9395	9400	9405	9410	9415	9420	9425	9430	9435	9440
88	9445	9450	9455	9460	9465	9469	9474	9479	9484	9489
89	9494	9499	9504	9509	9513	9518	9523	9528	9533	9538
90	9542	9547	9552	9557	9562	9566	9571	9576	9581	9586
91	9590	9595	9600	9605	9609	9614	9619	9624	9628	9633
92	9638	9643	9647	9652	9657	9661	9666	9671	9675	9680
93	9685	9689	9694	9699	9703	9708	9713	9717	9722	9727
94	9731	9736	9741	9745	9750	9754	9759	9763	9768	9773
95	9777	9782	9786	9791	9795	9800	9805	9809	9814	9818
96	9823	9827	9832	9836	9841	9845	9850	9854	9859	9863
97	9868	9872	9877	9881	9886	9890	9894	9899	9903	9908
98	9912	9917	9921	9926	9930	9934	9939	9943	9948	9952
99	9956	9961	9965	9969	9974	9978	9983	9987	9991	9996

x	e^x	e^{-x}	x	e^x	e^{-x}
0.00	1.0000	1.0000	1.5	4.4817	0.2231
0.01	1.0101	0.9901	1.6	4.9530	0.2019
0.02	1.0202	0.9802	1.7	5.4739	0.1827
0.03	1.0305	0.9705	1.8	6.0496	0.1653
0.04	1.0408	0.9608	1.9	6.6859	0.1496
0.05	1.0513	0.9512	2.0	7.3891	0.1353
0.06	1.0618	0.9418	2.1	8.1662	0.1225
0.07	1.0725	0.9324	2.2	9.0250	0.1108
0.08	1.0833	0.9231	2.3	9.9742	0.1003
0.09	1.0942	0.9139	2.4	11.023	0.0907
0.10	1.1052	0.9048	2.5	12.182	0.0821
0.11	1.1163	0.8958	2.6	13.464	0.0743
0.12	1.1275	0.8869	2.7	14.880	0.0672
0.13	1.1388	0.8781	2.8	16.445	0.0608
0.14	1.1503	0.8694	2.9	18.174	0.0550
0.15	1.1618	0.8607	3.0	20.086	0.0498
0.16	1.1735	0.8521	3.1	22.198	0.0450
0.17	1.1853	0.8437	3.2	24.533	0.0408
0.18	1.1972	0.8353	3.3	27.113	0.0369
0.19	1.2092	0.8270	3.4	29.964	0.0334
0.20	1.2214	0.8187	3.5	33.115	0.0302
0.21	1.2337	0.8106	3.6	36.598	0.0273
0.22	1.2461	0.8025	3.7	40.447	0.0247
0.23	1.2586	0.7945	3.8	44.701	0.0224
0.24	1.2712	0.7866	3.9	49.402	0.0202
0.25	1.2840	0.7788	4.0	54.598	0.0183
0.30	1.3499	0.7408	4.1	60.340	0.0166
0.35	1.4191	0.7047	4.2	66.686	0.0150
0.40	1.4918	0.6703	4.3	73.700	0.0136
0.45	1.5683	0.6376	4.4	81.451	0.0123
0.50	1.6487	0.6065	4.5	90.017	0.0111
0.55	1.7333	0.5769	4.6	99.484	0.0101
0.60	1.8221	0.5488	4.7	109.95	0.0091
0.65	1.9155	0.5220	4.8	121.51	0.0082
0.70	2.0138	0.4966	4.9	134.29	0.0074
0.75	2.1170	0.4724	5.0	148.41	0.0067
0.80	2.2255	0.4493	5.5	244.69	0.0041
0.85	2.3396	0.4274	6.0	403.43	0.0025
0.90	2.4596	0.4066	6.5	665.14	0.0015
0.95	2.5857	0.3867	7.0	1096.6	0.0009
1.0	2.7183	0.3679	7.5	1808.0	0.0006
1.1	3.0042	0.3329	8.0	2981.0	0.0003
1.2	3.3201	0.3012	8.5	4914.8	0.0002
1.3	3.6693	0.2725	9.0	8103.1	0.0001
1.4	4.0552	0.2466	10.0	22026	0.00005

Table F: Natural logarithm of x

x	ln x	x	ln x	x	ln x
		4.5	1.5041	9.0	2.1972
0.1	− 2.3026	4.6	1.5261	9.1	2.2083
0.2	− 1.6094	4.7	1.5476	9.2	2.2192
0.3	− 1.2040	4.8	1.5686	9.3	2.2300
0.4	− 0.9163	4.9	1.5892	9.4	2.2407
0.5	− 0.6931	5.0	1.6094	9.5	2.2513
0.6	− 0.5108	5.1	1.6292	9.6	2.2618
0.7	− 0.3567	5.2	1.6487	9.7	2.2721
0.8	− 0.2231	5.3	1.6677	9.8	2.2824
0.9	− 0.1054	5.4	1.6864	9.9	2.2925
1.0	0.0000	5.5	1.7047	10	2.3026
1.1	0.0953	5.6	1.7228	11	2.3979
1.2	0.1823	5.7	1.7405	12	2.4849
1.3	0.2624	5.8	1.7579	13	2.5649
1.4	0.3365	5.9	1.7750	14	2.6391
1.5	0.4055	6.0	1.7918	15	2.7081
1.6	0.4700	6.1	1.8083	16	2.7726
1.7	0.5306	6.2	1.8245	17	2.8332
1.8	0.5878	6.3	1.8405	18	2.8904
1.9	0.6419	6.4	1.8563	19	2.9444
2.0	0.6931	6.5	1.8718	20	2.9957
2.1	0.7419	6.6	1.8871	25	3.2189
2.2	0.7885	6.7	1.9021	30	3.4012
2.3	0.8329	6.8	1.9169	35	3.5553
2.4	0.8755	6.9	1.9315	40	3.6889
2.5	0.9163	7.0	1.9459	45	3.8067
2.6	0.9555	7.1	1.9601	50	3.9120
2.7	0.9933	7.2	1.9741	55	4.0073
2.8	1.0296	7.3	1.9879	60	4.0943
2.9	1.0647	7.4	2.0015	65	4.1744
3.0	1.0986	7.5	2.0149	70	4.2485
3.1	1.1314	7.6	2.0281	75	4.3175
3.2	1.1632	7.7	2.0412	80	4.3820
3.3	1.1939	7.8	2.0541	85	4.4427
3.4	1.2238	7.9	2.0669	90	4.4998
3.5	1.2528	8.0	2.0794	100	4.6052
3.6	1.2809	8.1	2.0919	110	4.7005
3.7	1.3083	8.2	2.1041	120	4.7875
3.8	1.3350	8.3	2.1163	130	4.8675
3.9	1.3610	8.4	2.1282	140	4.9416
4.0	1.3863	8.5	2.1401	150	5.0106
4.1	1.4110	8.6	2.1518	160	5.0752
4.2	1.4351	8.7	2.1633	170	5.1358
4.3	1.4586	8.8	2.1748	180	5.1930
4.4	1.4816	8.9	2.1861	190	5.2470

n	k	p=.05	.10	.15	.20	.25	.30	.35	.40	.45	.50
2	0	.9025	.8100	.7225	.6400	.5625	.4900	.4225	.3600	.3025	.2500
	1	.0950	.1800	.2550	.3200	.3750	.4200	.4550	.4800	.4950	.5000
	2	.0025	.0100	.0225	.0400	.0625	.0900	.1225	.1600	.2025	.2500
3	0	.8574	.7290	.6141	.5120	.4219	.3430	.2746	.2160	.1664	.1250
	1	.1354	.2430	.3251	.3840	.4219	.4410	.4436	.4320	.4084	.3750
	2	.0071	.0270	.0574	.0960	.1406	.1890	.2389	.2880	.3341	.3750
	3	.0001	.0010	.0034	.0080	.0156	.0270	.0429	.0640	.0911	.1250
4	0	.8145	.6561	.5220	.4096	.3164	.2401	.1785	.1296	.0915	.0625
	1	.1715	.2916	.3685	.4096	.4219	.4116	.3845	.3456	.2995	.2500
	2	.0135	.0486	.0975	.1536	.2109	.2646	.3105	.3456	.3675	.3750
	3	.0005	.0036	.0115	.0256	.0469	.0756	.1115	.1536	.2005	.2500
	4	.0000	.0001	.0005	.0016	.0039	.0081	.0150	.0256	.0410	.0625
5	0	.7738	.5905	.4437	.3277	.2373	.1681	.1160	.0778	.0503	.0312
	1	.2036	.3280	.3915	.4096	.3955	.3602	.3124	.2592	.2059	.1562
	2	.0214	.0729	.1382	.2048	.2637	.3087	.3364	.3456	.3369	.3125
	3	.0011	.0081	.0244	.0512	.0879	.1323	.1811	.2304	.2757	.3125
	4	.0000	.0004	.0022	.0064	.0146	.0284	.0488	.0768	.1128	.1562
	5	.0000	.0000	.0001	.0003	.0010	.0024	.0053	.0102	.0185	.0312
6	0	.7351	.5314	.3771	.2621	.1780	.1176	.0754	.0467	.0277	.0156
	1	.2321	.3543	.3993	.3932	.3560	.3025	.2437	.1866	.1359	.0938
	2	.0305	.0984	.1762	.2458	.2966	.3241	.3280	.3110	.2780	.2344
	3	.0021	.0146	.0415	.0819	.1318	.1852	.2355	.2765	.3032	.3125
	4	.0001	.0012	.0055	.0154	.0330	.0595	.0951	.1382	.1861	.2344
	5	.0000	.0001	.0004	.0015	.0044	.0102	.0205	.0369	.0609	.0938
	6	.0000	.0000	.0000	.0001	.0002	.0007	.0018	.0041	.0083	.0156
7	0	.6983	.4783	.3206	.2097	.1335	.0824	.0490	.0280	.0152	.0078
	1	.2573	.3720	.3960	.3670	.3115	.2471	.1848	.1306	.0872	.0547
	2	.0406	.1240	.2097	.2753	.3115	.3177	.2985	.2613	.2140	.1641
	3	.0036	.0230	.0617	.1147	.1730	.2269	.2679	.2903	.2918	.2734
	4	.0002	.0026	.0109	.0287	.0577	.0972	.1442	.1935	.2388	.2734
	5	.0000	.0002	.0012	.0043	.0115	.0250	.0466	.0774	.1172	.1641
	6	.0000	.0000	.0001	.0004	.0013	.0036	.0084	.0172	.0320	.0547
	7	.0000	.0000	.0000	.0000	.0001	.0002	.0006	.0016	.0037	.0078
8	0	.6634	.4305	.2725	.1678	.1001	.0576	.0319	.0168	.0084	.0039
	1	.2793	.3826	.3847	.3355	.2670	.1977	.1373	.0896	.0548	.0312
	2	.0515	.1488	.2376	.2936	.3115	.2965	.2587	.2090	.1569	.1094
	3	.0054	.0331	.0839	.1468	.2076	.2541	.2786	.2787	.2568	.2188
	4	.0004	.0046	.0185	.0459	.0865	.1361	.1875	.2322	.2627	.2734
	5	.0000	.0004	.0026	.0092	.0231	.0467	.0808	.1239	.1719	.2188
	6	.0000	.0000	.0002	.0011	.0038	.0100	.0217	.0413	.0703	.1094
	7	.0000	.0000	.0000	.0001	.0004	.0012	.0033	.0079	.0164	.0312
	8	.0000	.0000	.0000	.0000	.0000	.0001	.0002	.0007	.0017	.0039

Table H: Poisson probabilities $e^{-m}m^k/k!$

m	k=0	1	2	3	4	5	6	7	8	9	10
0.1	.9048	.0905	.0045	.0002	.0000						
0.2	.8187	.1637	.0164	.0011	.0001	.0000					
0.3	.7408	.2222	.0333	.0033	.0002	.0000					
0.4	.6703	.2681	.0536	.0072	.0007	.0001	.0000				
0.5	.6065	.3033	.0758	.0126	.0016	.0002	.0000				
0.6	.5488	.3293	.0988	.0198	.0030	.0004	.0000				
0.7	.4966	.3476	.1217	.0284	.0050	.0007	.0001	.0000			
0.8	.4493	.3595	.1438	.0383	.0077	.0012	.0002	.0000			
0.9	.4066	.3659	.1647	.0494	.0111	.0020	.0003	.0000			
1.0	.3679	.3679	.1839	.0613	.0153	.0031	.0005	.0001	.0000		
1.1	.3329	.3662	.2014	.0738	.0203	.0045	.0008	.0001	.0000		
1.2	.3012	.3614	.2169	.0867	.0260	.0062	.0012	.0002	.0000		
1.3	.2725	.3543	.2303	.0998	.0324	.0084	.0018	.0003	.0001	.0000	
1.4	.2466	.3452	.2417	.1128	.0395	.0111	.0026	.0005	.0001	.0000	
1.5	.2231	.3347	.2510	.1255	.0471	.0141	.0035	.0008	.0001	.0000	
1.6	.2019	.3230	.2584	.1378	.0551	.0176	.0047	.0011	.0002	.0000	
1.7	.1827	.3106	.2640	.1496	.0636	.0216	.0061	.0015	.0003	.0001	.0000
1.8	.1653	.2975	.2678	.1607	.0723	.0260	.0078	.0020	.0005	.0001	.0000
1.9	.1496	.2842	.2700	.1710	.0812	.0309	.0098	.0027	.0006	.0001	.0000
2.0	.1353	.2707	.2707	.1804	.0902	.0361	.0120	.0034	.0009	.0002	.0000
2.2	.1108	.2438	.2681	.1966	.1082	.0476	.0174	.0055	.0015	.0004	.0001
2.4	.0907	.2177	.2613	.2090	.1254	.0602	.0241	.0083	.0025	.0007	.0002
2.6	.0743	.1931	.2510	.2176	.1414	.0735	.0319	.0118	.0038	.0011	.0003
2.8	.0608	.1703	.2384	.2225	.1557	.0872	.0407	.0163	.0057	.0018	.0005
3.0	.0498	.1494	.2240	.2240	.1680	.1008	.0504	.0216	.0081	.0027	.0008
3.2	.0408	.1304	.2087	.2226	.1781	.1140	.0608	.0278	.0111	.0040	.0013
3.4	.0334	.1135	.1929	.2186	.1858	.1264	.0716	.0348	.0148	.0056	.0019
3.6	.0273	.0984	.1771	.2125	.1912	.1377	.0826	.0425	.0191	.0076	.0028
3.8	.0224	.0850	.1615	.2046	.1944	.1477	.0936	.0508	.0241	.0102	.0039
4.0	.0183	.0733	.1465	.1954	.1954	.1563	.1042	.0595	.0298	.0132	.0053
5.0	.0067	.0337	.0842	.1404	.1755	.1755	.1462	.1044	.0653	.0363	.0181
6.0	.0025	.0149	.0446	.0892	.1339	.1606	.1606	.1377	.1033	.0688	.0413
7.0	.0009	.0064	.0223	.0521	.0912	.1277	.1490	.1490	.1304	.1014	.0710
8.0	.0003	.0027	.0107	.0286	.0573	.0916	.1221	.1396	.1396	.1241	.0993
9.0	.0001	.0011	.0050	.0150	.0337	.0607	.0911	.1171	.1318	.1318	.1186
10.0	.0000	.0005	.0023	.0076	.0189	.0378	.0631	.0901	.1126	.1251	.1251

m	k=11	12	13	14	15	16	17	18	19	20	21
5.0	.0082	.0043	.0013	.0005	.0002						
6.0	.0225	.0113	.0052	.0022	.0009	.0003	.0001				
7.0	.0452	.0264	.0142	.0071	.0033	.0014	.0006	.0002	.0001		
8.0	.0722	.0481	.0296	.0169	.0090	.0045	.0021	.0009	.0004	.0002	.0001
9.0	.0970	.0728	.0504	.0324	.0194	.0109	.0058	.0029	.0014	.0006	.0003
10.0	.1137	.0948	.0729	.0521	.0347	.0217	.0128	.0071	.0037	.0019	.0009

t	$\phi(t)$	$\Phi(t)$
0.0	0.398 942	0.500 000
0.1	.396 952	.539 828
0.2	.391 043	.579 260
0.3	.381 388	.617 911
0.4	.368 270	.655 422
0.5	.352 065	.691 462
0.6	.333 225	.725 747
0.7	.312 254	.758 036
0.8	.289 692	.788 145
0.9	.266 085	.815 940
1.0	.241 971	.841 345
1.1	.217 852	.864 334
1.2	.194 186	.884 930
1.3	.171 369	.903 200
1.4	.149 727	.919 243
1.5	.129 518	.933 193
1.6	.110 921	.945 201
1.7	.094 049	.955 435
1.8	.078 950	.964 070
1.9	.065 616	.971 283
2.0	.053 991	.977 250
2.1	.043 984	.982 136
2.2	.035 475	.986 097
2.3	.028 327	.989 276
2.4	.022 395	.991 802
2.5	.017 528	.993 790
2.6	.013 583	.995 339
2.7	.010 421	.996 533
2.8	.007 915	.997 445
2.9	.005 953	.998 134
3.0	.004 432	.998 650
3.1	.003 267	.999 032
3.2	.002 384	.999 313
3.3	.001 723	.999 517
3.4	.001 232	.999 663
3.5	.000 873	.999 767
3.6	.000 612	.999 841
3.7	.000 425	.999 892
3.8	.000 292	.999 928
3.9	.000 199	.999 952
4.0	.000 134	.999 968
4.1	.000 089	.999 979
4.2	.000 059	.999 987
4.3	.000 039	.999 991
4.4	.000 025	.999 995
4.5	.000 016	.999 997

Table K: Random numbers

```
88 47 43 73 86    36 96 47 36 61    46 98 63 71 62    33 26 16 80 45
70 74 24 67 62    42 81 14 57 20    42 53 32 37 32    27 07 36 07 51
85 76 62 27 66    56 50 26 71 07    32 90 79 78 53    13 55 38 58 59
64 56 85 99 26    96 96 68 27 31    05 03 72 93 15    57 12 10 14 21
12 59 56 35 64    38 54 82 46 22    31 62 43 09 90    06 18 44 32 53

16 22 77 94 39    49 54 43 54 82    17 37 93 23 78    87 35 20 96 43
84 42 17 53 31    57 24 55 06 88    77 04 74 47 67    21 76 33 50 25
63 01 63 78 59    16 95 55 67 19    98 10 50 71 75    12 86 73 58 07
33 21 12 34 29    78 64 56 07 82    52 42 07 44 38    15 51 00 13 42
57 60 86 32 44    09 47 27 96 54    49 17 46 09 62    90 52 84 77 27

18 18 07 92 46    44 17 16 58 09    79 83 86 19 62    06 76 50 03 10
26 62 38 97 75    84 16 07 44 99    83 11 46 32 24    20 14 85 88 45
23 42 40 64 74    82 97 77 77 81    07 45 32 14 08    32 98 94 07 72
52 36 28 19 95    50 92 26 11 97    00 56 76 31 38    80 22 02 53 53
37 85 94 35 12    83 39 50 08 30    42 34 07 96 88    54 42 06 87 98

70 29 17 12 13    40 33 20 38 26    13 89 51 03 74    17 76 37 13 04
56 62 18 37 35    96 83 50 87 75    97 12 25 93 47    70 33 24 03 54
99 49 57 22 77    88 42 95 45 72    16 64 36 16 00    04 43 18 66 79
16 08 15 04 72    33 27 14 34 09    45 59 34 68 49    12 72 07 34 45
31 16 93 32 43    50 27 89 87 19    20 15 37 00 49    52 85 66 60 44

68 34 30 13 70    55 74 30 77 40    44 22 78 84 26    04 33 46 09 52
74 57 25 65 76    59 29 97 68 60    71 91 38 67 54    13 58 18 24 76
27 42 37 86 53    48 55 90 65 72    96 57 69 36 10    96 46 92 42 45
00 39 68 29 61    66 37 32 20 30    77 84 57 03 29    10 45 65 04 26
29 94 98 94 24    68 49 69 10 82    53 75 91 93 30    34 25 20 57 27

16 90 82 66 59    83 62 64 11 12    67 19 00 71 74    60 47 21 29 68
11 27 94 75 06    06 09 19 74 66    02 94 37 34 02    76 70 90 30 86
35 24 10 16 20    33 32 51 26 38    79 78 45 04 91    16 92 53 56 16
38 23 16 86 38    42 38 97 01 50    87 75 66 81 41    40 01 74 91 62
31 96 25 91 47    96 44 33 49 13    34 86 82 53 91    00 52 43 48 85

66 67 40 67 14    64 05 71 95 86    11 05 65 09 68    76 83 20 37 90
14 90 84 45 11    75 73 88 05 90    52 27 41 14 86    22 98 12 22 08
68 05 51 18 00    33 96 02 75 19    07 60 62 93 55    59 33 82 43 90
20 46 78 73 90    97 51 40 14 02    04 02 33 31 08    39 54 16 49 36
64 19 58 97 79    15 06 15 93 20    01 90 10 75 06    40 78 78 89 62

05 26 93 70 60    22 35 85 15 13    92 03 51 59 77    59 56 78 06 83
07 97 10 88 23    09 98 42 99 64    61 71 62 99 15    06 51 29 16 93
68 71 86 85 85    54 87 66 47 54    73 32 08 11 12    44 95 92 63 16
26 99 61 65 53    58 37 78 80 70    42 10 50 67 42    32 17 55 85 74
14 65 52 68 75    87 59 36 22 41    26 78 63 06 55    13 08 27 01 50
```

Solutions to Odd Numbered Problems

Chapter 1

1.3.3. 1.9%.

1.3.5. $I_1 = (0.902) I_0$, $I_2 = (0.814) I_0$, $I_n = (0.902)^n I_0$.

1.3.7. 71.2% and 28.8%.

1.3.9. $(0.15) (0.08) = 0.012$ or 1.2%.

1.4.1. Either 90% by simply adding 20%, or 76% by treating the original proportion of misclassified persons, that is, 30% as 100%.

1.5.1. $a + \dfrac{b}{c} = 29/6 = 4.83...$, $\dfrac{a+b}{c} = 3/2 = 1.5$,

$\dfrac{a}{b} + c = 34/5 = 6.8$, $\dfrac{a}{b+c} = 4/11 = 0.36...$.

1.5.3. $x^2 + xy + yx - y^2$ (distributive law twice),

$x^2 + xy + xy - y^2$ (commutative law of multiplication),

$x^2 + 2xy - y^2$ (associative law of addition).

1.5.5. a) $(17 \times 19) + (13 \times 19) = (17 + 13) \times 19 = 30 \times 19 = 570$,
b) $25 \times 17 \times 4 = 25 \times 4 \times 17 = 100 \times 17 = 1700$,
c) $33 \times 125 \times 5 \times 8 = 33 \times 1000 \times 5 = 165000$.

1.5.7. a) $6!/4! = 5 \times 6 = 30$,
b) $97!/98! = 1/98$.

1.6.1. a) $(-1)^2 + (-2)^2 + (-3)^2 = 1 + 4 + 9 = 14$,
b) $(-1)(-2)(-3) = -6$,
c) $1 \times 4 \times 9 = 36$,
d) $(5) (200) (8) (125) = 1000 \times 1000 = 1000000$.

1.6.3. b), d), e) true.
a) $3/4 = 0.75$, c) $(-6) < 5$, f) $-5 < -1 < 0$.

1.6.5. $w < 20$, $20 \leqq w < 22$, $22 \leqq w < 24$, $24 \leqq w$.

1.6.7. a) $x(8 - x) = 0$, $x_1 = 0$, $x_2 = 8$,
b) $x(px - 1) = 0$, $x_1 = 0$, $x_2 = 1/p$,
c) $x(1 - x)(1 + x) = 0$, $x_1 = 0$, $x_2 = 1$, $x_3 = -1$.

1.7.1. a) $x > 3$, b) $y < 7$, c) $u > 5$, d) $p \geq -1$,
e) $|s| < 2$, or: $-2 < s < +2$,
f) $|t| > \sqrt{3}$, or: $t > \sqrt{3}$ or $t < -\sqrt{3}$.

1.8.1. $\left(\dfrac{a+b}{2}\right)^2 \geq ab$ is equivalent to $a^2 + 2ab + b^2 \geq 4ab$ or $(a-b)^2 \geq 0$
which is obvious.

1.9.1. a) $\displaystyle\sum_{i=1}^{5} x_i$, b) $\displaystyle\sum_{i=0}^{k} z_i$, c) $\displaystyle\sum_{j=3}^{n} a_j$, d) $\displaystyle\sum_{k=1}^{4} a_k^2$, e) $\displaystyle\sum_{i=1}^{N} (a_i + b_i)^2$.

1.9.3. Apply associative and commutative laws of addition.

1.9.5. $(x_1 + a) + (x_2 + a) + \cdots + (x_n + a) = (x_1 + x_2 + \cdots + x_n) + na$.

1.9.7. a) $\bar{x} = 76.7$,
b) $+1.1$, -1.0, -4.4, $+4.7$, -0.4,
c) $\displaystyle\sum_{i=1}^{5} (x_i - \bar{x}) = 0$,
d) 43.82.

1.9.9. a) $\Sigma(x_i^2 - 2\bar{x}x_i + \bar{x}^2) = \Sigma x_i^2 - 2\bar{x}\Sigma x_i + n\bar{x}^2$.
Replace Σx_i by $n\bar{x}$ and simplify.
b) Replace $\bar{x}$ by $(\Sigma x_i)/n$ in previous formula.
c) Replace only one factor $\bar{x}$ by $(\Sigma x_i)/n$.

1.10.5. p^{-1}, $(a+b)^{-1}$, $2p^{-5}$, $5(x-z)^{-1}$, $(u-v)(u+v)^{-1}$.

1.10.7. 12 cm.

1.10.9. a) 5.1×10^{18} kg,
b) 1.1×10^{18} kg.

1.10.11. $120\,000$ years.

1.10.13. a) $10\,\text{kW}/4.19 = 2.4\,\text{kcal/sec}$.
b) $100\,\text{m}^3$ water are of mass 10^5 kg.
Rate of temperature increase $= 2.4 \times 10^{-5}\,°\text{C/sec}$.

1.10.15. Between 1.2×10^5 kg and 5.6×10^5 kg.

1.10.17. Let the energy to heat $1\,\text{m}^3$ of water by $1°\,\text{C}$ be "1 unit". Then the power station produces 300 units/sec. This energy per sec is distributed evenly over $200\,\text{m}^3$. Thus $1\,\text{m}^3$ obtains 300 units$/200 = 1.5$ units per sec. Hence, the temperature increases by $1.5°$ C.

1.10.19. $1\,\text{m}^3$ water contains $10^3 \times 35\,\text{ng} = 35 \times 10^{-6}$ g. Total surface water: $20 \times 5 \times 10^{12}\,\text{m}^3 = 10^{14}\,\text{m}^3$. Amount of PCB's: $10^{14} \times 35 \times 10^{-6}\,\text{g} = 3500\,\text{t}$ (metric tons).

1.10.21. 15.1×10^{13} m or 1000 times the distance earth-sun.

1.10.23. 7.2 ng (nanogram).

1.10.25. a) xy/a, b) a^2/b, c) $(a+2b)/(2-a)$, d) $x-y$.

1.10.27. a) $(x+y)/xy$, b) $(t-2)/2t$, c) $(u-1)/u^2$, d) $(5y-6x)/3x^2y^2$.

1.11.1. $7^{1/2}$, $10^{1/3}$, $A^{1/2}$, $(a+b)^{1/2}$, $(1-x)^{1/3}$, $3^{-1/2}$, $p^{-1/3}$.

1.11.3. $a^{9/4}$.

1.11.5. Factor 100.

1.11.7. $|a-b|$ or $b-a$.

1.12.1. $8.1, 4.0, 18.4, 20.8, 0.7, 0.2$.

1.12.3. a) 27.13, b) 27.

1.12.5. $a^2-b^2 \approx 0.10$, $(a+b)(a-b)=0.102$.

1.12.7. 7.9%, 17%, 20%.

1.14.1. $(a-b)^2(a+b)^2 = [(a-b)(a+b)]^2 = (a^2-b^2)^2 = a^4+b^4-2a^2b^2$.

Chapter 2

2.3.1. $S \subset R \subset P \subset U$, $S \subset T \subset P \subset U$, $R \cap T = S$.

2.4.1. $C \subset D$ or $D = C \cup \{5\}$.

2.4.3. a) $\{x|x<5\}$, b) $\{15/4\}$, c) $\{-9\}$, d) $\{10^{1/2}, -10^{1/2}\}$,
e) $\{x|x \geq 6\}$, f) $\{t|t \neq 0\}$.

2.5.1. $A = (a_4, a_5)$, $\bar{A} = \{a_0, a_1, a_2, a_3\}$,
$B = \{a_2, a_3, a_4, a_5\}$, $\bar{B} = \{a_0, a_1\}$.

2.7.1. Two points of intersection, one point of tangency, empty set.

2.7.3. For two planes: a line, a plane, empty set.
For three planes: one point, one line, a plane, empty set.

2.7.5. 607.

2.7.7. $A \setminus B = A \cap \bar{B}$, $B \setminus A = B \cap \bar{A}$.

2.7.9. $A_1 \cap A_2 = $ line joining P_1 and P_2 if $P_1 \neq P_2$.
$A_1 \cap A_2 = A_1$ if $P_1 = P_2$.

2.10.1. c) $[(A \vee B) \wedge D] \vee [C \wedge E]$,
d) $[(A \vee B) \wedge F \wedge (H \vee I)] \vee [\{(C \wedge G) \vee (D \wedge G) \vee E\} \wedge K]$.

2.10.3. a) The sum of two positive numbers is positive.
b) The product of two negative numbers is positive.
c) If the sum of the squares of two numbers is greater than zero, then one of the two numbers is different from zero.

Chapter 3

3.2.1. $A \times B = \{(0, 0), (0, 2), (0, 4), (1, 0), (1, 2), (1, 4)\}$,
$B \times A = \{(0, 0), (0, 1), (2, 0), (2, 1), (4, 0), (4, 1)\}$.

3.3.1. $A \times A$ contains 25 pairs, 19 of them satisfy the inequality.

3.3.3. $R = \{(P, Z), (P, 0), (P, D), (Z, 0), (Z, C), (Z, D),$
$(0, C), (C, 0), (0, 0), (C, C), (0, D), (C, D)\}$.

3.3.5. 1) $520 \, \text{mh}^{-1}$,   2) $290 \, \text{mh}^{-1}$,
3) $490 \, \text{mh}^{-1}$,   4) $390 \, \text{mh}^{-1}$,   5) $320 \, \text{mh}^{-1}$.

3.4.1. Uniqueness of the association in one direction.
Domain: $\{1, 2, ..., 20\}$,   Range: $\{0, 1\}$.

3.4.3. Only b), c), d) define functions.

3.5.1. $y = ax$ where $a = 2.8 \, \text{percent/kR}$.
Domain $\{x | 0 \leq x \leq 6\text{kR}\}$,   Range $\{y | 0 \leq y \leq 16.8\%\}$.
Angle is not meaningful, depends on units chosen.

3.5.3. $v = at$ where $a = 9.81 \, \text{m/sec}^2$. $v_1 = 0.981 \, \text{m/sec}$, $v_2 = 1.962 \, \text{m/sec}$,

3.5.5. Width of the gap $= 10 \, \text{m}/2\pi = 1.6 \, \text{m}$.

3.6.3. Start at $x = 0$, $y = -20$. Then plot $\Delta x = 1$, $\Delta y = 8.8$.

3.6.5. All lines pass through the point $x = 2$, $y = 3$.

3.6.7. $\Delta Q = 1.3 \, \text{mg}$, $\Delta t = 25 \, \text{sec}$, $\Delta Q/\Delta t = 0.052 \, \text{mg/sec}$, $b = 87.6 \, \text{mg}$.

3.6.9. $l = l_0 + aF$.

3.6.11. Yes. $\Delta y/\Delta x = 1.5$.

3.6.13. $y = ax + b$ where $a = 7.44$ and $b = 9 \, \text{mm}$.

3.6.15. $y = 1.8x + 32$,
$x = 36.0° \, C$ $36.1° \, C$ $36.2° \, C...$
$y = 96.8° \, F$ $96.98° \, F$ $97.16° \, F...$

3.6.17. b) Range $= \{N | 95 \leq N \leq 115\}$.

3.6.19. The difference δ is a function of v_0 for a particular person if for each value of v_0 only one measurement is made. The function is nonlinear.

Chapter 4

4.1.1. Range: $\{y | y \geq 0\}$ if $a > 0$,
$\{y | y \leq 0\}$ if $a < 0$.

4.1.3. Point set contained between parabola and straight line.

4.2.3. 36.8%

4.3.1. $x' = x - 1$, $x = x' + 1$, $y' = y + 2$, $y = y' - 2$,
$y' = -x'^2 + 4x' + 7$.

4.4.1.

x	y	Δy	$\Delta^2 y$
-1	12		
		-4	
0	8		2
		-2	
1	6		2
		0	
2	6		2
⋮	⋮	⋮	⋮

4.4.3. $\Delta f(x) = 3x^2 + 3x + 1$, $\Delta^2 f(x) = 6x + 6$, $\Delta^3 f(x) = 6$.

4.5.1. $v = 1.185$, 1.167, 1.111, $1.019 \,(\text{cm sec}^{-1})$.

4.6.5. a) $x(x - 2p) = 0$, $x_1 = 0$, $x_2 = 2p$,
      b) $x(qx - 3) = 0$, $x_1 = 0$, $x_2 = 3/q$,
      c) $x(rx + r + 3) = 0$, $x_1 = 0$, $x_2 = -(r + 3)/r$.

4.6.7. $y = \pm 2$.

4.6.9. $u = 9$.

4.6.11. $(x - 7)(x + 3s) = x^2 + (3s - 7)x - 21s = 0$.

4.6.13. $a = 5.24$, $b = 0.76$.

4.6.15. $D = m^2 - 6m - 27$. Real solutions if $D \geq 0$, that is, $m \leq -3$ or $m \geq 9$.

4.6.17. Discriminant $(N - 1)^2 + 2N = N^2 + 1 > 0$. For $N = 0$ one root $\lambda = \tfrac{1}{2}$.

4.7.1. Time interval between two cars:
in the first case 2.52 sec; 1400 cars/hour.
in the second case 1.80 sec; 2000 cars/hour.
Police regulation increases capacity of road.

Chapter 5

5.1.1. $l' = 10$ is also a period.

5.2.1. $\pi/6$, $\pi/4$, $\pi/3$, $2\pi/3$, $3\pi/4$, $3\pi/2$, $5\pi/2$.

5.4.1. a) $\sin\alpha = \sin(\pi - \alpha)$,   b) $\sin(\pi + \alpha) = -\sin\alpha$,
      c) $\cos(-\alpha) = \cos\alpha$,   d) $\cos\alpha = -\cos(\pi - \alpha)$.

5.5.1. Let x-axis point toward sun. Then $x = 142$ m, $y = 531$ m.

5.5.3. $x_1 = 8.0$ m, $y_1 = 26.3$ m, $x_2 = -18.1$ m, $y_2 = 4.8$ m, $x_3 = -8.6$ m, $y_3 = 30.1$ m.

5.5.5. a) $\alpha_1 = 30°$, $\alpha_2 = 150°$,
      b) $30° \leq \beta \leq 150°$.

5.5.7. a) $25.0°$,   b) $143.1°$,   c) $149.0°$,   d) $54.5°$.

5.6.1. a) $2/\tan\theta = 0.85$ m,
     b) $2\tan\theta = 4.71$ m.

5.6.3. $h = d\sin\alpha = 0.9$ cm, 2.5 cm, 3.8 cm.

5.8.1. $r = 2/\sin\alpha (0 < \alpha < \pi)$.

5.9.1. Sine wave with acrophase $t = 0$ and amplitude 2.

Chapter 6

6.1.1. a), b), d), e) are arithmetic; c), d), f), j) are geometric.

6.1.3. $N_0 \cdot 2^{12}$. The 25 % level of $N_0 \cdot 2^{12}$ is reached in 20 hours.

6.1.5. $N_0/16$. 125 hours.

6.2.3. $\Delta y = aq^{x+1} - aq^x = aq^x(q-1)$.

6.3.1. a), c), d)   monotone,
     b), e),   not monotone.

6.4.1. a) $x = \log y/\log 2$ $(y > 0)$,
     b) $x = \log(y/a)$ $(y > 0)$,
     c) $t = \log 2r/\log 5$ $(r > 0)$,
     d) $x = \log(Q/2)/\log w$ $(Q > 0)$.

6.5.5. 4.5×10^9, 5.5×10^9, 6.7×10^9.

6.5.7. a) 5.13 %,   b) 14 years.

6.5.9. 8.3 and 13.9 weeks.

6.5.11. 125 g, 312 g, 781 g, etc.

6.5.13. 0.64 %.

6.6.1. $I_2/I_1 = 31.6$.

6.6.3. a) 4.4,   b) 7.1,   c) 7.6.

6.6.5. $L = 10(\log(c/I_0) - 2\log r)$

6.7.1. $OA_1 = aq$, $OA_2 = aq^2$, etc. where $q = 1/\cos\alpha$.

6.7.3. Logarithmic spiral.

Chapter 7

7.2.9. $a \approx 121$ µg/10 ml, $q \approx 0.955$.

7.3.1. Let $D = \log d$, $M = \log m$. Then $\Delta M/\Delta D = 1.01$, $M = 1.01\, D + 0.53$.

7.3.5. a) exponential function, b) and c) power functions.

Chapter 8

8.1.1. a) 8,   b) $-1/3$,   c) $2/7$,   d) a/b.

8.1.3. a) 2,   b) 2,   c) $-1/3$,   d) $2x$.

8.1.5. $a_n = \dfrac{n}{\sqrt{n^2+1}} = \dfrac{1}{\sqrt{1+1/n^2}} \to 1$.

8.1.7. $\dfrac{(\sqrt{5+h}-\sqrt{5})(\sqrt{5+h}+\sqrt{5})}{h(\sqrt{5+h}+\sqrt{5})} = \dfrac{5+h-5}{h(\sqrt{5+h}+\sqrt{5})}$

$$= \dfrac{1}{\sqrt{5+h}+\sqrt{5}} \to \dfrac{1}{2\sqrt{5}}\ .$$

8.1.9. a), b), e) tend to zero, c) and d) tend to infinity.

8.2.1. a) e^3,   b) a^{-2}.

8.2.3. $\dfrac{\sin 5h}{h} = \dfrac{5\sin k}{k} \to 5$.

8.3.1. a) $728/486$,   b) $2\dfrac{1-11^{-6}}{1-11^{-1}}$,   c) $43/64$.

8.3.3. a) $1/(1-r)$,   b) $2c$,   c) $s/(s-1)$.

8.3.5. $U = 1/(1-w)^2$

8.3.7. $q = 1-0.023 = 0.977$,
     a) $M_n = M(1-q^{n+1})/(1-q)$,
     b) $M/(1-q) = 43.5\ M$.

8.3.9. $q = 1-p/100$. Total accumulation of poison $= d + dq + \cdots + dq^n$
     $= d(1-q^{n+1})/(1-q)$.

8.3.11. $\left|\dfrac{a_{n+1}}{a_n}\right| = \left|\dfrac{a^{n+1}}{(n+1)!} \cdot \dfrac{n!}{a^n}\right| = \left|\dfrac{a}{n+1}\right| \to 0$,

$\left|\dfrac{a}{n+1}\right| < q < 1$ from a certain n on.

8.4.1. a) tends to $+\infty$,   b) tends to $-\infty$.

8.4.3. For $t \to 0$, $i \to \infty$.   For $t \to \infty$, $i \to b$.

8.4.5. Let E be the energy of sound per second emitted by the bat and I_1 be the intensity received by the moth. Then $I_1 \propto E/r^2$. Let I_2 be the intensity of the echo. We may assume that $I_2 \propto I_1$. Finally, let I_3 be the intensity of sound absorbed by the bat's ears. Since the echo is equivalent to a new source of sound, we get $I_3 \propto I_2/r^2 \propto I_1/r^2 \propto E/r^4$. (Notice that doubling r reduces I_3 to $1/16$.)

8.5.3. Let $\{f_n\}$ be the sequence of fractions. Then $f_{n+1} = \dfrac{1}{1+f_n}$.

For $b_n = 1/f_n$, this is equivalent to $b_{n+1} = \dfrac{1}{b_n} + 1$, cf. Eq. (8.5.7).

Chapter 9

9.1.1. $v_1 = 31\ \text{km}/7.5\ \text{h} = 4.1\ \text{km}\,\text{h}^{-1}$,

$v_2 = 14\ \text{km}/3\ \text{h} = 4.7\ \text{km}\,\text{h}^{-1}$, $v_2 > v_1$.

9.1.3. a) $\Delta N/\Delta t = 90$,   b) $\Delta N/\Delta t = 540$,   c) $\Delta N/\Delta t = 450$.

9.1.5. Density $= \dfrac{(10+h)^2/3 - 10^2/3}{h} = \dfrac{20+h}{3}$ ($= 7.00,\ 6.70,\ 6.67$, resp.)

9.2.1. $\Delta y/\Delta x = \dfrac{a(x+h)^2 - ax^2}{h} = 2ax + ah \to 2ax$.

9.2.3. a) $(-2)\,x^{-3}$,   b) $5w^4$,   c) $\tfrac{2}{3}r^{-1/3}$,   d) $(-3)\,t^{-4}$,
      e) $\tfrac{1}{3}Q^{-2/3}$,   f) $-\tfrac{1}{2}p^{-3/2}$.

9.2.5. a) $dv/dt = a - b/t^2$,   b) $dU/dz = 2az + \tfrac{1}{2}bz^{-1/2} - \tfrac{1}{2}cz^{-3/2}$.

9.2.7. $\dfrac{dN}{dt} = 52 + 4t = 72\ h^{-1}$.

9.2.9. a) 0,   b) -0.12,   c) -0.06.

9.2.11. $\Delta y/\Delta x = \dfrac{\cos(x+h) - \cos x}{h} = -\dfrac{2}{h}\sin\!\left(x + \dfrac{h}{2}\right)\sin\dfrac{h}{2}$

$= -\sin(x+k) \cdot \dfrac{\sin k}{k} \to -\sin x$.

9.2.13. a) $y' = (x-3) + (x+7) = 2x + 4$,
      b) $y' = \sin x + x\cos x$,
      c) $z' = -\cos t - (1-t)\sin t$,
      d) $Q' = \cos^2\alpha - \sin^2\alpha$,
      e) $n' = \tfrac{1}{3}x^{-2/3}(1 - 2x) - 2x^{1/3}$,
      f) $f' = \tfrac{a}{2}y^{-1/2}\sin y + a\sqrt{y}\cos y$.

9.2.15. a) $\cos^2 x - \sin^2 x$,   b) $6t^2 - 6t + 10$,   c) $\cos u - u\sin u$,
      d) $-\tfrac{3}{2}w^{1/2} + \tfrac{1}{2}w^{-1/2}$.

9.2.17. a) $2(x+5)$,   b) $4u(u^2 - 3)$,   c) $-(t-2)^{-2}$,
      d) $2(1-v)^{-2}$,   e) $-\tfrac{3}{2}(4 - 3t)^{-1/2}$,   f) $4\cos(4\alpha - 5)$.

9.2.19. a) $u' = -4(x-1)^{-5}$,   b) $E' = -\dfrac{1}{w^2} - \dfrac{1}{(w-1)^2}$,

      c) $z' = \tfrac{1}{2}(t-1)^{-1/2} - \tfrac{1}{2}(t+1)^{-3/2}$,
      d) $f' = 2\alpha\cos\alpha(\cos\alpha - \alpha\sin\alpha)$.

9.2.21. a) $dy/dx = \dfrac{3x^2 - 2x^3}{(1-x)^2}$,

b) $dS/dt = \dfrac{ad - bc}{(ct+d)^2}$,

c) $\dfrac{dQ}{d\alpha} = \dfrac{(1+\cos\alpha)(-\cos\alpha) - (1-\sin\alpha)(-\sin\alpha)}{(1+\cos\alpha)^2}$

$\qquad = \dfrac{-\cos\alpha + \sin\alpha - 1}{(1+\cos\alpha)^2}$,

d) $dP/dt = (3+t^4)(1+t^2)^{-2}$,

e) $df/du = pu^2(pu^3+q)^{-2/3}$,

f) $dh/d\phi = \dfrac{2\cos 3\phi\cos 2\phi + 3\sin 2\phi\sin 3\phi}{\cos^2 3\phi}$.

9.2.23. $y' = 2x$, $\quad x = \sqrt{y-1}$ $\;(y>1)$,

$\dfrac{dx}{dy} = \dfrac{1}{2\sqrt{y-1}} = \dfrac{1}{2x} = \left(\dfrac{dy}{dx}\right)^{-1}$.

9.3.1. a) $y = 2x^3 + C$, $\quad$ b) $y = 4x^2 - 7x + C$, $\quad$ c) $u = \frac{a}{2}t^2 + bt + C$,

d) $y = \frac{5}{4}x^4 + C$, $\quad$ e) $W = t^2 - 8t + C$, $\quad$ f) $U = U_0 x + \sin x + C$,

g) $y = \frac{1}{3}\sin t + C$, $\quad$ h) $U = \frac{1}{2}\sin 2x + C$, $\quad$ i) $K = -1/u + C$.

9.5.1. ≈ 2.8.

9.5.3. a) $1/2$, $\quad$ b) 10, $\quad$ c) $a\dfrac{t^3}{3} + b\dfrac{t^2}{2} + ct$, $\quad$ d) 4, $\quad$ e) 2,

f) 2.

9.5.5. a) $-\frac{1}{5}x^{-5} + C$, $\quad$ b) $\frac{3}{5}t^{2/3} + C$,

c) $\frac{u^2}{2} + \frac{2}{3}u^3 + \frac{3}{4}u^4 + C$,

d) $\frac{10}{11}s^{\frac{11}{10}} + C$,

e) $-A\cos\theta + B\sin\theta + C$,

f) $-\dfrac{1}{\sqrt{2}}Q^{-1/2} + C$.

9.5.7. a) $\dfrac{1}{x+2}$, $\quad$ b) $\dfrac{ax-1}{x+1}$,

c) $\dfrac{t-3}{\sin t}$, $\quad$ d) $\sqrt{s^3 + 1}$.

9.6.1. a) $y'' = -6x$, $\quad$ b) $u'' = 40z^3 - 18z$, $\quad$ c) $W'' = 6t^{-3}$,

d) $-\frac{1}{2}s^{-3/2}$.

9.6.3. $v = at$, $\quad s = \frac{a}{2}t^2$.

9.6.5. a) convex downward, b) and c) convex upward, d) not convex,
e) convex upward for $x < 0$ and convex downward for $x > 0$.

9.6.7. a) maximum at $x = -3^{-1/2}$, minimum at $x = +3^{-1/2}$,
b) minimum at $t = -2$, c) maximum at $p = \frac{1}{2}$.

9.6.9. Necessary, but not sufficient. Under unusual conditions not
even necessary.

9.6.11. The condition is sufficient, but not necessary (since c need not
be the hypotenuse).

9.6.13. The condition is necessary, but not sufficient.

9.6.15. The condition $f''(x) = 0$ is necessary, but not sufficient for a
point of inflection.

9.7.1. a) absolute minimum at $x = 3/2$, absolute maximum at $x = 5$.
b) absolute minimum at $t = -2$, absolute maximum at $t = 3$.
c) maximum at $v = 1$, minimum at $v = 3$.
d) relative maximum at $x = -1$, relative minimum at $x = +1$,
absolute maximum at $x = 3$, absolute minimum at $x = -3$.

9.7.3. $dQ/ds = (s^2 + 2s - 15)(s + 1)^{-2} = 0$ for $s = 3$ and $s = -5$. For
$s \downarrow 3$ is $dQ/ds > 0$, for $s \uparrow 3$ is $dQ/ds < 0$. $Q(0) = 5$, $Q(3) = -4$
absolute minimum.

9.7.5. a) at $x = 5$ cm, b) $\int_0^{10} (10x - x^2)\, dx = 167$ mg.

9.7.7. $x = a/2$.

9.7.9. $dP/dv = -\dfrac{W^2}{2\varrho S v^2} + \tfrac{3}{2}\varrho A v^2 = 0$, $v = \left(\dfrac{W^2}{3\varrho^2 SA}\right)^{1/4}$.

9.7.11. $S = 2\pi r h + 2\pi r^2$, $V = \pi r^2 h$,

$S = 2\dfrac{V}{r} + 2\pi r^2$, $dS/dr = -2\dfrac{V}{r^2} + 4\pi r = 0$,

$r = \left(\dfrac{V}{2\pi}\right)^{1/3}$, $h : r = V : \pi r^3 = V : \dfrac{V}{2} = 2$, $h = 2r$.

9.8.1. a) $\dfrac{1}{6}\int_0^6 \left(\dfrac{x}{2} + 3\right) dx = 9/2$,

b) $\int_2^3 at^2\, dt = 19a/3$,

c) $\dfrac{1}{\pi}\int_{-\pi/2}^{\pi/2} \cos \alpha\, d\alpha = 2/\pi$.

9.9.1. $\delta S \approx 8\pi r \cdot \delta r$,
$\delta V \approx 4\pi r^2 \cdot \delta r = S \cdot \delta r$.

9.10.5. a) $\dfrac{1}{\omega}\sin\omega x \ (\omega \neq 0)$, b) $-\dfrac{1}{32}(1-8u)^4$,

c) $\dfrac{2}{3q}(p+qs)^{3/2}$, d) $\dfrac{1}{20}(4t+3)^5\Big|_0^2 = \dfrac{11^5-3^5}{20}$,

e) $\frac{1}{3}\sin 3\theta|_0^{\pi/2} = -\frac{1}{3}$, f) $-\frac{1}{5}(5x-1)^{-1}|_1^3 = 1/28$.

9.10.7. a) $\int x\sin x\,dx = -x\cos x + \sin x + C$,

b) $\int t\cos\omega t\,dt = \omega^{-2}(\omega t\sin\omega t + \cos\omega t) + C$,

c) $\int u(u+1)^{1/2}\,du = \frac{2}{3}u(u+1)^{3/2} - \frac{4}{15}(u+1)^{5/2} + C$.

Chapter 10

10.3.1. a) 6.931, b) 4.189, c) $2\ln 100 = 9.2104$.

10.4.1. a) 7.5244, b) 0.3935, c) 0.49658.

10.4.7. a) $\frac{1}{2}$, b) 0, c) $-a$.

10.5.1. a) 5.65, b) 0.253, c) 0.850.

10.5.3. a) $\exp(x\ln 2)$, b) $\exp(u\ln 10)$, c) $\exp(s\ln 4.43)$,
d) $\exp(-3.08886t)$, e) $\exp(-0.43124x)$.

10.5.5. $33.115e^{2x}$.

10.6.1. $\log 10 = 1$, $\ln 10 = 2.3026$, $\ln 10 = 2.3026\log 10$, etc.

10.7.1. a) $3e^{3x}$, b) $-2e^{1-2u}$, c) $(-t)\exp(-\frac{1}{2}t^2)$, d) $5/(5x+4)$,
e) $2v/(v^2-2)$, f) $-1/(s^2+s)$, g) $e^{t/2}(1+t/2)$,
h) $1+\ln 3u$, i) $1/[r(1-r)]$.

10.7.3. Use the chain rule with $u = f(x)$.

10.7.5. a) $e^u + C$, b) $\frac{1}{2}e^{2t} + C$, c) $-e^{-x} + C$, d) $\ln w + C$,
e) $\ln(x+1) + C$, f) $\frac{1}{2}\ln(2t+5) + C$.

10.7.7. $\delta y \approx (1/x)\,\delta x$,

10.7.9. Let $u = -x^2$, a) $e^u > 0$, b) $1/e^u \to 0$,
c) $y' = -2xe^u$, $y'' = -2e^u(1-2x^2)$,
$y' = 0$ implies $x = 0$ with $y'' < 0$,
d) $y'' = 0$ implies $1 - 2x^2 = 0$.

10.7.11. a) $y(0) = c(e^0 - e^0) = 0$,
b) $e^{-at} > e^{-bt}$ because of $b > a$,
e) $y'' = c(a^2 e^{-at} - b^2 e^{-bt}) = 0$, $t = (b-a)^{-1}\ln\left(\dfrac{b}{a}\right)^2$.

10.8.1. Use Eq. (10.8.3) with $a = 3, -3, \frac{1}{2}$. Thus

    a) e^3,   b) e^{-3},   c) $e^{1/2}$.

    d) $\dfrac{1}{h} \ln(1 + 3h) = \dfrac{3}{k} \ln(1 + k) \to 3$.

10.9.1. $dT/dt = -ak\exp(-kt)$.

10.9.3. a) $8.06\,\text{d}$,   b) $1.87\,\text{h}$.

10.9.5. $\delta N/N_0 \approx -\lambda e^{-\lambda t_1} \delta t$.

10.9.7. $dN/dt = -abk\exp(-b \cdot \exp kt)\exp kt$.

10.10.1. $e^{-x} \approx e^0 - x$.

10.10.3. $\sin x \approx x - \dfrac{x^3}{3!}$, $\cos x \approx 1 - \dfrac{x^2}{2}$.

10.10.5. $\ln(1 + x) = x - \dfrac{x^2}{2} + \dfrac{x^3}{3} - \dfrac{x^4}{4} + \cdots$,

    $\dfrac{d}{dx} \ln(1 + x) = 1 - x + x^2 - x^3 + x^4 - \cdots$

                  $= 1/(1 + x)$ for $|x| < 1$.

10.10.7. a) $\varDelta t = \dfrac{\ln 2}{\ln\left(1 + \dfrac{p}{100}\right)}$,

    b) $\varDelta t = \dfrac{100 \ln 2}{p} \approx \dfrac{70}{p}$.

10.11.1. a) $\sinh(-x) = \frac{1}{2}(e^{-x} - e^x)$,
      b) $\cosh(-x) = \frac{1}{2}(e^{-x} + e^x)$,
      d) $\sinh(x + y) = \frac{1}{2}(e^{x+y} - e^{-x-y})$.

10.11.3. $\sinh x = x + \dfrac{x^3}{3!} + \dfrac{x^5}{5!} + \cdots$,

    $\cosh x = 1 + \dfrac{x^2}{2!} + \dfrac{x^4}{4!} + \cdots$.

10.11.5. a) $e^x(x - 1) + C$,   b) $e^x(x + k - 1) + C$,

    c) $\dfrac{x^3}{3}(\ln x - \frac{1}{3}) + C$,   d) $(-\frac{1}{2})\ln(5 - 2z)|_1^2 = \ln 2$,

    e) $\dfrac{1}{\ln a} e^{x\ln a}\Big|_0^1 = (a - 1)/\ln a$,

    f) $\dfrac{1}{a} \ln(at + b)\Big|_a^{2a} = \dfrac{1}{a} \ln \dfrac{2a^2 + b}{a^2 + b}$ $(a > 0)$.

Chapter 11

11.1.1. $y' = 2ax + b = \dfrac{ax^2 + bx}{x} + ax = \dfrac{y}{x} + ax$.

11.3.1. a) $\dfrac{dy}{y} = \dfrac{dx}{x}$, $\quad \ln|y| = \ln|x| + C$, $\quad y = C_1 x$, $\quad C_1 \in \mathbb{R}$,

b) $\dfrac{dy}{y^2} = a\, dx$, $\quad -\dfrac{1}{y} = ax + C$, $\quad y = -\dfrac{1}{ax + C}$,

c) $\dfrac{dy}{y} = kt\, dt$, $\quad \ln|y| = k\dfrac{t^2}{2} + C$, $\quad y = C_1 e^{kt^2/2}$,

d) $\dfrac{du}{u^2} = t\, dt$, $\quad -\dfrac{1}{u} = \dfrac{t^2}{2} + C$, $\quad u = -\dfrac{1}{t^2/2 + C}$.

11.3.3. $\left(\dfrac{2}{v} - 1\right) dv = k\, dt$, $\quad 2\ln|v| - v = kt + C$;

$v = 1$, $\quad t = 0$ implies $C = -1$.

11.3.5. $N = N_1 e^{c(t - t_1)}$.

11.3.7. $\lambda > 0.5$ if t is measured in years.

11.3.9. a) $\dfrac{dy}{dt} = 0.76 y$, $\quad$ b) $\dfrac{dy}{dt} = 2\sqrt{y}$, $\quad$ c) $\dfrac{dy}{dt} = y - 1$.

11.3.11. $I(x) = I_0 e^{-\mu x}$, $\quad \dfrac{dI}{dx} = -\mu I_0 e^{-\mu x}$, $\quad \dfrac{dI}{dx} = -\mu I$.

11.4.3. $\dfrac{dN}{dt} = (\lambda - \mu) N + v$, $\quad N = N_0 e^{(\lambda - \mu)t} + \dfrac{v}{\lambda}(e^{(\lambda - \mu)t} - 1)$.

11.4.5. $\dfrac{dz}{z - 1} = dx (z > 1)$, $\quad \ln(z - 1) = x + C$, $\quad C = 0$, $\quad z = 1 + e^x$.

11.4.7. $Q\, dQ = (x - 3)\, dx$,

$\dfrac{Q^2}{2} = \dfrac{1}{2}(x - 3)^2 + C$, $\quad C = 1/2$, $\quad Q^2 = (x - 3)^2 + 1$.

11.5.1. a) $y = 2 + \dfrac{3}{1 + ke^{3t}}$, $\quad$ b) $u = 3 - \dfrac{3}{1 + ke^{-3t/2}}$,

c) $z = 1 + \dfrac{2}{1 + ke^{2t}}$, $\quad$ d) $y = 2 - \dfrac{4}{1 + ke^{4x}}$.

11.5.3. a) $y = \dfrac{2}{1 + e^{-t}}$, $\quad$ b) $y = \dfrac{4}{1 + \frac{1}{3}e^{-4t}}$,

c) $y = \dfrac{1}{1 + e^{-t/3}}$, $\quad$ d) $y = \dfrac{4}{1 + 3^{-t}}$.

11.6.1. $V = cL^3$, $\quad \dfrac{dV}{dt} = 3cL^2 \dfrac{dL}{dt}$,

$$\frac{1}{V}\frac{dV}{dt} = \frac{3}{L}\frac{dL}{dt}.$$

11.6.3. $\dfrac{dy}{y} = \dfrac{m\,dx}{x}$, $\quad \ln|y| = m\ln x + C$,

$\quad y = Ax^m$, $\quad A = -\dfrac{1}{2^m}$.

11.6.5. a) $y_2 - y_1 = A(e^{kt_2} - e^{kt_1})$,

b) $\dfrac{\Delta y}{y} = \dfrac{A(e^{kt_2} - e^{kt_1})}{e^{kt_1}}$,

c) $\dfrac{\Delta y}{\Delta t} = \dfrac{A(e^{kt_2} - e^{kt_1})}{t_2 - t_1}$,

d) $\dfrac{dy}{dt} = kAe^{kt_1}$,

e) $\dfrac{1}{y}\dfrac{dy}{dt} = k$.

11.7.1. $x = Ae^{\lambda t}$, $\quad y = Be^{\lambda t}$,

$\left.\begin{matrix} (7-\lambda)A - 4B = 0 \\ -9A + (7-\lambda)B = 0 \end{matrix}\right\}$ $\lambda_1 = 1$, $\quad \lambda_2 = 13$.

$\dfrac{A_1}{B_1} = \dfrac{2}{3}$, $\quad \dfrac{A_2}{B_2} = -\dfrac{2}{3}$, $\quad A_1 = 2k$, $\quad B_1 = 3k$,

$A_2 = 2m$, $\quad B_2 = -3m$,
$x = 2ke^t + 2me^{13t}$,
$y = 3ke^t - 3me^{13t}$.

11.7.3. The discriminant of $\lambda^2 - (a+d)\lambda + (ad - bc) = 0$ is $(a+d)^2 - 4(ad - bc) = (a-d)^2 + 4bc > 0$.

11.7.5. a) Population growth increases poverty, poverty increases population growth.

b) Burning fossil fuels increases SO_2 pollution, public concern grows; to decrease pollution, energy production has to be modified.

c) Low standard of living may reduce health and production. Lower production decreases the standard of living.

d) The more cars, the higher the density of traffic, the more roads have to be constructed. More roads stimulate the purchase of cars.

e) Continuous use of drugs may lead to habituation; this could imply sickness and call for more drugs.

f) Destruction of forests could cause degradation of soil. To continue agriculture, more forests are destroyed.

11.9.1. $d^2x/dt^2 = -A\omega^2 \cos\omega t - B\omega^2 \sin\omega t$,
$\omega^2 = k$, $\omega_1 = +\sqrt{k}$, $\omega_2 = -\sqrt{k}$,
A, B being arbitrary constants.

Chapter 12

12.2.1. a) $\partial f/\partial x = a$, $\partial f/\partial y = -b$,

b) $\partial g/\partial u = 2u - 2v$, $\partial g/\partial v = -2u + 2v$,

c) $\partial \Phi/\partial s = \partial \Phi/\partial t = e^{s+t}$,

d) $\partial h/\partial x = a\exp(ax + by + cz)$,   etc.

e) $\partial \Psi/\partial \alpha = 3\cos(3\alpha + \beta)$, $\partial \Psi/\partial \beta = \cos(3\alpha + \beta)$,

f) $\partial Q/\partial v = w/v$, $\partial Q/\partial w = \ln v$,

g) $\partial F/\partial r = -(r - 2s)^{-2}$, $\partial F/\partial s = 2(r - 2s)^{-2}$,

h) $\partial G/\partial s = -abt/(bs - ct)^2$, $\partial G/\partial t = abs/(bs - ct)^2$.

12.2.3. a) $f_{xx} = 2a$, $f_{xy} = b$, $f_{yy} = 2c$,

b) $h_{uu} = n(n-1) u^{n-2} v^n$, $h_{uv} = n^2 (uv)^{n-1}$,
$h_{vv} = n(n-1) u^n v^{n-2}$,

c) $Q_{vv} = -w/v^2$, $Q_{vw} = 1/v$, $Q_{ww} = 0$,

d) $S_{xx} = S_{yy} = S_{zz} = 2$, $S_{xy} = S_{xz} = S_{yz} = 0$,

e) $\Phi_{ss} = 0$, $\Phi_{st} = ae^{at}$, $\Phi_{tt} = a^2 se^{at}$,

f) $\Psi_{\alpha\alpha} = -A\sin\alpha$, $\Psi_{\alpha\beta} = 0$, $\Psi_{\beta\beta} = -B\sin\beta$.

12.2.5. $Q_x = x(x^2 + y^2)^{-1/2} = xQ^{-1}$, $Q_y = yQ^{-1}$,
$Q_{xx} = Q^{-1} - xQ^{-2}Q_x = Q^{-1} - x^2Q^{-3}$,
$Q_{yy} = Q^{-1} - y^2Q^{-3}$,
$Q_{xx} + Q_{yy} = 2Q^{-1} - (x^2 + y^2) Q^{-3} = Q^{-1}$.

12.3.1. $\left. \begin{array}{l} \partial z/\partial x = 18x - 6y = 0 \\ \partial z/\partial y = -6x + 4y - 6 = 0 \end{array} \right\}$ $x = 1$, $y = 3$, $z = 2$.

12.3.3. Let $15 + 4x - 3y + 5xy - x^2 - 2y^2 = S$,
$\partial S/\partial x = 4 + 5y - 2x$, $\partial S/\partial y = -3 + 5x - 4y$,
$\left. \begin{array}{l} \partial z/\partial x = (4 + 5y - 2x) \exp S = 0 \\ \partial z/\partial y = (-3 + 5x - 4y) \exp S = 0 \end{array} \right\}$ $x = -\frac{1}{17}$, $y = -\frac{14}{17}$.

12.3.5. In the first and third quadrant x and y are of equal sign, hence $z > 0$. In the second and fourth quadrant x and y are of unequal sign, hence $z < 0$.

12.3.7. Let $x =$ amount of sulfur, $y =$ number of absences.

$$\Sigma x_i = 71, \quad \Sigma y_i = 265, \quad \Sigma x_i y_i = 4244, \quad \Sigma x_i^2 = 1103.$$

$$\left.\begin{array}{r} 1103a + 71b = 4244 \\ 71a + 5b = 265 \end{array}\right\} a = 5.074, \quad b = -19.049,$$

$$y = (5.074) x - 19.049.$$

12.3.9. $x = py + q, \quad \varepsilon_i = py_i + q - x_i.$
Let $S = \Sigma \varepsilon_i^2$. Then $\partial S/\partial p = 0$ implies
$p\Sigma y_i^2 + q\Sigma y_i = \Sigma x_i y_i, \quad \partial S/\partial q = 0 \quad$ implies
$p\Sigma y_i + mq = \Sigma x_i \quad$ or

$$\left.\begin{array}{r} 16571p + 265q = 4244 \\ 265p + 5q = 71 \end{array}\right\} p = 0.1904, \quad q = 4.1078,$$

$$x = (0.1904) y + 4.1078 \quad \text{or} \quad y = (5.252) x - 21.57.$$

Chapter 13

13.2.1. $\{A, B\}$: blood pressure below 150 mm,
$\{B, C\}$: blood pressure above 120 mm,
$\{A, C\}$: blood pressure below 120 mm or above 150 mm,
$\{A, B, C\}$: blood pressure arbitrary.

13.2.3. Intersection $\{BB\}$,
Union $\{BB, BG, GB\}$.

13.2.5. A and B, $\quad$ A and D, $\quad$ C and D.

13.3.1. 0.7 and 0.3.

13.3.3. a) $19/37$, $\quad$ b) $6/37$, $\quad$ c) $11/37$.

13.3.5. $11/20$.

13.3.7. $\{ABC, ACB, BAC, BCA, CAB, CBA\}$, Probability $1/6$.

13.4.1. a) 0.77, $\quad$ b) 0.54.

13.4.3. Use a Venn diagram.

13.5.1. The reduced outcome space is $\{BG, GB, GG\}$. All three events
have the same probability. Hence $P(GG|\text{not } BB) = 1/3$.

13.5.3. a) 0.0067, $\quad$ b) 0.891.

13.5.5. a) $39\% + 48\% - 15\% = 72\%$ have either antigen, 28% have no
antigen.
b) 33% of total population have B, but not A. Conditional
probability is $33/48 = 0.69$ or 69%.

13.5.7. $(3/8)(2/7) = 3/28$.

13.5.9. a) 0.044, b) 0.0925 .

13.6.1. a) $(2/6)(2/6) = 1/9$,   b) $2(1/6)(1/6) = 1/18$,
        c) $3(1/6)(1/6) = 1/12$.

13.6.3. a) $1/4$,   b) $1/4$,   c) $1/4$,   d) $1/4$,   e) 0 .

13.6.5. $p_1 = r^2$, $p_2 = 2pq$, $p_3 = p^2 + 2pr$, $p_4 = q^2 + 2qr$.

13.7.1. Number of words $25^2 = 625$.
        a) yes,   b) no .

13.7.3. For each bar there exist two possibilities. Number of symbols
        at most $2^7 = 128$.

13.7.5. 10, 20, 56, 9, 1, 330 .

13.7.7. $\binom{10}{1} = \binom{9}{0} + \binom{9}{1} = 10$, $\binom{10}{2} = \binom{9}{1} + \binom{9}{2} = 45$,   etc.

13.7.9. $\binom{38}{3} = 8436$.

13.7.11. $\binom{10}{4} = 210$.

13.7.13. $8!/(5!\,3!) = \binom{8}{3} = 56$.

13.7.15. a) $\binom{10}{3}$,   b) $\binom{10}{4}$.

13.7.17. a) $1 + 4x + 6x^2 + 4x^3 + x^4$,
         b) $1 - 4x + 6x^2 - 4x^3 + x^4$,
         c) $32 + 80p + 80p^2 + 40p^3 + 10p^4 + p^5$,
         d) $z^3 + \frac{3}{2}z^2 + \frac{3}{4}z + \frac{1}{8}$,
         e) $12a^5 + 40a^3 + 12a$.

13.7.19. a) $\dfrac{n+1}{n+1-k}$,   b) $\dfrac{1}{k}m(m-1)\cdots(m-k+1)$,

         c) $\binom{N+1}{5}$.

13.7.21. a) $a^{n-1} + \binom{n-1}{1}a^{n-2}b + \binom{n-1}{2}a^{n-3}b^2 + \cdots + b^{n-1}$,

         b) $a^{2n} + \binom{2n}{1}a^{2n-1}b + \binom{2n}{2}a^{2n-2}b^2 + \cdots + b^{2n}$.

13.8.1. a) $\binom{5}{2}\big/2^5 = 10/32$,   b) $1/32$,   c) $31/32$,   d) $16/32$.

13.8.3. The probability of no "double six" is $(35/36)^{24} = 0.5086$. It is more advantageous to bet that this case occurs.

13.8.5. a) $0.822^3 = 0.555$,   b) $1 - 0.555 = 0.445$.

13.8.7. a) $\binom{4}{1}\left(\frac{1}{4}\right)\left(\frac{3}{4}\right)^3 = 0.422$,   b) $1 - \left(\frac{3}{4}\right)^4 = 0.684$.

13.8.9. $P(GGGG\,|\,\text{at least } GG) = \dfrac{1/16}{6/16 + 4/16 + 1/16} = \dfrac{1}{11}$.

13.8.11. $15/16$.

13.9.1. c) $\mu = E(X) = (0.5)\,1 + (0.3)\,5 + (0.2)\,10 = 4.0$,
$\sigma^2 = E(X - \mu)^2 = (0.5)\,3^2 + (0.3)\,1^2 + (0.2)\,6^2 = 12.0$,
$\sigma = 3.46$.

13.9.3. $E(X) = 39$, $\sigma = 2.93$.

13.9.5. $E(S_6) = (0.7351)\,0 + (0.2321)\,1 + (0.0305)\,2$

$$+ (0.0021)\,3 + (0.0001)\,4 = 0.2998,$$

$E(S_6) = 6(0.05) = 0.30$. Difference caused by rounding errors.

13.9.7. $E(X) = 5(0.5) = 2.5$, $\operatorname{Var}(X) = 5(0.5)\,(0.5) = 1.25$.

13.9.9. a) $E(N_1) = 240(0.25) = 60$, $E(N_2) = 240(0.5) = 120$,
$E(N_3) = E(N_1)$.

b) $\operatorname{Var}(N_1) = 240(0.25)\,(0.75) = 45$, $\sigma_1 = 6.71$,
$\operatorname{Var}(N_2) = 240(0.50)\,(0.50) = 60$, $\sigma_2 = 7.75$,
$\operatorname{Var}(N_3) = \operatorname{Var}(N_1)$, $\sigma_3 = \sigma_1$.

13.10.1. $p_0 = (0.5)^0\, e^{-0.5}/0! = 0.6065$,
$p_1 = (0.5)^1\, e^{-0.5}/1! = 0.3033$,
$p_2 = (0.5)^2\, e^{-0.5}/2! = 0.0758$,
$p_3 = (0.5)^3\, e^{-0.5}/3! = 0.0126$,   etc.

13.10.3. $E(X) = (0.1353)\,0 + (0.2707)\,1 + (0.2707)\,2 + \cdots$

$$+ (0.0002)\,9 = 1.9994,$$

$E(X) = m = 2$ (Difference due to rounding errors).
$\operatorname{Var} X = E(X - m)^2 = (0.1353)\,2^2 + (0.2707)\,1^2 + (0.2707)\,0^2 + \cdots$

$$+ (0.0002)\,7^2 = 1.9972,$$

$\operatorname{Var} X = m = 2$ (Difference due to rounding errors).

13.10.5. a) $\sigma = \sqrt{m} = 2.10$,
b) $1 - p_0 = 1 - (4.4)^0\, e^{-4.4}/0! = 0.9877$.

13.10.7. $\sigma = \sqrt{m} = 2.83$. Deviations of $1, 2, 3$ occur frequently, larger deviations are rare.

13.10.9. $m = 5$. $p_0 = 0.0067$, $p_1 = 0.0337$, $p_2 = 0.0842$, etc. (Table H).

13.11.1. $x = \pm 1$.

Chapter 14

14.1.1.

$$\begin{array}{cccc} & BB & Bb & bb \\ AA & \begin{pmatrix} n_{00} & n_{01} & n_{02} \\ Aa & n_{10} & n_{11} & n_{12} \\ aa & n_{20} & n_{21} & n_{22} \end{pmatrix} \end{array}$$

14.2.1. a) $(8, 8, 4, 10)$,   b) $(2, 6, -2, 10)$,
c) $(0, 0, 0, 0)$,   d) $(-6, -2, -6, 0)$.

14.2.3.
a) $\begin{pmatrix} 1 & 0 & 1 \\ 5 & 6 & 5 \end{pmatrix}$,   b) $\begin{pmatrix} 5 & -4 & 1 \\ -5 & 4 & 3 \end{pmatrix}$,

c) $\begin{pmatrix} -1 & 2 & 1 \\ 10 & 7 & 6 \end{pmatrix}$,   d) $\begin{pmatrix} 11 & -8 & 3 \\ -5 & 14 & 11 \end{pmatrix}$.

14.2.5. a) 51,   b) 13,   c) 19,
d) 51,   e) 57,   f) 89,   g) 5.

14.2.7. $f'q = 0(23) + 1(17) + 2(8) + 3(6) + 4(8) + 5(2) = 93$.

14.2.9. $E(X) = (p_1, p_2, ..., p_m) \begin{pmatrix} x_1 \\ x_2 \\ \vdots \\ x_m \end{pmatrix}$.

14.2.11. a) $\begin{pmatrix} 15 \\ 15 \end{pmatrix}$,   b) $(5, 1)$,

c) $\begin{pmatrix} 15 & -3 \\ 12 & 19 \end{pmatrix}$,   d) $\begin{pmatrix} 41 & 36 \\ -3 & -4 \end{pmatrix}$.

14.2.13. a) $\begin{pmatrix} a+b+c+d \\ -a+b-c+d \end{pmatrix}$ b) $(a+b+c+d, -a+b-c+d)$.

14.2.15. $u'Su = 3x^2 - xy + y^2$.

14.2.17. $A(B+C) = AB + AC = \begin{pmatrix} a+3b & 3a+4b \\ c+3d & 3c+4d \end{pmatrix}$.

14.2.19. $A^2 = \begin{pmatrix} 1 & 2 \\ 0 & 1 \end{pmatrix} \begin{pmatrix} 1 & 2 \\ 0 & 1 \end{pmatrix} = \begin{pmatrix} 1 & 4 \\ 0 & 1 \end{pmatrix}$, $A^n = \begin{pmatrix} 1 & 2n \\ 0 & 1 \end{pmatrix}$.

14.2.21. $U^n = \begin{pmatrix} 1^n & 0 & 0 \\ 0 & 2^n & 0 \\ 0 & 0 & 3^n \end{pmatrix}$.

14.2.23. $\frac{1}{10} \begin{pmatrix} 8 & 5 \\ 6 & 5 \end{pmatrix} \begin{pmatrix} 5 & -5 \\ -6 & 8 \end{pmatrix} \neq \begin{pmatrix} 1 & 0 \\ 0 & 1 \end{pmatrix}$.

14.2.25. $A'A = \frac{1}{25} \begin{pmatrix} 3 & -4 \\ 4 & 3 \end{pmatrix} \begin{pmatrix} 3 & 4 \\ -4 & 3 \end{pmatrix} = \begin{pmatrix} 1 & 0 \\ 0 & 1 \end{pmatrix}$.

14.2.27. $x = 17$, $y = -7$.

14.2.29. a) $ab' = 15$,   b) $a'b = \begin{pmatrix} 3 \\ 1 \\ 1 \end{pmatrix} (4, 2, 1) = \begin{pmatrix} 12 & 6 & 3 \\ 4 & 2 & 1 \\ 4 & 2 & 1 \end{pmatrix}$.

14.2.31. $AB = \begin{pmatrix} a_1x_1 + a_2x_2 + a_3x_3 & a_1y_1 + a_2y_2 + a_3y_3 \\ b_1x_1 + b_2x_2 + b_3x_3 & b_1y_1 + b_2y_2 + b_3y_3 \end{pmatrix}$,

$B'A' = \begin{pmatrix} a_1x_1 + a_2x_2 + a_3x_3 & b_1x_1 + b_2x_2 + b_3x_3 \\ a_1y_1 + a_2y_2 + a_3y_3 & b_1y_1 + b_2y_2 + b_3y_3 \end{pmatrix} = (AB)'$.

14.3.1. $n_{01} = 34000$, $n_{11} = 300$, $n_{21} = 240$.

14.3.3. a) $TO = \begin{pmatrix} p & q & 0 \\ \frac{1}{2}p & \frac{1}{2} & \frac{1}{2}q \\ 0 & p & q \end{pmatrix} \begin{pmatrix} p^2 & 2pq & q^2 \\ p^2 & 2pq & q^2 \\ p^2 & 2pq & q^2 \end{pmatrix}$

$= \begin{pmatrix} p^2(p+q) & 2pq(p+q) & q^2(p+q) \\ p^2(p+q) & 2pq(p+q) & q^2(p+q) \\ p^2(p+q) & 2pq(p+q) & q^2(p+q) \end{pmatrix} = \begin{pmatrix} p^2 & 2pq & q^2 \\ p^2 & 2pq & q^2 \\ p^2 & 2pq & q^2 \end{pmatrix} = O$,

b) $T^2 = \begin{pmatrix} p & q & 0 \\ \frac{1}{2}p & \frac{1}{2} & \frac{1}{2}q \\ 0 & p & q \end{pmatrix} \begin{pmatrix} p & q & 0 \\ \frac{1}{2}p & \frac{1}{2} & \frac{1}{2}q \\ 0 & p & q \end{pmatrix} = \begin{pmatrix} p^2 + \frac{1}{2}pq & pq + \frac{1}{2}q & \frac{1}{2}q^2 \\ \frac{1}{2}p^2 + \frac{1}{4}p & pq + \frac{1}{4} & \frac{1}{4}q + \frac{1}{2}q^2 \\ \frac{1}{2}p^2 & \frac{1}{2}p + pq & \frac{1}{2}pq + q^2 \end{pmatrix}$

$= \begin{pmatrix} \frac{1}{2}p + \frac{1}{2}p^2 & \frac{1}{2}q + pq & \frac{1}{2}q^2 \\ \frac{1}{4}p + \frac{1}{2}p^2 & \frac{1}{4} + pq & \frac{1}{4}q + \frac{1}{2}q^2 \\ \frac{1}{2}p^2 & \frac{1}{2}p + pq & \frac{1}{2}q + \frac{1}{2}q^2 \end{pmatrix} = \frac{1}{2}T + \frac{1}{2}O$,

c) $T^3 = T(T^2) = T(\frac{1}{2}T + \frac{1}{2}O) = \frac{1}{2}T^2 + \frac{1}{2}TO$

$= \frac{1}{2}(\frac{1}{2}T + \frac{1}{2}O) + \frac{1}{2}O = \frac{1}{4}T + \frac{3}{4}O$.

14.3.5. $P = \begin{pmatrix} 0.4 & 0.6 \\ 0.2 & 0.8 \end{pmatrix}$, $P^2 = \begin{pmatrix} 0.28 & 0.72 \\ 0.24 & 0.76 \end{pmatrix}$.

14.4.1. $a_1 + a_2 + a_3 = \begin{pmatrix} 4 \\ 6 \end{pmatrix}$.

14.4.3. $a + b = \begin{pmatrix} 12 \\ 3 \end{pmatrix}$, $a - b = \begin{pmatrix} 2 \\ -7 \end{pmatrix}$, $-a = \begin{pmatrix} -7 \\ 2 \end{pmatrix}$,

$-b = \begin{pmatrix} -5 \\ -5 \end{pmatrix}$, $\frac{1}{2}a = \begin{pmatrix} 7/2 \\ -1 \end{pmatrix}$, $-\frac{3}{5}b = \begin{pmatrix} -3 \\ -3 \end{pmatrix}$.

14.4.5. $\frac{1}{2}(a + b) = \begin{pmatrix} 5 \\ -1 \end{pmatrix}$.

14.4.7. a) $|a| = 5$, $|b| = \sqrt{53}$,

b) $a'b = -13$, $\cos\alpha = \dfrac{-13}{5\sqrt{53}} = -0.3571$, $\alpha = 110.9^0$.

14.4.9. $u'w = 30 - 30 = 0$; u and w are orthogonal.

14.4.11. $F_1 + F_2 + F_3 = \begin{pmatrix} 0 \\ -2 \\ 10 \end{pmatrix}$, $F = \begin{pmatrix} 0 \\ 2 \\ -10 \end{pmatrix}$.

14.4.13. $v'w = -10 + 6 + 4 = 0$; v and w are orthogonal.

14.4.15. $|a| = \sqrt{14}$, $|b| = \sqrt{41}$, $a'b = 7$,
$\cos\alpha = 0.2922$, $\alpha = 73.0^0$.

14.4.17. $a'b = |a|\,|b|\cos\pi = -|a|\,|b|$.

14.5.1. $\frac{1}{3}(a_1 + a_2 + a_3) = (17/3, 3, 3)'$.

14.5.3. $|F| = 2|F_1|\cos\dfrac{\alpha}{2}$.

14.6.1. a) 7 ,   b) 14 ,   c) -102 .

14.6.3. a) $x_1 = -9/17$,   b) $x = \frac{1}{2}(-26r - 15s)$,
$x_2 = +38/17$. $y = -5s - 8r$.

14.6.5. a) $4\begin{vmatrix} 7 & 5 \\ -2 & 8 \end{vmatrix} = 264$,   b) $4\begin{vmatrix} 1 & 1 \\ 7 & 0 \end{vmatrix} + 5\begin{vmatrix} -2 & 1 \\ 3 & 7 \end{vmatrix} = -113$.

14.6.7. $x_1 = -10/7$, $x_2 = -103/7$, $x_3 = -140/7$.

14.6.9. a) 236 (multiply first line by -2 and add to third line),
b) 102 (subtract first column from third column),
c) -475 (multiply first line by 2 and add to second line),
d) 8645 (factor out 13 and 7).

14.6.11. $\begin{vmatrix} 1 & a & a^2 \\ 1 & b & b^2 \\ 1 & c & c^2 \end{vmatrix} = \begin{vmatrix} 0 & a-b & a^2-b^2 \\ 0 & b-c & b^2-c^2 \\ 1 & c & c^2 \end{vmatrix}$

$= (a-b)(b-c) \begin{vmatrix} 1 & a+b \\ 1 & b+c \end{vmatrix} = (a-b)(b-c)(c-a).$

14.6.13. $\det(\lambda A) = \lambda^3 \begin{vmatrix} 4 & 2 & 1 \\ 3 & -4 & 5 \\ 2 & 2 & 2 \end{vmatrix} = -50\lambda^3.$

14.7.1. $\begin{pmatrix} x & y \\ u & v \end{pmatrix} = \begin{pmatrix} 5 & -7 \\ -2 & 3 \end{pmatrix}.$

14.7.3. $A^{-1} = \frac{1}{6} \begin{pmatrix} 4 & 1 & -2 \\ 6 & 3 & -6 \\ -12 & -3 & 12 \end{pmatrix}.$

14.7.5. a) $A^{-1} = \begin{pmatrix} 1 & -4 & 6 \\ 0 & 1 & -3 \\ 0 & 0 & 1 \end{pmatrix}$, b) $B^{-1} = -\frac{1}{6} \begin{pmatrix} -2 & 0 & 0 \\ 2 & -6 & 0 \\ 5 & -15 & 3 \end{pmatrix}.$

14.7.7. $A^2 = \begin{pmatrix} 41 & 55 \\ 55 & 74 \end{pmatrix}$, $(A^2)^{-1} = \frac{1}{9} \begin{pmatrix} 74 & -55 \\ -55 & 41 \end{pmatrix}$,

$A^{-1} = \frac{1}{3} \begin{pmatrix} 7 & -5 \\ -5 & 4 \end{pmatrix}$, $(A^{-1})^2 = \frac{1}{9} \begin{pmatrix} 74 & -55 \\ -55 & 41 \end{pmatrix}.$

14.7.9. $y = Ax,$ $A^{-1}y = A^{-1}Ax,$

$x = A^{-1}y,$ $A^{-1} = \frac{1}{6} \begin{pmatrix} -2 & 4 & 1 \\ -6 & 6 & 3 \\ 12 & -12 & -3 \end{pmatrix}.$

14.8.1. $a - b - c = O$. Therefore, a, b, and c are linearly dependent.

14.8.3. a) $x = t$ (arbitrary), $y = 3t - 7$,
b) no solution,
c) $x = t$ (arbitrary), $y = 3t$,
d) $x = -3$, $y = -14$,
e) $x = y = 0$,
f) $x = y = 0$.

14.8.5. $\begin{vmatrix} \lambda & 12 \\ 3 & \lambda \end{vmatrix} = 0,$ $\lambda_1 = 6,$ $\lambda_2 = -6.$

14.9.1. a) $\lambda_1 = 3,$ $x_1 = (1, -5),$
       $\lambda_2 = 7,$ $x_2 = (1, -1),$
   b) $\lambda_1 = 1,$ $x_1 = (1, -2),$
       $\lambda_2 = 9,$ $x_2 = (3, 2),$
   c) $\lambda_1 = 3,$ $x_1 = (2, -1),$
       $\lambda_2 = 11,$ $x_2 = (2, -5).$

14.9.3. $\lambda_1 = 2,$ $x_1 = (1, -2, 3),$
       $\lambda_2 = 1,$ $x_2 = (0, 1, -2),$
       $\lambda_3 = 9,$ $x_3 = (0, 3, 2).$

14.9.5. $\begin{vmatrix} p_1 - 1 & q_1 \\ p_2 & q_2 - 1 \end{vmatrix} = \begin{vmatrix} -q_1 & q_1 \\ p_2 & -p_2 \end{vmatrix} = 0.$

14.9.7. $\begin{vmatrix} -\lambda & 1 \\ \frac{1}{4} & \frac{1}{2} - \lambda \end{vmatrix} = \lambda^2 - \frac{\lambda}{2} - \frac{1}{4} = 0, \quad \lambda = \dfrac{1 \pm \sqrt{5}}{4}.$

14.9.9. $\det(M - \lambda I) = \lambda(\lambda^3 - 48\lambda - 12) = 0.$

Chapter 15

15.2.1. a) $5, 53.1°,$ b) $5, -53.1°,$ c) $\sqrt{2}, 225°,$ d) $5\sqrt{2}, -45°.$

15.3.1. a) $1 + 12i,$ b) $-12 + \frac{15}{2}i,$ c) $-10,$
   d) $+27,$ e) $-16,$ f) $41 + 61i,$
   g) $-10 - 49i,$ h) $-5 + 12i,$ i) $-9 - 40i,$ k) $73.$

15.3.3. a) $\frac{1}{29}(2 - 5i),$ b) $\dfrac{a - bi}{a^2 + b^2},$ c) $\frac{1}{10}(3 + i),$
   d) $-\frac{1}{10}(1 + 13i).$

15.4.1. a) $5e^{0.9273i},$ b) $5e^{-0.9273i},$
   c) $2\sqrt{2}e^{i\pi/4},$ d) $\sqrt{17}e^{2.8966i},$
   e) $\sqrt{145}e^{3.2247i},$ f) $10e^{2.2143i}.$

15.4.3. a) $e^{\frac{5}{6}\pi i},$ b) $e^{\frac{\pi}{6}i},$
   c) $e^{\frac{4\pi i}{3}},$ d) $e^{\frac{\pi}{2}i} = i.$

15.5.1. a) $x = \pm 3i,$ b) $x = -3 \pm 4i,$ c) $\lambda = 3 \pm 3i,$
   d) $p = -6 \pm 5i,$ e) $u = \frac{1}{4}(5 \pm \sqrt{23}i),$ f) $s = 1 \pm 7i.$

15.5.3. a) $3 \pm i,$ b) $5 \pm 3i.$

15.6.1. a) $\omega = \sqrt{3/2},$ $x = a\cos\omega t + b\sin\omega t,$
   b) $x = e^{-t/2}(A_1 e^{it} + A_2 e^{-it}).$

References

Abbott, B.C., Brady, A.J.: Amphibian muscle. In: Physiology of the amphibia. Moore, J.A. (Ed.). Chap. 6, pp. 329—370. London-New York: Academic Press 1964.

Ackerman, E.: Biophysical science. Englewood Cliffs, N.J.: Prentice-Hall 1962.

Adler, I.: Simplifying the formula. Letter publ. in Science 153, 936 (1966).

Alexander, R.M.: Animal mechanics. Seattle: University of Washington Press 1968.

Allen, E.S.: Six-place tables. New York-San Francisco-Toronto-London: McGraw-Hill 1947.

Anderson, K.P.: Manual of sampling and statistical methods for fisheries biology. Part II, Chap. 5: Computations. FAO Fisheries Technical Paper No. 26, Suppl. 1. Rome: Food and Agriculture Organization of the United Nations 1965.

Arbib, M.A.: Brains, machines and mathematics. New York-San Francisco-Toronto-London: McGraw-Hill 1964.

Atkins, G.L.: Multicompartment models for biological systems. London: Methuen 1969.

Bailey, N.T.J.: The mathematical theory of epidemics. New York-London: Hafner. London: Griffin 1957.

— The elements of stochastic processes with applications to the natural sciences. London-New York-Sydney: Wiley & Sons 1964.

— The mathematical approach to biology and medicine. London-New York-Sydney: Wiley & Sons 1967.

Bak, T.A., Lichtenberg, J.: Mathematics for scientists. New York-Amsterdam: Benjamin 1966.

Bakir, F. et al.: Methylmercury poisoning in Iraq. Science 181, 230—241 (1973).

Bandoni, R.J., Koske, R.E.: Monolayers and microbial dispersal. Science 183, 1079—1081 (1974).

Barlow's Tables: see Comrie.

Barnett, V.D.: The Monte Carlo solution of a competing species problem. Biometrics 18, 76—103 (1962).

Bartlett, M.S.: The use of transformations. Biometrics 3, 39—52 (1947).

— Stochastic population models in ecology and epidemiology. London: Methuen, London-New York-Sydney: Wiley & Sons 1960.

Batschelet, E.: Statistical methods for the analysis of problems in animal orientation and certain biological rhythms. Washington, D.C.: Amer. Inst. Biol. Sciences 1965.

— Striebel, H.R.: Nomogramm zur Bestimmung der reellen und komplexen Wurzeln einer Gleichung vierten Grades. Z. angew. Math. Phys. 3, 156—159 (1952).

Beier, W.: Biophysik. Eine Einführung in die physikalische Analyse elementarer biologischer Strukturen und Vorgänge. 2nd edition. Leipzig: VEB G. Thieme 1962.

Benedict, F.G.: Vital energetics. Report No. 503. Washington, D.C.: Carnegie Institute 1938.

Beroza, M., Knipling, E.F.: Gypsy moth control with the sex attractant pheromone. Science 177, 19—27 (1972).

Bertalanffy, L. von: see von Bertalanffy, L.

Beyer, W.H. (Ed.): Handbook of tables for probability and statistics. Cleveland, Ohio: Chemical Rubber 1966.

Blaxter, K.L., Graham, N.M., Wainman, F.W.: Some observations on the digestibility of food by sheep, and on related problems. Brit. J. Nutr. 10, 69—91 (1956).

Bliss, C. I.: Statistics in biology. Vol. 2. New York-San Francisco-Toronto-London: McGraw-Hill 1970.

— Blevins, D. L.: The analysis of seasonal variation in measles. Amer. J. Hyg. **70**, 328—334 (1959).

Blume, J.: Nachweis von Perioden durch Phasen- und Amplitudendiagramm mit Anwendungen aus der Biologie, Medizin und Psychologie. Köln-Opladen: Westdeutscher Verlag 1965.

Boag, J. W.: Maximum likelihood estimates of the proportion of patients cured by cancer therapy. J. Roy. Stat. Soc., Ser. B, **11**, 15—53 (1949).

Bourret, J. A., Lincoln, R. G., Carpenter, B. H.: Fungal endogenous rhythms expressed by spiral figures. Science **166**, 763—764 (1969).

Brian, M. V.: Social insect populations. London-New York: Academic Press 1965.

Brody, S.: Bioenergetics and growth. With special reference to the efficiency complex in domestic animals. New York-Amsterdam-London: Reinhold 1945.

Brohmer, P.: Fauna von Deutschland. 9th edition. Heidelberg: Quelle & Meyer 1964.

Brown, F. A., Jr.: A unified theory for biological rhythms. Rhythmic duplicity and the genesis of circa periodisms. In: Circadian clocks. Proc. of the Feldafing Summer School 1964. Aschoff, J. (Ed.). Amsterdam: North-Holland 1965.

Cajori, F.: A history of mathematical notations. Vols. 1 and 2. Chicago: Open Court 1928 and 1929.

Calloway, N. O.: Ages of experimental animals. Letter publ. in Science **150**, 1771 (1965).

Cannings, C., Edwards, A. W. F.: Natural selection and the de Finetti diagram. Annals Human Genetics **31**, 421—428 (1968).

Cavalli-Sforza, L. L., Edwards, A. W. F.: Phylogenetic analysis. Models and estimation procedures. Amer. J. Hum. Genet. **19**, 233—257 (1967).

Chapman, D. G.: Stochastic models in animal population ecology. Proc. Fifth Berkeley Symposium on Mathematical Statistics and Probability, **4**, 147—162. Berkeley, Calif.: Univ. of California Press 1967.

Chiang, C. L.: Competition and other interactions between species. In: Statistics and mathematics in biology. Kempthorne, O., Bancroft, T. A., Gowen, J. W., Lush, J. L. (Eds.). Chap. 14. Reprinted from 1954 edition. New York-London: Hafner 1964.

— Introduction to stochastic processes in biostatistics. London-New York-Sydney: Wiley & Sons 1968.

Chiarappa, L., Chiang, H. C., Smith, R. F.: Plant pests and diseases: Assessment of crop losses. Science **176**, 769—773 (1972).

Cole, K. S.: Theory, experiment, and the nerve impulse. In: Theoretical and mathematical biology. Waterman, T. H. and Morowitz, H. J. (Eds.), pp. 136—171. New York-Toronto-London: Blaisdell 1965.

— Membranes, ions and impulses. A chapter of classical biophysics. Berkeley-Los Angeles: University of California Press 1968.

Comrie, L. J. (Ed.): Barlow's tables of squares, cubes, square roots, cube roots, and reciprocals of all integers up to 12,500. London: Spon. New York: Chemical Publ. Co. 1962.

Comroe, J. H.: Physiology of respiration. An introductory text. Chicago: Year Book 1965.

Consolazio, C. F., Johnson, R. E., Pecora, L. J.: Physiological measurements of metabolic functions in man. New York-San Francisco-Toronto-London: McGraw-Hill 1963.

Copp, D. H., Cockcroft, D. W., Kueh, Y.: Calcitonin from ultimobranchial glands of dogfish and chickens. Science **158**, 924–925 (1967).

Crow, J. F., Kimura, M.: An introduction to population genetic theory. New York, Evanston, London: Harper & Row 1970.

D'Ancona, U.: The struggle for existence. Leiden: Brill 1954.

Davis, D. S.: Nomography and empirical equations. 2nd edition. New York-Amsterdam-London: Reinhold 1962.

Davis, F. S., Wayland, J. R., Merkle, M. G.: Ultrahighfrequency electromagnetic fields for weed control: Phytotoxicity and selectivity. Science 173, 535—537 (1971).

Davis, H. T., Fisher, V. J.: Tables of mathematical functions, Vol. 3: Arithmetical tables. San Antonio, Texas: The Principia Press of Trinity University 1962.

Davson, H.: A textbook of general physiology, 3rd edition. Boston: Little, Brown 1964.

Defares, J. G., Sneddon, I. N., Wise, M. E.: An Introduction to the mathematics of medicine and biology. 2nd edition. Amsterdam-London: North-Holland 1973.

de Finetti, B.: Considerazioni matematiche sull' ereditarietà mendeliana. Metron 6, 3—41 (1926).

Diem, K., Lentner, C. (Eds.): Scientific tables. 7th edition. Basel: Ciba-Geigy (1970).

Dingle, H.: Migration strategies of insects. Science 175, 1327—1335 (1972).

Dohan, F. C.: Air pollutants and incidence of respiratory disease. Arch. Environ. Health 3, 387—395 (1961).

Dormer, K. J.: Shoot organization in vascular plants. London: Chapman and Hall 1972.

Dwight, H. B.: Tables of integrals and other mathematical data. New York: MacMillan 1961.

Elias, H.: Three-dimensional structure identified from single sections. Science 174, 993—1000 (1971).

— Hennig, A.: Stereology of the human renal glomerulus. In: Quantitative methods in morphology. Weibel, E. R., Elias, H. (Eds.), pp. 130—166. Berlin-Heidelberg-New York: Springer 1967.

Engelhardt, W.: Was lebt in Tümpel, Bach und Weiher? Stuttgart: Kosmos, Franck'sche Verlagshandlung 1962.

Ewens, W. J.: Population genetics. London: Methuen 1969.

Feinstein, A. R.: Clinical judgment. Baltimore: The Williams & Wilkins Co. 1967.

Feller, W.: On the logistic law of growth and its empirical verifications in biology. Acta Biotheor. (Leiden) 5, 51—64 (1940).

— Diffusion processes in genetics. In: Proceedings of the Second Berkeley Symposium on mathematical statistics and probability. Neyman, J. (Ed.), pp. 227—246. Berkeley: University of California Press 1951.

— An introduction to probability theory and its applications. Vol. 1, 3rd edition. London-New York-Sydney: Wiley & Sons 1968.

Ficken, R. W., Ficken, M. S., Hailman, J. P.: Temporal pattern shifts to avoid acoustic interference in singing birds. Science 183, 762—763 (1974).

Fischmeister, H. F.: Apparative Hilfsmittel der Stereologie. In: Quantitative methods in morphology. Weibel, E. R., Elias, H. (Eds.), pp. 221—249. Berlin-Heidelberg-New York: Springer 1967.

Fisher, R. A.: The theory of inbreeding. 2nd edition. Edinburgh-London: Oliver and Boyd 1965.

Fraenkel, G. S., Gunn, D. L.: The orientation of animals. Kineses, taxes and compass reactions. New York: Dover 1961.

Frisch, Karl von: see von Frisch, K.

Fyhn, H. J., Petersen, J. A., Johansen, K.: Heart activity and high-pressure circulation in Cirripedia. Science 180, 513—515 (1973).

Gani, J.: Stochastic formulations for life tables, age distributions and mortality curves. In: The mathematical theory of the dynamics of biological populations. M. S. Bartlett and R. W. Hiorns (Eds.), pp. 291—302. London and New York: Academic Press 1973.

Gause, G. F.: The struggle for existence. Baltimore: Williams and Wilkins 1934. New York: Hafner 1964 and 1969.

Gebelein, H., Heite, H. J.: Statistische Urteilsbildung. Berlin-Göttingen-Heidelberg: Springer 1951.

Geigy: Scientific tables. See Diem, K.

Gelbaum, B. R., March, J. G.: Mathematics for the social and behavioral sciences. Probability, calculus and statistics. Philadelphia-London: Saunders.

General Electric Company: Tables of the individual and cumulative terms of Poisson distribution. New York: Van Nostrand 1962.

George, F. H.: The brain as a computer. Oxford-London-Edinburgh-New York: Pergamon Press 1961.

Gnedenko, B. V., Khinchin, A. Y.: An elementary introduction to the theory of probability. San Francisco-London: Freeman 1961.

Goel, N. S., Maitra, S. C., Montroll, E. W.: On the Volterra and other nonlinear models of interacting species. New York-London: Academic Press 1971.

—, Richter-Dyn, N.: Stochastic models in biology. New York-San Francisco-London: Academic Press 1974.

Goldberg, S.: Probability, an introduction. Englewood Cliffs: Prentice-Hall 1960.

Goldin, A., Venditti, J. M., Mantel, N., Kline, I., Gang, M.: Evaluation of combination chemotherapy with three drugs. Cancer Research 28, 950—960 (1968).

Good, I. J.: A power law and a logarithmic law in athletics. Amer. Statistician 1971, 54.

Goodman, L. A.: The analysis of population growth when the birth and death rates depend upon several factors. Biometrics 25, 659—681 (1969).

Goodwin, B. C.: Temporal organization in cells. A dynamic theory of cellular control processes. London-New York: Academic Press 1963.

Gowen, J. W.: Humoral and cellular elements in natural and acquired resistance to typhoid. Amer. J. Human Genet. 4, 285—302 (1952).

— Effects of X-rays of different wave lengths on viruses. In: Statistics and mathematics in biology. Kempthorne, O., Bancroft, T. A., Gowen, J. W., Lush, J. L. (Eds.). Reprint of 1954 edition, Chap. 39, pp. 495—510. New York-London: Hafner 1964.

Gradštejn, I. S., Ryžik, I. M.: Table of integrals, series and products. London-New York: Academic Press 1965.

Grande, F., Taylor, H. L.: Adaptive changes in the heart, vessels, and patterns of control under chronically high loads. In: Handbook of physiology. Field, J. (Ed.). Section 2, Vol. 3, pp. 2615—2677. Washington, D. C.: Amer. Physiol. Soc. 1965.

Gray, Sir James: Animal locomotion. New York: Norton 1968.

Griffith, J. S.: Mathematical neurobiology. An introduction to the mathematics of the nervous system. London and New York: Academic Press 1971.

Gröbner, W., Hofreiter, N.: Integraltafel. Two parts, 4th edition. Wien-New York: Springer 1965 and 1966.

Grodins, F. S.: Control theory and biological systems. New York-London: Columbia University Press 1963.

Grossman, S. I., Turner, J. E.: Mathematics for the biological sciences. New York: Macmillan. London: Collier Macmillan 1974.

Guelfi, J.: Initiation mathématique à la physique médicale et à la biologie. 2nd edition. Paris: Masson 1966.

Hadeler, K. P.: Mathematik für Biologen. Berlin-Heidelberg-New York: Springer 1974.

Halberg, F.: Chronobiology. Ann. Rev. Physiol. 31, 675—725 (1969).

— Engeli, M., Hamburger, C., Hillman, D.: Spectral resolution of low-frequency, small-amplitude rhythms in excreted ketosteroid; probable androgen-induced circaseptan desynchronization. Acta Endocr. (Kbh.) Suppl. 103, 54 pp. (1965).

— Reinberg, A.: Rythmes circadiens et rythmes de basses fréquences en physiologie humaine. J. Physiol. (Paris) 59, 117—200 (1967).

Halberg, F., Tong, Y. L., Johnson, E. A.: Circadian system phase — an aspect of temporal morphology; procedures and illustrative examples. Proc. International Congress of Anatomists. In: The cellular aspects of biorhythms. Symposium on Rhythmic Research, pp. 20—48. Berlin-Heidelberg-New York: Springer 1967.

Hamilton, W. F.: Falling drop method: Density determination of biologic fluids. In: Medical physics. Glasser, O. (Ed.), pp. 427—429. Chicago: Year Book 1947.

Hammond, K. R., Householder, J. E.: Introduction to the statistical method. Foundations and use in the behavioral sciences. Reprint. New York: Knopf 1963.

Harvey, G. R., Steinhauer, W. G., Teal, J. M.: Polychlorobiphenyls in North Atlantic ocean water. Science **180**, 643—644 (1973).

Hays, W. L.: Statistics for psychologists. New York-Chicago-San Francisco-Toronto-London: Holt, Rinehart, and Winston 1963.

Heinz, E.: Probleme bei der Diffusion kleiner Substanzmengen innerhalb des menschlichen Körpers. Biochem. Z. **319**, 482—492 (1949).

Hennig, A.: Fehlerbetrachtungen zur Volumenbestimmung aus der Integration ebener Schnitte. In: Quantitative methods in morphology. Weibel, E. R., Elias, H. (Eds.), pp. 99—129. Berlin-Heidelberg-New York: Springer 1967.

Hodges, J. L., Lehmann, E. L.: Basic concepts of probability and statistics. San Francisco: Holden-Day 1964.

Hodgkin, A. L., Huxley, A. F.: A quantitative description of membrane current and its application to conduction and excitation in nerve. J. Physiol. **117**, 500—544 (1952).

Holm, L. G., Weldon, L. W., Blackburn, R. D.: Aquatic weeds. Science **166**, 699—709 (1969).

Hoyle, G.: Neuromuscular mechanism of a locust skeletal muscle. Proc. Roy. Soc. London, B, **143**, 343—367 (1955).

Huff, D.: How to lie with statistics. New York: Norton 1954.

Hughes, G. M.: Locomotion: Terrestrial. In: The Physiology of Insecta. Rockstein, M. (Ed.). Vol. 2, Chap. 4, pp. 227—254. London-New York: Academic Press 1965.

Huntley, H. E.: The divine proportion. A study in mathematical beauty. New York: Dover 1970.

Huxley, J. S.: Problems of relative growth. London: Methuen 1932.

Inman, D. L., Brush, B. M.: The coastal challenge. Science **181**, 20—32 (1973).

Inman, R. E., Ingersoll, R. B., Levy, E. A.: Soil. A natural sink for carbon monoxide. Science **172**, 1229—1231 (1971).

Iosifescu, M., Tăutu, P.: Stochastic processes and applications in biology and medicine. Vol. I (theory), Vol. II (models). Biomathematics Vol. 3 and 4. Berlin-Heidelberg-New York: Springer 1973.

Jacquard, A.: The genetic structure of populations. Berlin-Heidelberg-New York: Springer 1974.

Jacquez, J. A.: Compartmental analysis in biology and medicine. Kinetics of distribution of tracer-labeled materials. Amsterdam-London-New York: Elsevier 1972.

Jardine, N., Sibson, R.: Mathematical taxonomy. London-New York: Wiley 1971.

Jarman, M.: Examples of quantitative zoology. London: Edward Arnold 1970.

Jenks, G. F., Brown, D. A.: Three-dimensional map construction. Science **154**, 857—864 (1966).

Jerison, H. J.: Brain evolution: New light on old principles. Science **170**, 1224—1225 (1970).

Johnson, F. H., Eyring, H., Polissar, M. J.: The kinematic basis of molecular biology. London-New York-Sydney: Wiley & Sons 1954.

Kamke, E.: Differentialgleichungen. Lösungsmethoden und Lösungen. Vol. 1, 3rd edition. New York: Chelsea 1948.

Karlin, S.: Sex and infinity. A mathematical analysis of the advantages and disadvantages of genetic recombination. In: The mathematical theory of the dynamics of biological populations. M. S. Bartlett and R. W. Hiorns (Eds.), pp. 155—194. London and New York: Academic Press 1973.

Karsten, K. G.: Charts and graphs. Englewood Cliffs: Prentice-Hall 1925.

Kemeny, J. G., Snell, J. L.: Mathematical models in the social sciences. Boston: Ginn 1962.

Kempthorne, O.: An intoduction to genetic statistics. London-New York-Sydney: Wiley & Sons 1957.

Kendall, M. G., Stuart, A.: The advanced theory of statistics. Vol. 3. London: Griffin 1966.

Kerlinger, F. N.: Foundations of behavioral research. Educational and psychological inquiry. New York-Chicago-San Francisco-Toronto-London: Holt, Rinehart, and Winston 1964.

Keyfitz, N.: Introduction to the mathematics of population. Reading-Menlo Park-London-Don Mills: Addison-Wesley 1968.

Khazanie, R. G.: An indication of the asymptotic nature of the Mendelian Markov process. J. Appl. Prob. **5**, 350—356 (1968).

Kimura, M.: Diffusion models in population genetics. J. appl. Prob. **1**, 177—232 (1964).

— Stochastic processes in population genetics, with special reference to distribution of gene frequencies and probability of gene fixation. In: Mathematical topics in population genetics. Ken-ichi Kojima (Ed.). Biomathematics, Vol. 1, pp. 178—209. Berlin-Heidelberg-New York: Springer 1970.

— Ohta, T.: Theoretical aspects of population genetics. Princeton, N. J.: Princeton University Press 1971.

King, K., Jr., Hare, P. E.: Amino acid composition of planktonic Foraminifera: A paleo-biochemical approach to evolution. Science **175**, 1461—1463 (1972).

Kingman, J. F. C.: Markov population processes. J. Appl. Prob. **6**, 1—18 (1969).

Kleerekoper, H.: Orientation through chemo-reception in fishes. In: Animal orientation and navigation, NASA-Symposium at Wallops Station, Va. S. R. Galler *et al.* (Eds.), pp. 459—468. U.S. Government Printing Office, Washington, D. C. 1972.

Klotter, K.: General properties of oscillating systems. Cold Spr. Harb. Symp. quant. Biol. **25**, 185—187 (1960).

Kostitzin, V. A.: Mathematical biology. London: Harrap 1939.

Kramer, H. H.: Recombination in selfed chromosome interchange heterozygotes. In: Statistics and mathematics in biology. Kempthorne, O., Bancroft, T. A., Gowen, J. W., Lush, J. L. (Eds.). Chap. 40, pp. 511—522. Reprinted. New York-London: Hafner 1964.

Kummer, B.: Biomechanik des Säugetierskelets. In: Handbuch der Zoologie, Vol. 8, part 6, pp. 1—80. Berlin: De Gruyter 1959.

Kynch, G. J.: Mathematics for the chemist. London-New York: Academic Press 1955.

Lampert, F., Bahr, G. F., Rabson, A. S.: Herpes simplex virus: Dry mass. Science **166**, 1163—1164 (1969).

Lang, H. J.: Über das Lichtrückenverhalten des Guppy *(Lebistes reticulatus)* in farbigen und farblosen Lichtern. Z. vergl. Physiol. **56**, 296—340 (1967).

Latham, J. L.: Elementary reaction kinetics. London: Butterworths 1964.

Laue, R.: Elemente der Graphentheorie und ihre Anwendung in den biologischen Wissenschaften. Leipzig: Akademische Verlagsgesellschaft Geest & Portig 1970.

Lee, R. E., Jr.: The size of suspended particulate matter in air. Science **178**, 567—575 (1972).

Lefort, G.: Mathématiques pour les sciences biologiques et agronomiques. Vol. 1 (text) and Vol. 2 (exercises). Paris: Colin 1967.

Leigh, E. G., Jr., Lewontin, R. C., Pavlidis, T.: Some mathematical problems in biology. Providence: Amer. Math. Soc. 1968.

Leslie, P. H.: On the use of matrices in certain population mathematics. Biometrika **33**, 183—212 (1945).

Leslie, P. H.: Some further notes on the use of matrices in population mathematics. Biometrika **35**, 213—245 (1948).

— An analysis of the data for some experiments carried out by Gause with populations of the protozoa, *Paramecium Aurelia* and *Paramecium caudatum*. Biometrika **44**, 314—327 (1957).

— The properties of a certain lag type of population growth and the influence of an external random factor on a number of such populations. Physiol. Zool. **32**, 151—159 (1959).

— Gower, J. C.: The properties of a stochastic model for the predator-prey type of inter-action between two species. Biometrika **47**, 219—234 (1960).

Levene, H., Pavlovsky, O., Dobzhansky, T.: Interaction of the adaptive values in poly-morphic experimental populations of *Drosophila pseudoobscura*. Evolution **8**, 335—349 (1954).

Levens, A. S.: Nomography. 2nd edition. London-New York-Sydney: Wiley & Sons 1959.

— Graphics. With an introduction to conceptual design. 1st ed. 1962, 2nd enlarged ed. 1968. London-New York-Sydney: Wiley & Sons 1962 and 1968.

— Graphical methods in research. London-New York-Sydney: Wiley & Sons 1965.

Levin, B. R.: A model for selection in systems of species competition. In: Concepts and models of biomathematics. Simulation techniques and methods. Heinmets, F. (Ed.), pp. 237—275. New York: Dekker 1969.

Levins, R.: Evolution in changing environments. Some theoretical explorations. Princeton: Princeton University Press 1968.

Lewis, E. G.: On the generation and growth of a population. Sankhya **6**, 93—96 (1942).

Li, C. C.: Population genetics. 2nd impression. London-Chicago: The University of Chicago Press 1958.

— Human genetics. Principles and methods. New York-San Francisco-Toronto-London: McGraw-Hill 1961.

Lillestøl, J.: Another approach to some Markov chain models in population genetics. J. Appl. Prob. **5**, 9—20 (1968).

Liss, A., Maniloff, J.: Isolation of mycoplasmatales viruses and characterization of MVL 1, MVL 52, and MVG 51. Science **173**, 725—727 (1971).

Lopez, A.: Problems in stable population theory. Princeton: Office of Population Research 1961.

Lotka, A. J.: Elements of mathematical biology. Unabridged republication of: Elements of physical biology. New York: Dover 1956.

Luce, R. D.: Individual choice behavior. A theoretical analysis. London-New York-Sydney: Wiley & Sons 1959.

Luce, R. D., Galanter, E.: Discrimination. In: Handbook of mathematical psychology. Luce, R. D., Bush, R. R., Galanter, E. (Eds.). Chap. 4, Vol. 1. London-New York-Sydney: Wiley & Sons 1963a.

— — Psychophysical scaling. In: Handbook of mathematical psychology. Luce, R. D., Bush, R. R., Galanter, E. (Eds.). Chap. 5, Vol. 1. London-New York-Sydney: Wiley & Sons 1963b.

MacArthur, R. H.: Geographical Ecology. Patterns in the distribution of species. New York-Evanston-London: Harper & Row 1972.

—, Connell, J. H.: The biology of populations. London-New York-Sydney: Wiley & Sons 1966.

McBrien, V. O.: Introductory analysis. New York: Appleton-Century-Crofts 1961.

McDonald, D. A.: Blood flow in arteries. London: Arnold 1960.

McNaughton, S. J.: Photosynthetic system II: Racial differentiation in *Typha latifolia*. Science **156**, 1363 (1967).

Magar, M. E.: Data analysis in biochemistry and biophysics. New York and London: Academic Press 1972.

Matis, J. H., Hartley, H. O.: Stochastic compartmental analysis: Model and least squares estimation from time series data. Biometrics **27**, 77—102 (1971).

Meredith, W. M.: Basic mathematical and statistical tables for psychology and education. New York-San Francisco-Toronto-London: McGraw-Hill 1967.

Metzler, C. M.: Usefullness of the two-compartment open model in pharmacokinetics. J. Amer. Statist. Assoc. **66**, 49—54 (1971).

Miescher, K.: Zur Frage der Alternsforschung. Experientia (Basel) **11**, special reprint with additions, 1—40 (1955).

Milhorn, H. T.: The application of control theory to physiological systems. Philadelphia-London: Saunders 1966.

Mills, D. R., Kramer, F. R., Spiegelman, S.: Complete nucleotide sequence of a replicating RNA molecule. Science **180**, 916—927 (1973).

Mitchell, H. C., Thaemert, J. C.: Three dimensions in fine structure. Science **148**, 1480—1482 (1965).

Mitchell, R. L., Anderson, I. C.: Catalase photoinactivation. Science **150**, 74 (1965).

Montroll, E. W.: Some statistical aspects of the theory of interacting species. In: Some mathematical questions in biology III. J. D. Cowan (Ed.), pp. 99—143. Providence, R. I.: American Mathematical Society 1972.

Moran, P. A. P.: The statistical processes of evolutionary theory. Oxford: Clarendon Press 1962.

Morowitz, H. J.: Entropy for biologists. An introduction to thermodynamics. London-New York: Academic Press 1970.

Morscher, E.: Development and clinical significance of the anteversion of the femoral neck. Wiederherstellungschir. u. Traum. (Reconstr. Surg. Traumat.) **9**, 107—125 (1967).

Mosimann, J. E.: Elementary probability for the biological sciences. New York: Appleton-Century-Crofts 1968.

Mosteller, F., Rourke, R. E. K., Thomas, G. B.: Probability with statistical applications. Reading-Menlo Park-London-Don Mills: Addison-Wesley 1961.

Nahikian, H. M.: A modern algebra for biologists. Chicago-London: University of Chicago Press 1964.

National Bureau of Standards. Table of natural logarithms. Vols. 1—4. Federal Works Agency. Work Projects Administration for the City of New York 1941.

— Tables of probability functions. Vol. 2. Federal Works Agency. Work Projects Administration for the City of New York 1942.

— Tables of the exponential function e^x. 2nd edition. Washington, D. C.: U.S. Government Printing Office 1947.

— Tables of the binomial probability distribution. Washington, D. C.: U.S. Government Printing Office 1950.

— Tables of 10^x (Antilogarithms to the base 10). Washington, D. C.: U.S. Government Printing Office 1953.

Neel, J. V., Schull, W. J.: Human heredity. 3rd impression. Chicago-London: University of Chicago Press 1958.

Nelson, E.: Gesetzmäßigkeiten der Gestaltwandlung im Blütenreich. Ihre Bedeutung für das Problem der Evolution. Chernex-Montreux: E. Nelson 1954.

Northrop, J. H., Kunitz, M., Herriot, R. M.: Crystalline enzymes. 2nd edition. New York: Cambridge University Press 1948.

Ostrander, C. C.: A mathematical study of the genus Pentremites. In: Numerical taxonomy. A. J. Cole (Ed.), pp. 165—180. London and New York: Academic Press 1969.

Patrick, W. H., Jr., Gotoh, S., Williams, B. G.: Strengite dissolution in flooded soils and sediments. Scienne **179**, 564—565 (1973).

Patten, B. C.: Systems ecology: A course sequence in mathematical ecology. BioScience **16**, 593—598 (1966).

Pauwels, F.: Gesammelte Abhandlungen zur funktionellen Anatomie des Bewegungsapparates. Berlin-Heidelberg-New York: Springer 1965.

Pavlidis, T.: Biological oscillators: Their mathematical analysis. New York-London: Academic Press 1973.

Pearce, C.: A new deterministic model for the interaction between predator and prey. Biometrics **26**, 387—392 (1970).

Pearson, E. S., Hartley, H. O. (Eds.): Biometrika tables for statisticians, Vol. 1, 3rd edition. Cambridge: University Press 1966.

Peirce, B. O.: A short table of integrals. 3rd edition. Boston: Ginn 1929.

Pennycuick, C. J.: The mechanics of bird migration. Ibis **111**, 525—556 (1969).

Peskar, B., Spector, S.: Serotonin. Radioimmunoassay. Science **179**, 1340—1341 (1973).

Peters, R. M.: The mechanical basis of respiration. An approach to respiratory pathophysiology. Boston: Little, Brown 1969.

Pielou, E. C.: An introduction to mathematical ecology. London-New York-Sydney: Wiley & Sons 1969.

Pittendrigh, C. S., Bruce, V. G.: An oscillator model for biological clocks. In: Rhythmic and synthetic processes in growth. Rudnick, D. (Ed.), pp. 75—109. Princeton: Princeton University Press 1957.

Platt, D. R.: Natural history of the hognose snakes *Heterodon platyrhinos* and *Heterodon nasicus*. University of Kansas Publications, Museum of Natural History **18**, No. 4, 253—420 (1969).

Platt, R. B., Griffiths, J. F.: Environmental measurements and interpretation. New York-Amsterdam-London: Reinhold 1964.

Pollard, J. H.: On the use of the direct matrix product in analysing certain stochastic population models. Biometrika **53**, 397—415 (1966).

— Mathematical models for the growth of human populations. Cambridge: University Press 1973.

Radinsky, L.: Relative brain size: A new measure. Science **155**, 836—837 (1967).

Rains, D. W.: Light-enhanced potassium absorption by corn leaf tissue. Science **156**, 1382—1383 (1967).

Randall, J. E.: Elements of biophysics. Chicago: Year Book 1958.

— Elements of biophysics, 2nd edition. Chicago: Year Book 1962.

Rashevsky, N.: Mathematical biophysics. Physico-mathematical foundations of biology. Vols. I and II, 3rd edition. New York: Dover 1960.

— Mathematical principles in biology and their applications. Springfield: Thomas 1961.

— Some medical aspects of mathematical biology. Springfield: Thomas 1964.

Reinke, W. A., Taylor, C. E., Immerwahr, G. E.: Nomograms for simplified demographic calculations. Public Health Reports **84**, No. 5, 431—444 (1969).

Ricci, B.: Physiological basis of human performance. Philadelphia: Lea and Febiger 1967.

Riggs, D. S.: Control theory and physiological feedback mechanisms. Baltimore: Williams and Wilkins 1970.

— The mathematical approach to physiological problems. A critical primer. Baltimore: Williams and Wilkins 1963. Paperback edition, Cambridge-London: M.I.T. Press 1970.

Roach, S. A.: The theory of random clumping. London: Methuen 1968.

Robertson, A.: A theory of limits in artificial selection with many linked loci. In: Mathematical topics in population genetics. Ken-ichi Kojima (Ed.). Biomathematics, Vol. 1, pp. 246—288. Berlin-Heidelberg-New York: Springer 1970.

Romig, H. G.: 50—100 binomial tables. London-New York-Sydney: Wiley & Sons 1953.

Rosen, R.: Optimality principles in biology. London: Butterworths 1967.

Rosenfeld, A.: Picture processing by computer. London-New York: Academic Press 1969.

Rouvray, D. H.: The search for useful topological indices in chemistry. Amer. Scientist **61**, 729—735 (1973).

Rubinow, S. I.: Mathematical problems in the biological sciences. Philadelphia: Soc. Indust. Appl. Math. 1973.

Ruch, T. C., Patton, H. D.: Physiology and biophysics. Philadelphia-London: Saunders 1965.

Salzer, H. E., Levine, N.: Tables of sines and cosines to ten decimal places at thousandths of a degree. Oxford-London-Edinburgh-New York: Pergamon Press. New York: Macmillan 1962.

Saunders, L., Fleming, R.: Mathematics and statistics for use in pharmacy, biology, and chemistry. London: The Pharmaceutical Press 1957.

Schaffer, H. E., Mettler, L. E.: Teaching models in population genetics. BioScience **20**, 1304—1310 (1970).

Schanz, F.: Wachstumsansprüche der Cladophoracee *Rhizoclonium hieroglyphicum* Kütz. in Reinkultur. Ph. D. dissertation, Univ. Zürich 1974.

Schaub, H.: Über die Grossforaminiferen im Untereocaen von Campo (Ober-Aragonien). Eclogae Geol. Helv. **59**, 355—378 (1966).

Schips, M.: Mathematik und Biologie. Leipzig-Berlin: Teubner 1922.

Schmid, C. F.: Handbook of graphic presentation. New York: Ronald Press 1954.

Schmidt-Nielsen, K.: How animals work. Cambridge: at the University Press 1972.

— Locomotion. Energy cost of swimming flying, and running. Science **177**, 222—228 (1972a).

Schüepp, O.: Meristeme. Wachstum und Formbildung in den Teilungsgeweben höherer Pflanzen. Basel-Stuttgart: Birkhäuser 1966.

Schuler, R., Kreuzer, F.: Properties and performance of membrane-covered rapid polarographic oxygen catheder electrodes for continuous oxygen recording *in vivo*. In: Progress in respiration research. Herzog, H. (Ed.). Vol. 3, pp. 64—78. Basel-New York: Karger 1969.

Seal, H. L.: Multivariate statistical analysis for biologists. London-New York-Sydney: Wiley & Sons 1964.

Searle, S. R.: Matrix algebra for the biological sciences (including applications in statistics). London-New York-Sydney: Wiley & Sons 1966.

Selby, S. M. (Ed.): Standard mathematical tables. Student edition. 16th edition. Cleveland: Chemical Rubber 1968.

Setlow, R. B., Pollard, E. C.: Molecular biophysics. Reading-Menlo Park-London-Don Mills: Addison-Wesley 1962.

Severinghaus, J. W., Stupfel, M.: Respiratory physiologic studies during hypothermia. From: The physiology of induced hypothermia, pp. 52—57. Proceedings of a Symposium 1955. Dripps, R. D. (Ed.). Washington, D.C.: Publication No. 451 of the National Academy of Sciences 1956.

Simpson, G. G., Roe, A., Lewontin, R. C.: Quantitative zoology. Revised edition. New York-Burlingame: Harcourt, Brace 1960.

Singh, I. J., Gunberg, D. L.: Quantitative histology of changes with age in rat bone cortex. J. Morphology **133**, 241—252 (1971).

Skellam, J. G.: Seasonal periodicity in theoretical population ecology. Proc. Fifth Berkeley Symposium on mathematical statistics and probability, Vol. 4, pp. 179—205. Berkeley-Los Angeles: University of California Press 1967.

Slijper, E. J.: Riesen und Zwerge im Tierreich. Hamburg-Berlin: Parey 1967.

Slobodkin, L. B.: Growth and regulation of animal populations. New York-Chicago-San Francisco-Toronto-London: Holt, Rinehart and Winston 1961.

Smith, C. A. B.: Biomathematics, the principles of mathematics for students of biological and general science. Vol. 1: Algebra, geometry, calculus. Vol. 2: Numerical methods, matrices, probability, statistics. 4th edition. London: Griffin 1966 and 1969.

Smith, J. M.: Mathematical ideas in biology. London-New York: Cambridge University Press 1968.

Sollberger, A.: Biological rhythm research. Amsterdam-London-New York: Elsevier 1965.

Solomon, A. K.: Compartmental methods of kinetic analysis. In: Mineral metabolism. Comar, C. L., Bronner, F. (Eds.). Vol. 1, part A, Chap. 5. London-New York: Academic Press 1960.

Sparrow, A. H., Underbrink, A. G., Rossi, H. H.: Mutations induced in *Tradescantia* by small doses of X-rays and neutrons: Analysis of dose-response curves. Science **176**, 916—918 (1972).

Steinhaus, H.: Mathematical snapshots. Revised edition. New York: Oxford University Press 1960.

Stevens, B.: Chemical kinetics. For general students of chemistry. London: Chapman & Hall 1965.

Stevens, S. S.: Mathematics, measurement, and psychophysics. In: Handbook of experimental psychology. S. S. Stevens (Ed.). New York: Wiley & Sons, 1—49 (1951).

— Neural events and the psychophysical law. Science **170**, 1043—1050 (1970).

Stibitz, G. R.: Mathematics in medicine and the life sciences. Chicago: Year Book 1966.

Strehler, B. L.: Time, cells, and aging. 2nd printing. London-New York: Academic Press 1963.

Stuhlman, O.: An introduction to biophysics. London-New York-Sydney: Wiley & Sons 1943.

Stumpff, K.: Grundlagen und Methoden der Periodenforschung. Berlin: Springer 1937.

Sunderman, F. W., Boerner, F.: Normal values in clinical medicine. Philadelphia-London: Saunders 1949.

Sved, J. A., Mayo, O.: The evolution of dominance. In: Mathematical topics in population genetics. Ken-ichi Kojima (Ed.). Biomathematics, Vol. **1**, 289—316. Berlin-Heidelberg-New York: Springer 1970.

Swanson, D. A.: Magma supply rate at Kilauea volcano, 1952—1971. Science **175**, 169—170 (1972).

Swerdloff, R. S., Pozefsky, T., Tobin, J. D., Andres, R.: Influence of age on the intravenous tolbutamide response test. Diabetes **16**, 161—170 (1967).

Sykes, Z. M.: On discrete stable population theory. Biometrics **25**, 285—293 (1969).

Tarski, A.: Introduction to logic and to the methodology of deductive sciences. 3rd edition. New York: Oxford University Press 1965.

Taunton-Rigby, A., Sher, S. E., Kelley, P. R.: Lysergic acid diethylamide: Radioimmunoassay. Science **181**, 165—166 (1973).

Taylor, T. R., Aitchison, J., McGirr, E. M.: Doctors as decision-makers: A computer-assisted study of diagnosis as a cognitive skill. Brit. Med. J., 35—40 (1971, July issue).

Teissier, G.: Relative growth. In: The physiology of crustacea. Waterman, T. H. (Ed.). Vol. 1, Chap. 16. London-New York: Academic Press 1960.

Tenney, S. M., Remmers, J. E.: Comparative quantitative morphology of the mammalian lung: diffusion area. Nature (London) **197**, 54—56 (1963).

Thews, G.: Nomogramme zum Säure-Basen-Status des Blutes und zum Atemgastransport. Berlin-Heidelberg-New York: Springer 1971.

Thompson, d'Arcy W.: On growth and form. First edition. Cambridge: At the University Press 1917.

— On growth and form. Abridged edition. Cambridge: University Press 1961.

Thrall, R. M., Mortimer, J. A., Rebman, K. R., Baum, R. F. (Eds.): Some mathematical models in biology. Revised Edition. Report No. 40241-R-7 prepared at the University of Michigan 1967.

Timoféeff-Ressovsky, N. W., Zimmer, K. G.: Das Trefferprinzip in der Biologie. Leipzig: Hirzel 1947.

Tóth, L. F.: What the bees know and what they do not know. Bull. Amer. math. Soc. **70**, 468—481 (1964).

Tucker, V. A., Schmidt-Koenig, K.: Flight speeds of birds in relation to energetics and wind directions. The Auk **88**, 97—107 (1971).

Turner, J. R. G.: Changes in mean fitness under natural selection. In: Mathematical topics in population genetics. Ken-ichi Kojima (Ed.). Biomathematics, Vol. **1**, 32—78. Berlin-Heidelberg-New York: Springer 1970.

Volterra, V.: Leçons sur la théorie mathématique de la lutte pour la vie. Paris: Gauthier-Villars 1931.

von Bertalanffy, L.: Theoretische Biologie. Vol. 2: Stoffwechsel, Wachstum. 2nd edition. Bern: Francke 1951.

von Frisch, K.: The dance language and orientation of bees. Cambridge: Belknap Press of Harvard University Press 1967.

Vorobyov, N. N.: The Fibonacci numbers. Boston: Heath 1961.

Wagner, G.: Ergänzung zu Variationsstatistik. Mitt. Ver. Schweiz. Naturwissenschafts-lehrer **2**, 5—6 (1957).

Waltman, P.: Deterministic threshold models in the theory of epidemics. Lecture Notes in Biomathematics 1. Berlin-Heidelberg-New York: Springer 1974.

Waterman, T. H.: The analysis of spacial orientation. Ergebn. Biol. **26**, 98—117 (1963).
— Morowitz, H. J. (Ed.): Theoretical and mathematical biology. New York-Toronto-London: Blaisdell 1965.

Watt, K. E. F.: Ecology and resource management. A quantitative approach. New York-San Francisco-Toronto-London: McGraw-Hill 1968.

Weibel, E. R.: Morphometry of the human lung. Berlin-Göttingen-Heidelberg: Springer 1963.
— Elias, H.: Introduction to stereologic principles. In: Quantitative methods in morpho-logy. Weibel, E. R., Elias, H. (Eds.), pp. 89—98. Berlin-Heidelberg-New York: Springer 1967.

Welch, L. F., Adams, W. E., Carmon, J. L.: Yield response surfaces, isoquants, and economic fertilizer optima for coastal Bermuda grass. Agron. J. **55**, 63—67 (1963).

Went, F. W.: The size of man. Amer. Scientist **56**, 400—413 (1968).

Wever, R.: Virtual synchronization towards the limits of the range of entrainment. J. theor. Biol. **36**, 119—132 (1972).

Weyl, H.: Symmetry. Princeton: Princeton University Press 1952.

Wilbur, H. M., Collins, J. P.: Ecological aspects of amphibian metamorphosis. Science **182**, 1305—1314 (1973).

Wilbur, K. M., Owen, G.: Growth. In: Physiology of Mollusca. Wilbur, K. M., Yonge, C. M. (Eds.). Vol. 1, Chap. 7. London-New York: Academic Press 1964.

Williams, M., Lissner, H. R.: Biomechanics of human motion. Philadelphia-London: Saunders 1962.

Williamson, M.: The analysis of biological populations. London: Arnold 1972.

Wilson, R.: Tax the integrated pollution exposure. Science **178**, 182—183 (1972).

Winfree, A. T.: Spiral waves in chemical activity. Science **175**, 634—635 (1972).

Woodwell, G. M.: Radioactivity and fallout: The model pollution. BioSience **19**, 884—887 (1969).

Worthing, A. G., Geffner, J.: Treatment of experimental data. 8th printing. London-New York-Sydney: Wiley & Sons. London: Chapman & Hall 1959.

Wright, S.: The interpretation of multivariate systems. Chap. 2. In: Statistics and mathe-matics in biology. Kempthorne, O., Bancroft, T. A., Gowen, J. W., Lush, J. L. (Eds.). Reprinted. New York-London: Hafner 1964.

— Evolution and the genetics of populations. Vol. 1: Genetic and biometric foundations. Vol. 2: The theory of gene frequencies. Chicago-London: University of Chicago Press 1968 and 1969.

Author and Subject Index

Bio-mathematics

Managing Editors: K. Krickeberg, S. A. Levin

Forthcoming Volumes

Springer-Verlag
Berlin
Heidelberg
New York

Volume 8

A. T. Winfree

The Geometry of Biological Time

1979. Approx. 290 figures. Approx. 580 pages
ISBN 3-540-09373-7

The widespread apperance of periodic patterns
in nature reveals that many living organisms are
communities of biological clocks. This land-
mark text investigates, and explains in mathematical terms, periodic processes in living
systems and in their non-living analogues. Its
lively presentation (including many drawings),
timely perspective and unique bibliography will
make it rewarding reading for students and re-
searchers in many disciplines.

Volume 9

W. J. Ewens

Mathematical Population Genetics

1979. 4 figures, 17 tables. Approx. 330 pages
ISBN 3-540-09577-2

This graduate level monograph considers the
mathematical theory of population genetics,
emphasizing aspects relevant to evolutionary
studies. It contains a definitive and comprehensive discussion of relevant areas with references
to the essential literature. The sound presentation and excellent exposition make this book a
standard for population geneticists interested in
the mathematical foundations of their subject
as well as for mathematicians involved with
genetic evolutionary processes.

Volume 10

A. Okubo

Diffusion and Ecological Problems:
Mathematical Models

1979. Approx. 114 figures. Approx. 300 pages
ISBN 3-540-09620-5

This is the first comprehensive book on mathematical models of diffusion in an ecological
context. Directed towards applied mathematicians, physicists and biologists, it gives a
sound, biologically oriented treatment of the
mathematics and physics of diffusion.

Journal of

Mathematical Biology

ISSN 0303-6812   Title No. 285

Editorial Board: H. J. Bremermann, Berkeley, CA;
F. A. Dodge, Yorktown Heights, NY; K. P. Hadeler,
Tübingen; S. A. Levin, Ithaca, NY; D. Varjú, Tübingen

Advisory Board: M. A. Arbib, Amherst, MA;
E. Batschelet, Zürich; W. Bühler, Mainz; B. D. Coleman,
Pittsburgh, PA; K. Dietz, Tübingen; W. Fleming, Provi-
dence, RI; D. Glaser, Berkeley, CA; N. S. Goel, Bing-
hamton, NY; J. N. R. Grainger, Dublin; F. Heinmets,
Natick, MA; H. Holzer, Freiburg i. Br.; W. Jäger, Heidel-
berg; K. Jänich, Regensburg; S. Karlin, Stanford, CA;
S. Kauffman, Philadelphia, PA; D. G. Kendall, Cam-
bridge; N. Keyfitz, Cambridge, MA; B. Khodorov,
Moscow; E. R. Lewis, Berkeley, CA; D. Ludwig, Van-
couver; H. Mel, Berkeley, CA; H. Mohr, Freiburg i. Br.;
E. W. Montroll, Rochester, NY; A. Oaten, Santa Barbara,
CA; G. M. Odell, Troy, NY; G. Oster, Berkeley, CA;
A. S. Perelson, Providence, RI; T. Poggio, Tübingen;
K. H. Pribram, Stanford, CA; S. I. Rubinow, New York,
NY; W. v. Seelen, Mainz; L. A. Segel, Rehovot;
W. Seyffert, Tübingen; H. Spekreijse, Amsterdam;
R. B. Stein, Edmonton; R. Thom, Bures-sur-Yvette;
Jun-ichi Toyoda, Tokyo; J. J. Tyson, Blacksburgh, VA;
J. Vandermeer, Ann Arbor, MI.

Springer-Verlag
Berlin
Heidelberg
New York

The **Journal of Mathematical Biology** publishes papers
in which mathematics leads to a better understanding of
biological phenomena, mathematical papers inspired by
biological research and papers which yield new experimental data bearing on mathematical models. The scope
is broad, both mathematically and biologically and
extends to relevant interfaces with medicine, chemistry,
physics and sociology. The editors aim to reach an
audience of both mathematicians and biologists.

Subscription information and sample copy upon request.